Fundamental Concepts of Abstract Algebra

GERTURDE EHRLICH
The University of Maryland

DOVER PUBLICATIONS, INC.
Mineola, New York

Bibliographical Note

This Dover edition, first published in 2011, is an unabridged republication of the work originally published in 1991 by PWS-Kent Publishing Company, Boston.

Library of Congress Cataloging-in-Publication Data

Ehrlich, Gertrude.
Fundamental concepts of abstract algebra / Gertrude Ehrlich. — Dover ed.
p. cm.
Originally published: Boston : PWS-Kent Pub., 1991.
Includes bibliographical references and index.
ISBN-13: 978-0-486-48589-8 (pbk.)
ISBN-10: 0-486-48589-7 (pbk.)
1. Algebra, Abstract. I. Title.

QA162.E37 2011
512'.02—dc23

2011019127

Manufactured in the United States by Courier Corporation
48589701
www.doverpublications.com

"All right," said the Cat; and this time it vanished quite slowly, beginning with the end of the tail, and ending with the grin, which remained some time after the rest of it had gone.

"Well! I've often seen a cat without a grin," thought Alice, "but a grin without a cat! It's the most curious thing I ever saw in all my life!"

(Lewis Carroll, *Alice in Wonderland*)

Preface

This book is intended as a text in abstract algebra for undergraduate mathematics majors. It contains ample material for a two-semester sequence, including a thorough treatment of field theory during the second semester. Some knowledge of the rudiments of linear algebra is assumed from the start, and a fair number of examples are drawn from this area. In Sections 10–17 of Chapter 3, finite-dimensional vector spaces are treated in an abstract algebraic setting. These sections may be used for review, for reference, or to enhance the linear algebra background of the students. Some possible two-semester sequences are described below.

The exercise list in each section of each chapter starts with an exercise consisting of True-or-False statements, selected with the aim of inducing a thoughtful review of materials covered in the section, and dispelling misconceptions at an early stage.

Possible Two-Semester Sequences:
Model 1 (including a full treatment of Galois Theory)
First Semester:
- Chapter 1 (1–3)
- Chapter 2 (1–11; 12, 13 if time permits)
- Chapter 3 (1–9; 10 if time permits)

Second Semester:
- Chapter 2 (14, 15)
- Chapter 4 (1–17)

Model 2 (including some field theory, but not Galois theory, and an introduction to abstract linear algebra)
First Semester:
- Chapter 1 (1–3)
- Chapter 2 (1–11)
- Chapter 3 (1–8)

Second Semester:
- Chapter 2 (12–15)
- Chapter 3 (9–17)
- Chapter 4 (1–5)

Acknowledgements

I am deeply indebted to my colleagues, Professors John Horvath, Adam Kleppner, and David Schneider, who taught from the manuscript and made many valuable comments; to numerous students, notably Greg Grant and Aaron Naiman (champion error finders); and to Kristi Aho Kraft and Virginia Sauber Vargas who typed a large portion of the manuscript, with much skill and forbearance.

I would also like to express my appreciation to the following reviewers: Daniel D. Anderson (University of Iowa), Ida Z. Arms (Indiana University of Pennsylvania), Robert Beezer (University of Puget Sound), Patrick J. Costello (Eastern Kentucky University), John C. Higgins (Brigham Young University), Frederick Hoffman (Florida Atlantic University), William J. Keane (Boston College), Daniel B. Shapiro (Ohio State University), Anita A. Solow (Grinnell College), John R. Stallings (University of California-Berkeley), Paul M. Weichsel (University of Illinois), and Anne Ludington Young (Loyola College).

Glossary of Symbols

$\mathbb{Z}, \mathbb{Q}, \mathbb{R}, \mathbb{C}$	The set of all integers, rational numbers, real numbers, and complex numbers, respectively.
$\mathbb{Z}^+, \mathbb{Q}^+, \mathbb{R}^+$	The set of all *positive* integers, rational numbers, and real numbers, respectively.
$\mathbb{Q}^*, \mathbb{R}^*, \mathbb{C}^*$	The set of all *non-zero* rational numbers, real numbers, and complex numbers, respectively.
$M_n(A)$	The set of all $n \times n$ matrices with entries in A.
$\mathrm{GL}_n(\mathbb{R})$	The set of all non-singular $n \times n$ matrices with real entries.
$\mathrm{SL}_n(\mathbb{R})$	The set of all $n \times n$ matrices with real entries and with determinant equal to 1.
$\subset$	Is a subset of.
$\supset$	Contains as a subset.
$\in$	Is an element of.
$f: A \to B$	f is a function from A to B.

Contents

1 Preliminaries 1

1.1 Introduction *1*
1.2 Sets, Relations, and Functions *4*
1.3 The Integers *16*

2 Groups 27

2.1 Binary Operations *27*
2.2 Groups *33*
2.3 Subgroups, Cyclic Groups, and the Order of an Element *43*
2.4 Isomorphism *54*
2.5 Cosets and Lagrange's Theorem *60*
2.6 Permutation Groups *66*
2.7 Cayley's Theorem; Geometric Groups *77*
2.8 Normal Subgroups *81*
2.9 Homomorphism, Factor Groups, and the Fundamental Theorem of Homomorphism for Groups *87*
2.10 Further Isomorphism Theorems; Simple Groups *101*
2.11 Automorphisms and Invariant Subgroups *104*
2.12 Direct Products of Groups *108*
2.13 The Structure of Finite Abelian Groups *112*
2.14 Solvable Groups *119*
2.15 Primary Groups and the Sylow Theorems *124*

3 *Rings, Modules, and Vector Spaces* 132

3.1 Rings and Subrings *132*
3.2 Ring Homomorphisms, Ideals, Residue Class Rings, and Simple Rings *139*
3.3 Fundamental Theorem of Homomorphism for Rings *148*
3.4 Maximal and Prime Ideals *150*
3.5 Polynomial Rings *156*
3.6 Principal Ideal Domains *163*
3.7 Euclidean Domains *174*
3.8 Fields of Quotients of Integral Domains *179*
3.9 Polynomials over Unique Factorization Domains *183*
3.10 Groups with Operators; Modules *190*
3.11 Vector Spaces, Subspaces, and Linear Independence *194*
3.12 Basis and Dimension *199*
3.13 Linear Transformations *207*
3.14 Coordinate Vectors, Matrices, and Determinants *211*
3.15 Representation of Linear Transformations by Matrices *217*
3.16 Non-Singular Matrices, Change of Basis, and Similarity *223*
3.17 Eigenvalues and Diagonalization *229*

4 *Fields* 238

4.1 Subfields, Extensions, Prime Fields, and Characteristic *238*
4.2 Adjunctions; Algebraic and Transcendental Elements *242*
4.3 Finding an Extension in Which a Given Polynomial Has a Zero *246*
4.4 Algebraic and Transcendental Extensions; Degree of an Extension; Finite Fields *250*
4.5 Classical Constructions I *257*
4.6 Extension of Isomorphisms *263*
4.7 Normal Extensions *266*
4.8 Separable Extensions *273*
4.9 Galois Extensions and Galois Groups *280*
4.10 The Fundamental Theorem of Galois Theory *285*
4.11 Roots of Unity *291*
4.12 Radical Extensions I *297*
4.13 Radical Extensions II *305*
4.14 Transcendence Sets and the Definition of a General Polynomial of Degree n *311*

4.15 Symmetric Functions and the Unsolvability of a General Polynomial of Degree $n > 4$ **315**
4.16 Solution of a General Polynomial of Degree $n \leq 4$ by Radicals **320**
4.17 Classical Constructions II; The Fundamental Theorem of Algebra **328**

Bibliography 334

Index 336

1

Preliminaries

The following historical remarks are intended to be read at the outset of your study, and then reread from time to time as your knowledge of the subject increases.

1.1

Introduction

Abstraction Is Power

Much of the power of mathematics stems from its abstractness, which is the source of its universality. Mathematics was abstract from its very beginnings in prehistoric times. Primitive humans, confronted (through millions of years) with examples of sets such as these,

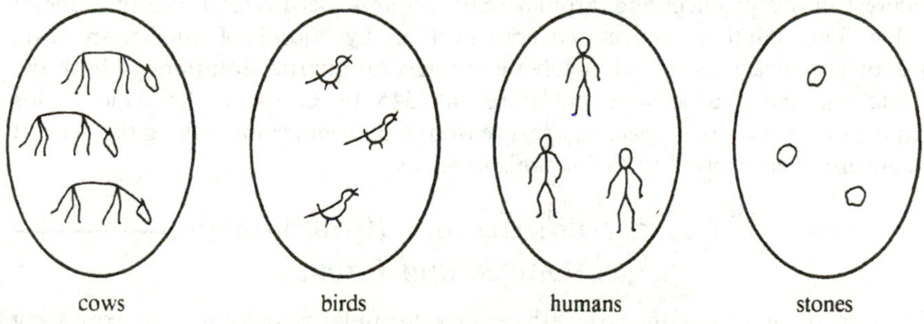

Figure 1

gradually learned to *abstract from* the differences of such sets, and to recognize a common property: "threeness." In this way, through many acts of abstraction, they created the natural numbers. The abstract concepts 1, 2, 3, 4, 5, 6, ... enabled them to deal more efficiently with concrete problems such as: if to , I add [cow pictograms], how many cows will I have? If to ○ ○ , I add ○○○, how many stones? For, they could now solve, once and for all, the problem "2 + 3 = ?", and the number of resulting objects would be the same, no matter whether the sets being combined consisted of cows or stones.

Such acts of abstraction, resulting in greater universality and in increased power to solve problems, pervade the history of mathematics. We shall make them our central theme as we develop some of the basic concepts of that portion of mathematics now known as abstract algebra.

Ancient Origins of Algebra

Algebra had its origin in ancient times. The earliest known written record of problems we would now classify as algebraic is contained in Babylonian tablets and Egyptian papyri dating from around 1700 B.C. The solution of quadratic equations was known to the Babylonians at that time, and is believed also to have been known in ancient China. Based on Hindu and Greek sources, Mohammed ibn Mûsâ al-Khowârizmî of Baghdad, in 825 A.D., wrote a textbook called *Al-jebr*, which, following its translation during the twelfth century, was to have considerable influence on the subsequent development of algebra in Europe. (The word *algebra* is derived from the title of this book, and the word *algorithm* from its author's name.)

Polynomial Equations Through the Sixteenth Century

The basic problem of classical algebra was the solution of polynomial equations. Since the general linear and quadratic equations had been solved in pre-Christian times, the algebraic problem awaiting consideration through the Dark Ages and the Middle Ages was the solution of the general polynomial equation of degree three or more. The intellectual revival in Renaissance Italy brought about renewed interest in this problem and, around 1510, Scipione del Ferro solved the general cubic. This solution was rediscovered in 1535 by Niccola Tartaglia. In 1540, Lodovico Ferrari succeeded in solving the general quartic. Solutions of both the cubic and the quartic were published in 1545 by Girolamo Cardano in his expository work *Ars Magna*. For this reason, the formulas for solving the general cubic are often referred to as *Cardan's formulas*.

The Nineteenth Century Breakthrough: Abel, Galois, and Groups

Thus, by the middle of the sixteenth century, formulas were known for expressing the roots of the general polynomial equation of degree $n \leq 4$ in terms of the coefficients of the equation, using finitely many elementary algebraic operations and root extractions. Put more briefly, it was known how to solve the general equation of degree $n \leq 4$ *by radicals*. The question remaining open was: Can the general polynomial equation of degree $n \geq 5$ be solved by radicals? Little progress was made on this problem until the late eighteenth and early nineteeth centuries.

The work of Joseph Louis Lagrange (1736–1813) on the quintic, the proof by Karl Friedrich Gauss (1777–1855) in 1799 of the Fundamental Theorem of Algebra (which states that every polynomial equation with complex coefficients *has* a root in the complex number field), and the discovery by Augustin Louis Cauchy (1789–1857) of permutation groups set the stage for the great breakthrough. During the first third of the nineteenth century, two young geniuses, both tragically short-lived, settled the question of solvability by radicals for equations of degree $n \geq 5$. In 1824, the Norwegian Niels Henrik Abel (1802–1829), proved that the general polynomial equation of degree $n \geq 5$ is not solvable by radicals. Independently, a very young Frenchman, Evariste Galois (1811–1832), discovered the general theory of solvability for polynomial equations (now known as *Galois Theory*) which implies, in particular, the unsolvability of the general polynomial equation of degree $n \geq 5$ by radicals. The essential feature of Galois' discovery was the association between a polynomial and a certain group of permutations on its roots.

The significance of this discovery lies far beyond the solution of one ancient problem. Indeed, little use is made of the existing formulas for solving the general cubic and quartic, and no one seriously laments the non-existence of formulas for solving the general equation of degree $n \geq 5$. For most practical purposes, there are numerical methods quite adequate to the task of *approximating* the roots of such equations to any desired degree of accuracy.

The Birth of Abstract Algebra: Further Uses of Groups

The significance of Galois' discovery lies in the introduction of the group concept as an important tool in mathematics. Abstract algebra was born at that time! After Galois' results had finally been published by Liouville in 1846, the use of the group concept began to spread within mathematics. In 1872, Felix Klein (1849–1925), published his Erlanger Program, proposing to formulate all of geometry as the study of invariants under groups of transformations. In 1893, Marius Sophus Lie (1842–1899), published his three-volume work on continuous groups of transformations. His theory forms a fundamental part of the theory of continuous functions. As time progressed, group theory thus found its way into geometry, analysis, topology, and other areas of mathematics where it made profound contributions. During the twentieth century, the use of group theory began to transcend the boundaries of mathematics, as groups became essential tools in such diverse fields of physics as crystallography, quantum mechanics, and elementary particle theory.

Other Algebraic Structures and Their Uses

Abstract algebra deals with algebraic structures. Simplest among the important algebraic structures is the group. A group consists of a set and a binary operation defined on the set, subject to certain requirements. All other important algebraic structures are basically groups in which additional operations or relations interact with the group operation. Among them are rings (including fields), linear (or vector) spaces, and algebras. These structures were introduced during the late nineteenth and early twentieth centuries. Abstract algebra as we now know it was profoundly influenced by such mathematicians as Richard Dedekind (1831-1916), who was first to formulate the notion of an ideal; Ernst Steinitz (1871-1928), who developed the abstract theory of fields; and Emmy Noether (1882-1935), whose contributions to abstract ideal theory and non-commutative algebras set the tone for algebraic research during this century.

The impact of abstract algebra on other areas of mathematics as well as on physical sciences was not confined to groups. For example, normed linear spaces (i.e., linear spaces in which a notion of length is defined) form the basis of modern (functional) analysis. A particular kind of normed linear space, known as a *Hilbert space*, plays an important role in modern physics. Galois Theory itself has found application in algebraic coding theory which is used in the construction of error-correcting codes for electronic communication systems.

The Science and Art of Mathematics

As exemplified by the history of algebra, the history of mathematics tends to advance from specific problems to the generalizations arising from their solution. The greater abstraction thus attained makes possible the solution of wider classes of specific problems. As mathematics develops, the pure mathematics of one era often becomes applied mathematics—sometimes decades or even centuries later. It would, however, be a mistake to conclude that the sole justification for mathematical research is the probable future applicability of its results to science or technology. Mathematics is a science in its own right, as well as an art, fueled by the desire to discover, and by the urge to create.

Recommended reading: [20], [21], [22], [25], [26].

1.2

Sets, Relations, and Functions

Our introduction to set theory will be largely informal. We leave the term *set* undefined, remarking only on its intuitive content: a collection, an aggregate, a bunch of objects. These objects, the members of a set, are called its *elements*. We write "$a \in A$" to signify that a is an element of the set A. (A set consisting of just one element is sometimes called a *singleton*; a set consisting of just two elements is called a *pair*.) It proves convenient to include among sets an *empty set*: a set without elements. Two sets are *equal* if they consist of the same elements. Equivalently, two sets are equal if no element of either set fails to be in the other. As a consequence, there cannot be more than one empty set—for, if A and B are both empty sets, than neither contains an element which fails to be in the other.

Notation: The symbol "$\varnothing$" denotes the empty set.

If A and B are sets, then B *is a subset of* A if every element of B is an element of A. We write "$B \subset A$" to signify that B is a subset of A.

It is easy to verify the following statements (see Exercise 1.2.2):

(1) $\varnothing \subset A$ for any set A;

(2) $A \subset A$ for any set A;

(3) If A, B are sets such that $A \subset B$ and $B \subset A$, then $A = B$;

(4) If A, B, C are sets such that $A \subset B$ and $B \subset C$, then $A \subset C$.

Notation: We use braces to specify the elements of a set A, or of a subset B of A, either by listing the elements explicitly, or by stating a condition the elements must satisfy.

Example 1:

$\{1, 2, 3\}$ is the set consisting of 1, 2, 3.

Example 2:

$\{x \in \mathbb{Z} \mid x > 0\}$ is the set of all positive integers. For any two sets A and B, we denote by "$A \setminus B$" the set $\{x \in A \mid x \notin B\}$. In particular, if $B \subset A$, then $A \setminus B$ is called the *complement* of B in A.

The elements of a set may themselves be sets. In such a case, it is important not to confuse the relations of set membership ($\in$) and set inclusion ($\subset$). If we think of the United Nations as a set of nations, and of each nation as a set of citizens, then the United Nations is a set whose elements are the member nations. Thus, for example, U.S. $\in$ U.N, but U.S. $\not\subset$ U.N. Also, if Jones is a U.S. citizen, then Jones $\in$ U.S., but Jones $\notin$ U.N. (This illustrates that, unlike $\subset$, $\in$ is not transitive.)

Every set, A, has a *power set*, P_A: the set whose elements are the subsets of A. For example, if $A = \{1, 2, 3\}$, then

$$P_A = \{\varnothing, \{1\}, \{2\}, \{3\}, \{1, 2\}, \{1, 3\}, \{2, 3\}, \{1, 2, 3\}\}.$$

It is easy to see that a finite set of n elements ($n \geq 0$) has 2^n subsets (see Exercise 1.2.5). (For this reason, the power set of a set A is often denoted by "2^A".)

In discussing sets, we generally assume a "universe of discourse"—a set $\mathscr{U}$ to which the elements of all sets under discussion belong.

Now let $\mathscr{C}$ be a non-empty set of sets. Then the *union* of the sets in $\mathscr{C}$ is the set defined by

$$\bigcup_{X \in \mathscr{C}} X = \{x \in \mathscr{U} \mid x \in X \quad \text{for } \textit{some} \quad X \in \mathscr{C}\}.$$

It consists of all elements (in $\mathscr{U}$) which belong to *any one* of the sets of $\mathscr{C}$.

The *intersection* of the sets in $\mathscr{C}$ is the set defined by

$$\bigcap_{X \in \mathscr{C}} X = \{x \in \mathscr{U} \mid x \in X \quad \text{for } \textit{all} \quad X \in \mathscr{C}\}.$$

It consists of all elements (in $\mathscr{U}$) which belong to *all* of the sets of $\mathscr{C}$.

For example, if $X_1 = \{1, 2\}$, $X_2 = \{2\}$, $X_3 = \{2, 3\}$, then $\bigcup_{i=1}^{3} X_i = \{1, 2, 3\}$, while $\bigcap_{i=1}^{3} X_i = \{2\}$.

The union of two sets A and B (see Figure 2) is denoted by "$A \cup B$," and their intersection by "$A \cap B$."

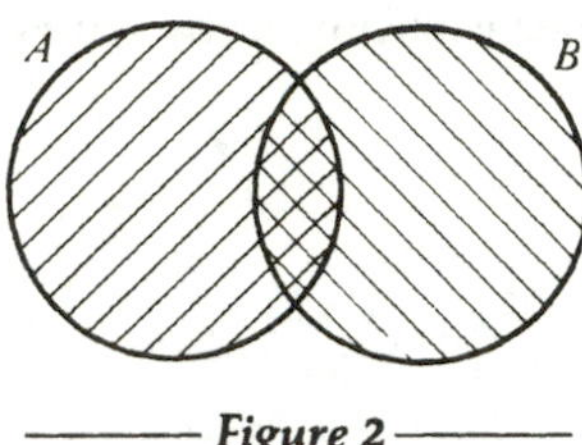

Figure 2

Given non-empty sets A and B, we can form the set $A \times B$ consisting of all ordered pairs (a, b), where $a \in A$, $b \in B$. One can give formal definitions of *ordered pair*—for example, the definition due to N. Wiener: $(a, b) = \{\{a\}, \{a, b\}\}$—which, in effect, selects two elements, $a \in A$ and $b \in B$, and then distinguishes one of them, say, a, as being "it" in some sense, notationally designated as "first." This makes it possible to prove the most important fact about ordered pairs: $(a_1, b_1) = (a_2, b_2)$ if and only if $a_1 = a_2$ and $b_1 = b_2$. Another approach (which we adopt here) is to introduce ordered pairs as undefined entities, merely postulating that $(a_1, b_1) = (a_2, b_2)$ if and only if $a_1 = a_2$ and $b_1 = b_2$.

The set $A \times B$ is called the *Cartesian product* of A and B, exemplified readily by the Cartesian plane $\mathbb{R} \times \mathbb{R}$ (see Figure 3).

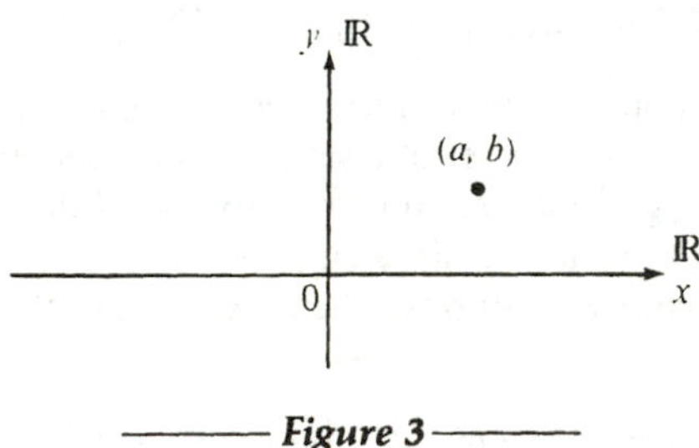

Figure 3

Other graphic examples of Cartesian products are the sets $\mathbb{R} \times \mathbb{Z}$, $\mathbb{Z} \times \mathbb{R}$, $\mathbb{Z} \times \mathbb{Z}$, which we ask you to visualize presently. For further practice, let $A = \{1, 2\}$, $B = \{1, 2, 3\}$. Write out the elements of $A \times A$, $A \times B$, and $B \times B$.

The notions of "ordered pair" and "Cartesian product of two sets" can be generalized: for finitely many sets $A_1, \ldots, A_n$ $(n \geq 1)$, we introduce *ordered n-tuples* $(a_1, \ldots, a_n)$, postulating that

$$(a_1, \ldots, a_n) = (b_1, \ldots, b_n)$$
$$(a_i, b_i \in A_i \quad \text{for} \quad i = 1, \ldots, n)$$

if and only if $a_i = b_i$ for each $i = 1, \ldots, n$.

The set of all ordered *n*-tuples $(a_1, \ldots, a_n)$ $(a_i \in A_i, i = 1, \ldots, n)$ is the *Cartesian product* $A_1 \times \ldots \times A_n$ of the sets $A_1, \ldots, A_n$.

Binary Relations

Let A be a non-empty set. A binary relation on A selects certain ordered pairs from $A \times A$. (For example, the relation $<$ on $\mathbb{Z}$ selects the ordered pairs (a, b) such that $a < b \quad (a, b \in Z)$.) For simplicity, we identify each binary relation with the set of ordered pairs it selects:

Definition 1.2.1: If A is a non-empty set and R is a non-empty subset of $A \times A$, then R is a *binary relation on A*.

Example:

The relation $<$ on the set $A = \{1, 2, 3\}$ is the set $\{(1, 2), (1, 3), (2, 3)\}$, while the relation $\leq$ on A is the set $\{(1, 1), (1, 2), (1, 3), (2, 2), (2, 3), (3, 3)\}$.

Since it is clumsy to write, e.g., $(a, b) \in <$ instead of $a < b$, we continue to use the customary notation: if R is a binary relation on A and $(a, b) \in R$, we write aRb.

Several special properties which a given binary relation may or may not have will be of interest to us.

Definition 1.2.2: Let R be a binary relation on a set A. Then R is

(a) *reflexive* if aRa holds for each $a \in A$;

(b) *symmetric* if aRb implies bRa $(a, b \in A)$;

(c) *transitive* if aRb and bRc implies aRc $(a, b, c \in A)$;

(d) *anti-symmetric* if aRb and bRa implies $a = b$ $(a, b \in A)$;

(e) *trichotomous* if for each $a, b \in A$, exactly one of the following holds: aRb, $a = b$, bRa.

If R is reflexive, symmetric, and transitive, then R is an *equivalence relation* on A.

If R is reflexive, anti-symmetric, and transitive, then R is a *partial order relation* on A.

If R is transitive and trichotomous, then R is a (*total*) *order relation* on A.

Example 1:

Equality is an equivalence relation on any set.

Example 2:

Similarity is an equivalence relation on the set of all triangles.

Example 3:

For each positive integer n, similarity of matrices is an equivalence relation on the set of all $n \times n$ matrices with real entries. (Recall: if A, B are $n \times n$ matrices with real entries, then matrix A is similar to matrix B if there exists a non-singular $n \times n$ matrix P with real entries such that $P^{-1}AP = B$.)

Example 4:

$\leq$ is a partial order relation on the set of all real numbers (or rationals, or integers).

Example 5:

$\subset$ is a partial order relation on the power set of any set. (See Exercise 1.2.2.)

Example 6:

$<$ is a total order relation on the set of all real numbers (or rationals, or integers).

Equivalence relations play a fundamental role in abstract algebra. Let us examine them more closely.

Definition 1.2.3: Let R be an equivalence relation on a set A. For each $a \in A$, let $\bar{a} = \{x \in A \mid xRa\}$. The set $\bar{a}$ is *the equivalence class of a modulo R*. The set whose elements are the equivalence classes $\bar{a}$ $(a \in A)$ is the *factor set of A modulo R*.

Lemma: If R is an equivalence relation on A, then, for $a, b \in A$, $aRb \Leftrightarrow \bar{a} = \bar{b}$. (In words: a bears the relation R to b if and only if the equivalence class of a with respect to R is equal to the equivalence class of b with respect to R.)

Proof: Suppose aRb. Let $y \in \bar{a}$. Then yRa. From yRa and aRb, by transitivity, we have yRb. Hence $y \in \bar{b}$. But then $\bar{a} \subset \bar{b}$. By symmetry of R, bRa. Hence, similarly, $\bar{b} \subset \bar{a}$. It follows that $\bar{a} = \bar{b}$.

Conversely, suppose $\bar{a} = \bar{b}$. By reflexivity, we have aRa, hence $a \in \bar{a}$. But then $a \in \bar{b}$, and so aRb, ∎

Definition 1.2.4: Let A be a non-empty set. A set $\mathscr{P}$ of non-empty subsets of A is a *partition of A* if

(1) $A = \bigcup_{X \in \mathscr{P}} X$

and

(2) if $X, Y \in \mathscr{P}$, then either $X = Y$ or $X \cap Y = \varnothing$.

(Informally, if $\mathscr{P}$ is a partition of A, we often say: the sets belonging to $\mathscr{P}$ *partition* A. The sets in $\mathscr{P}$ are sometimes called the *cells* of the partition $\mathscr{P}$.)

Theorem 1.2.1 (Fundamental Partition Theorem): Let A be a non-empty set, and let R be an equivalence relation on A. Then the equivalence classes with respect to R form a partition of A.

Proof: Let $\mathscr{P} = \{\bar{x} \mid x \in A\}$. Since R is reflexive, $a \in \bar{a}$ for each $a \in A$. Hence $A \subset \bigcup_{x \in A} \bar{x}$. On the other hand, $\bigcup_{x \in A} \bar{x} \subset A$, and so $A = \bigcup_{x \in A} \bar{x}$. Thus $\mathscr{P}$ satisfies Condition 1 of Definition 1.2.4.

For $x, y \in A$, suppose $\bar{x} \cap \bar{y} \neq \varnothing$. Then there is some $z \in A$ such that $z \in \bar{x}$ and $z \in \bar{y}$. But then zRx and zRy. By the Lemma, we have $\bar{z} = \bar{x}$ and $\bar{z} = \bar{y}$. It follows that $\bar{x} = \bar{y}$, and so $\mathscr{P}$ satisfies Condition 2 of Definition 1.2.4. ■

The converse of this theorem holds also: given any partition $\mathscr{P}$ of a set A, there exists an equivalence relation R on A such that $\mathscr{P}$ is the set of all equivalence classes with respect to R (Exercise 1.2.4).

Example 1:

On the set $\mathbb{R}$ of all real numbers, define R by aRb if $|a| = |b|$. Clearly R is an equivalence relation on $\mathbb{R}$. For $x \in R$, $\bar{x} = \{-x, x\}$. Every real number belongs to one and only one of the sets $\{-x, x\}$.

Example 2:

On the set $\mathbb{C}$ of all complex numbers, define R by: zRw if $|z| = |w|$. (Note: if $z = a + bi$, $a, b \in \mathbb{R}$, then $|z| = \sqrt{a^2 + b^2}$.) R is an equivalence relation on $\mathbb{C}$. In the complex plane, the equivalence classes are represented by the concentric circles with center at the origin and arbitrary non-negative real radius (see Figure 4).

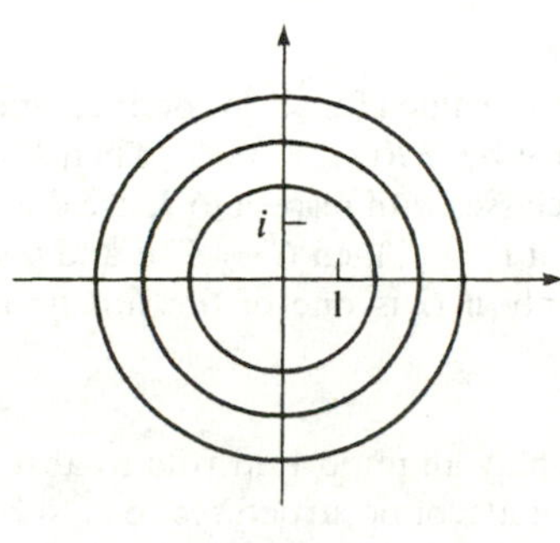

Figure 4

Example 3:

On the set $M_n(\mathbb{R})$ of all $n \times n$ matrices with real entries, define R by: ARB if $\det A = \det B$. Then R is an equivalence relation on $M_n(\mathbb{R})$. The equivalence class of any matrix A in $M_n(\mathbb{R})$ consists of all matrices X in $M_n(\mathbb{R})$ such that $\det X = \det A$. Every $n \times n$ matrix belongs to exactly one of these classes.

Example 4:

On the set $M_n(\mathbb{R})$, define R by: ARB if $B = P^{-1}AP$ for some non-singular $n \times n$ matrix P in $M_n(\mathbb{R})$ (i.e., if A is similar to B). The equivalence class of a matrix A consists of all matrices similar to A. These classes partition $M_n(\mathbb{R})$. (If matrix A represents the linear operator L on $\mathbb{R}^n$ relative to some basis for $\mathbb{R}^n$, then the equivalence class of A consists of all matrices representing L with respect to different bases for $\mathbb{R}^n$.)

Example 5:

On $\mathbb{R}$, define R by: $\alpha R \beta$ if $\alpha - \beta$ is an integral multiple of 2π. Then R is an equivalence relation on $\mathbb{R}$. If we associate with each real number α the angle of α radians whose initial side is the positive x-axis, then the equivalence classes with respect to R may be depicted by half-lines radiating from the origin.

Example 6:

On the set $\mathbb{R} \times \mathbb{R}$ of all ordered pairs of real numbers, define R by: $(a, b)\ R\ (c, d)$ if $a^2 + b^2 = c^2 + d^2$ $(a, b, c, d \in R)$. Then R is an equivalence relation on $\mathbb{R} \times \mathbb{R}$. The equivalence classes may be depicted by concentric circles about the origin, in the Cartesian plane.

Example 7:

Let A be the set of all real-valued functions defined and continuously differentiable on $\mathbb{R}$. Define R on A by: FRG if $F' = G'$. Then R is an equivalence relation on A. The equivalence classes with respect to R are the indefinite integrals $\int f(x)\, dx$ (f continuous), for: if $F' = f$, then $G' = F'$ if and only if $G(x) = F(x) + C$ for all $x \in R$, i.e., if and only if G is one of the functions in the set of functions denoted by "$\int f(x)\, dx$."

Partial order relations play an important role in abstract algebra since certain interesting substructures of algebraic structures (e.g., subgroups, ideals, etc.) form special partially ordered sets, called lattices, under inclusion.

Definition 1.2.5: Let L be a set, R a partial order relation on L, and let a, $b \in L$.

(1) An element $\gamma \in L$ is a *greatest lower bound* (*infimum*) of a and b with respect to the partial order relation R if

(i) $\gamma R a$ and $\gamma R b$,

and

(ii) cRa and cRb $(c \in L)$ implies $cR\gamma$.

(2) An element $\lambda \in L$ is a *least upper bound* (*supremum*) of a and b with respect to the partial order relation R if

(i) $aR\lambda$ and $bR\lambda$,

and

(ii) aRc and bRc $(c \in L)$ implies λRc.

In general, there is no guarantee of the existence of a greatest lower bound, or a least upper bound, of two elements of a partially ordered set. However, in case of existence, the following corollary guarantees uniqueness.

Corollary: If R is a partial order relation on a set L, then two elements $a, b \in L$ have at most one greatest lower bound and at most one least upper bound in L, with respect to R.

Proof: Let $a, b \in L$. Suppose γ_1 and γ_2 in L are both greatest lower bounds for a and b. By condition (i) of (1) (Definition 1.2.5), applied to γ_1, we have $\gamma_1 Ra$ and $\gamma_1 Rb$. Hence, by condition (ii) of (1), applied to γ_2, we have $\gamma_1 R\gamma_2$. On the other hand, by condition (i) of (1), applied to γ_2, we have $\gamma_2 Ra$ and $\gamma_2 Rb$. Hence, by condition (ii) of (1), applied to γ_1, we have $\gamma_2 R\gamma_1$. Since R is anti-symmetric, it follows that $\gamma_1 = \gamma_2$.

A similar proof applies to least upper bounds. ■

Notation: We write glb(a, b) and lub(a, b) to designate, respectively, the greatest lower bound and the least upper bound of two elements a and b.

Definition 1.2.6: Let L be a set, R a partial order relation on L. Then L forms a *lattice* with respect to R if every pair of elements of L has a greatest lower bound and a least upper bound in L, with respect to R.

Example:

The power set P_X of a set X forms a lattice with respect to the relation $\subset$ (set inclusion), with glb$(A, B) = A \cap B$ and lub$(A, B) = A \cup B$, for each $A, B \in P_X$. (See Exercise 1.2.8.)

A convenient way to depict a lattice, L, is by means of a *lattice diagram*, in which straight line segments connect any element of L to its immediate successors and predecessors in L.

For example, the power set P_X of the set $X = \{1, 2, 3\}$, considered as a lattice under set inclusion, may be represented by the following lattice diagram (Figure 5):

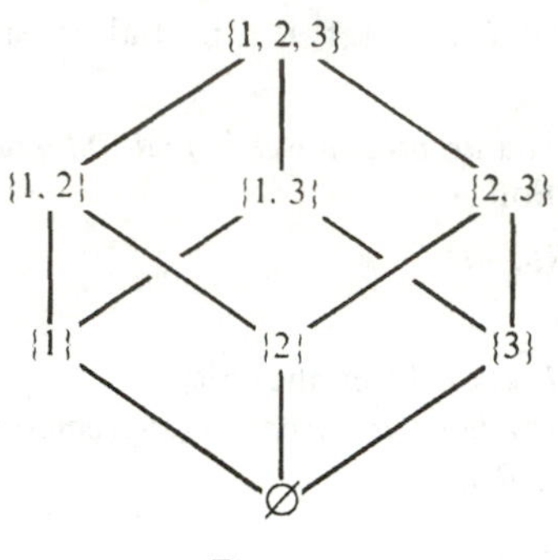

Figure 5

Functions

A function f from a set A to a set B associates with each element of A exactly one element of B. Thus, f selects a set of ordered pairs from $A \times B$. For simplicity, we define a function to *be* the set of all ordered pairs it selects (its graph!).

Definition 1.2.7: Let A, B be non-empty sets. A *function* (or *mapping*) *from A to B* is a subset f of $A \times B$ such that, for each $a \in A$, there is exactly one $b \in B$ for which $(a, b) \in f$. The set A is the *domain*, the set B the *codomain*, of f.

Notation: We write $f : A \to B$ to signify that f is a function with domain A and codomain B. Given $f : A \to B$, we write $b = f(a)$, or $f : a \mapsto b$, if $(a, b) \in f$.

If $f : A \to B$, then the *image* (or *range*) of f is the set

$$\operatorname{Im} f = \{b \in B \mid b = f(a) \quad \text{for some} \quad a \in A\}.$$

If $f : A \to B$, then f is *onto* (or *surjective*) if $\operatorname{Im} f = B$, and f is 1-1 (or *injective*) if $a_1 \neq a_2$ implies $f(a_1) \neq f(a_2)$. (Equivalently, f is 1-1 if $f(a_1) = f(a_2)$ implies $a_1 = a_2$.) If $f : A \to B$ is both 1-1 and onto, then f is a *bijection.*

If A, B, C are sets and $f : A \to B$, $g : B \to C$ are functions, then $g \circ f : A \to C$ is the function defined by

$$(g \circ f)(a) = g(f(a))$$

for each $a \in A$. The function $g \circ f$ is called the *composite* of f and g.

Theorem 1.2.2: Composition of functions is associative, i.e., if $f : A \to B$, $g : B \to C$, and $h : C \to D$, then

$$h \circ (g \circ f) = (h \circ g) \circ f.$$

Proof: It is easy to check that $h \circ (g \circ f)$ and $(h \circ g) \circ f$ are both functions with domain A and codomain D. To complete the proof, we need show only that they are the same set of ordered pairs, i.e., that they assign the same image to each element of their (common) domain. Thus, let $a \in A$. Then

$$(h \circ (g \circ f))(a) = h((g \circ f)(a)) = h(g(f(a)))$$

and

$$((h \circ g) \circ f)(a) = (h \circ g)(f(a)) = h(g(f(a))).$$

Hence the two functions $h \circ (g \circ f)$ and $(h \circ g) \circ f$ are equal. ∎

On any set A, we can define the function ι_A which sends every element of A to itself. The function $\iota_A : A \to A$ such that $\iota_A(a) = a$ for each $a \in A$ is the *identity function* on A. (For example, $\iota_{\mathbb{R}} : \mathbb{R} \to \mathbb{R}$ is the function whose graph is the line $y = x$.)

Definition 1.2.8: If $f : A \to B$, then $g : B \to A$ is a *left inverse function* for f if $g \circ f = \iota_A$; $g : B \to A$ is a *right inverse function* for f if $f \circ g = \iota_B$; and $g : B \to A$ is a *2-sided inverse function* (or simply, an *inverse function*) for f if $g \circ f = \iota_A$ and $f \circ g = \iota_B$. (In particular, in case $A = B$, this last condition reads simply $g \circ f = \iota_A = f \circ g$.)

Theorem 1.2.3: Let $f : A \to B$. Then (1) f is 1-1 if and only if f has a left inverse function; (2) f is onto if and only if f has a right inverse function; (3) f is 1-1 onto (a bijection) if and only if f has a 2-sided inverse function.

A function $f : A \to B$ has at most one 2-sided inverse function.

Proof: (1) Suppose $f : A \to B$ has a left inverse function $g : B \to A$. If $f(a_1) = f(a_2)$ $(a_1, a_2 \in A)$, then $g(f(a_1)) = g(f(a_2))$, so $(g \circ f)(a_1) = (g \circ f)(a_2)$, $\iota_A(a_1) = \iota_A(a_2)$, $a_1 = a_2$. Thus, f is 1-1.

Conversely, suppose f is 1-1. Define $g : B \to A$ by $g(b) = a$ if $f(a) = b$; $g(b) = a_0$ if $b \in B \setminus \operatorname{Im} f$, where a_0 is an arbitrary element chosen from A. Then, for each $a \in A$, $g(f(a)) = g(b) = a$ (where $b = f(a)$), and so $g \circ f = \iota_A$. Thus, g is a left inverse function of f.

(2) Suppose $f : A \to B$ has a right inverse function $g : B \to A$. Let $b \in B$. Then $b = \iota_B(b) = (f \circ g)(b) = f(g(b))$; hence $b \in \operatorname{Im} f$. But then $\operatorname{Im} f = B$, and so f is onto.

Conversely, suppose f is onto. For each $b \in B$, let $A_b = \{a \in A \mid f(a) = b\}$. Since f is onto, $A_b \neq \varnothing$ for each $b \in B$, and we may select*, for each $b \in B$, an element $a_b \in A_b$. Now define $g : B \to A$ by $g(b) = a_b$, for each $b \in B$. Then $(f \circ g)(b) = f(g(b)) = f(a_b) = b$, for each $b \in B$, and so $f \circ g = \iota_B$. Thus, g is a right inverse function for f.

(3) If f has a 2-sided inverse function $g : B \to A$, then f is 1-1 onto, by (1) and (2).

Conversely, if f is 1-1 onto, we need only define $g : B \to A$ by $g(b) = a$ if $f(a) = b$. Then $(f \circ g)(b) = f(g(b)) = f(a) = b$, for each $b \in B$ $(a = g(b))$, whence $f \circ g = \iota_B$, and $(g \circ f)(a) = g(f(a)) = a$ for each $a \in A$ $(b = f(a))$, whence $g \circ f = \iota_A$. Thus, g is a 2-sided inverse function of f.

---

* A more formal discussion of set theory would include the Axiom of Choice, which legitimizes such selections. It states that, given any set $\mathscr{C}$ of non-empty sets, there is a set consisting of elements, one chosen from each of the sets in $\mathscr{C}$.

Finally, suppose $f : A \to B$ has a left inverse $g_1 : B \to A$ and a right inverse $g_2 : B \to A$. Then $g_1 = g_1 \circ (f \circ g_2) = (g_1 \circ f) \circ g_2 = g_2$. From this, it follows immediately that a function $f : A \to B$ can have at most one 2-sided inverse function. ■

Recommended reading: [42], [44].

Exercises 1.2

True or False

1. If R is the binary relation defined on the set $\mathbb{Z}$ of all integers by xRy if $x + y = 0$, then R is an equivalence relation on $\mathbb{Z}$.
2. If R is the binary relation defined on $M_2(\mathbb{R})$ by ARB if $B = PAQ$, where P, Q are non-singular matrices in $M_2(\mathbb{R})$, then R is an equivalence relation on $M_2(\mathbb{R})$.
3. If $f : A \to B$, and R is defined on A by: xRy if $f(x) = f(y)$, then R is an equivalence relation on A.
4. There is a set that has the same number of elements as its power set.
5. If $<$ is an order relation on a set X, then $\leq$ (i.e., $<$ or $=$) is a partial order relation on X.

1.2.1. Let $A = \{1, 2, 3\}$, $B = \{2, 3, 4\}$. List the elements of each of the following sets: $A \cup B$, $A \cap B$, $A \times B$, $B \times A$, P_A, P_B.

1.2.2. Prove: *(1)* If A is a set, then $\varnothing \subset A$;
(2) If A is a set, then $A \subset A$;
(3) If A, B are sets such that $A \subset B$ and $B \subset A$, then $A = B$;
(4) If A, B, C are sets such that $A \subset B$ and $B \subset C$, then $A \subset C$.

1.2.3. Give an example to prove that a binary relation that is symmetric and transitive need not be reflexive. (It is very tempting to believe otherwise: from aRb and bRa, by transitivity, one has aRa! Why does this *not* prove that symmetry and transitivity imply reflexivity?)

1.2.4. Prove the converse of the Fundamental Partition Theorem: If $\mathscr{P}$ is any partition of a non-empty set A, then there exists an equivalence relation R on A such that the equivalence classes with respect to R are the cells of the partition $\mathscr{P}$.

1.2.5. Prove that a set A consisting of n elements (n a positive integer) has 2^n subsets. (Hint: Imagine yourself making up each subset by deciding, for each element of A, whether it is to be put into the subset or left out of it.)

1.2.6. Prove that the composite of two 1-1 maps is 1-1; the composite of two onto maps is onto; and the composite of two bijections is a bijection.

1.2.7. Let $f : X \to Y$ and let $Z = \text{Im } f$. Define functions $g : X \to Z$ and $h : Z \to Y$ such that $f = h \circ g$.

1.2.8. Prove that the power set P_X of any set X forms a lattice with respect to the relation $\subset$, with $\text{glb}(A, B) = A \cap B$ and $\text{lub}(A, B) = A \cup B$, for each A, $B \in P_X$.

1.2.9. *(1)* Draw a lattice diagram for the power set P_X of the set $X = \{1, 2, 3, 4\}$ under the partial order relation $\subset$.

(2) Let $L = \{1, 2, 3\}$. Show that L forms a lattice with respect to the relation $\leq$. Draw a lattice diagram depicting L.

1.2.10. Let X be a set and let A, B, C be subsets of X.

(1) Prove that $A \cup (B \cap C) = (A \cup B) \cap (A \cup C)$
and $A \cap (B \cup C) = (A \cap B) \cup (A \cap C)$. (Distributive Laws)

(2) Prove that $X \setminus (A \cap B) = (X \setminus A) \cup (X \setminus B)$
and $X \setminus (A \cup B) = (X \setminus A) \cap (X \setminus B)$. (De Morgan's Laws)

1.2.11. Let $\mathbb{Z}^+$ be the set of all positive integers and let $\Phi : \mathbb{Z}^+ \to \mathbb{Z}^+$ be defined by: $\Phi(k) = k + 1$ for each $k \in \mathbb{Z}^+$.

(1) Prove that Φ is 1-1 but not onto.

(2) Define infinitely many left inverse functions for Φ.

1.2.12. Determine whether each of the following functions has a right, left, or two-sided inverse function and find the two-sided inverse function if there is one; otherwise, find at least two one-sided inverse functions, or explain why none exist.

(a) $f : \mathbb{R}^+ \cup \{0\} \to \mathbb{R}^+ \cup \{0\}$ defined by $f(x) = \sqrt{x}$ $(x \geq 0$ in $\mathbb{R})$;

(b) $g : \mathbb{R}^+ \cup \{0\} \to \mathbb{R}$ defined by $g(x) = \sqrt{x}$ $(x \geq 0$ in $\mathbb{R})$;

(c) $h : \mathbb{R} \to \mathbb{R}$ defined by $h(x) = x^2$ $(x \in \mathbb{R})$;

(d) $k : \mathbb{R} \to \mathbb{R}^+ \cup \{0\}$ defined by $k(x) = x^2$ $(x \in \mathbb{R})$.

1.2.13. Let $f : A \to B$ be a function. On A define a binary relation R by: aRb if $f(a) = f(b)$.

(1) Prove that R is an equivalence relation on A.

(2) For any element $b \in \operatorname{Im} f$, let $f^{-1}(b) = \{a \in A \mid f(a) = b\}$.

This set is called the *inverse image* of b with respect to f. (The notation is in no way intended to suggest the existence of an inverse function, f^{-1}, for f).

Prove that the equivalence classes with respect to the relation R defined in (1) are precisely the inverse images of the elements of $\operatorname{Im} f$.

1.2.14. Despite the informality of our treatment of set theory, we try in this exercise to give you a taste of the proof that *there is no "set of all sets."*

Indeed, suppose there *is* a set S of which every set is an element. Then S has a subset K consisting of all those sets X that are *not* elements of themselves (i.e., such that $X \notin X$). Now ask whether $K \in K$. The result will be tantalizing!

1.2.15. If A, B are non-empty sets, then A is defined to be *equipotent* to B if there exists a bijection from A to B. Prove that equipotence is an equivalence relation on any non-empty set of non-empty sets. (If two sets are equipotent, we say that they have "the same cardinality.")

1.2.16. Let A be a non-empty set and let P_A be the power set of A.

(1) Prove that A is equipotent to a subset of P_A (see the definition in the preceding exercise).

(2) Prove that there is *no* mapping of A *onto* P_A (hence, certainly A is not equipotent to P_A). (Hint: Suppose there exists a mapping $f : A \to P_A$ that maps A *onto* P_A. Let $K = \{a \in A \mid a \notin f(a)\}$. Then $K \in P_A$. Since f is onto, $K = f(k)$ for some $k \in A$. Now ask whether $k \in K$.)

1.2.17. *(1)* If A, B are sets, define $A \precsim B$ by: either $A = \varnothing$ or A is equipotent to a subset of B. Prove that $\precsim$ is a reflexive and transitive relation on any non-empty set of sets. ($\precsim$ is also anti-symmetric and is thus a partial order relation, by a theorem in set theory known as the Cantor-Schroeder-Bernstein Theorem.)

(2) If A, B are sets, define $A \prec B$ by: $A \preceq B$ holds, but $B \preceq A$ does not hold. Use Exercise 1.2.16 to prove that, for any set A, $A \prec P_A$. (This result implies that, for every set, there is a "larger" set, in the sense of the relation $\prec$, i.e., there is no largest cardinality, either finite or transfinite.)

1.3 The Integers

Among the indispensable tools in our study of abstract algebra will be certain properties of the integers—in particular: well-ordering of the positive integers, induction, and the division algorithm. Also indispensable, mainly as motivation for several important algebraic notions, will be the relation of congruence.

We begin with a list of axioms (a list that can be expressed much more compactly once algebraic structures have been developed—see Exercise 3.2.26).

Axioms for $\mathbb{Z}$

Let $\mathbb{Z}$ be a set of elements, to be called *integers*. Assume that $\mathbb{Z}$ has at least two elements, denoted by "0" and "1," with $0 \neq 1$.

***Axioms regarding* $+$:**

A_0: For each $(a, b) \in \mathbb{Z} \times \mathbb{Z}$, there is exactly one integer $a + b \in \mathbb{Z}$.

A_1: $a + b = b + a$ for all $a, b \in \mathbb{Z}$.

A_2: $a + (b + c) = (a + b) + c$ for all $a, b, c \in \mathbb{Z}$.

A_3: $a + 0 = a$ for all $a \in \mathbb{Z}$.

A_4: For each $a \in \mathbb{Z}$, there is an integer $-a \in \mathbb{Z}$ such that $a + (-a) = 0$.

***Axioms regarding* $\cdot$:**

M_0: For each $(a, b) \in \mathbb{Z} \times \mathbb{Z}$, there is exactly one integer $a \cdot b \in \mathbb{Z}$.

M_1: $a \cdot b = b \cdot a$ for all $a, b \in \mathbb{Z}$.

M_2: $a \cdot (b \cdot c) = (a \cdot b) \cdot c$ for all $a, b, c \in \mathbb{Z}$.

M_3: $a \cdot 1 = a$ for all $a \in \mathbb{Z}$.

***Distributivity of* $\cdot$ *over* $+$:**

D: $a \cdot (b + c) = a \cdot b + a \cdot c$ for all $a, b, c \in \mathbb{Z}$.

***Axioms regarding* $<$:**

O_1: For each $a, b \in \mathbb{Z}$, exactly one of the following holds:

$$a < b$$
$$a = b$$
$$b < a.$$

O_2: For $a, b, c \in \mathbb{Z}$, $a < b$ and $b < c$ implies $a < c$. (Note: O_1 and O_2 state that $<$ is an order relation on $\mathbb{Z}$.)

O_3: If $a, b, c \in \mathbb{Z}$ and $a < b$, then $a + c < b + c$.

O_4: If $a, b, c \in \mathbb{Z}$, with $a < b$ and $0 < c$, then $a \cdot c < b \cdot c$.

Notations: Write $\mathbb{Z}^+ = \{a \in \mathbb{Z} | 0 < a\}$. (The elements of $\mathbb{Z}^+$ are called *positive* integers.)

Write $a \leq b$ if either $a < b$ or $a = b$.

Write $a - b$ for $a + (-b)$.

Well-ordering axiom:

WO: Every non-empty subset of $\mathbb{Z}^+$ has a least element, i.e., if $M \subset \mathbb{Z}^+$, $M \neq \varnothing$, then there is an integer $s \in M$ such that $s \leq m$ for all $m \in M$.

The use of these axioms may be illustrated by proving various very basic properties of the integers. We begin with several lemmas whose proofs are elementary and may be helpful to readers who have not "proved" before. *More impatient readers may wish to skip these lemmas, with the understanding that the results contained in them will be assumed in proving the theorems that follow.*

Lemma 1: If $a \in \mathbb{Z}$ and $0 < a$, then $-a < 0$.

Proof: From $0 < a$, by O_3 we have $0 + (-a) < a + (-a)$. Hence, using A_1, A_3, and A_4, we obtain $-a < 0$. ■

Lemma 2: If $z \in \mathbb{Z}$ and $a + z = a$ for all $a \in \mathbb{Z}$, then $z = 0$.

Proof: In particular, $0 + z = 0$, and so $z = 0$, by A_1 and A_3. ■

Lemma 3: If $a, b \in \mathbb{Z}$ such that $a + b = 0$, then $b = -a$.

Proof: By A_0 and A_2, $((-a) + a) + b = -a + (a + b)$. Thus, if $a + b = 0$, then $0 + b = -a + 0$, by A_1 and A_4. But then $b = -a$, by A_3. ■

Lemma 4: For each $a \in \mathbb{Z}$, $-(-a) = a$.

Proof: Let $a \in \mathbb{Z}$. By A_4 and A_1, $0 = a + (-a) = -a + a$. By Lemma 3, it follows that $-(-a) = a$. ■

Lemma 5: For all $a \in \mathbb{Z}$, $a \cdot 0 = 0 = 0 \cdot a$.

Proof: Let $a \in \mathbb{Z}$. By A_3 and D, we have $a \cdot 0 = a \cdot (0 + 0) = a \cdot 0 + a \cdot 0$. But then, by A_4, A_2, and A_3,

$$0 = -(a \cdot 0) + a \cdot 0 = -(a \cdot 0) + (a \cdot 0 + a \cdot 0) = (-(a \cdot 0) + a \cdot 0) + a \cdot 0 = 0 + a \cdot 0 = a \cdot 0.$$

Thus, using A_1, we have $a \cdot 0 = 0 = 0 \cdot a$. ■

Lemma 6: For $a, b \in \mathbb{Z}$, $a \cdot (-b) = -(a \cdot b) = (-a) \cdot b$, and $(-a) \cdot (-b) = a \cdot b$.

Proof: Let $a, b \in \mathbb{Z}$. By D, A_4 and Lemma 5, we have $a \cdot b + a \cdot (-b) = a \cdot (b + (-b)) = a \cdot 0 = 0$. Hence, by Lemma 3, $a \cdot (-b) = -(a \cdot b)$. A similar proof shows that $(-a) \cdot b = -(a \cdot b)$. Using these results and Lemma 4, we have $(-a) \cdot (-b) = -(a \cdot (-b)) = -(-(a \cdot b)) = ab$. ∎

Lemma 7: $0 < 1$.

Proof: By O_1, exactly one of the following holds: $0 < 1$; $0 = 1$; $1 < 0$. By hypothesis, $0 \neq 1$. Suppose $1 < 0$. Then, by Lemma 1, $0 < -1$. Hence, by O_4, $0 \cdot (-1) < (-1) \cdot (-1)$. But then, by Lemmas 5 and 6, $0 < 1$, contrary to our hypothesis that $1 < 0$. By O_1, we conclude: $0 < 1$ must hold. ∎

Notation: Henceforth, we shall use the notations "$a < b$" and "$b > a$" interchangeably.

Up to this point, we have made no use of the well-ordering axiom for $\mathbb{Z}^+$. Indeed, all of the results contained in Lemmas 1–7 are valid more generally for rational numbers and, even more generally, for real numbers. (They would, in fact, hold in any ordered integral domain—see Exercises 3.2.24–3.2.26.) A drastic change occurs as soon as we make use of the well-ordering axiom: the discrete structure of the integers begins to emerge.

Theorem 1.3.1: There is no integer a such that $0 < a < 1$. (Equivalently, every positive integer is ≥ 1, or: 1 is the least positive integer.)

Proof: By Lemma 7, $0 < 1$. Suppose there is an integer a such that $0 < a < 1$. Let $M = \{a \in \mathbb{Z} | 0 < a < 1\}$. Then $M \subset \mathbb{Z}^+$ and $M \neq \varnothing$; hence, M contains a least element, m_0, and we have $0 < m_0 < 1$. Multiplying the double inequality by m_0, we obtain $0 < m_0^2 < m_0 < 1$. But this contradicts the status of m_0 as least in M! It follows that there is *no* integer a such that $0 < a < 1$. ∎

Corollary: Let $k \in \mathbb{Z}$.

(1) There is no integer such that $k < a < k + 1$.

(2) If $T_k = \{a \in \mathbb{Z} | k < a\}$, then every non-empty subset of T_k has a least element. (In particular, every non-empty set of non-negative integers has a least element.)

We leave the proof as an exercise (see Exercise 1.3.1).

Theorem 1.3.2 (First Induction Principle): If $M \subset \mathbb{Z}^+$ such that

(a) $1 \in M$

and, for $n \in \mathbb{Z}$,

(b) $n \in M \Rightarrow n + 1 \in M$,

then $M = \mathbb{Z}^+$.

Proof: Suppose $M \neq \mathbb{Z}^+$. Let $\bar{M} = \mathbb{Z}^+ \backslash M$. Then $\bar{M} \neq \varnothing$ and $\bar{M} \subset \mathbb{Z}^+$. By well-ordering of $\mathbb{Z}^+$, $\bar{M}$ has a least element, s. Now, $s \neq 1$ since $1 \in M$ (by (a)), hence $1 \notin \bar{M}$. So $s = m + 1$ where $0 < m$. Since s is *least* in $\bar{M}$, and $m < s$, we have $m \in M$. By (b), $s = m + 1 \in M$. Contradiction!

It follows that $M = \mathbb{Z}^+$. ■

This theorem is often stated in propositional, rather than set-theoretic form. Let $P(n)$ be a proposition regarding a positive integer n such that

(a) $P(1)$ is true

and

(b) if $P(n)$ is true, then $P(n + 1)$ is true.

Then $P(n)$ is true for all positive integers n.

Letting M be the truth set of P (i.e., the set of all positive integers n such that $P(n)$ is true), one easily reconciles these two versions of the First Induction Principle.

Theorem 1.3.3 (Second Induction Principle): For each positive integer n, let $I_n = \{k \in \mathbb{Z}^+ | k \leq n\}$. Let $M \subset \mathbb{Z}^+$ be such that

(a) $1 \in M$

and

(b) $I_n \subset M \Rightarrow n + 1 \in M$.

Then $M = \mathbb{Z}^+$.

Proof: Let $K = \{n \in \mathbb{Z}^+ | I_n \subset M\}$. By Theorem 1.3.1, $I_1 = \{1\}$; hence, by (a), $I_1 \subset M$. But then $1 \in K$.

Suppose $n \in K$. Then $I_n \subset M$, and so, by (b), we have that $n + 1 \in M$. By the Corollary of Theorem 1.3.1, $I_{n+1} \subset M$, whence $n + 1 \in K$. But then, by the First Induction Principle, $K = \mathbb{Z}^+$. From this, we conclude that $I_n \subset M$ for each $n \in \mathbb{Z}^+$. But then (since $n \in I_n$ for each $n \in \mathbb{Z}^+$), we have $\mathbb{Z}^+ \subset M$. Since $M \subset \mathbb{Z}^+$, it follows that $M = \mathbb{Z}^+$. ■

As another illustration of a proof using well-ordering, we prove the following important theorem.

Theorem 1.3.4 Division Algorithm: Let $a, b \in \mathbb{Z}$, $b \neq 0$. Then there exist $q, r \in \mathbb{Z}$ such that $a = bq + r$, where $0 \leq r < |b|$. The integers q and r are uniquely determined for given a and b. (Recall that a is called the *dividend*, b the *divisor*, q the *quotient*, and r the *remainder*.)

Proof: *Existence.* Case 1. First suppose $b > 0$ and $a \geq 0$. Let $M = \{a - bx | x \in \mathbb{Z}, a - bx \geq 0\}$. Since $a = a - b \cdot 0 \geq 0$, we have $a \in M$; hence $M \neq \varnothing$. By the Well-Ordering Axiom for $\mathbb{Z}^+$ and Theorem 1.3.1, Corollary, M has a least element r. Then $r = a - bq$ for some $q \in \mathbb{Z}$, and so $a = bq + r$, with $0 \leq r$. We need to show that $r < b = |b|$. Suppose $r \geq b$. Then $r = b + s$ for some integer $s \geq 0$; hence $s = r - b = a - b(q + 1) \in M$. But $s = r - b < r$, and r is the least element of M. Contradiction! It follows that $0 \leq r < b = |b|$, as required.

Case 2. Next, suppose $b > 0$ and $a < 0$. By Lemma 1, $-a > 0$; hence, by Case 1, there are integers $\bar{q}$ and $\bar{r}$ such that

$$-a = b\bar{q} + \bar{r}, \quad 0 \leq \bar{r} < b = |b|.$$

If $\bar{r} = 0$, we have $a = bq + r$, where $q = -\bar{q}$, $r = 0$. If $\bar{r} > 0$, then $a = b(-\bar{q}) - \bar{r} = b(-\bar{q} - 1) + b - \bar{r}$. Let $q = -\bar{q} - 1$, $r = b - \bar{r}$. Then $a = bq + r$ and, from $0 < \bar{r} < b$, we conclude that $0 < r < b = |b|$.

Case 3. Finally, suppose $b < 0$, $a \in \mathbb{Z}$. Then $-b > 0$; hence, by Cases 1 and 2, there are integers $\bar{q}$ and $\bar{r}$ such that

$$a = (-b)\bar{q} + r, \quad 0 \leq \bar{r} < -b = |b|.$$

But then $a = bq + r$, $0 \leq r < |b|$, where $q = -\bar{q}$ and $r = \bar{r}$.

Uniqueness. For a, b in $\mathbb{Z}$, $(b \neq 0)$, suppose $a = bq_1 + r_1 = bq_2 + r_2$, $0 \leq r_1 < |b|$, $0 \leq r_2 < |b|$ $(q_1, q_2, r_1, r_2 \in \mathbb{Z})$. Suppose $r_1 \neq r_2$, say, $r_1 < r_2$. Then $0 < b(q_1 - q_2) = r_2 - r_1 < |b|$. But then $0 < |b|\,|q_1 - q_2| < |b|$. However, $q_1 - q_2 \neq 0$ (else $r_1 = r_2$); hence $|q_1 - q_2| \geq 1$ (by Theorem 1.3.1) and $|b|\,|q_1 - q_2| \geq |b| \cdot 1 = |b|$. Contradiction. It follows that $r_1 = r_2$. But then, from $bq_1 + r_1 = bq_2 + r_2$, we conclude that $bq_1 = bq_2$. Since $b \neq 0$, this implies that $q_1 = q_2$ (see Exercise 1.3.4). ∎

An important binary relation on $\mathbb{Z}$ is the relation of divisibility.

Definition 1.3.1: If a, b are integers such that $a = bq$ for some integer q, then b *divides* a (or: *b is a divisor of a*).

Notation: We write $b|a$ to signify that b divides a. (Careful! Do not confuse the relation symbol "|" with the operation symbol "/". For example, $3|6$ because $3 \cdot 2 = 6$, i.e., $6/3 = 2$).

Note that, by this definition, $0|0$ (which is uninteresting, but does no harm), while $0|a$ is impossible for all non-zero integers a.

Definition 1.3.2: An integer $p > 1$ such that the only divisors of p are ± 1 and $\pm p$ is called a *prime*. An integer $n > 1$ that is not a prime is a *composite integer.*

Definition 1.3.3: If $a, b \in \mathbb{Z}$, then an integer d is a *greatest common divisor* of a and b if

(1) $d|a$ and $d|b$ (i.e., d is a common divisor of a and b)

and

(2) if c is an integer such that $c|a$ and $c|b$, then $c|d$ (i.e., every common divisor of a and b divides d).

Remark: From Definition 1.3.3, it follows immediately that any two greatest common divisors d_1, d_2 of a pair of integers a, b must divide each other, so that $d_1 = \pm d_2$ (see Exercise 1.3.7.) This implies that two integers can have *at most one positive* greatest common divisor. In the following theorem, we prove the existence of a (unique) positive greatest common divisor, for any pair of integers a, b not both zero.

Theorem 1.3.5: Every pair of integers a, b not both 0 has a unique positive greatest common divisor d. There exist integers s, t such that $d = as + bt$.

Proof: Consider the set $M = \{ax + by \mid x, y \in \mathbb{Z}, ax + by > 0\}$. $M \neq \emptyset$, for: since either a or b is non-zero, $\pm a$ or $\pm b$ is a positive integer contained in M (e.g., if $a < 0$, then $-a = a(-1) + b \cdot 0 \in M$). Hence M contains a least positive integer $d = ax_0 + by_0$ $(x_0, y_0 \in \mathbb{Z})$. By Theorem 1.3.4, there are integers q, r such that $a = dq + r$, $0 \le r < d$, and $r = a - dq = a - (ax_0 + by_0)q = a(1 - x_0q) + b(-y_0q)$. Thus, if $r > 0$, then $r \in M$, contrary to the hypothesis that d is least in M. We conclude that $r = 0$, and that $d \mid a$. Similarly, $d \mid b$, and so d satisfies Condition 1 of Definition 1.3.3.

Does d satisfy Condition 2 of Definition 1.3.3? Suppose c is an integer such that $c \mid a$ and $c \mid b$. Then $a = ch$, $b = ck$ for some $h, k \in \mathbb{Z}$. Hence $d = ax_0 + by_0 = (ch)x_0 + (ck)y_0 = c(hx_0 + ky_0)$, and so $c \mid d$.

It follows that d is a greatest common divisor of a and b. By the Remark following Definition 1.3.3, d is the *only* positive greatest common divisor of a and b. ■

Notation: For a, b not both 0, we write "gcd(a, b)" to designate the positive greatest common divisor of a and b.

Remark: Our definitions imply that 0 is the (unique) greatest common divisor of 0 and 0. This does no harm and is somewhat useful in generalizations to principal ideal domains (see Chapter 3).

Definition 1.3.4: Two integers a, b are *relatively prime* if $\gcd(a, b) = 1$.

Example:

$\gcd(6, 35) = 1$; hence 6 and 35 are relatively prime.

Corollary (of Definition 1.3.4 and Theorem 1.3.5): Two integers a, b are relatively prime if and only if there exist integers s, t such that $1 = sa + tb$.

We leave the proof as an exercise (Exercise 1.3.10).

Remark: Definitions 1.3.3 and 1.3.4, Theorem 1.3.5, and the preceding Corollary can be extended to an arbitrary finite number of integers $a_1, \ldots, a_n$ (see Exercise 1.3.11).

An important consequence of Theorem 1.3.5 is the following theorem:

Theorem 1.3.6: Let a, b, c be integers, with $\gcd(a, c) = 1$. If $c \mid ab$, then $c \mid b$.

Proof. There exist integers s, t such that

$$1 = as + ct,$$

whence

$$b = (as + ct)b = (as)b + (ct)b = (ab)s + c(tb).$$

If $c|ab$, then $ab = ch$ for some $h \in \mathbb{Z}$, and so $b = (ch)s + c(tb) = c(hs) + c(tb) = c(hs + tb)$. Thus $c|b$. ■

An immediate consequence of Theorem 1.3.6 is Euclid's Lemma on primes.

Lemma (Euclid): If p is a prime and a, b are integers such that $p|ab$, then $p|a$ or $p|b$.

Proof: It suffices to prove that, if $p|ab$ and $p\nmid a$, then $p|b$. Thus, suppose $p\nmid a$. Then $\gcd(p, a) = 1$. Hence, by Theorem 1.3.6, $p|b$. ■

By a simple induction argument, Euclid's Lemma can be extended to any finite number of factors: if p is a prime such that $p|a_1 \cdot \cdots \cdot a_m$, where the a_i are integers, then $p|a_j$ for some j, $1 \leq j \leq m$. It is in this form that Euclid's Lemma serves as a lemma for the Fundamental Theorem of Arithmetic, which we are now ready to prove.

Theorem 1.3.7 (Fundamental Theorem of Arithmetic): Every positive integer $n > 1$ is equal to a product of primes. This factorization is unique up to rearrangement of the prime factors.

Proof: *Existence.* We use the Second Induction Principle (Theorem 1.3.3). Let M be the set consisting of 1 and all positive integers that are products of primes. Then $1 \in M$, by hypothesis. Let n be an integer such that $n > 1$ and $I_{n-1} \subset M$. If n is prime, then clearly $n \in M$. If n is composite, then there are integers h, k such that $n = hk$, $1 < h < n$, $1 < k < n$. But then, by Theorem 1.3.1, Corollary, we have $h, k \in I_{n-1}$, and each is a product of primes, hence so is n. By the Second Induction Principle, we conclude that $M = \mathbb{Z}^+$, and so every positive integer $n > 1$ is a product of primes.

Uniqueness. We again use the Second Induction Principle. Let K be the set consisting of 1 and all positive integers whose prime factorization is unique (up to rearrangement of factors). Then $1 \in K$ by hypothesis. For $n > 1$, suppose that $I_{n-1} \subset K$, and that n has prime factorizations

$$n = p_1 \ldots p_s = q_1 \ldots q_t \quad (s, t \in \mathbb{Z}^+), \tag{1}$$

where the p_i and the q_j are primes, not necessarily distinct. If n is prime, then clearly $s = t = 1$ and $p_1 = n = q_1$. If n is composite, then, by Euclid's Lemma, since $p_1|q_1 \ldots q_t$, we have $p_1|q_j$ for some j, $1 \leq j \leq t$; but then $p_1 = q_j$ since $p_1 \neq 1$ and q_j is prime. Without loss of generality, we may label the prime factors on the right-hand side of (1) in such a way that $p_1 = q_1$. Then, since n is not prime, we have $1 < n/p_1 = p_2 \ldots p_s = q_2 \ldots q_t < n$. Since $I_{n-1} \subset K$, we conclude that $s - 1 = t - 1$ and, after suitable rearrangement and relabeling of the factors, $p_i = q_i$ for each $i = 2, \ldots, s$. But then $s = t$ and $p_i = q_i$ for each $i = 1, \ldots, s$, and so $n \in K$. By the Second Induction Principle, it follows that $K = \mathbb{Z}^+$; but then, for all integers $n > 1$, the prime factorization is unique (up to rearrangement of factors). ■

Corollary: Every positive integer $n > 1$ may be expressed in one and only one way as $n = p_1^{\alpha_1} \dots p_m^{\alpha_m}$, where $m \geq 1$, the p_i are primes, with $p_1 < p_2 < \cdots < p_m$, and $\alpha_i \geq 1$ in $\mathbb{Z}$, for each $i = 1, \dots, m$.

We now embark on our final venture in $\mathbb{Z}$ by introducing an equivalence relation which will prove to be the prototype of an important relation in group theory.

Let n be a positive integer. On $\mathbb{Z}$, define a binary relation as follows:

Definition 1.3.5: Let $a, b \in \mathbb{Z}$. Then a *is congruent to* b modulo n ($a \equiv b \bmod n$) if $n | a - b$.

Theorem 1.3.8: For n any positive integer, $\equiv$ mod n is an equivalence relation on $\mathbb{Z}$.

Proof: *Reflexivity*: If $a \in \mathbb{Z}$, then $a \equiv a \bmod n$ since $n | a - a$. *Symmetry*: If $a, b \in \mathbb{Z}$ and $a \equiv b \bmod n$, then $n | a - b$; hence $n | b - a$, and so $b \equiv a \bmod n$. *Transitivity*: if $a, b, c \in \mathbb{Z}$, with $a \equiv b \bmod n$ and $b \equiv c \bmod n$, then $n | a - b$, $n | b - c$. But then, since $a - c = a - b + b - c$, we have $n | a - c$, and so $a \equiv c \bmod n$. ■

For every integer k, the equivalence class of k with respect to congruence modulo n is the set

$$\bar{k} = \{x \in \mathbb{Z} | x \equiv k \bmod n\} = \{k + nq | q \in \mathbb{Z}\}.$$

$\bar{k}$ is called the *residue class of* k modulo n.

Theorem 1.3.9: Let n be a positive integer. Then every integer is congruent modulo n to exactly one of the integers $0, 1, \dots, n - 1$.

Proof: Let $k \in \mathbb{Z}$. By the Division Algorithm, $k = nq + r$ for some $q, r \in \mathbb{Z}$, $0 \leq r < n$. Since $k - r = nq$, $k \equiv r \bmod n$. If also $k \equiv r_1 \bmod n$, $0 \leq r_1 < n$, then $k - r_1 = nq_1$ for some $q_1 \in \mathbb{Z}$, and so $k = nq_1 + r_1$. By the uniqueness of quotient and remainder, we have $q_1 = q$ and $r_1 = r$. ■

Recall that, if R is an equivalence relation on a set A, then for $a, b \in A$, $aRb \Leftrightarrow \bar{a} = \bar{b}$ (Lemma, p. 8). Here, we have $a \equiv b \bmod n \Leftrightarrow \bar{a} = \bar{b} \Leftrightarrow n | a - b$. Using this fact, we immediately obtain the following corollary.

Corollary: If $k \in \mathbb{Z}$, then there is exactly one $r \in \mathbb{Z}$, $0 \leq r < n$, such that $\bar{k} = \bar{r}$.

Thus for each positive integer n, we have exactly n residue classes

$$\begin{aligned} \bar{0} &= \{nq | q \in \mathbb{Z}\} \\ \bar{1} &= \{nq + 1 | q \in \mathbb{Z}\} \\ &\vdots \\ \overline{n-1} &= \{nq + (n-1) | q \in \mathbb{Z}\}. \end{aligned}$$

These residue classes partition $\mathbb{Z}$: every integer belongs to exactly one of the residue classes $\bar{0}, \bar{1}, \dots, \overline{n-1}$.

In Chapter 2, we shall return to consider the set $\bar{\mathbb{Z}}_n = \{\bar{0}, \bar{1}, \ldots, \overline{n-1}\}$ and define on it the structure of a group. See Figure 6.

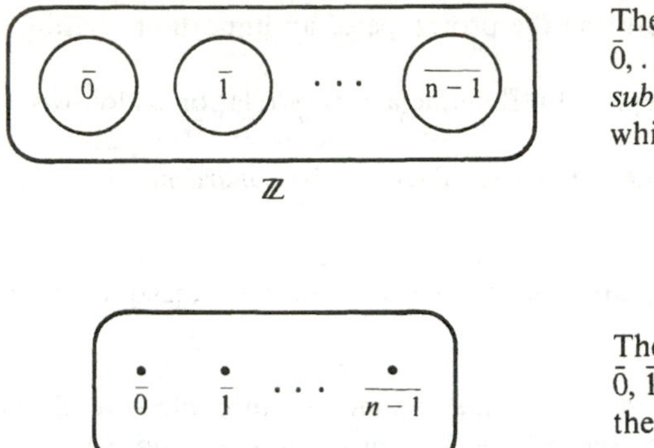

The residue classes $\bar{0}, \ldots, \overline{n-1}$ are *subsets* of $\mathbb{Z}$ which partition $\mathbb{Z}$.

The residue classes $\bar{0}, \bar{1}, \ldots, \overline{n-1}$ are the *elements* of $\bar{\mathbb{Z}}_n$.

Figure 6

Recommended reading: [1], [31], [42].

Exercises 1.3

True or False

1. If $M \subset \mathbb{Z}^+$ has the property $n \in M \Rightarrow n + 1 \in M$, then $M = \mathbb{Z}^+$.
2. If $P(n)$ is a proposition regarding a positive integer n such that $P(1)$ is true, $P(2)$ is true, and $P(3)$ is true, then $P(n)$ is true for all positive integers n.
3. In an induction proof, you first assume what you're trying to prove, then prove it.
4. The remainder obtained when -37 is divided by 5 is -2.
5. The integers 28 and 64 are congruent modulo 1, 2, 3, 4, 6, 9, 12, 18, and 36.

1.3.1. Prove the Corollary of Theorem 3.1.

1.3.2. If $a, b \in \mathbb{Z}^+$, prove that $a + b$ and ab are in $\mathbb{Z}^+$.

1.3.3. Prove: if $ab = 0 \quad (a, b \in \mathbb{Z})$, then either $a = 0$ or $b = 0$.

1.3.4. Prove: if $bx = by \quad (b \neq 0, x, y \in \mathbb{Z})$, then $x = y$.

1.3.5. Assuming as known the definition of absolute value, and the result that $|ab| = |a| \cdot |b| \quad (a, b \in \mathbb{Z})$, prove: if $a|b$, $b \neq 0$, then $|a| \leq |b|$.

1.3.6. Prove that the only divisors of 1 are ± 1.

1.3.7. Prove: if $a, b \in \mathbb{Z}$ such that $a|b$ and $b|a$, then $a = \pm b$.

1.3.8. Use induction to prove each of the following summation formulas:

(a) $1 + 2 + \cdots + n = n(n + 1)/2$, for each $n \in \mathbb{Z}^+$;

(b) $1^2 + 2^2 + \cdots + n^2 = n(n + 1)(2n + 1)/6$, for each $n \in \mathbb{Z}^+$.

CAUTION: Before embarking on an induction proof, be sure you understand fully that the second requirement of the Induction Principle (Theorem 1.3.2) is a *conditional statement.* You must prove:

$$\text{IF} \quad n \in M, \quad \text{THEN} \quad n + 1 \in M;$$

or, in propositional form:

$$\text{IF} \quad P(n) \quad \text{is true, THEN} \quad P(n+1) \quad \text{is true.}$$

1.3.9. Use induction to prove that a finite set of n elements has 2^n subsets (recall Exercise 1.2.5).

1.3.10. Prove that two integers a and b are relatively prime (in the sense of Definition 1.3.4) if and only if there are integers s, t such that $sa + tb = 1$.

1.3.11. Extend Definition 1.3.3 of *greatest common divisor*, Definition 1.3.4 of *relatively prime*, Theorem 1.3.5, and the Corollary on p. 21 to an arbitrary finite set of integers $a_1, \ldots, a_k$ $(k \geq 1)$.

1.3.12. For $a, b \in \mathbb{Z}$, $m \in \mathbb{Z}$ is called a *least common multiple* of a and b if

(1) $a|m$ and $b|m$

and

(2) if t is any integer such that $a|t$ and $b|t$, then $m|t$.

(a) Prove: if m_1, m_2 are least common multiples of a and b, then $m_1 = \pm m_2$. Conclude that two integers a, b have at most one positive least common multiple.

(b) If $a, b \in \mathbb{Z}$, with either a or b equal to 0, prove that 0 is the only least common multiple of a and b.

(c) If $a, b \in \mathbb{Z}$, $a \neq 0$, $b \neq 0$, prove that the least positive integer m such that $a|m$ and $b|m$ is a least common multiple of a and b, according to the definition given previously. (By (a), m is the unique positive least common multiple of a and b; we denote it by $\operatorname{lcm}(a, b)$.)

1.3.13. Let a, b be positive integers, $p_1, \ldots, p_s$ $(s \geq 1)$ distinct primes such that

$$a = p_1^{\alpha_1} \ldots p_s^{\alpha_s}, \quad \alpha_i \geq 0 \quad \text{in} \quad \mathbb{Z}\ (i = 1, \ldots, s),$$

$$b = p_1^{\beta_1} \ldots p_s^{\beta_s}, \quad \beta_i \geq 0 \quad \text{in} \quad \mathbb{Z}\ (i = 1, \ldots, s).$$

(Note that any two positive integers $a > 1$, $b > 1$ can be expressed in this way, for the same set of primes p_i, since zero exponents are allowed.)

Prove that

$$\gcd(a, b) = \prod_{i=1}^{s} p_i^{\gamma_i}$$

$$\operatorname{lcm}(a, b) = \prod_{i=1}^{s} p_i^{\delta_i},$$

where $\gamma_i = \min\{\alpha_i, \beta_i\}$, $\delta_i = \max\{\alpha_i, \beta_i\}$ $(i = 1, \ldots, s)$.

1.3.14. Let a, b, c be integers such that $\gcd(a, b) = 1$, $a|c$ and $b|c$. Prove that $ab|c$.

1.3.15. For n a positive integer, $a, b, c, d \in \mathbb{Z}$, prove: if $a \equiv b \bmod n$ and $c \equiv d \bmod n$, then $a + c \equiv b + d \bmod n$ and $ac \equiv bd \bmod n$. Conclude: if $a \equiv b \bmod n$, then $a + c \equiv b + c \bmod n$ and $ac \equiv bc \bmod n$.

1.3.16. Let n be a positive integer. Prove the following cancellation laws:
(1) if $a + c \equiv b + c \bmod n$, then $a \equiv b \bmod n$;
(2) if $ac \equiv bc \bmod n$ *and* $\gcd(c, n) = 1$, then $a \equiv b \bmod n$.
Give an example to show that the hypothesis $\gcd(c, n) = 1$ in (2) is necessary.

1.3.17. A useful property of the real number system is the *Archimedean property*: if μ, ν are real numbers, $\nu > 0$, then there is a positive integer k such that $k\nu \geq \mu$.
(1) Use the Archimedean property to prove that, for each $\mu, \nu \in \mathbb{R}$, $\nu > 0$, there is an integer q and a real number r such that $\mu = \nu q + r$, $0 \leq r < \nu$. Prove that q and r are uniquely determined for each $\mu, \nu > 0$ in $\mathbb{R}$.
(2) Given $\nu > 0$, define a binary relation "$\equiv \bmod \nu$" on $\mathbb{R}$ by: $x \equiv y \bmod \nu$ if $x - y = q\nu$ for some integer q. Prove that $\equiv \bmod \nu$ is an equivalence relation on $\mathbb{R}$.
(3) Let $\nu = 2\pi$. What are the equivalence classes with respect to $\equiv \bmod 2\pi$? Find a graphic representation of these equivalence classes. (For example, how might you represent the class $\overline{\pi/4}$ to which $\pi/4$ belongs?)

1.3.18. Let a, b be integers, not both equal to 0. Prove that $\text{lcm}(a, b) = ab/\gcd(a, b)$.

1.3.19. A set A is defined to be *infinite* if there is a 1-1 mapping of A onto some proper subset of A. Prove that the sets $\mathbb{Z}$ and $\mathbb{Z}^+$ are infinite sets.

1.3.20. A set A is defined to be *finite* if either $A = \varnothing$ or there is a positive integer n such that the initial segment $I_n = \{1, 2, \ldots, n\}$ can be mapped 1-1 onto A. Using induction, prove that a finite set is not infinite, i.e., there is no 1-1 mapping of a finite set onto a proper subset of itself.

1.3.21. Use induction to prove that every subset of a finite set is finite.

1.3.22. A set B is defined to be *denumerable* if there is a 1-1 mapping of $\mathbb{Z}^+$ onto B. Prove that every infinite set A has a denumerable subset. (Hint: the Axiom of Choice is needed here.)

1.3.23. Use the preceding exercise and the definitions of *finite* and *infinite* given in Exercises 1.3.20 and 1.3.19 to prove: if a set is not finite, then it is infinite. (Thus, combining this exercise and Exercise 1.3.20, we have the result: a set is infinite if and only if it is not finite.)

1.3.24. (1) Prove that the relation $|$ (is a divisor of) is a partial order relation on $\mathbb{Z}^+$. Is $|$ a partial order relation on $\mathbb{Z}$?
(2) With respect to the partial order relation $|$ on $\mathbb{Z}^+$, prove that, for $a, b \in \mathbb{Z}^+$, $\text{glb}(a, b) = \gcd(a, b)$, and $\text{lub}(a, b) = \text{lcm}(a, b)$.
(3) Conclude that $\mathbb{Z}^+$ forms a lattice with respect to $|$, with gcd and lcm serving, respectively, as glb and lub.

1.3.25. A subset S of a lattice L is said to form a *sublattice* of L if S forms a lattice with respect to the partial order of L, restricted to S, with glb and lub defined as in L. Prove that the positive divisors of any positive integer n form a sublattice of the lattice $\mathbb{Z}^+$ described in Exercise 1.3.24. Draw a lattice diagram for the sublattice whose elements are the positive divisors of 24.

2

Groups

2.1 Binary Operations

The portion of mathematics currently classified as "algebra" generally deals with mathematical structures involving binary operations.

Definition 2.1.1: A *binary operation* is a function whose domain is a set of ordered pairs. In particular, if $\circ : A \times A \to A$, then $\circ$ is a *binary operation on A*.

Notation: If $\circ\,(a, b) = c$, we write $c = a \circ b$.

Example 1:

Addition and multiplication are binary operations on $\mathbb{Z}$, on $\mathbb{Q}$, on $\mathbb{R}$, and on $\mathbb{C}$.

Example 2:

Subtraction is a binary operation on $\mathbb{Z}$, but not on $\mathbb{Z}^+$.

Example 3:

If V is a vector space over $\mathbb{R}$, then
(a) vector addition is a binary operation on V;
(b) scalar multiplication is a function from $\mathbb{R} \times V$ to V; hence it is a binary operation with domain $\mathbb{R} \times V$.
(c) An inner product on V (e.g., the dot product on $V = \mathbb{R}^n$) is a function from $V \times V$ to $\mathbb{R}$; hence it is a binary operation with domain $V \times V$, but not a binary operation *on* V.

Definition 2.1.2: Let $\circ$ be a binary operation. Then $\circ$ is *commutative* if $a \circ b = b \circ a$ for all a, b for which both $a \circ b$ and $b \circ a$ are defined; $\circ$ is *associative* if $(a \circ b) \circ c = a \circ (b \circ c)$ for all a, b, c for which both $(a \circ b) \circ c$ and $a \circ (b \circ c)$ are defined.

In particular, if $\circ$ is a binary operation *on a set A*, then $\circ$ is *commutative* if $a \circ b = b \circ a$ for all $a, b \in A$; and $\circ$ is *associative* if $(a \circ b) \circ c = a \circ (b \circ c)$ for all $a, b, c \in A$.

Example 1:

Addition and multiplication on $\mathbb{Z}$, $\mathbb{Q}$, $\mathbb{R}$, or $\mathbb{C}$ are commutative and associative.

Example 2:

Subtraction on $\mathbb{Z}$, $\mathbb{Q}$, $\mathbb{R}$, or $\mathbb{C}$ is neither commutative nor associative.

Example 3:

Multiplication of matrices (for example, over $\mathbb{R}$) is associative but not commutative. Addition of matrices is commutative and associative.

Example 4:

For $n \geq 1$, let $M_n(\mathbb{R})$ be the set of all $n \times n$ matrices over $\mathbb{R}$. Then addition and multiplication of matrices are binary operations *on* $M_n(\mathbb{R})$. Addition is commutative and associative. Multiplication is associative but not commutative except in the case where $n = 1$.

Example 5:

If A is a non-empty set, and F_A is the set of all functions $f: A \to A$, then composition of functions is an associative binary operation on F_A. (See Theorem 1.2.2). Composition is not commutative except in the case where A consists of just one element. (See Exercise 2.1.8.)

Example 6:

If P_A is the power set of a set A, then $\cup$ and $\cap$ (see Chapter 1, Section 2) are binary operations on P_A. Both are commutative as well as associative.

Example 7:

Again, let P_A be the power set of a set A. On P_A, define a binary operation Δ by: $X \,\Delta\, Y = (X \cup Y) \backslash (X \cap Y)\,(X, Y \in P_A)$. Then Δ is both commutative and associative (see Figure 1). The set $X \,\Delta\, Y$ is called the *symmetric difference* of X and Y.

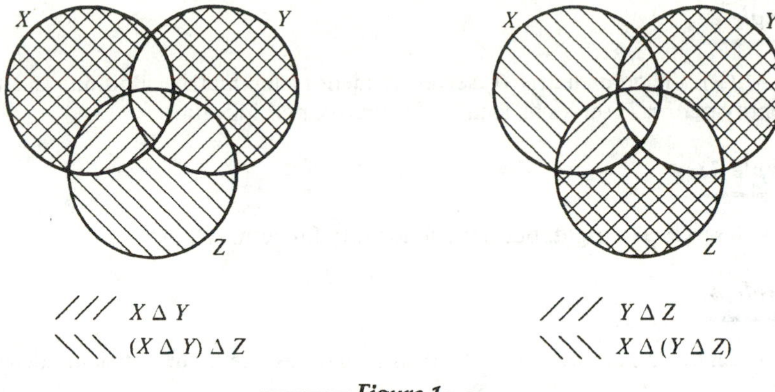

Figure 1

Example 8:

On the unit interval [0, 1], define a binary operation $\circ$ by: $p \circ q = 1 - pq$. Then $\circ$ is obviously commutative, but $\circ$ is not associative. For: $(p \circ q) \circ r = (1 - pq) \circ r = 1 - (1 - pq)r = 1 - r + pqr$, while $p \circ (q \circ r) = 1 - p(q \circ r) = 1 - p(1 - qr) = 1 - p + pqr \quad (p, q, r \in [0, 1])$.

This operation computes probabilities! If P, Q, R are independent events occurring with probabilities p, q, r, respectively, then $p \circ q = 1 - pq$ is the probability that P and Q will not both occur; $(p \circ q) \circ r = (1 - r) + pqr$ is the probability that either R will fail to occur, or the three events P, Q, R will all occur; $p \circ (q \circ r) = (1 - p) + pqr$ is the probability that either P will fail to occur, or the three events will all occur.

In the familiar number systems $\mathbb{Z}$, $\mathbb{Q}$, $\mathbb{R}$, $\mathbb{C}$, zero is a special element with respect to addition: $a + 0 = a = 0 + a$ for any number a in the system. Analogously, 1 is a special element for multiplication: $a \cdot 1 = a = 1 \cdot a$, for any number a in the system. Both are identity elements, in the sense of the following definition.

Definition 2.1.3: Let $\circ$ be a binary operation on a set A, and let $e \in A$. Then e is a *right identity for* $\circ$ if $a \circ e = a$ for all $a \in A$; e is a *left identity for* $\circ$ if $e \circ a = a$ for all $a \in A$; e is a *two-sided identity for* $\circ$ if $e \circ a = a = a \circ e$ for all $a \in A$. (If we omit the adjectives *right*, *left*, or *two-sided* in discussing an identity element e, it will be understood that e is a two-sided identity.)

Note: if $\circ$ is commutative, then every identity element for $\circ$ is, of course, two-sided.

Example 1:

In $\mathbb{Z}$, $\mathbb{Q}$, $\mathbb{R}$, or $\mathbb{C}$, 0 is an identity for $+$ and 1 is an identity for $\cdot$.

Example 2:

In $M_n(\mathbb{R})$, the zero matrix 0_n serves as identity for matrix addition, and the identity matrix I_n serves as identity for matrix multiplication.

Example 3:

In $\mathbb{Z}$, 0 serves as a right, but not left, identity for subtraction.

Example 4:

If $\circ$ is defined on $\mathbb{R}^+$ by $a \circ b = b^a$, then 1 serves as a left, but not right, identity for $\circ$.

Example 5:

In the power set P_A of any set A, $\varnothing$ serves as identity for $\cup$ and for Δ (symmetric difference—see Example 7, p. 28); A serves as identity for $\cap$.

Example 6:

In the set F_A of all functions $f: A \to A$, where A is a non-empty set, the identity function, ι_A, serves as identity for composition.

Example 7:

One-sided identities need not be unique. Let

$$A = \left\{ \begin{pmatrix} a & 0 \\ b & 0 \end{pmatrix} \middle| \, a, b \in \mathbb{R} \right\}.$$

Since

$$\begin{pmatrix} a & 0 \\ b & 0 \end{pmatrix}\begin{pmatrix} c & 0 \\ d & 0 \end{pmatrix} = \begin{pmatrix} ac & 0 \\ bc & 0 \end{pmatrix}$$

for all $a, b, c, d \in \mathbb{R}$, matrix multiplication (restricted to A) serves as a binary operation on A. For each $k, a, b \in \mathbb{R}$, we have

$$\begin{pmatrix} a & 0 \\ b & 0 \end{pmatrix}\begin{pmatrix} 1 & 0 \\ k & 0 \end{pmatrix} = \begin{pmatrix} a & 0 \\ b & 0 \end{pmatrix}$$

$$\begin{pmatrix} 1 & 0 \\ k & 0 \end{pmatrix}\begin{pmatrix} a & 0 \\ b & 0 \end{pmatrix} = \begin{pmatrix} a & 0 \\ ka & 0 \end{pmatrix}.$$

Thus, for each $k \in \mathbb{R}$, the matrix $\begin{pmatrix} 1 & 0 \\ k & 0 \end{pmatrix}$ serves as a right, but not left, identity for matrix multiplication (restricted to A).

Theorem 2.1.1: Let $\circ$ be a binary operation on a set A. If $e \in A$ is a left identity for $\circ$ and $e' \in A$ is a right identity for $\circ$, then $e' = e$. Hence A contains *at most one* two-sided identity for $\circ$.

Proof: $e = e \circ e' = e'$. ■

Thus, if there *is* a (two-sided) identity for $\circ$, in A, then this identity is unique, i.e., it is the only (two-sided) identity and, indeed, the only right or left identity, for $\circ$ in A.

An identity element in a set A is characterized by its behavior toward *every* element of the set, with respect to a given binary operation. This is in sharp contrast to the behavior of inverse elements, which *belong to* individual elements of the set.

Definition 2.1.4: Let A be a set, $\circ$ a binary operation on A, and let $e \in A$ be an identity element for $\circ$ (right, left, or two-sided). Let $a, b \in A$.

(1) If $b \circ a = e$, then b is a *left inverse of a* with respect to $\circ$ and the identity e.

(2) If $a \circ b = e$, then b is a *right inverse of a* with respect to $\circ$ and the identity e.

(3) If $a \circ b = e = b \circ a$, then b is a *(two-sided) inverse of a* with respect to $\circ$ and the identity e.

Of course, if $\circ$ is commutative, any inverse is necessarily two-sided. The term *inverse*, used without a modifying adjective, will be understood to mean *two-sided inverse*.

Example 1:

In $\mathbb{Z}$, $\mathbb{Q}$, $\mathbb{R}$, or $\mathbb{C}$, $-a$ serves as inverse for a with respect to $+$ and the identity 0.

Example 2:

In $\mathbb{Q}^*$, $\mathbb{R}^*$, or $\mathbb{C}^*$, $1/a$ serves as inverse for $a \neq 0$, with respect to $\cdot$ and the identity 1.

Example 3:

In the set $M_n(\mathbb{R})$, $-A = (-a_{ij})_{n\times n}$ serves as inverse of the matrix $A = (a_{ij})_{n\times n}$ with respect to matrix addition and the identity 0_n. If $A \in M_n(\mathbb{R})$ is non-singular, then the matrix A^{-1} serves as (two-sided) inverse for A with respect to matrix multiplication and the identity I_n.

Example 4:

In the power set P_A of a set A, with respect to $\cup$ and the identity $\varnothing$, only $\varnothing$ has an inverse. In fact, $\varnothing$ is its own inverse since $\varnothing \cup \varnothing = \varnothing$. With respect to $\cap$ and the identity A, only A has an inverse. In fact, A is its own inverse since $A \cap A = A$. With respect to Δ (symmetric difference) and the identity $\varnothing$, every $X \in P_A$ is its own inverse since $X \,\Delta\, X = (X \cup X)\backslash(X \cap X) = X\backslash X = \varnothing$.

Somewhat analogous to Theorem 2.1.1 regarding identity elements, we have the following theorem regarding inverse elements. But note that associativity (not required in Theorem 2.1.1) is an essential ingredient here.

Theorem 2.1.2: Let A be a set, $\circ$ an associative binary operation on A, and let $e \in A$ be a two-sided identity for $\circ$. Let $a \in A$. If $b_1 \in A$ is a left inverse of a, and $b_2 \in A$ is a right inverse of a, with respect to $\circ$ and the identity e, then $b_1 = b_2$. Hence an element of A can have *at most one* two-sided inverse in A, with respect to $\circ$ and e.

Proof: $b_1 = b_1 \circ (a \circ b_2) = (b_1 \circ a) \circ b_2 = b_2$. ■

Thus, if an element $a \in A$ has a two-sided inverse, this inverse is unique, i.e., it is the one and only two-sided inverse of a.

Remark: To refresh your memory on matrices and linear algebra, cf. [27], [30], or refer to Chapter 3.

Exercises 2.1

True or False

1. If $\circ$ is the binary operation defined on $\mathbb{R}^+$ by: $a \circ b = b^a$, then every element of $\mathbb{R}^+$ has a left inverse with respect to $\circ$ and the left identity 1.
2. The binary operation $\circ$ defined on $\mathbb{R}$ by $a \circ b = 2a + b$ is commutative.
3. The binary operation $\circ$ defined on $\mathbb{R}$ by $a \circ b = 2ab$ is associative.
4. If e is a two-sided identity with respect to a binary operation $\circ$ on a set A, then at least one element of A has a two-sided inverse with respect to $\circ$ and the identity e.
5. If $\circ$ is a binary operation defined on a set A and e is a two-sided identity with respect to $\circ$, then $a \circ a = a$ for at least one $a \in A$.

2.1.1. Let $S = \left\{\begin{pmatrix} a & b \\ 0 & 0 \end{pmatrix} \middle| a, b \in \mathbb{R}\right\}$.

(1) Prove that matrix multiplication (restricted to S) is a binary operation on S.

(2) Prove that $\begin{pmatrix} 1 & 0 \\ 0 & 0 \end{pmatrix}$ serves as a left, but not right, identity for $\cdot$. In S, find infinitely many left identities for $\cdot$. Does S contain a two-sided identity for $\cdot$?

(3) Find all matrices in S that have left inverses with respect to $\cdot$ and the identity $\begin{pmatrix} 1 & 0 \\ 0 & 0 \end{pmatrix}$.

2.1.2. Let E be a 2×2 matrix such that $E^2 = E$, $E \neq 0_2$, $E \neq I_2$.

(1) Let $H = \{EX \mid X \in M_2(\mathbb{R})\}$. Prove that matrix multiplication (restricted to H) is a binary operation on H and that $E \in H$ serves as a left, but not right, identity for $\cdot$.

(2) Let $K = \{XE | X \in M_2(\mathbb{R})\}$. Translate the statements in (1) regarding H to corresponding statements about K. Prove your statements about K.

(3) Relate this exercise to Exercise 2.1.1.

2.1.3. On $\mathbb{R}$, define $\circ$ by $a \circ b = a + b - ab$ $(a, b \in \mathbb{R})$. Prove that $\circ$ is a commutative and associative binary operation on $\mathbb{R}$. Prove that 0 serves as identity with respect to $\circ$. Find all real numbers that have inverses with respect to $\circ$ and the identity 0.

2.1.4. Let $F_{\mathbb{R}}$ be the set of all real-valued functions defined on $\mathbb{R}$, and let $\circ$ be composition of functions. Note that $\circ$ is a binary operation on $F_{\mathbb{R}}$. Verify that the identity function $\imath_{\mathbb{R}}: \mathbb{R} \to \mathbb{R}$ defined by $\imath_{\mathbb{R}}(x) = x$ for each $x \in \mathbb{R}$ serves as two-sided identity with respect to $\circ$. Prove that the function $g: \mathbb{R} \to \mathbb{R}$ defined by $g(x) = e^x$ $(x \in \mathbb{R})$ has a left, but not right, inverse with respect to $\circ$ and the identity $\imath_{\mathbb{R}}$. Give an example of a function $h: \mathbb{R} \to \mathbb{R}$ that has a right, but not left, inverse with respect to $\circ$ and the identity $\imath_{\mathbb{R}}$.

2.1.5. In $M_2(\mathbb{R})$, find all matrices that are their own inverses with respect to matrix multiplication and the identity $I_2 = \begin{pmatrix} 1 & 0 \\ 0 & 1 \end{pmatrix}$.

2.1.6. On $M_2(\mathbb{R})$, define $\circ$ by $A \circ B = AB - BA$ $(A, B \in M_2(\mathbb{R}))$. Prove that $\circ$ is a non-commutative, non-associative binary operation on $M_2(\mathbb{R})$.

2.1.7. Let $F_\infty(\mathbb{R})$ be the set of all infinitely many times differentiable functions $f: \mathbb{R} \to \mathbb{R}$. On $F_\infty(\mathbb{R})$, define $*$ by $f * g = fg' + gf'$. Prove that $*$ is a commutative, non-associative binary operation on $F_\infty(\mathbb{R})$.

2.1.8. Let A be a non-empty set, and let F_A be the set of all functions $f: A \to A$. Prove that composition of functions is a non-commutative binary operation on F_A except in the case where A consists of just one element.

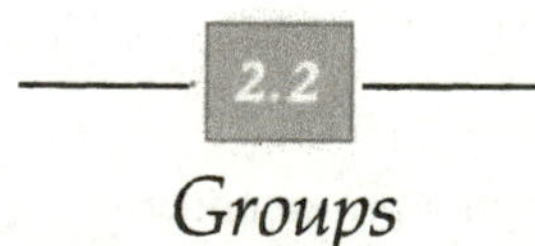

Groups

We are now ready to encounter the first class of algebraic structures which we will study in some detail: groups. To define a group, we require two basic ingredients: a set G, and a binary operation $\circ$ defined on G. Both the set and the binary operation must be the same if two groups are to be equal. This suggests that we consider ordered pairs $\langle G, \circ \rangle$, where G is a set, $\circ$ a binary operation on G. If $\circ$ is associative, then $\langle G, \circ \rangle$ is a *semigroup*. If G contains a two-sided identity, then $\langle G, \circ \rangle$ is a *groupoid*. For $\langle G, \circ \rangle$ to be a group, we require associativity, the existence of a two-sided identity, and the existence of two-sided inverses for all the elements of G.

Definition 2.2.1: A *group* is an ordered pair $\langle G, \circ \rangle$, where G is a set, $\circ$ a binary operation on G, satisfying the following conditions:

(1) $\circ$ is associative;

(2) there is a (two-sided) identity $e \in G$, for the operation $\circ$;

(3) for every element $a \in G$, there is an element $b \in G$ such that b is a (two-sided) inverse of a, with respect to $\circ$ and the identity e. (More comfortably, if $\langle G, \circ \rangle$ is a group, we usually say: "G forms a group under $\circ$," or simply "G is a group" if there is no doubt as to the operation.)

If $\langle G, \circ \rangle$ is a group such that $\circ$ is commutative, then $\langle G, \circ \rangle$ is an *abelian group*. The number of elements of the set G is called the *order* of the group $\langle G, \circ \rangle$.

Remark 1: By Theorems 2.1.1 and 2.1.2, if $\langle G, \circ \rangle$ is a group, then G contains *exactly one* identity e, and every element of G has *exactly one* inverse with respect to $\circ$ and e.

Remark 2: By Definition 2.1.1, the assertion in Definition 2.2.1 that $\circ$ is a binary operation on G guarantees that, for each $a, b \in G$, there is exactly one element $a \circ b$ in G. (The existence, for each $a, b \in G$, of an element $a \circ b$ in G is often described by saying that G "is closed" under $\circ$.)

Notation: While, in Definition 2.2.1, we use the neutral notation "$\circ$" for the group operation, henceforth the group operation will usually be denoted by either "$\cdot$" or "$+$". Additive notation is practically never used unless the operation is commutative.

If the group operation is denoted by "$\circ$" or by "$\cdot$", we shall generally use "e" or "1" to denote the identity and "a^{-1}" to denote the inverse of $a \in G$. If the group operation is denoted by "$+$", we shall generally use "0" to denote the identity, and "$-a$" to denote the inverse of $a \in G$.

Example 1:

$\langle \mathbb{Z}, + \rangle$, $\langle \mathbb{Q}, + \rangle$, $\langle \mathbb{R}, + \rangle$, $\langle \mathbb{C}, + \rangle$, $\langle \mathbb{Q}^*, \cdot \rangle$, $\langle \mathbb{R}^*, \cdot \rangle$, $\langle \mathbb{C}^*, \cdot \rangle$, are all infinite abelian groups.

Note that $\mathbb{Z} \setminus \{0\}$ does not form a group under $\cdot$.

Example 2:

If V is a vector space over $\mathbb{R}$, then $\langle V, + \rangle$ is an infinite abelian group, where "$+$" denotes addition of vectors.

Example 3:

For n a positive integer, let G be the set of all non-singular $n \times n$ matrices over $\mathbb{R}$, and let $\cdot$ be matrix multiplication. Then $\langle G, \cdot \rangle$ is an infinite group, non-abelian except in the case where $n = 1$. (We shall henceforth denote this group by "$GL_n(\mathbb{R})$." It is called the *general linear group* associated with n.)

A large class of non-abelian groups is provided by the following theorem:

Theorem 2.2.1: Let A be a non-empty set, and let S_A be the set of all bijections $\alpha : A \to A$. Then S_A forms a group under composition. S_A is non-abelian unless A is either a singleton or a pair. If A is infinite, so is S_A. If A is a finite set of n elements, then the order of S_A is $n!$.

Proof: For simplicity in future use, we denote composition by $\cdot$, or omit the operation symbol altogether. Let $\alpha, \beta \in S_A$. Then the composite function $\alpha\beta$ is a bijection from A to A. (See Exercise 1.2.6.) Hence $\cdot$ acts as a binary operation on S_A. By Theorem 1.2.2, $\cdot$ is associative. For each $\alpha \in S_A$, and each $a \in A$, $(\alpha\, \iota_A)(a) = \alpha(\iota_A(a)) = \alpha(a)$, and $(\iota_A \alpha)(a) = \iota_A(\alpha(a)) = \alpha(a)$, whence $\alpha \iota_A = \alpha = \iota_A \alpha$. Thus ι_A serves as a two-sided identity element for $\cdot$. By Theorem 1.2.3, each $\alpha \in S_A$ has a two-sided inverse function β which is itself a bijection from A to A since it has α as its inverse function. Since $\alpha\beta = \iota_A = \beta\alpha$, β serves as the inverse element of α with respect to composition and the identity ι_A. We have proved that $\langle S_A, \cdot \rangle$ is a group.

If A has either just one element or just two elements, then S_A is clearly abelian (Exercise 2.2.4). If A has more than two elements, let x, y, z be distinct elements of A. Define mappings α, β as follows: $\alpha(x) = y$, $\alpha(y) = x$, $\alpha(t) = t$ for each $t \in A$ different from both x and y; $\beta(x) = y$, $\beta(y) = z$, $\beta(z) = x$, $\beta(t) = t$ for each $t \in A$ different from x, y, and z. Then $\alpha\beta \neq \beta\alpha$. For: $(\alpha\beta)(x) = \alpha(\beta(x)) = \alpha(y) = x$, while $(\beta\alpha)(x) = \beta(\alpha(x)) = \beta(y) = z \neq x$. Thus S_A is not abelian.

The proof of the assertions regarding the order of S_A is left as an exercise (see Exercise 2.2.6). ■

Definition 2.2.2: A bijection of a finite set A is called a *permutation of A*. (Thus, if A is finite, S_A is the group of all permutations of A.) In particular, if $A = \{1, 2, \ldots, n\}$, then S_A is denoted by "S_n" and is called *the symmetric group of degree n*.

> ***Note:*** S_n is a group of order $n!$. The phrase "*of degree n*" reflects the association between S_n and a general polynomial of degree n in Galois Theory (see Chapter 1, Section 1 and Chapter 4, Sections 14 and 15). The group concept found its origin in permutation groups, and, despite the ubiquity of groups within mathematics as well as in some of its applications, groups remain intimately linked to permutations. In fact, as we shall see (Theorem 2.7.1, Cayley's Theorem) every group is structurally equivalent ("isomorphic") to a group of permutations.

In Section 2.6, we study permutation groups in detail. Here we confine ourselves to an examination of the operations in S_3. For the time being, we denote a permutation σ on the set $X_3 = \{1, 2, 3\}$ by the symbol

$$\sigma = \begin{pmatrix} 1 & 2 & 3 \\ b_1 & b_2 & b_3 \end{pmatrix},$$

where b_i is the image under σ of $i \in X_3$. The elements of S_3 are the permutations

$$A = \begin{pmatrix} 1 & 2 & 3 \\ 1 & 2 & 3 \end{pmatrix}$$

$$E = \begin{pmatrix} 1 & 2 & 3 \\ 2 & 3 & 1 \end{pmatrix}$$

$$J = \begin{pmatrix} 1 & 2 & 3 \\ 3 & 1 & 2 \end{pmatrix}$$

$$N = \begin{pmatrix} 1 & 2 & 3 \\ 2 & 1 & 3 \end{pmatrix}$$

$$S = \begin{pmatrix} 1 & 2 & 3 \\ 1 & 3 & 2 \end{pmatrix}$$

$$W = \begin{pmatrix} 1 & 2 & 3 \\ 3 & 2 & 1 \end{pmatrix}.$$

(The capital letter designations are in agreement with notation used in Section 2.6.)

For each $i \in X_3$, and $\sigma, \tau \in S_3$, we apply τ before σ to compute the product σ, τ. For example, in S_3, we have

$$\mathrm{JN} = \begin{pmatrix} 1 & 2 & 3 \\ 3 & 1 & 2 \end{pmatrix}\begin{pmatrix} 1 & 2 & 3 \\ 2 & 1 & 3 \end{pmatrix} = \begin{pmatrix} 1 & 2 & 3 \\ 1 & 3 & 2 \end{pmatrix} = \mathrm{S}.$$

The products computed in this fashion can be tabulated in a group table (known as a *Cayley table*) as follows:

	A	*E*	*J*	*N*	*S*	*W*
A	*A*	*E*	*J*	*N*	*S*	*W*
E	*E*	*J*	*A*	*W*	*N*	*S*
J	*J*	*A*	*E*	*S*	*W*	*N*
N	*N*	*S*	*W*	*A*	*E*	*J*
S	*S*	*W*	*N*	*J*	*A*	*E*
W	*W*	*N*	*S*	*E*	*J*	*A*

(The computations are made much easier by the use of cycle notation, which we develop in Section 2.6.)

Observe that every element of S_3 occurs exactly once in each row and in each column of the table. (Theorem 2.2.6 will explain why this must always be the case.) Inverses of individual elements can be easily read from the table and the lack of commutativity of the group is evident at a glance from the fact that the table fails to be symmetric with respect to its principal (i.e., upper left to lower right) diagonal.

We proceed next to a class of finite abelian groups, derived from the group $\langle \mathbb{Z}, + \rangle$: the groups of residue classes modulo an integer $n > 0$. These groups will provide our first encounter with the notion of a quotient structure—one of the basic concepts of abstract algebra.

Theorem 2.2.2: Let n be a positive integer. On the set $\bar{\mathbb{Z}}_n$ of all residue classes modulo n (see Theorems 1.3.8 and 1.3.9 and the Corollary of 1.3.9), define $\oplus_n$ by: $\bar{h} \oplus_n \bar{k} = \overline{h+k}$ $(h, k \in \mathbb{Z})$. Then $\oplus_n$ is a binary operation on $\bar{\mathbb{Z}}_n$ and $\langle \bar{\mathbb{Z}}_n, \oplus_n \rangle$ is an abelian group.

Proof: We need to prove that our prescription for forming $\bar{h} \oplus \bar{k}$ defines a binary operation on $\bar{\mathbb{Z}}_n$, i..e., a function from $\bar{\mathbb{Z}}_n \times \bar{\mathbb{Z}}_n$ to $\bar{\mathbb{Z}}_n$.

There is no doubt that any two residue classes $\bar{h}$ and $\bar{k}$ can be combined under $\oplus_n$ to produce another residue class $\overline{h+k}$, but there should be, initially, some doubt regarding the uniqueness of the resulting class. Recall that $\bar{a} = \bar{b}$ if and only if $a \equiv b \bmod n$. (Thus, e.g., for $n = 6$, $\bar{4} = \overline{10}$.) Could the "circle-sum" be dependent on the particular numbers in terms of which the classes we "add" are expressed? (For example, if $n = 6$, is $\bar{4} \oplus_6 \bar{3} = \overline{10} \oplus_6 \overline{-9}$? Yes, in this case, since $\overline{4+3} = \bar{7} = \bar{1}$, and $\overline{10 + (-9)} = \bar{1}$.) We must show in general that the "circle-sum" of two classes does not depend on the representation of the classes. Thus, suppose, for $h_1, h_2, k_1, k_2 \in \mathbb{Z}$, we have $\bar{h}_1 = \bar{h}_2, \bar{k}_1 = \bar{k}_2$. Then $h_1 \equiv h_2 \bmod n$ and $k_1 \equiv k_2 \bmod n$, whence $h_1 + k_1 \equiv h_2 + k_2 \bmod n$ (see Exercise 1.3.15). But then $\overline{h_1 + k_1} = \overline{h_2 + k_2}$. We may now conclude that $\oplus_n$ is a binary operation on $\bar{\mathbb{Z}}_n$.

The operation $\oplus_n$ is commutative since, for all $h, k \in \mathbb{Z}$, $\bar{h} \oplus_n \bar{k} = \overline{h+k} = \overline{k+h} = \bar{k} \oplus_n \bar{h}$, and associative since, for all h, k, l in $\mathbb{Z}$,

$$\begin{aligned}(\bar{h} \oplus_n \bar{k}) \oplus_n \bar{l} = \overline{h+k} \oplus_n \bar{l} &= \overline{(h+k)+l} = \overline{h+(k+l)} \\ &= \bar{h} \oplus_n \overline{k+l} = \bar{h} \oplus_n (\bar{k} \oplus_n \bar{l}).\end{aligned}$$

The residue class $\bar{0}$ serves as identity since

$$\bar{0} \oplus_n \bar{h} = \overline{0+h} = \bar{h} = \bar{h} \oplus_n \bar{0}$$

for all $h \in \mathbb{Z}$. For each $h \in \mathbb{Z}$,

$$\overline{-h} \oplus_n \bar{h} = \overline{-h+h} = \bar{0} = \bar{h} \oplus_n \overline{-h},$$

whence every element of $\bar{\mathbb{Z}}_n$ has an inverse in $\bar{\mathbb{Z}}_n$ with respect to $\oplus_n$ and the identity 0. We have proved that $\langle \bar{\mathbb{Z}}_n, \oplus_n \rangle$ is an abelian group. ■

Corollary: For each positive integer n, there is a group of order n.

Proof: By Theorem 1.3.9, for each $n \in \mathbb{Z}^+$, $\bar{\mathbb{Z}}_n = \{\bar{0}, \bar{1}, \ldots, \overline{n-1}\}$; hence $\bar{\mathbb{Z}}_n$ forms a group of order n. ■

Remarks: The elements of $\bar{\mathbb{Z}}_n$ are the equivalence classes, in $\mathbb{Z}$, with respect to the equivalence relation $\equiv \bmod n$. (Thus, $\bar{\mathbb{Z}}_n$ is the factor set of $\mathbb{Z}$ with respect to the equivalence relation $\equiv \bmod n$.) Specifically, the elements of $\bar{\mathbb{Z}}_n$ are the sets

$$\begin{aligned}\bar{0} &= \{nq \mid q \in \mathbb{Z}\} \\ \bar{1} &= \{nq + 1 \mid q \in \mathbb{Z}\} \\ &\vdots \\ \overline{n-1} &= \{nq + (n-1) \mid q \in \mathbb{Z}\}.\end{aligned}$$

The factor set $\bar{\mathbb{Z}}_n$ forms a group under the operation $\oplus_n$: $\langle \bar{\mathbb{Z}}_n, \oplus_n \rangle$ is our first example of a factor group, or quotient group (see Definition 2.9.3)—one type of quotient structure. Forming quotient structures is an act of abstraction that can provide useful insights regarding a given structure. For example, in forming $\langle \mathbb{Z}_n, \oplus_n \rangle$, we *abstract from* the differences between the (infinitely many) elements that constitute each class: *each of the equivalence classes* $\bar{h}$ $(h \in \mathbb{Z})$ *becomes a single element of* $\mathbb{Z}_n$. The group $\langle \mathbb{Z}_n, \oplus_n \rangle$ retains some of the properties of $\langle \mathbb{Z}, + \rangle$ (e.g., commutativity) and loses others (e.g., infiniteness).

"Factoring out" multiples of n to form $\langle \mathbb{Z}_n, \oplus_n \rangle$ allows us to focus on certain features of $\mathbb{Z}$ itself. For example, for $n = 11$, the fact that $\bar{6} \oplus_{11} \overline{10} = \bar{5}$ tells us: if an integer that is 6 more than a multiple of 11 is added (in $\mathbb{Z}$) to an integer that is 10 more than a multiple of 11, we get an integer that is 5 more than a multiple of 11.

In the simplest case, where $n = 2$, we have $\bar{0}$ equal to the set of all even integers, $\bar{1}$ equal to the set of all odd integers, and the Cayley table for $\langle \mathbb{Z}_2, \oplus_2 \rangle$. (Table a) merely reflects the familiar behavior of evenness and oddness under addition in $\mathbb{Z}$, somewhat crudely represented by Table b.

$\oplus_2$	$\bar{0}$	$\bar{1}$
$\bar{0}$	$\bar{0}$	$\bar{1}$
$\bar{1}$	$\bar{1}$	$\bar{0}$

a.

+	even	odd
even	even	odd
odd	odd	even

b.

We next deduce some simple but quite basic consequences of Definition 2.2.1. We have already commented on the uniqueness of the identity element, and of inverse elements, in a group. Two further basic results regarding inverses are contained in the following theorem.

Theorem 2.2.3: Let $\langle G, \cdot \rangle$ be a group.

(1) If $a \in G$, then $(a^{-1})^{-1} = a$.

(2) If $a, b \in G$, then $(ab)^{-1} = b^{-1}a^{-1}$.

Proof: *(1)* From $a^{-1}a = e = aa^{-1}$, we conclude that a is a two-sided inverse of a^{-1}; by the uniqueness of inverses, this implies that $a = (a^{-1})^{-1}$.

(2) Since $(ab)(b^{-1}a^{-1}) = ((ab)b^{-1})a^{-1} = (a(bb^{-1}))a^{-1} = (ae)a^{-1} = aa^{-1} = e$ and, similarly, also $(b^{-1}a^{-1})(ab) = e$, we conclude, again invoking uniqueness of inverses, that $b^{-1}a^{-1} = (ab)^{-1}$. ∎

Note that in additive notation, these results read

(1) $-(-a) = a$ for each $a \in G$;

(2) $-(a + b) = (-b) + (-a) \quad (a, b \in G)$.

(It is important to learn to abstract from notation, among other things.)

Theorem 2.2.4 (Cancellation Laws): Let $\langle G, \cdot \rangle$ be a group, and let $a, b, x \in G$.

(1) Left-cancellation: if $xa = xb$, then $a = b$.

(2) Right-cancellation: if $ax = bx$, then $a = b$.

Proof: ***(1)*** If $xa = xb$, then

$$x^{-1}(xa) = x^{-1}(xb)$$
$$(x^{-1}x)a = (x^{-1}x)b$$
$$ea = eb$$
$$a = b.$$

(2) We leave the proof as an exercise see Exercise 2.2.2). ■

(For practice: write the statement and the proof of this theorem and the theorems that follow in additive notation.)

To illustrate the power of cancellation laws, we prove the following theorem which states that the requirements (2) and (3) of Definition 2.2.1 can be judiciously weakened.

Theorem 2.2.5: Let G be a set, $\cdot$ an associative binary operation on G. If G contains a left identity e_l for $\cdot$, and every element $a \in G$ has a left inverse a_l with respect to $\cdot$ and the left identity e_l, then $\langle G, \cdot \rangle$ is a group. (Another version of this theorem may be obtained by assuming the existence for a right identity e_r relative to which every element of G has a right inverse. However, there is no correct version that *mixes* left and right—see Exercise 2.2.7.)

Proof: We first prove that left-cancellation holds in G. For $a, b, x \in G$, suppose $xa = xb$. Let x_l be a left inverse of x with respect to e_l. Then

$$x_l(xa) = x_l(xb)$$
$$(x_lx)a = (x_lx)b$$
$$e_la = e_lb$$
$$a = b.$$

We next prove that the left identity e_l is, in fact, two-sided. Let $a \in G$, and let a_l be a left inverse of a with respect to e_l. Then

$$e_l = e_le_l$$
$$a_la = (a_la)e_l$$
$$a_la = a_l(ae_l).$$

By left-cancellation, $a = ae_l$. Thus the left identity e_l is also a right identity. Henceforth, call it e.

Again, let $a \in G$, and let a_l be a left inverse of a with respect to e. We have

$$a_la = e;$$

hence,

$$(a_la)a_l = ea_l = a_le$$
$$a_l(aa_l) = a_le.$$

By left-cancellation, $aa_l = e$. Thus, a_l is a two-sided inverse of a, with respect to the two-sided identity e.

It follows that $\langle G, \cdot \rangle$ is a group. ■

The following theorem shows that linear equations can be solved in a group. It explains why, in a Cayley table, every group element occurs once and only once in each row and in each column.

Theorem 2.2.6: Let $\langle G, \cdot \rangle$ be a group, and let $a, b \in G$. Then

(1) there is a unique element $x \in G$ such that $ax = b$;

(2) there is a unique element $y \in G$ such that $ya = b$.

Proof: *(1) Existence*. let $x = a^{-1}b$. Then $ax = a(a^{-1}b) = (aa^{-1})b = eb = b$. *Uniqueness*. if $x_1, x_2 \in G$ such that $ax_1 = ax_2 = b$, then (by left-cancellation) $x_1 = x_2$.

(2) We leave the proof as an exercise. ■

The existence conditions in (1) and (2) of Theorem 2.2.6 are powerful enough to make an associative system into a group.

Theorem 2.2.7: Let $\cdot$ be an associative binary operation on a non-empty set G. Suppose that, for each $a, b \in G$, there exist $x, y \in G$ such that $ax = b$, $ya = b$. Then $\langle G, \cdot \rangle$ is a group.

We leave the proof as an exercise which we particularly urge you to do. This exercise serves as a kind of litmus test for your understanding up to this point. There is a trap waiting to catch the unwary (see Exercise 2.2.12).

Generalized Products; Exponentiation

Let $\langle G, \cdot \rangle$ be a group. Then the binary operation $\cdot$ allows us to combine two elements at a time. Products of n factors, for any $n \geq 1$, may be defined recursively as follows: let $\{a_i\}_{i \in \mathbb{Z}^+}$ be a sequence of elements in G. Then

$$\prod_{i=1}^{1} a_i = a_1$$

$$\prod_{i=1}^{n+1} a_i = \left(\prod_{i=1}^{n} a_i\right) a_{n+1} \quad \text{for each} \quad n \in \mathbb{Z}^+.$$

The empty product of $\{a_i\}_{i \in \mathbb{Z}^+}$ is defined to be equal to 1, the identity of G. (The validity of recursive definitions is based on a theorem derived from the Induction Principle; cf. [44], p. 48.)

It can be shown that *generalized associativity* holds, i.e., roughly stated: parentheses can be inserted arbitrarily in a product without changing its value (Exercise 2.2.16).

Exponentiation in a group $\langle G, \cdot \rangle$ can also be defined recursively. For $a \in G$,

$$\begin{aligned} a^1 &= a \\ a^{n+1} &= a^n \cdot a \quad \text{for each} \quad n \in \mathbb{Z}^+ \\ a^0 &= e \\ a^{-n} &= (a^n)^{-1} \quad \text{for each} \quad n \in \mathbb{Z}^+. \end{aligned} \tag{1}$$

It can be proved, making use of induction (see Exercise 2.2.15), that the following laws of exponents are valid in any group $\langle G, \cdot \rangle$:

For each $a \in G$,

$$a^h \cdot a^k = a^{h+k} \quad \text{for all} \quad h, k \in \mathbb{Z}; \qquad (2)$$
$$(a^h)^k = a^{hk} \quad \text{for all} \quad h, k \in \mathbb{Z}.$$

If G is abelian, then for all $a, b \in G$,

$$(ab)^k = a^k b^k \quad \text{for all} \quad k \in \mathbb{Z}. \qquad (3)$$

Of course, in additive notation, $\prod_{i=1}^{n} a_i$ is replaced by $\sum_{i=1}^{n} a_i$, and exponents are replaced by coefficients. We urge you to carry out the translation of the definitions in detail. Specifically, the definition of exponents is replaced, in additive notation, by

$$\begin{aligned} 1a &= a \\ (n+1)a &= na + a \\ 0a &= 0_G \quad \text{(the identity of } \langle G, + \rangle) \qquad (1') \\ (-n)a &= -na \quad \text{for each} \quad n \in \mathbb{Z}^+ \end{aligned}$$

and the laws of exponents become the laws of coefficients:

$$\left.\begin{aligned} ha + ka &= (h+k)a \\ h(ka) &= (hk)a \end{aligned}\right\} \quad \text{for all} \quad h, k \in \mathbb{Z}, \quad a \in G. \qquad (2')$$

If $+$ is commutative (as is usually the case when additive notation is used), then, for $a, b \in G$,

$$k(a + b) = ka + kb \quad \text{for each} \quad k \in \mathbb{Z}. \qquad (3')$$

Exercises 2.2

True or False

1. Every finite group is abelian.
2. If $\circ$ is an associative binary operation on a set A such that both right and left cancellation laws are valid in A, then $\langle A, \circ \rangle$ is a group.
3. There is an empty group.
4. The elements of $\mathbb{Z}_n$ (n a positive integer) are integers.
5. If a, b are elements of a group $\langle G, \cdot \rangle$, then $x = ba^{-1}$ is a solution of the equation $ax = b$.

2.2.1. Compute

(a) $\alpha\beta$ where $\alpha = \begin{pmatrix} 1 & 2 & 3 & 4 \\ 2 & 1 & 4 & 3 \end{pmatrix}$, $\beta = \begin{pmatrix} 1 & 2 & 3 & 4 \\ 4 & 1 & 2 & 3 \end{pmatrix}$ in S_4.

(b) N^3JW^2 in S_3 (see the table on p. 36).

(c) α^{-1} where $\alpha = \begin{pmatrix} 1 & 2 & 3 & 4 & 5 & 6 \\ 3 & 2 & 4 & 1 & 6 & 5 \end{pmatrix}$ in S_6.

(d) $\begin{pmatrix} 2 & 3 \\ 5 & 1 \end{pmatrix}^{-1}$ in $GL_2(\mathbb{R})$.

(e) $5 \cdot \bar{2} \oplus_6 2 \cdot \bar{5}$ in $\mathbb{Z}_6$.

(f) $-(\overline{13} \oplus_7 \overline{-45})$ in $\mathbb{Z}_7$.

(g) $\left(\begin{pmatrix} 1 & 2 \\ 2 & 1 \end{pmatrix}\begin{pmatrix} 3 & 1 \\ 2 & 3 \end{pmatrix}\begin{pmatrix} 1 & 2 \\ 0 & 1 \end{pmatrix}\right)^{-1}$ in $GL_2(\mathbb{R})$.

2.2.2. Prove the right-cancellation law (Theorem 2.2.4).

2.2.3. Let $\langle G, \cdot \rangle$ be a group. Prove that there is exactly one element $x \in G$ such that $x \cdot x = x$.

2.2.4. Prove that every group of order 1 or 2 is abelian. Conclude that the group S_A of all bijections on a set A consisting of ≤ 2 elements is abelian.

2.2.5. Prove that every group of order 3 or 4 is abelian.

2.2.6. Let A be a set and let S_A be the set of all bijections on A.

Prove:

(1) if A has n elements (n a positive integer), then S_A has $n!$ elements;

(2) if A is an infinite set, so is S_A.

2.2.7. On the set $S = \{1, 2\}$, define $\alpha : S \to S$ and $\beta : S \to S$ by: $\alpha(1) = \alpha(2) = 1$; $\beta(1) = \beta(2) = 2$. Let $G = \{\alpha, \beta\}$, and let $\cdot$ be composition of maps. Prove that $\cdot$ is an associative binary operation on G; G contains a right identity for $\cdot$, relative to which every element of G has a left inverse. Prove that $\langle G, \cdot \rangle$ is not a group.

2.2.8. Let G be a group of even order. Prove that G contains at least one element $a \neq e$ such that a is its own inverse. (Hint: If $x \in G$, $x \neq x^{-1}$, then x and x^{-1} form a pair.)

2.2.9. Let $\langle G, \cdot \rangle$ be a group such that every element of G is its own inverse. Prove that G is abelian. (Hint: given $a, b \in G$, use the fact that each of the elements a, b and ab is its own inverse.)

2.2.10. Let $M_2(\mathbb{R})$ be the set of all 2×2 matrices over $\mathbb{R}$, and let $A, B \in M_2(\mathbb{R})$.

(1) Does $\begin{pmatrix} 1 & 2 \\ 2 & 3 \end{pmatrix} A = \begin{pmatrix} 1 & 2 \\ 2 & 3 \end{pmatrix} B$ imply $A = B$?

(2) Does $\begin{pmatrix} 1 & 2 \\ 3 & 6 \end{pmatrix} A = \begin{pmatrix} 1 & 2 \\ 3 & 6 \end{pmatrix} B$ imply $A = B$?

Justify your answer fully. (If you answer "no" to one of the questions, you must give a specific counterexample.)

2.2.11. In the set $M_2(\mathbb{R})$ of all 2×2 matrices over $\mathbb{R}$, does every equation $AX = B$ ($A, B \in M_2(\mathbb{R})$) have a solution X in $M_2(\mathbb{R})$? Prove or disprove.

2.2.12. Prove Theorem 2.2.7: let G be a non-empty set, $\cdot$ an associative binary operation on G. If, for each $a, b \in G$, there are $x, y \in G$ such that $ax = b$, $ya = b$, then $\langle G, \cdot \rangle$ is a group. (Hint: use Theorem 2.2.5.)

2.2.13. *(1)* Let G be a non-empty *finite* set and let $\cdot$ be an associative binary operation on G such that both left and right cancellation laws hold for $\cdot$. Prove that $\langle G, \cdot \rangle$ is a group. (See Exercises 1.3.19 and 1.3.20.)

(2) Give an example of an *infinite* set G, an associative binary operation $\cdot$ on G satisfying right and left cancellation laws, such that $\langle G, \cdot \rangle$ is *not* a group.

2.2.14. Give an example of a finite set G and a binary operation $\cdot$ on G satisfying the right, but not left, cancellation law. Is $\langle G, \cdot \rangle$ a group?

2.2.15. Prove the laws of exponents in a group $\langle G, \cdot \rangle$; i.e., prove: if $a \in G$, $h, k \in \mathbb{Z}$, then

(1) $a^h \cdot a^k = a^{h+k}$

(2) $(a^h)^k = a^{hk}$. (Hint: First deal with positive exponents, using induction.)

2.2.16. Prove the generalized associative law, i.e., prove: if $\langle G, \cdot \rangle$ is a group, $a_1, \ldots, a_n \in G$ $(n \geq 1)$, and $1 \leq h_1 < h_2 < \cdots < h_s = n$,

$$P_1 = \prod_{i=1}^{h_1} a_i, \quad P_2 = \prod_{i=h_1+1}^{h_2} a_i, \ldots, \quad P_s = \prod_{i=h_{s-1}+1}^{h_s} a_i,$$

then $\prod_{j=1}^{s} P_j = \prod_{i=1}^{n} a_i$.

2.3

Subgroups, Cyclic Groups, and the Order of an Element

By Definition 2.1.1, if A is a set and $\circ$ is a binary operation on A, then $a \circ b \in A$ for each $a, b \in A$.

Definition 2.3.1: If $\circ$ is a binary operation on a set A and $B \subset A$, then B is *closed under* $\circ$ if $x \circ y \in B$ for each $x, y \in B$. In this case, if we define $\circ_B$ on B by: $x \circ_B y = x \circ y$ for each $x, y \in B$, then $\circ_B$ is a binary operation *on* B. We call it the operation induced by $\circ$ on B, and usually drop the subscript "B"; i.e., we use "$\circ$" to denote both the given operation on A and the operation induced by it on B.

Example:

Consider the binary operation $+$ on $\mathbb{Q}$. Since $\mathbb{Z}$ is closed under addition of rational numbers, $+$ induces a binary operation $+_{\mathbb{Z}}$ on $\mathbb{Z}$. We usually simply denote it by $+$: it is the familiar addition of integers. On the other hand, let $B = \{1, 2, 3\} \subset \mathbb{Z}$. Here, B is not closed under addition, hence $+$ does not induce a binary operation on B.

All this is preliminary to introducing the rather simple concept of a subgroup.

Definition 2.3.2: Let $\langle G, \circ \rangle$ be a group, and let $H \subset G$. Then H is a *subgroup of* G if $\circ$ induces a binary operation, $\circ$, on H such that $\langle H, \circ \rangle$ is a group.

Notation: We write $H < G$ to indicate that H is a subgroup of G.

Definition 2.3.2 is all right as far as it goes, but it does not provide us with an explicit procedure for checking whether a given subset of a group is a subgroup.

The following theorem serves this purpose.

Theorem 2.3.1: Let $\langle G, \circ \rangle$ be a group, and let $H \subset G$. Then $H < G$ if and only if the following three conditions hold:

(1) $a \circ b \in H$ for each $a, b \in H$ (i.e., H is closed under $\circ$);

(2) $e \in H$;

(3) $a^{-1} \in H$ for each $a \in H$.

Proof: Suppose $H < G$. Then $\circ$ induces a binary operation on H, i.e., H is closed under $\circ$. H contains an identity element, e_H, and we do not know to start with that $e_H = e$. However, from $e_H = e_H \circ e_H = e_H \circ e$, by left-cancellation in G, we readily conclude: $e_H = e$. Thus, $e \in H$. Every element a of the group $\langle H, \circ \rangle$ has an inverse element, a', in H. From $a \circ a' = e = a \circ a^{-1}$ (where a^{-1} is the inverse of a in G), we conclude by left-cancellation that $a' = a^{-1}$. Thus, if $a \in H$, then $a^{-1} \in H$. We have proved that (1), (2), and (3) hold.

Conversely, suppose H is a subset of G satisfying (1), (2), and (3). Since the induced operation $\circ$ on H inherits associativity from the given operation $\circ$ on G, we conclude that $\langle H, \circ \rangle$ satisfies all the conditions of Definition 2.2.1. But then $H < G$. ■

Corollary: Let $\langle G, \circ \rangle$ be a group, and let $H \subset G$. Then $H < G$ if and only if

(1) $H \neq \varnothing$

and

(2) $a \circ b^{-1} \in H$ for each $a, b \in H$.

We leave the proof as an exercise (see Exercise 2.3.2).

While the corollary provides a more elegant characterization of subgroups, in practice the checklist of properties (1), (2), and (3) of Theorem 3.3.1 frequently provides the most natural procedure for determining whether a given subset of a group is a subgroup.

Example 1:

If $\langle G, \circ \rangle$ is any group, then G and $E = \{e\}$ obviously form subgroups of G. (A subgroup different from G or E is called a *proper subgroup* of G. E is called the *trivial subgroup* of G.)

Example 2:

In the sequence of groups $\langle \mathbb{Z}, + \rangle$, $\langle \mathbb{Q}, + \rangle$, $\langle \mathbb{R}, + \rangle$, $\langle \mathbb{C}, + \rangle$, each group is a subgroup of all that follow.

Example 3:

$\{-1, 1\}$ and $\mathbb{Q}^+$ are subgroups of $\langle \mathbb{Q}^*, \cdot \rangle$.

Example 4:

$\{-1, 1\}$, $\mathbb{Q}^+$, and $\mathbb{R}^+$ are subgroups of $\langle \mathbb{R}^*, \cdot \rangle$.

Example 5:

$H = \{z \in \mathbb{C} \mid |z| = 1\}$ is a subgroup of $\langle \mathbb{C}^*, \cdot \rangle$.

Example 6:

Let A be a non-empty set. If $a \in A$, and $H_a = \{\alpha \in S_A \mid \alpha(a) = a\}$, then $H_a < S_A$.

Example 7:

For n a positive integer, let $G = GL_n(\mathbb{R})$ (the group of all non-singular $n \times n$ matrices over $\mathbb{R}$), and let

$$H = \{A \in G \mid \det A = 1\}.$$

Then $H < G$.

(The subgroup H is generally denoted by $SL_n(\mathbb{R})$—the "special linear group" determined by n, over $\mathbb{R}$.)

Example 8:

Any subspace of a vector space V forms a subgroup of the group $\langle V, + \rangle$.

It should be amply clear by now that a subset S of a group G has to be very special to be a subgroup of G. However, as we shall see very shortly, every subset of a group G *generates* a subgroup of G.

Theorem 2.3.2: Let $\langle G, \circ \rangle$ be a group, and let $\mathscr{C}$ be a set of subgroups of G. Then $\bigcap_{H \in \mathscr{C}} H < G$.

Proof: We use the Corollary of Theorem 2.3.1. Since each $H \in \mathscr{C}$ contains the identity, e, of G, we have $e \in \bigcap_{H \in \mathscr{C}} H$, and so $\bigcap_{H \in \mathscr{C}} H \neq \varnothing$.

Suppose $a, b \in \bigcap_{H \in \mathscr{C}} H$. Then $a, b \in H$, for each $H \in \mathscr{C}$; hence $a \circ b^{-1} \in H$, for each $H \in \mathscr{C}$. But then $a \circ b^{-1} \in \bigcap_{H \in \mathscr{C}} H$. It follows that $\bigcap_{H \in \mathscr{C}} H < G$. ∎

Consider now a group G, and a subset S of G. In general, S will not *be* a subgroup of G. But S is *contained in* at least one subgroup of G, namely G itself. What is the "smallest" subgroup of G that contains S? In what sense "smallest"?

Definition 2.3.3: Let $\langle G, \circ \rangle$ be a group, and let $S \subset G$. Then the *subgroup* of G which is *generated by* S is

$$[S] = \bigcap_{\substack{H < G \\ S \subset H}} H.$$

In particular, if $S = \{a\}$, then $[S] = \bigcap_{\substack{H < G \\ a \in H}} H$ is called the *cyclic subgroup of G generated by a*, and is denoted simply by $[a]$.

By Theorem 2.3.2, $[S]$ is a subgroup of G for *every* subset S of G (including $S = \varnothing$, in which case $[S] = \{e\} = [e]$). The subgroup $[S]$ is contained in every subgroup H of G that contains the set S. This is the sense in which $[S]$ is "smallest." [For G a finite group, $[S]$ will indeed be the subgroup of *smallest order* that contains S. But for G infinite, the characterization of $[S]$ as an intersection is essential. For example, in $\langle \mathbb{Z}, + \rangle$, let $S = \{2\}$. Then the only subgroups of $\mathbb{Z}$ which contain $S = \{2\}$ are $\mathbb{Z}$ and $2\mathbb{Z}$ (the subgroup consisting of all even integers). Thus, $[S] = \mathbb{Z} \cap 2\mathbb{Z} = 2\mathbb{Z}$. But note that $\mathbb{Z}$ and $2\mathbb{Z}$ contain "the same number of elements," in the sense that their elements can be matched up one-to-one under the bijection $f\colon \mathbb{Z} \to 2\mathbb{Z}$ defined by $f(a) = 2a$, for each $a \in \mathbb{Z}$.]

As was the case with the definition of *subgroup*, Definition 2.3.2 is conceptually elegant, but impractical for daily use. We need a more explicit characterization of the subgroup $[S]$ generated by a subset S, a characterization that will allow us to recognize the elements of $[S]$.

Theorem 2.3.3: Let $\langle G, \cdot \rangle$ be a group.

(1) If $S \subset G$, $S \neq \varnothing$, then $[S] = \{a_1^{s_1} \dots a_m^{s_m} | a_i \in S, s_i \in \mathbb{Z}, m \geq 1\}$, where the a_i are not necessarily distinct. (Equivalently, $[S]$ is the set of all products of elements of S and their inverses.)

(2) In particular, for $a \in G$, the subgroup $[a]$ generated by a is given by: $[a] = \{a^k | k \in \mathbb{Z}\}$.

Proof: *(1)* Write $P = \{a_1^{s_1} \dots a_m^{s_m} | a_i \in S,\ s_i \in \mathbb{Z},\ m \geq 1\}$. Running through the checklist in Theorem 2.3.1, we obtain immediately that $P < G$. For each $a \in S$, $a = a^1 \in P$. Hence $S \subset P$. Thus, P is one of the subgroups of G which contain S. By Definition 2.3.2, $[S]$ is the intersection of all such subgroups; hence $[S] \subset P$.

On the other hand, since $[S]$ is a subgroup of G, and $S \subset [S]$, all products and inverses of elements of S are in $[S]$. This implies that $P \subset [S]$. But then $[S] = P$.

(2) If $S = \{a\}$, the set P of (1) reduces to $P = \{a^k | k \in \mathbb{Z}\}$, and we have $[a] = \{a^k | k \in \mathbb{Z}\}$. ∎

Again, note the additive version:

(1′) $[S] = \{s_1 a_1 + \dots + s_m a_m | a_i \in S,\ s_i \in \mathbb{Z},\ m \geq 1\}$

(2′) $[a] = \{ka | k \in \mathbb{Z}\}$.

Example 1:

Let $S = \{2, 3\} \subset \mathbb{Z}$. By Theorem 2.3.3, the subgroup of $\langle \mathbb{Z}, + \rangle$ generated by S is $[S] = \{2s_1 + 3s_2 | s_1, s_2 \in \mathbb{Z}\}$. Since $1 = 2 \cdot (-1) + 3 \cdot 1$, we have $1 \in [S]$. Hence (by Theorem 2.3.1), $h1 \in [S]$ for each $h \in \mathbb{Z}$. But then $[S] = \mathbb{Z}$.

Example 2:

On the other hand, if $T = \{4, 6\}$, then

$$[T] = \{4s_1 + 6s_2 | s_1, s_2 \in \mathbb{Z}\} = \{2(2s_1 + 3s_2) | s_1, s_2 \in \mathbb{Z}\}.$$

Since every integer can be expressed in the form $2s_1 + 3s_2$ for some $s_1, s_2 \in \mathbb{Z}$ (see Example 1), $[T]$ is simply the set $2\mathbb{Z}$ of all even integers.

Note that, even though each of the preceding subgroups was defined in terms of more than one generator, each has turned out to be generated by a single one of its elements:

$$[S] = [1] \quad \text{and} \quad [T] = [2].$$

This is no accident since, in fact, every subgroup of $\mathbb{Z}$ has this property (see Theorem 2.3.5).

Definition 2.3.4: A group G is *cyclic* if there is some $a \in G$ such that $G = [a]$.

From Theorem 2.3.3, we have: if $\langle G, \cdot \rangle$ is cyclic, then $G = [a] = \{a^k | k \in \mathbb{Z}\}$, for some $a \in G$. (In additive notation: $G = \{ka | k \in \mathbb{Z}\}$.)

Example 1:

$\mathbb{Z} = [1] = [-1]$; hence it is cyclic.

Example 2:

For each positive integer n, $\mathbb{Z}_n = [\bar{1}]$; hence it is cyclic. In fact, (see Exercise 2.3.13), $\mathbb{Z}_n = [\bar{k}]$ for each $k \in \mathbb{Z}$ such that k and n are relatively prime.

Example 3:

If G is any group, then the subgroup $[a]$ generated by any element a of G is, of course, a cyclic group.

We shall see shortly that $\langle \mathbb{Z}, + \rangle$ and the groups $\langle \mathbb{Z}_n, \oplus_n \rangle$ form the structural prototypes of *all* cyclic groups.

Theorem 2.3.4: Every cyclic group is abelian.

Proof: This follows immediately from Theorem 3.3.3. For, if $G = [a]$, then $G = \{a^k | k \in \mathbb{Z}\}$. Thus, if $x, y \in G$, then $x = a^h$, $y = a^k$, for some $h, k \in \mathbb{Z}$. But then $xy = a^h a^k = a^{h+k} = a^{k+h} = a^k a^h = yx$. ■

The converse is false! "Most" abelian groups are *not* cyclic. Familiar examples of non-cyclic abelian groups include $\langle \mathbb{Q}, + \rangle$, $\langle \mathbb{Q}^*, + \rangle$, $\langle \mathbb{R}, + \rangle$, $\langle \mathbb{R}^*, + \rangle$, $\langle \mathbb{C}, + \rangle$, $\langle \mathbb{C}^*, + \rangle$ (see Exercise 2.3.14). However, cyclic groups (as we shall see in

Section 2.12) do play a central role in the structure theory of finitely generated abelian groups, for which they form the basic building blocks.

Theorem 2.3.5: Every subgroup of a cyclic group is cyclic.

Proof: The essential tools in this proof are the Well-Ordering Property of $\mathbb{Z}^+$, and the Division Algorithm.

Let $G = [a]$ be a cyclic group, and let $H < G$. If $H = \{e\}$, then $H = [e]$; hence it is cyclic. If $H \neq \{e\}$, there is some non-zero integer s such that $a^s \in H$. Then also $a^{-s} = (a^s)^{-1} \in H$, and one of s, $-s$ is positive. Let $T = \{t \in \mathbb{Z}^+ | a^t \in H\}$. We have just shown that $T \neq \varnothing$. By well-ordering of $\mathbb{Z}^+$, T has a least element, t_0. We want to prove that $H = [a^{t_0}]$. Of course, since $a^{t_0} \in H$ (by the definition of T), all powers of a^{t_0} are in H; hence $[a^{t_0}] \subset H$. Thus, only the opposite inclusion remains to be established. Let $h \in H$. Since $G = [a]$, there is some integer k such that $h = a^k$. By the Division Algorithm,

$$k = t_0 q + r$$

for some $q, r \in \mathbb{Z}$, $0 \leq r < t_0$.

Hence $a^r = a^{k - t_0 q} = a^k (a^{t_0})^{-q} \in H$. Thus, if $r > 0$, then $r \in T$. But t_0 is the least integer in T, so $0 < r < t_0$ is impossible. We conclude that $r = 0$, whence $k = t_0 q$, and $h = a^k = (a^{t_0})^q \in [a^{t_0}]$. It follows that $H \subset [a^{t_0}]$. But then H $= [a^{t_0}]$, a cyclic group. ■

Corollary: Every subgroup of $\mathbb{Z}$ is of the form $k\mathbb{Z} = \{kq | q \in \mathbb{Z}\}$ for some $k \in \mathbb{Z}$, $k \geq 0$.

Theorem 2.3.6: Let $G = [a]$ be a cyclic group.

(1) If G is finite, of order n, then $G = \{e, a, \ldots, a^{n-1}\}$ with $a^n = e$, and $n = \min \{t \in \mathbb{Z}^+ | a^t = e\}$.

(2) If G is infinite, then the powers of a are all distinct, i.e., for $h \neq k$ $(h, k \in \mathbb{Z})$, $a^h \neq a^k$. In particular, $a^h \neq e$ for $h \neq 0$.

Proof: *(1)* By Theorem 2.3.3, $G = \{a^k | k \in \mathbb{Z}\}$. Since G is a finite set, the powers a^k are *not* all distinct. Hence for some $h < k$ $(h, k \in \mathbb{Z})$, we have $a^h = a^k$, and $a^{k-h} = e$. Let

$$T = \{t \in \mathbb{Z}^+ | a^t = e\}.$$

Then $T \neq \varnothing$ since $k - h > 0$. Hence there is a least positive integer l in T. Clearly $\{e, a, \ldots, a^{l-1}\} \subset G$. The elements $e, a, \ldots, a^{l-1}$ are all distinct—else, for some $i, j \in \mathbb{Z}$, $0 \leq i < j < l$, $a^i = a^j$, and $a^{j-i} = e$. But $0 < j - i \leq j < l$, contrary to the choice of l as least in T. We want to prove that $G = \{e, a, \ldots, a^{l-1}\}$ (which will imply that $l = n$), and thus must prove the opposite inclusion

$$G \subset \{e, a, \ldots, a^{l-1}\}.$$

Let $g \in G$. Then $g = a^s$ for some $s \in \mathbb{Z}$. By the Division Algorithm, $s = lq + r$ for some $q, r \in \mathbb{Z}$, $0 \leq r < l$. Hence

$$a^s = a^{lq+r} = (a^l)^q a^r = e^q a^r = a^r.$$

But then $g = a^s \in \{e, a, \ldots, a^{l-1}\}$. It follows that $G = \{e, a, \ldots, a^{l-1}\}$. Since $|G| = n$, and the elements $e, a, \ldots, a^{l-1}$ are all distinct, we conclude that $l = n$,

$$G = \{e, a, \ldots, a^{n-1}\},$$

and $a^n = e$, as promised.

Note that this result makes the finite cyclic group $G = [a]$ cyclic in a very graphic sense (see Figure 2). If its elements are arranged in a circle,

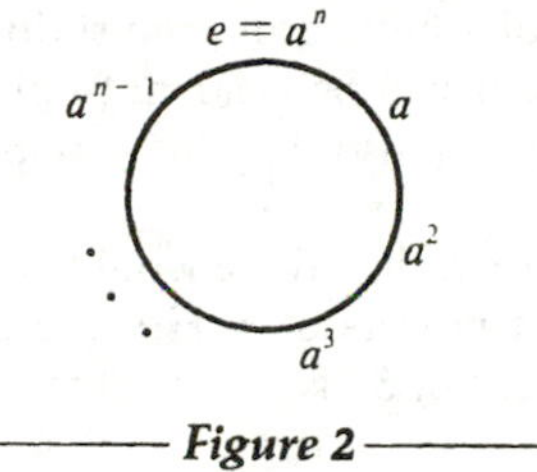

Figure 2

then, moving clockwise along the circle, we see that each element is a times its predecessor.

(2) If $G = [a]$ is infinite, suppose $a^h = a^k$ for some $h, k \in \mathbb{Z}$, $h < k$. As in the proof of (1), it follows that there is a least positive integer l such that $a^l = e$ and that $\{e, a, \ldots, a^{l-1}\} = G$. But this is impossible since G is infinite. Hence the powers of a are all distinct. In particular, $a^h \neq a^0 = e$ for $h \neq 0$. ■

A pleasant by-product of the preceding theorem is a very simple subgroup test for *finite* subsets of a group G.

Corollary: Let $\langle G, \cdot \rangle$ be a group, H a *finite* non-empty subset of G. If H is closed under the group operation $\cdot$, then $H < G$.

Proof: Since $H \neq \varnothing$, there is some $a \in H$. By closure of H under $\cdot$, using induction, we conclude that $a^k \in H$ for all positive integers k. Since H is finite, the powers $a^k (k \in \mathbb{Z}^+)$ are not all distinct; hence $a^i = a^j$ for some $i, j \in \mathbb{Z}^+$, with $i < j$. Let $t = j - i$. Then $t > 0$ and $a^t = a^{j-i} = e$. Thus, $e \in H$. From $a^t = e$, we have $a \cdot a^{t-1} = e = a^{t-1} \cdot a$; hence $a^{t-1} = a^{-1}$. If $t = 1$, then $a = e$; hence $a^{-1} = e \in H$. If $t > 1$, then $t - 1 > 0$ and $a^{-1} = a^{t-1} \in H$, by closure. Since a was chosen arbitrarily in H, we conclude that the inverse of every element of H is located in H. Thus, H satisfies the three subgroup conditions of Theorem 2.3.1, and so $H < G$. ■

Example 1:

Let us find the cyclic subgroups of S_3, using the Cayley table on p. 36. They are

$$[A] = \{A\}$$
$$[E] = \{A, E, E^2 = J\} = \{A, J, J^2 = E\} = [J]$$
$$[N] = \{A, N\}$$
$$[S] = \{A, S\}$$
$$[W] = \{A, W\}.$$

Note that the orders of these subgroups are 1, 2, and 3.

(There are, in fact, no other subgroups except for the whole group, S_3, itself. This could be established quite laboriously by showing that no other subset of S_3 is closed under the group operation. However, very soon, we will learn that the order of any subgroup of a finite group must divide the order of the group (Theorem 2.5.2). From this, it will be clear not only that any subgroup of S_3 must have order 1, 2, 3, or 6, but also that any group of order 1, 2, or 3 must be cyclic.)

S_3 is *not* cyclic. For, if it were, then there would be some element σ in S_3 such that S_3 consists of 6 distinct powers of σ. However, the maximum number of distinct powers of any element of S_3 is 3. (Or note: S_3 is not abelian!)

Example 2:

In $GL_2(\mathbb{R})$, (the group of all non-singular 2×2 matrices with real entries), we find the cyclic subgroups generated by the matrices $A = \begin{pmatrix} 0 & 1 \\ -1 & 0 \end{pmatrix}$ and $B = \begin{pmatrix} 1 & 1 \\ 0 & 1 \end{pmatrix}$. We have

$$A^2 = \begin{pmatrix} 0 & 1 \\ -1 & 0 \end{pmatrix}\begin{pmatrix} 0 & 1 \\ -1 & 0 \end{pmatrix} = \begin{pmatrix} -1 & 0 \\ 0 & -1 \end{pmatrix}$$
$$A^3 = \begin{pmatrix} 0 & 1 \\ -1 & 0 \end{pmatrix}\begin{pmatrix} -1 & 0 \\ 0 & -1 \end{pmatrix} = \begin{pmatrix} 0 & -1 \\ 1 & 0 \end{pmatrix}$$
$$A^4 = \begin{pmatrix} 1 & 0 \\ 0 & 1 \end{pmatrix} = I_2 \quad \text{(the } 2 \times 2 \text{ identity matrix).}$$

By Theorem 2.3.6, it follows that $[A] = \{I_2, A, A^2, A^3\}$. For B, we have $B^2 = \begin{pmatrix} 1 & 2 \\ 0 & 1 \end{pmatrix}$, $B^3 = \begin{pmatrix} 1 & 3 \\ 0 & 1 \end{pmatrix}, \ldots, B^n = \begin{pmatrix} 1 & n \\ 0 & 1 \end{pmatrix} \neq I_2$ for each positive integer n. By Theorem 2.3.6, it follows that the cyclic subgroup $[B]$ generated by B is infinite, consisting of the infinitely many distinct powers B^k, $k \in \mathbb{Z}$. Explicitly,

$$[B] = \left\{\begin{pmatrix} 1 & k \\ 0 & 1 \end{pmatrix} \middle| k \in \mathbb{Z}\right\}.$$

$\left(\text{Note that, for } n \in \mathbb{Z}^+, (B^n)^{-1} = B^{-n} = \begin{pmatrix} 1 & -n \\ 0 & 1 \end{pmatrix}.\right)$

Order of an Element

As illustrated by the preceding examples, Theorem 2.3.6 tells us a great deal about each element of an arbitrary group G. For, if $g \in G$, then g generates a cyclic subgroup $[g]$ of G and this subgroup is either finite cyclic, or infinite cyclic. Hence either

$$[g] = \{e, g, \ldots, g^{n-1}\}$$

for some positive integer n, or

$$[g] = \{g^k \mid k \in \mathbb{Z}\},$$

with the powers of g all distinct.

Definition 2.3.5: Let $\langle G, \cdot \rangle$ be a group, and let $g \in G$. Then the *order*, $o(g)$, *of* g, is equal to the order of the cyclic subgroup $[g]$ generated by g.

In Example 1 preceding, we thus have $o(A) = 1$, $o(E) = o(J) = 3$, and $o(N) = o(S) = o(W) = 3$. In Example 2, $o(A) = 4$, while $o(B)$ is infinite.

Example:

In $\langle \mathbb{Z}_6, \otimes_6 \rangle$, we have

$$[\bar{0}] = \{\bar{0}\}; \text{ hence } \bar{0} \text{ has order } 1;$$

$$[\bar{1}] = \{\bar{0}, \bar{1}, \bar{2}, \bar{3}, \bar{4}, \bar{5}\} = \mathbb{Z}_6; \text{ hence } \bar{1} \text{ has order } 6;$$

$$[\bar{2}] = \{\bar{0}, \bar{2}, \bar{4}\}; \text{ hence } \bar{2} \text{ has order } 3;$$

$$[\bar{3}] = \{\bar{0}, \bar{3}\}; \text{ hence } \bar{3} \text{ has order } 2;$$

$$[\bar{5}] = \{\bar{0}, \bar{5}, \bar{4}, \bar{3}, \bar{2}, \bar{1}\} = \mathbb{Z}_6; \text{ hence } \bar{5} \text{ has order } 6.$$

(We have listed the elements in the sequence in which they arise "cyclically," e.g., $0 \cdot \bar{5}, 1 \cdot \bar{5}, 2 \cdot \bar{5}, 3 \cdot \bar{5}, 4 \cdot \bar{5}, 5 \cdot \bar{5}$ in $[\bar{5}]$.)

Obviously, every element g of a group G of finite order will have finite order (since $[g]$ is a subset of the finite set G). On the other hand, if G has infinite order, an element $g \in G$, $g \neq e$, may have either finite or infinite order, as illustrated by Example 2. (The identity element, of course, always has order 1.)

You may have already observed some of the basic properties of the order of a group element.

Theorem 2.3.7: Let $\langle G, \cdot \rangle$ be a group, $a \in G$, and let n be a positive integer. Then the following statements are equivalent:

(1) a has finite order n;

(2) $a^n = e$ and $n \leq t$ for each positive integer t such that $a^t = e$;

(3) $a^n = e$ and $n \mid s$ for each integer s such that $a^s = e$.

Proof: $1 \Rightarrow 2$: If $o(a) = n$, then

$$[a] = \{e, a, \ldots, a^{n-1}\},$$

with $e, a, \ldots, a^{n-1}$ all distinct, and $a^n = e$. Thus n is the least positive integer t such that $a^t = e$.

$2 \Rightarrow 3$: Again, as in the proof of Theorem 2.3.5, if $n = \min\{t \in \mathbb{Z}^+ \mid a^t = e\}$, then $n \mid s$ for every integer s such that $a^s = e$.

(3) clearly implies (2), and (2) implies (1) since, in the proof of Theorem 2.3.5, we saw that, if $n = \min\{t \in \mathbb{Z}^+ \mid a^t = e\}$, then n is the order of G. ■

Corollary: Let $\langle G, \cdot \rangle$ be a group, and let $a \in G$. Then the following statements are equivalent:

(1) a has infinite order;

(2) there is no positive integer t such that $a^t = e$.

Remark: By Theorem 2.3.7 and its Corollary, finding the order of a group element, a, amounts to determining whether there exists a positive integer k such that $a^k = e$. If so, then $o(a)$ is the *least* such positive integer. If not, then $o(a)$ is infinite.

Example:

In $\langle \mathbb{C}^*, \cdot \rangle$, $o(-1) = 2$ since $-1 \neq 1$ and $(-1)^2 = 1$. What is the order of i? We have $i^2 = -1$, $i^3 = -i$, $i^4 = 1$. Thus, $o(i) = 4$. What is the order of $w = -\frac{1}{2} + \frac{\sqrt{3}}{2}i$? We have $w^2 = \frac{1}{4} - \frac{1}{2}\sqrt{3}i - \frac{3}{4} = -\frac{1}{2} - \frac{\sqrt{3}}{2}i$, $w^3 = ww^2 = \frac{1}{4} + \frac{3}{4} = 1$. Thus $o(w) = 3$.

A complex number z is called an n-th root of unity if $z^n = 1$ for some $n \in \mathbb{Z}^+$. If $z \in \mathbb{C}$ is an n-th root of unity, then z is a *primitive* n-th root of unity if n is the *least* positive integer such that $z^n = 1$, i.e., if $o(z) = n$ in $\mathbb{C}^*$. Thus, -1 is a primitive square root, w is a primitive cube root, and i a primitive 4-th root of unity. By Theorem 3.3.7, if z is a primitive n-th root of unity (i.e., if $o(z) = n$), then z is an s-th root of unity (i.e., $z^s = 1$) if and only if n divides s. For example, $w^s = 1$ if and only if s is a multiple of 3.

Which complex numbers have finite order in $\mathbb{C}^*$? By Theorem 3.3.7 and its Corollary, we know that $z \in \mathbb{C}^*$ *has finite order* if and only if there is a positive integer t such that $z^t = 1$, i.e., *if and only if z is a root of unity*. In this case, z is a primitive n-th root of unity, where $n = \min\{t \in \mathbb{Z}^+ \mid z^t = 1\} = o(z)$ in $\mathbb{C}^*$. (Roots of unity will be studied in detail in Chapter 4, Section 11.)

Exercises 2.3

True or False

1. Every abelian group is cyclic.
2. Every element of a group G is contained in a cyclic subgroup of G.
3. Every subset of a group G is contained in a cyclic subgroup of G.
4. A finite cyclic group has exactly one generator.
5. Every infinite cyclic group has exactly two generators.

2.3.1. Find the elements of $[\alpha]$, and find $o(\alpha)$, if

(a) $\alpha = 3$ in $\langle \mathbb{Z}, + \rangle$

(b) $\alpha = \bar{3}$ in $\langle \mathbb{Z}_{10}, \oplus_{10} \rangle$

(c) $\alpha = \bar{3}$ in $\langle \mathbb{Z}_9, \oplus_9 \rangle$

(d) $\alpha = \begin{pmatrix} 1 & 2 & 3 & 4 & 5 & 6 \\ 2 & 4 & 6 & 1 & 5 & 3 \end{pmatrix}$ in S_6

(e) $\alpha = \begin{pmatrix} 2 & 0 \\ 0 & 3 \end{pmatrix}$ in $GL_2(\mathbb{R})$

(f) $\alpha = \begin{pmatrix} 0 & 1 \\ 1 & 0 \end{pmatrix}$ in $GL_2(\mathbb{R})$

(g) $\alpha = \begin{pmatrix} 0 & -1 \\ 1 & -1 \end{pmatrix}$ in $GL_2(\mathbb{R})$

(h) $\alpha = \begin{pmatrix} 0 & -1 \\ 1 & 1 \end{pmatrix}$ in $GL_2(\mathbb{R})$

(i) $\alpha = 1 + i$ in $\mathbb{C}^*$

(j) $\alpha = -\frac{1}{2} + i\frac{\sqrt{3}}{2}$ in $\mathbb{C}^*$.

2.3.2. Prove the Corollary of Theorem 2.3.1.

2.3.3. Let G be a group such that every $a \in G$, $a \neq e$, has order 2. Prove that G is abelian. (Recall Exercise 2.2.9.)

2.3.4. Let $H < G$ and let $x \in G$. Define $xHx^{-1} = \{xhx^{-1} | h \in H\}$. Prove that $xHx^{-1} < G$. Prove that $|H| = |xHx^{-1}|$.

2.3.5. Prove: if G is a group, $x \in G$, then $o(x) = o(x^{-1})$.

2.3.6. *(a)* Let G be an abelian group. Prove that the elements of finite order in G form a subgroup of G. (This subgroup is called the *torsion subgroup* of G.)

(b) Do the elements of finite order in an arbitrary group form a subgroup? Prove or disprove.

(c) Do the elements of infinite order in any group form a subgroup? Prove or disprove.

2.3.7. Prove that, in an infinite cyclic group G, there are exactly two elements a, b such that $[a] = [b] = G$.

2.3.8. Let G be a group, and let A, B be subgroups of G. Prove that the set-theoretic union $A \cup B$ is a subgroup of G if and only if either $A \subset B$ or $B \subset A$. (Hint: if neither $A \subset B$ nor $B \subset A$, then there is an element $a \in A \backslash B$ and an element $b \in B \backslash A$. Consider the product ab.)

2.3.9. Prove that the subgroups of a group G form a lattice with respect to the partial order relation $\subset$ (set inclusion), with $\text{glb}(A, B) = A \cap B$, $\text{lub}(A, B) = [A \cup B]$, for A, B subgroups of G. (Caution: $[A \cup B]$ is the *subgroup of G generated by the set $A \cup B$*. The set $A \cup B$ itself is very rarely a subgroup—see the preceding exercise.)

2.3.10. Let G be a cyclic group of finite order n. Prove that, for each divisor m of n, G has exactly one subgroup of order m. Make a lattice diagram showing the subgroups of $\mathbb{Z}_{24}$.

2.3.11. In the group $GL_2(\mathbb{R})$, find the elements of the subgroup, G, generated by the two matrices

$$A = \begin{pmatrix} 0 & 1 \\ 1 & 0 \end{pmatrix} \quad \text{and} \quad B = -A = \begin{pmatrix} 0 & -1 \\ -1 & 0 \end{pmatrix}.$$

Draw a lattice diagram showing the subgroups of G.

2.3.12. In the multiplicative group $\langle \mathbb{C}^*, \cdot \rangle$ of all non-zero complex numbers, find the elements of the subgroup G generated by i and $-i$. Draw a lattice diagram showing the subgroups of G and compare this group with the group in Exercise 2.3.8.

2.3.13. Let $G = [a]$ be a cyclic group of finite order n. Prove that, for every integer k, the order of a^k is $n/\gcd(n, k)$. Conclude that $G = [a^k]$ if and only if $\gcd(n, k) = 1$, so that G has exactly $\phi(n)$ generators (where ϕ is the Euler Phi-function, defined by: $\phi(n) =$ the number of integers k such that $0 < k \leq n$ and $\gcd(k, n) = 1$).

2.3.14. Prove that each of the following groups fails to be cyclic: $\langle \mathbb{Q}, + \rangle$, $\langle \mathbb{Q}^*, \cdot \rangle$, $\langle \mathbb{R}, + \rangle$, $\langle \mathbb{R}^*, \cdot \rangle$, $\langle \mathbb{C}, + \rangle$, $\langle \mathbb{C}^*, \cdot \rangle$. (Hint: Remember Theorem 2.3.5.)

2.3.15. For n a positive integer, a complex number ζ is called an *n-th root of unity* if $\zeta^n = 1$. Prove that, for each positive integer n, the n-th roots of unity form a cyclic subgroup of $\langle \mathbb{C}^*, \cdot \rangle$. (Use de Moivre's Theorem.)

2.3.16. Let G be a group, and let $a, b \in G$ be two commuting elements of finite order.

(1) Prove: if $\gcd(o(a), o(b)) = 1$, then $o(ab) = o(a)\, o(b)$.

(2) Prove or disprove: in general, $o(ab) = \operatorname{lcm}(o(a), o(b))$.

Hint: be careful!

2.3.17. Let G be a group that has no proper subgroups. Prove that G is finite cyclic of prime order.

Isomorphism

Consider the Cayley table for $\langle \mathbb{Z}_3, \oplus_3 \rangle$ and the Cayley table for the subgroup W of $\mathbb{C}^*$ generated by $w = -\frac{1}{2} + i\frac{\sqrt{3}}{2}$:

$\oplus_3$	$\bar{0}$	$\bar{1}$	$\bar{2}$
$\bar{0}$	$\bar{0}$	$\bar{1}$	$\bar{2}$
$\bar{1}$	$\bar{1}$	$\bar{2}$	$\bar{0}$
$\bar{2}$	$\bar{2}$	$\bar{0}$	$\bar{1}$

$\mathbb{Z}_3$

$\cdot$	1	ω	ω^2
1	1	ω	ω^2
ω	ω	ω^2	1
ω^2	ω^2	1	ω

W

The groups $\mathbb{Z}_3$ and W are clearly not the same, but their tables bear a close resemblance to each other. In fact, under the correspondence

$$\begin{aligned}\bar{0} &\mapsto 1\\ \bar{1} &\mapsto \omega\\ \bar{2} &\mapsto \omega^2,\end{aligned}$$

the table for $\mathbb{Z}_3$ corresponds (in every entry) to the table for W, i.e., the sum of any two elements of $\mathbb{Z}_3$ corresponds to the product of the corresponding elements of W. For example, $\bar{1} \oplus_3 \bar{2} = \bar{0}$ in $\mathbb{Z}_3$. In W, the product of the corresponding elements is $\omega \cdot \omega^2$, which equals 1, the element of W corresponding to $\bar{0}$. This shows that the two groups, while different, are *structurally alike.* We have just illustrated one of the most important concepts in algebra: isomorphism. The word, derived from the Greek *isos* (equal) and *morphe* (form) means "structural likeness." We encounter it first in connection with groups, but will meet it again whenever we study a new type of algebraic structure.

Definition 2.4.1: Let $\langle G, \circ\rangle$, $\langle G', \circ'\rangle$ be groups. An *isomorphism of* $\langle G, \circ\rangle$, *onto* $\langle G', \circ'\rangle$ is a bijection

$$\phi : G \to G'$$

such that

$$\phi(x \circ y) = \phi(x) \circ' \phi(y)$$

for each $x, y \in G$.

If there exists an isomorphism from G onto G', we say that G *is isomorphic to* G', and write $G \cong G'$.

Remark: On any non-empty set of groups, $\cong$ is an equivalence relation (see Exercise 2.4.8). If $G \cong G'$, because of the symmetry of $\cong$, we simply say: "G and G' are isomorphic."

Example 1:

In the discussion preceding Definition 2.4.1, we observed that $\langle \mathbb{Z}_3, \oplus_3\rangle$ and $\langle W, \cdot\rangle$ (where $W = [\omega] \subset \mathbb{C}^*$) are isomorphic groups. If we define $\phi : \mathbb{Z}_3 \to W$ by

$$\begin{aligned}\phi(\bar{0}) &= 1\\ \phi(\bar{1}) &= \omega\\ \phi(\bar{2}) &= \omega^2\end{aligned}$$

then ϕ is an isomorphism:

$$\phi(\bar{h} \oplus \bar{k} = \phi(h) \cdot \phi(k)$$

for each h, k $(0 \le h, k \le 2)$.

ϕ *is not the only* isomorphism of $\mathbb{Z}_3$ onto W. If we define $\psi : \mathbb{Z}_3 \to W$ by

$$\begin{aligned}\psi(\bar{0}) &= 1\\ \psi(\bar{1}) &= \omega^2\\ \psi(1) &= \omega\end{aligned}$$

then ψ is also an isomorphism. This is apparent from the Cayley tables, with the table for W rearranged as shown:

$\oplus_3$	$\bar{0}$	$\bar{1}$	$\bar{2}$
$\bar{0}$	$\bar{0}$	$\bar{1}$	$\bar{2}$
$\bar{1}$	$\bar{1}$	$\bar{2}$	$\bar{0}$
$\bar{2}$	$\bar{2}$	$\bar{0}$	$\bar{1}$

$\mathbb{Z}_3$

$\cdot$	1	ω	ω^2
1	1	ω^2	ω
ω^2	ω^2	ω	1
ω	ω	1	ω^2

W

Example 2:

$\langle \mathbb{Z}_2, \oplus_2 \rangle$ and $H = \{-1, 1\} < \langle \mathbb{Q}^+, \cdot \rangle$ are isomorphic groups. The mapping $\phi : \mathbb{Z}_2 \to H$ defined by:

$$\phi(\bar{0}) = 1$$
$$\phi(\bar{1}) = -1$$

is an isomorphism of $\mathbb{Z}_2$ onto H, as can be readily seen from the Cayley tables

$\oplus_2$	$\bar{0}$	$\bar{1}$
$\bar{0}$	$\bar{0}$	$\bar{1}$
$\bar{1}$	$\bar{1}$	$\bar{0}$

and

$\cdot$	1	-1
1	1	-1
-1	-1	1

Example 3:

Let $\langle G, \circ \rangle = \langle \mathbb{R}, + \rangle$ and let $\langle G', \circ' \rangle = \langle \mathbb{R}^+, \cdot \rangle$. Define $\phi : \mathbb{R} \to \mathbb{R}^+$ by:

$$\phi(x) = e^x \quad \text{for each} \quad x \in \mathbb{R}.$$

Then ϕ is a bijection since it has inverse function $\psi : \mathbb{R}^+ \to \mathbb{R}$ defined by

$$\psi(y) = ln(y), \quad \text{for each} \quad y \in \mathbb{R}^+.$$

(See Theorem 1.2.3.)

For each $x, y \in \mathbb{R}$,

$$\phi(x + y) = e^{x+y} = e^x \cdot e^y = \phi(x) \cdot \phi(y).$$

We conclude that ϕ is an isomorphism of $\langle \mathbb{R}, + \rangle$ onto $\langle \mathbb{R}^+, \cdot \rangle$.

Example 4:

Getting bolder, we might ask whether $\langle \mathbb{R}, + \rangle$ and $\langle \mathbb{R}^*, \cdot \rangle$ are isomorphic as well. The answer is "no," but the non-existence of an isomorphism is sometimes harder to prove than the existence of an isomorphism. Students often make the mistake of merely proving that some particular mapping that comes to mind is not an isomorphism. This is obviously insufficient since what must be proved is that *no isomorphism exists.* A more fruitful approach is to look for a *structural property* (i.e, a property which isomorphism would preserve) that one of the groups has and the other fails to have.

In the case of $\langle \mathbb{R}, + \rangle$ and $\langle \mathbb{R}^*, \cdot \rangle$, observe that, in $\langle \mathbb{R}, + \rangle$, you can take *one half* of every number, i.e., if $a \in \mathbb{R}$, then there is a number $x \in \mathbb{R}$ such that $x + x = a$. When $+$ is translated to $\cdot$, how does "one half" translate? For $b \in \mathbb{R}^*$, it translates to a number $y \in \mathbb{R}^*$ such that $y \cdot y = b$, i.e., to a *square root* of b. But there are numbers in $\mathbb{R}^*$ (namely, all negative numbers) that do not have real square roots! We conclude that $\langle \mathbb{R}, + \rangle$ and $\langle \mathbb{R}^*, \cdot \rangle$, are *not* isomorphic.

A formal proof might look something like this: Suppose $\phi : \mathbb{R} \to \mathbb{R}^*$ is an isomorphism of $\langle \mathbb{R}, + \rangle$ and $\langle \mathbb{R}^*, \cdot \rangle$. Then there is some $a \in \mathbb{R}$ such that $\phi(a) = -1$. Let $x = a/2$. Then $\phi(a) = \phi(x + x) = \phi(x) \cdot \phi(x) = -1$. But this is impossible since -1 has no square root in $\mathbb{R}$. Contradiction! It follows that there exists no isomorphism of $\langle \mathbb{R}, + \rangle$ and $\langle \mathbb{R}^*, \cdot \rangle$.

Example 5:

For each of the groups $\langle \overline{\mathbb{Z}}_n, \oplus_n \rangle$ $(n \in \mathbb{Z}^+)$, whose elements are the *residue classes* modulo n, there is a corresponding group $\langle \mathbb{Z}_n, +_n \rangle$ whose elements are the integers $0, 1, 2, \ldots, n-1$, i.e., the *residues* (meaning: remainders) modulo n. If we define $+_n$ on $\mathbb{Z}_n = \{0, 1, \ldots, n-1\}$ by

$$h +_n k = \mathrm{r}$$

where r is the remainder obtained when $h + k$ is divided by n, then $+_n$ is a binary operation on $\mathbb{Z}_n$, and $\langle \mathbb{Z}_n, +_n \rangle$ is a group isomorphic to the residue class group $\langle \overline{\mathbb{Z}}_n, \oplus_n \rangle$. We leave the verification of the details as an exercise (see Exercise 2.4.12). A look at the Cayley tables for the two groups in the case of $n = 4$ should be convincing.

$\oplus_4$	$\bar{0}$	$\bar{1}$	$\bar{2}$	$\bar{3}$
$\bar{0}$	$\bar{0}$	$\bar{1}$	$\bar{2}$	$\bar{3}$
$\bar{1}$	$\bar{1}$	$\bar{2}$	$\bar{3}$	$\bar{0}$
$\bar{2}$	$\bar{2}$	$\bar{3}$	$\bar{0}$	$\bar{1}$
$\bar{3}$	$\bar{3}$	$\bar{0}$	$\bar{1}$	$\bar{2}$

$\overline{\mathbb{Z}}_4$

$+_4$	0	1	2	3
0	0	1	2	3
1	1	2	3	0
2	2	3	0	1
3	3	0	1	2

$\mathbb{Z}_4$

When two groups are isomorphic, we shall say that they represent *the same abstract group*. This immediately raises some interesting questions, such as:

(1) How many abstract groups are there of any given finite order?

(2) How many abstract abelian groups are there of any given finite order?

The answer to the first question is known only for very special orders. In Section 2.5, we provide a simple answer for the case of prime orders. The answer to the second question is known and appears in Section 2.13.

For the time being, we content ourselves with finding all abstract *cyclic* groups by showing that every cyclic group is isomorphic either to $\langle \mathbb{Z}, + \rangle$ or to one of the groups $\langle \bar{\mathbb{Z}}_n, \oplus_n \rangle$.

Theorem 2.4.1:

(1) Every infinite cyclic group is isomorphic to $\langle \mathbb{Z}, + \rangle$.

(2) Every cyclic group of finite order n is isomorphic to the group $\langle \bar{\mathbb{Z}}_n, \oplus_n \rangle$ (hence to $\langle \mathbb{Z}_n, +_n \rangle$).

Proof:

(1) Let$\langle G, \cdot \rangle$ be an infinite cyclic group. Then $G = [a] = \{a^k | k \in \mathbb{Z}\}$, for some $a \in G$. Define $\phi : \mathbb{Z} \to G$ by: $\phi(k) = a^k$, for each $k \in \mathbb{Z}$. Then ϕ is a well-defined map, clearly onto, and ϕ is 1-1 by Theorem 2.3.6. For $h, k \in \mathbb{Z}$,

$$\phi(h + k) = a^{h+k} = a^h \cdot a^k = \phi(h) \cdot \phi(k).$$

Hence ϕ is an isomorphism.

(2) Let$\langle G, \cdot \rangle$ be a finite cyclic group of order n. Then $G = \{a^0, a^1, \dots, a^{n-1}\}$, by Theorem 2.3.6. "Define" $\phi : \bar{\mathbb{Z}}_n \to G$ by: $\phi(\bar{k}) = a^k$, for each $k \in \mathbb{Z}$. We put quotes around the word *define* since it is not immediately clear that our prescription for ϕ does indeed define a function. Clearly, for each $k \in \mathbb{Z}$, we have $\bar{k} \in \bar{\mathbb{Z}}_n$ and $a^k \in G$. To show that ϕ is well-defined, we must prove: if $\bar{h} = \bar{k}$ in $\bar{\mathbb{Z}}_n$, then $a^h = a^k$. As a bonus, we will, at the same time, obtain that ϕ is 1-1. For, $\bar{h} = \bar{k} \Leftrightarrow h \equiv k \mod n \Leftrightarrow n | (k - h) \Leftrightarrow a^{k-h} = e$ (by Theorem 2.3.7) $\Leftrightarrow a^h = a^k$. We conclude that ϕ is a (well-defined) 1-1 map, clearly onto since every element of G is $a^k = \phi(\bar{k})$ for some $k \in \mathbb{Z}$.

For $h, k \in \mathbb{Z}$, we have

$$\phi(\bar{h} \oplus_n \bar{k}) = \phi(\overline{h + k}) = a^{h+k} = a^h \cdot a^k = \phi(\bar{h}) \cdot \phi(\bar{k}).$$

It follows that ϕ is an isomorphism of $\bar{\mathbb{Z}}_n$ onto G. ■

(We shall give a briefer and more elegant proof of this theorem in Section 2.9.)

Exercises 2.4

True or False

1. The additive group $\langle \mathbb{Q}, + \rangle$ is isomorphic to the multiplicative group $\langle \mathbb{Q}, \cdot \rangle$.

2. If G is a group, then some element, a, of G may occur more than once in one of the rows of the Cayley table of G.
3. The additive group of all integers is isomorphic to the additive group of all rational numbers.
4. If two groups are isomorphic, then they have the same order.
5. If two cyclic groups have the same order, then they are isomorphic.

2.4.1. Which of the following functions $f : \mathbb{R} \to \mathbb{R}$ are isomorphisms of the additive group of $\mathbb{R}$ onto itself?

(a) $f(x) = (e^x - e^{-x})/2$,
(b) $f(x) = 3x - 3$,
(c) $f(x) = x^3$.

2.4.2 Prove that if $\psi(x) = \ln x$ for each $x > 0$ in $\mathbb{R}$, then ψ is an isomorphism of the multiplicative group $\mathbb{R}^+$ of all positive real numbers onto the additive group $\mathbb{R}$ of all real numbers. (ψ is the inverse of the isomorphism ϕ in Example 3, p. 56.)

2.4.3. Let $\phi : G \to G'$ be an isomorphism. Prove that $\phi(e) = e'$ and $\phi(a^{-1}) = (\phi(a))^{-1}$ for each $a \in G$. (Do you need to use the hypothesis that ϕ is 1-1 or onto?)

2.4.4. Let $\phi : G \to G'$ be an isomorphism. Prove that, for each $a \in G$, $o(a) = o(\phi(a))$.

2.4.5. Prove that $\langle \mathbb{Z}, + \rangle$ is isomorphic to each of its non-trivial subgroups.

2.4.6. Prove that the additive group of all rational numbers is *not* isomorphic to $\langle \mathbb{Z}, + \rangle$.

2.4.7. If $H < G$ and $x \in G$, prove that the subgroup $xHx^{-1} = \{xhx^{-1} | h \in H\}$ is isomorphic to H.

2.4.8. Prove that the relation "is isomorphic to" is an equivalence relation on any non-empty set of groups.

(The equivalence classes are *isomorphism classes* of groups. We say that groups belonging to the same isomorphism class represent the same "abstract group.")

2.4.9. Prove that, given any finite set G, there is a binary operation $\cdot$ on G such that $\langle G, \cdot \rangle$ is a group.

(In Exercise 2.12.18, you are asked to extend this result to an arbitrary set.)

Note: In Exercise 1.2.14, we have indicated why there is no "set of all sets." From this exercise and Exercise 2.12.18, it follows that there is no "set of all groups"—hence, the cautious wording in Exercise 2.4.8.

2.4.10. Find two non-isomorphic groups of order 6. (*Prove* that they are not isomorphic.)

2.4.11. Prove that the additive group $\langle \mathbb{C}, + \rangle$ of all complex numbers is isomorphic to the additive group of the vector space $\mathbb{R}^2$.

2.4.12. Let n be a positive integer and let $\mathbb{Z}_n = \{0, 1, \ldots, n-1\}$. For $h, k \in \mathbb{Z}_n$, let $h +_n k = r$, where r is the remainder obtained when $h + k$ is divided by n. Prove that $+_n$ is a binary operation on $\mathbb{Z}_n$ such that $\langle \mathbb{Z}_n, +_n \rangle$ is a group. (Is $\langle \mathbb{Z}_n, +_n \rangle$ a subgroup of $\langle \mathbb{Z}, + \rangle$?)

2.4.13. Find all possible Cayley tables for a group of order 1, 2, 3, or 4. How many abstract groups of each of these orders are there?

2.5

Cosets and Lagrange's Theorem

We begin this section by generalizing the partitioning of $\mathbb{Z}$ into residue classes modulo a positive integer n to obtain partitionings of an arbitrary group G into equivalence classes determined by a subgroup H. This will lead immediately to a very simple, but very fundamental, theorem regarding subgroups of finite groups: the order of any subgroup of a finite group divides the order of the group. We end the section with a number of interesting consequences of this result, some of them well-known classical theorems in elementary number theory.

Recall the definition of the set $\mathbb{Z}_n$, for $n \in \mathbb{Z}^+$. We started with the equivalence relation $\equiv \bmod n$ defined by $a \equiv b \bmod n$ if $n|a - b$, and we let $\mathbb{Z}_n$ be the set of all equivalence classes with respect to $\equiv \bmod n$. Since $n|a - b$ if and only if $a - b \in n\mathbb{Z}$, we can reformulate the definition of $\equiv \bmod n$ in group-theoretic terms as follows:

$$a \equiv b \bmod n \quad \text{if and only if} \quad a + (-b) \in H,$$

where $H = n\mathbb{Z}$, the cyclic subgroup of $\langle \mathbb{Z}, + \rangle$ generated by n. The elements of $\mathbb{Z}_n$ are the residue classes $\bar{k} = \{nq + k | q \in \mathbb{Z}\}$. (By Theorem 1.3.9, for each $k \in \mathbb{Z}$, $\bar{k} = \bar{r}$ for some r, $0 \leq r \leq n - 1$; hence there are exactly n distinct residue classes for any given positive integer n.) With $H = n\mathbb{Z}$, it is natural to write $\bar{k} = H + k$, for each $k \in \mathbb{Z}$. The sets $H + k$ ($k \in \mathbb{Z}$) are the *cosets* of H in $\mathbb{Z}$. Since they are the equivalence classes with respect to $\equiv \bmod n$, the cosets of H in $\mathbb{Z}$ partition $\mathbb{Z}$.

We use this *coset decomposition* of $\mathbb{Z}$ as a model for defining cosets for arbitrary subgroups of arbitrary groups. In the general case, of course, we have to provide for non-commutativity of the group operation.

Definition 2.5.1: Let $\langle G, \cdot \rangle$ be a group, and let $H < G$. For each $a \in G$,

$$Ha = \{ha | h \in H\}$$

is the *right coset of H determined by a*;

$$aH = \{ah | h \in H\}$$

is the *left coset of H determined by a.*

(If $\cdot$ is commutative, then, for any $H < G$, $aH = Ha$ for all $a \in G$. We shall see that certain distinguished subgroups of non-commutative groups also have this property.)

Translated to additive notation, the sets in Definition 2.5.1 would read:

$$H + a = \{h + a | h \in H\}$$

$$a + H = \{a + h | h \in H\}$$

But recall that additive notation is generally used only for commutative operations; hence, "$H + a$" and "$a + H$" would generally denote the same set.

Example:

As we have already seen, for n a positive integer, the cosets of $H = n\mathbb{Z}$ are the residue classes $\bar{k} = H + k$, $k \in \mathbb{Z}$. The n distinct residue classes $H + 0$, $H + 1, \ldots, H + (n - 1)$ partition $\mathbb{Z}$.

Generalizing the equivalence relation which gave rise to the residue classes, we get, in general, two distinct relations.

Definition 2.5.2: Let $\langle G, \cdot \rangle$ be a group, and let $H < G$. For $a, b \in G$, define

$$a \equiv_r b \bmod H \quad \text{if} \quad ab^{-1} \in H$$

$$a \equiv_l b \bmod H \quad \text{if} \quad b^{-1}a \in H.$$

(Observe how this definition generalizes the relation $\equiv \bmod n$ in $\mathbb{Z}$. For $a, b \in \mathbb{Z}$, we have $a \equiv_r b \bmod n\mathbb{Z}$ if $a + (-b) \in n\mathbb{Z}$. This is the case precisely when $a \equiv b \bmod n$. Of course, "$a \equiv_l b \bmod n\mathbb{Z}$ if $(-b) + a \in n\mathbb{Z}$" says exactly the same thing, since $+$ is commutative.)

Theorem 2.5.1: Let $\langle G, \cdot \rangle$ be a group, and let $H < G$.

(1) The binary relations $\equiv_r \bmod H$ and $\equiv_l \bmod H$ are equivalence relations on G.

(2) The equivalence classes with respect to $\equiv_r \bmod H$ are the right cosets of H in G. The equivalence classes with respect to $\equiv_l \bmod H$ are the left cosets of H in G.

Proof: *(1)* $\equiv_r \bmod H$ is reflexive: for, if $a \in G$, then $aa^{-1} = e \in H$, hence $a \equiv_r a \bmod H$.

$\equiv_r \bmod H$ is symmetric: if $a \equiv_r b \bmod H$, then $ab^{-1} \in H$; hence $ba^{-1} = (ab^{-1})^{-1} \in H$, and so $b \equiv_r a \bmod H$.

$\equiv_r \bmod H$ is transitive: if $a \equiv_r b \bmod H$ and $b \equiv_r c \bmod H$, then $ab^{-1} \in H$ and $bc^{-1} \in H$; hence $(ab^{-1})(bc^{-1}) = ac^{-1} \in H$, and so $a \equiv_r c \bmod H$.

Thus, $\equiv_r \bmod H$ is an equivalence relation on G.

We leave the proof for $\equiv_l \bmod H$ as an exercise (see Exercise 2.5.4).

(2) For each $a \in G$, since $x \equiv_r a \bmod H \Leftrightarrow xa^{-1} \in H \Leftrightarrow xa^{-1} = h$ for some $h \in H \Leftrightarrow x = ha$ for some $h \in H$, we conclude that $\bar{a} = Ha$.

Again, we leave the left-hand version as an exercise (see Exercise 2.5.4). ■

Corollary 1: If $H < G$, then

(1) $Ha = Hb \Leftrightarrow ab^{-1} \in H$;

(2) $aH = bH \Leftrightarrow b^{-1}a \in H$

Proof: This follows immediately from Theorem 2.5.1 and the Lemma preceding the Fundamental Partition Theorem (Theorem 1.2.1) ■

IMPORTANT: Corollary 1 should be an often-used tool in dealing with coset decompositions!

From Theorem 1.2.1, we immediately obtain the following corollary.

Corollary 2: If $H < G$, then

(1) the right cosets of H in G partition G;

(2) the left cosets of H in G partition G.

Further Examples

Example 1:

In $\langle \mathbb{R}^*, \cdot \rangle$, let $H = \{-1, 1\} = \{x \in \mathbb{R}^* \mid |x| = 1\}$. The cosets of H in G are the sets $Ha = \{-a, a\}$, $a \in \mathbb{R}^*$.

Example 2:

In $\langle \mathbb{C}^*, \cdot \rangle$, let $H = \{z \in \mathbb{C} \mid |z| = 1\}$. The cosets of H in G are the sets $H\rho$, $\rho \in \mathbb{R}^+$. For: if $z \in \mathbb{C}^*$, then $z = \rho(\cos\theta + i\sin\theta)$, $\rho \in \mathbb{R}^+, \theta \in \mathbb{R}$. Since $|\cos\theta + i\sin\theta| = \sqrt{\cos^2\theta + \sin^2\theta} = 1$, $\cos\theta + i\sin\theta \in H$. But then $Hz = H\rho$, by Corollary 1 above, since $z\rho^{-1} = \cos\theta + i\sin\theta \in H$. $H\rho$ is the set of all $z \in \mathbb{C}$ such that $|z| = \rho$.

Example 3:

Let $G = GL_2(\mathbb{R})$, the group of all 2×2 non-singular matrices over $\mathbb{R}$, and let $H = SL_2(\mathbb{R})$, the subgroup of G consisting of all 2×2 matrices with determinant 1. Then the right cosets of H in G are the sets $H\begin{pmatrix} a & 0 \\ 0 & 1 \end{pmatrix}$ for $a \in \mathbb{R}^*$.

For, if $A \in G$ and $\det A = a \in \mathbb{R}$, then $a \neq 0$ and $HA = H\begin{pmatrix} a & 0 \\ 0 & 1 \end{pmatrix}$ since

$$\det\left[A\begin{pmatrix} a & 0 \\ 0 & 1 \end{pmatrix}^{-1}\right] = \det A \det\begin{pmatrix} a & 0 \\ 0 & 1 \end{pmatrix}^{-1} = a \cdot \frac{1}{a} = 1, \text{ whence } A\begin{pmatrix} a & 0 \\ 0 & 1 \end{pmatrix}^{-1} \in H,$$

and so $HA = H\begin{pmatrix} a & 0 \\ 0 & 1 \end{pmatrix}$. Similarly, the left cosets of H in G are the sets $\begin{pmatrix} a & 0 \\ 0 & 1 \end{pmatrix}H$, $a \in \mathbb{R}^*$.

Thus, as was the case in $\mathbb{Z}_n$, cosets can often be conveniently represented by *well-chosen* elements: for example, in $\mathbb{Z}_6$, we may prefer to write $\bar{2}$ in place of its equivalent $\overline{31286}$. In $\langle \mathbb{C}^*, \cdot \rangle$, with H as in Example 1, we may prefer $H5$ to its equivalent $H(\sqrt{22} + i\sqrt{3})$; and in Example 3, we may prefer $H\begin{pmatrix} 2 & 0 \\ 0 & 1 \end{pmatrix}$ to its equivalent $H\begin{pmatrix} 238 & 1859.25 \\ 16 & 125 \end{pmatrix}$.

We are now almost ready to make use of coset decompositions to obtain a very basic result on the structure of finite groups. Two simple lemmas will be helpful.

Lemma 1: Let G be a group and let $H < G$. Then, for each $a \in G$,

(1) $\phi_r : H \to Ha$ defined by $\phi_r(h) = ha$ for each $h \in H$ is a bijection;
and
(2) $\phi_l : H \to aH$ defined by $\phi_l(h) = ah$ for each $h \in H$ is a bijection.

Proof: (1) ϕ_r is a well-defined map, clearly onto; $\phi_r(h) = \phi_r(k) \Rightarrow ha = ka \Rightarrow h = k$. The proof of (2) is analogous. ■

Thus, the number of elements in any right (or left) coset of H in G is equal to the order of H.

Lemma 2: Let $H < G$, and let

$$\mathscr{C}_r = \{Ha \mid a \in G\}; \quad \mathscr{C}_l = \{aH \mid a \in G\}.$$

"Define" $\gamma : \mathscr{C}_r \to \mathscr{C}_l$ by: $\gamma(Ha) = a^{-1}H \quad (a \in G)$. Then γ is a bijection.

Proof: We need to prove that γ is well-defined and 1-1. (It is clearly onto.) Let $a, b \in G$. Then $Ha = Hb \Leftrightarrow ab^{-1} \in H \Leftrightarrow (ab^{-1})^{-1} = ba^{-1} \in H \Leftrightarrow (b^{-1})^{-1}\, a^{-1} \in H \Leftrightarrow a^{-1}H = b^{-1}H$. ■

Thus, a given subgroup has as many right cosets as it has left cosets.
Lemma 2 is not really needed to prove the theorem that follows, but it makes the definition of *index* more pleasant to state.

Definition 2.5.3: If $H < G$, then the number of right (or left) cosets of H in G is called the *index* of H in G. We write $[G : H]$ to denote the index of H in G.

Theorem 2.5.2 (Lagrange): (1) If $H < G$, then $|H| \cdot [G : H] = |G|$.
(2) If G is finite and $H < G$, then $|H|\,|\,|G|$ and $[G : H]\,|\,|G|$.

Proof: (1) Since the (say) right cosets of H in G partition G and are all the same size, we have $|G| = |H| \cdot [G : H]$. This is obvious in case G is finite, and true also when G is infinite if multiplication of transfinite cardinal numbers is defined in the customary way (cf. [43] or [44].)
(2) For G finite, (2) follows immediately from (1). ■

Usually, the term *Lagrange's Theorem* is applied specifically to the result that *the order of any subgroup of a finite group divides the order of the group.*

Theorem 2.5.2 yields a rich harvest of corollaries, including some famous classical theorems in number theory.

Corollary 1: The order of any element of a finite group divides the order of the group.

Proof: Let G be a finite group, $a \in G$. Then $o(a) = |[a]|$; hence it divides $|G|$. ■

Corollary 2: If G is a finite group of order n, then, for each $a \in G$, $a^n = e$.

Proof: By Corollary 1, for $a \in G$, $n = o(a) \cdot s$ for some $s \in \mathbb{Z}$. Hence

$$a^n = a^{o(a)\cdot s} = (a^{o(a)})^s = e^s = e. \qquad \blacksquare$$

Corollary 3: Every group of prime order is cyclic. (Hence, for each prime p, there is exactly one abstract group of order p.)

Proof: If G is a group of prime order p, then the only possible orders of subgroups of G are 1 and p. Hence the only possible subgroups of G are $\{e\}$ and G. But then, for each $a \neq e$, $a \in G$, we have $[a] = G$, and so G is cyclic. (By Theorem 2.4.1, G is isomorphic to $\mathbb{Z}_p$.) ■

Before proceeding to the number-theoretic corollaries, we need to introduce a second binary operation, $\circ_n$, on the set $\mathbb{Z}_n$ of all residue classes modulo a positive integer n.

For $h, k \in \mathbb{Z}$, "define" $\bar{h} \circ_n \bar{k} = \overline{hk}$. We leave it as an exercise to prove that $\circ_n$ is a (well-defined) binary operation on $\mathbb{Z}_n$. Clearly, $\langle \mathbb{Z}_n, \circ_n \rangle$ is not a group, for: $\bar{1}$ serves as identity for $\circ_n$, and there is *no* integer k such that $\bar{0} \circ_n \bar{k} = \bar{1}$. However, $\circ_n$ induces a binary operation, also to be denoted by $\circ_n$, on the subset of $\mathbb{Z}_n$ that consists of those residue classes $\bar{k}$ where k and n are relatively prime.

Theorem 2.5.3: For n a positive integer, let

$$\mathbb{Z}_n^* = \{\bar{k} \in \mathbb{Z}_n \mid \gcd(k, n) = 1\},$$

and let $\circ_n$ be defined by $\bar{k} \circ_n \bar{h} = \overline{kh}$, where $h, k \in \mathbb{Z}$, with $\gcd(k, n) = \gcd(h, n) = 1$. Then $\circ_n$ is a binary operation on $\mathbb{Z}_n^*$ such that $\langle \mathbb{Z}_n^*, \circ_n \rangle$ is a group.

We leave the proof as a "programmed" exercise (see Exercise 2.5.14).

The order of the group $\mathbb{Z}_n^*$ is equal to $\varphi(n)$, the number of elements in the set $\{k \in \mathbb{Z} \mid 0 < k \leq n, \gcd(k, n) = 1\}$. ($\varphi$ is the *Euler Phi-Function.*)

We continue with our list of corollaries of Lagrange's Theorem.

Corollary 4 (Euler's Theorem): Let n be a positive integer. If k is an integer such that $\gcd(k, n) = 1$, then $k^{\varphi(n)} \equiv 1 \bmod n$.

Proof: The group $\langle \mathbb{Z}_n^*, \circ_n \rangle$ has order $\varphi(n)$. If $k \in \mathbb{Z}$, with $\gcd(k, n) = 1$, then $\bar{k} \in \mathbb{Z}_n^*$. Hence $\bar{k}^{\varphi(n)} = \bar{1}$ in $\mathbb{Z}_n^*$, by Corollary 2. But then (by the definition of $\circ_n$), $\overline{k^{\varphi(n)}} = \bar{1}$, and so $k^{\varphi(n)} \equiv 1 \bmod n$. ■

Corollary 5 (Fermat's Little Theorem): If p is a prime, and k is an integer such that $p \nmid k$, then $k^{p-1} \equiv 1 \bmod p$.

Proof: This follows immediately from Euler's Theorem since, for p prime, $\varphi(p) = p - 1$, and the condition $p \nmid k$ is equivalent to $\gcd(p, k) = 1$. ■

This theorem, named for its discoverer, Pierre Auguste de Fermat (1601-1665), has, in recent years, become an important tool in the construction of public key codes (cf. [32]).

Exercises 2.5

True or False

1. If $H < G$ and $a \in G$, then $Ha < G$.
2. A finite subgroup of a group G must have finite index in G.
3. A group of order 212 may have a subgroup of order 7.
4. For every positive integer n, the group $\langle \mathbb{Z}, + \rangle$ has a subgroup of index n.
5. For every positive integer n, the group $\langle \mathbb{Z}, + \rangle$ has a subgroup of order n.

2.5.1. For each of the following groups G and subgroups H, list the elements of each right coset, and of each left coset, of H in G, and find the index $[G : H]$.
(a) $G = \bar{\mathbb{Z}}_{10}$, $H = [\bar{8}]$.
(b) $G = \bar{\mathbb{Z}}_{10}$, $H = [\bar{9}]$.
(c) $G = S_3$, $H = [J]$.
(d) $G = S_3$, $H = [S]$.
The notation in (c), (d) is as in the table on p. 36.

2.5.2. Let G be a group and let H be a subgroup of G. Prove that the only right or left coset of H that is a subgroup of G is H itself.

2.5.3. Let G be a group, H a subgroup of G. Prove: if $a \in G$, then $Ha = H$ if and only if $a \in H$; $aH = H$ if and only if $a \in H$.

2.5.4. Prove that, for $H < G$, the relation $\equiv_l \bmod H$ is an equivalence relation on G. Prove that the equivalence classes with respect to $\equiv_l \bmod H$ are the left cosets of H in G (see Theorem 2.5.1).

2.5.5. Prove that every non-trivial subgroup of $\langle \mathbb{Z}, + \rangle$ has finite index.

2.5.6. Let n be a positive integer and let m be a positive divisor of n. Prove that a cyclic group of order n has exactly one subgroup of index m (recall Exercise 2.3.10).

2.5.7. *(1)* Prove: if H is a finite subgroup of an infinite group G, then H has infinite index in G.
(2) Prove that $\langle \mathbb{Q}^*, \cdot \rangle$ has exactly one finite non-trivial subgroup (of infinite index, by (1)).

2.5.8. Prove that the additive group $\mathbb{Z}$ of all integers has infinite index in the additive group $\mathbb{Q}$ of all rational numbers.

2.5.9. Prove that the additive group $\mathbb{R}$ of all real numbers has infinite index in the additive group $\mathbb{C}$ of all complex numbers.

2.5.10. Prove that the multiplicative group $\langle \mathbb{R}^*, \cdot \rangle$ has infinite index in the multiplicative group $\langle \mathbb{C}^*, \cdot \rangle$

2.5.11. Let $G = GL_2(\mathbb{R})$, and let $H = \{cI_2 | c \in \mathbb{R}\}$. Note that $H < G$. Prove that $HA = AH$ for each $A \in G$.

2.5.12. Let $G = GL_2(\mathbb{R})$ and let $H = SL_2 \mathbb{R})$ (the subgroup of G consisting of all 2×2 matrices in G with determinant 1). Let $A = \begin{pmatrix} 1 & 2 \\ 3 & 4 \end{pmatrix}$.
(1) Prove that $HA = \{X \in G | \det X = -2\}$.
(2) Prove that $AH = \{X \in G | \det X = -2\}$.
(3) Conclude that $AH = HA$.
(4) Generalize this result to an arbitrary matrix $A \in G$.

2.5.13. In $GL_2(\mathbb{R})$, let $H = [M]$, where $M = \begin{pmatrix} 1 & 1 \\ 0 & 1 \end{pmatrix}$.

(1) Prove that $H = \left\{ \begin{pmatrix} 1 & k \\ 0 & 1 \end{pmatrix} \middle| k \in \mathbb{Z} \right\}$.

(2) For $A = \begin{pmatrix} 2 & 2 \\ 0 & 1 \end{pmatrix}$, what are the elements of the cosets AH and HA? Is $AH = HA$?

(3) Answer the questions in (2) with A replaced by $B = \begin{pmatrix} 1 & 0 \\ 1 & 1 \end{pmatrix}$.

2.5.14. Let n be a positive integer.

(1) On $\mathbb{Z}_n$, define $\bar{h} \circ_n \bar{k} = \overline{hk}$ $(h, k \in \mathbb{Z})$. Prove that $\circ_n$ is a binary operation on $\mathbb{Z}_n$. Prove that $\circ_n$ is associative and commutative, and that $\bar{1}$ serves as identity for $\circ_n$.

(2) Let $\mathbb{Z}_n^* = \{\bar{k} \in \mathbb{Z}_n \mid \gcd(k, n) = 1\}$. Prove that this definition makes sense, i.e., prove: if $\bar{h} = \bar{k}$ and $\gcd(k, n) = 1$, then $\gcd(h, n) = 1$.

(3) Prove that, for each $\bar{h}, \bar{k} \in \mathbb{Z}_n^*$, $\bar{h} \circ_n \bar{k} \in \mathbb{Z}_n^*$.

(4) Prove that, for each $\bar{h} \in \mathbb{Z}_n^*$, there is some $\bar{k} \in \mathbb{Z}_n^*$ such that $\bar{h} \circ_n \bar{k} = \bar{1}$.

(5) Conclude that $\langle \mathbb{Z}_n^*, \circ_n \rangle$ is a group.

(6) What is the order of the group $\mathbb{Z}_n^*$?

2.5.15. Let G be a finite abelian group, consisting of elements $a_1, \ldots, a_n$.

(1) Prove that $\left(\prod_{i=1}^{n} a_i \right)^2 = \prod_{i=1}^{n} a_i^2 = e$.

(2) If n is odd, prove that $\prod_{i=1}^{n} a_i = e$.

2.6

Permutation Groups

Closer acquaintance with the structure of permutation groups will provide us with a handy source of examples to illustrate the theory.

Recall that, if A is any set, then the set S_A of all bijections $\sigma: A \to A$ forms a group under composition (Theorem 2.2.1). If A is a finite set of n elements, then the bijections on A are called permutations of A, and the permutation group S_A is isomorphic (see Exercise 2.6.7) to S_n, the symmetric group of degree n, consisting of all permutations of the set $X_n = \{1, 2, \ldots, n\}$ (see Definition 2.2.2). It therefore suffices to study the symmetric groups S_n.

Example:

In S_8, let σ be the permutation on $X_8 = \{1, 2, 3, 4, 5, 6, 7, 8\}$ given by $\sigma = \begin{pmatrix} 1 & 2 & 3 & 4 & 5 & 6 & 7 & 8 \\ 3 & 1 & 2 & 4 & 7 & 8 & 6 & 5 \end{pmatrix}$ where each number in the first row is mapped under σ to the number underneath it in the second row. Closer inspection reveals that the effect of σ on X_8 can be described by means of cycles (represented by the circles shown in Figure 3, where each element is mapped under σ to the next one on its circle, proceeding clockwise).

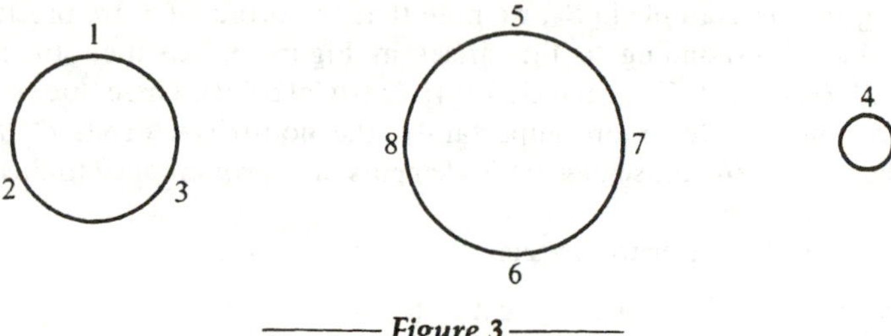

Figure 3

It is evident that the sets $\{1, 2, 3\}$, $\{5, 7, 6, 8\}$ and $\{4\}$ form a partition of the set X_8—a partition induced by the given permutation σ. If we define permutations σ_1, σ_2, and σ_3 (corresponding, respectively, to the circles in Figure 3) by:

$$\sigma_1 = \begin{pmatrix} 1 & 2 & 3 & 4 & 5 & 6 & 7 & 8 \\ 3 & 1 & 2 & 4 & 5 & 6 & 7 & 8 \end{pmatrix}$$

$$\sigma_2 = \begin{pmatrix} 1 & 2 & 3 & 4 & 5 & 6 & 7 & 8 \\ 1 & 2 & 3 & 4 & 7 & 8 & 6 & 5 \end{pmatrix}$$

$$\sigma_3 = \begin{pmatrix} 1 & 2 & 3 & 4 & 5 & 6 & 7 & 8 \\ 1 & 2 & 3 & 4 & 5 & 6 & 7 & 8 \end{pmatrix} = \iota_{X_8},$$

then $\sigma = \sigma_1 \sigma_2 \sigma_3 = \sigma_1 \sigma_2$ since σ_3 is just the identity permutation. Thus, we have expressed σ as a product of permutations, each of which permutes a subset of X_8 cyclically.

To generalize this example, since a certain partition of X_8 was the key to success, we look for a suitable equivalence relation.

Definition 2.6.1: For $n \geq 1$, let $\sigma \in S_n$. On X_n, define a binary relation $\sim_\sigma$ by $a \sim_\sigma b$ if $b = \sigma^k a$ for some $k \in \mathbb{Z}$.

Example:

In the preceding example in S_8, we have $1 \sim_\sigma 2$, $1 \sim_\sigma 3$, $2 \sim_\sigma 3$; $5 \sim_\sigma 7$, $5 \sim_\sigma 8$, etc. In short, elements of X_8 which are on the same circle bear the relation $\sim_\sigma$ to each other.

Theorem 2.6.1: For $n \geq 1$, if $\sigma \in S_n$, then $\sim_\sigma$ (as defined in Definition 2.6.1) is an equivalence relation on the set X_n.

Proof: The relation $\sim_\sigma$ is reflexive since, for each $k \in X_n$, $\sigma^0(a) = a$. The relation $\sim_\sigma$ is symmetric, for: if $a \sim_\sigma b$ $(a, b \in X_n)$, then $b = \sigma^k a$ for some $k \in \mathbb{Z}$; hence $a = (\sigma^k)^{-1} b = \sigma^{-k} b$, and so $b \sim_\sigma a$. The relation $\sim_\sigma$ is transitive, for: if $a \sim_\sigma b$ and $b \sim_\sigma c$ $(a, b, c \in X_n)$, then $b = \sigma^h a$ and $c = \sigma^k b$ $(h, k \in \mathbb{Z})$, and so $c = \sigma^k \sigma^h a = \sigma^{k+h} a$. But then $a \sim_\sigma c$. ■

Definition 2.6.2: For $n \geq 1$, let $\sigma \in S_n$. Then the equivalence classes with respect to $\sim_\sigma$ are the *orbits of* σ. An orbit consisting of just one element of X_n is called a *trivial* orbit; an orbit consisting of more than one element of X_n is called a *non-trivial* orbit.

Returning to our example in S_8, we note that the orbits of σ are precisely the subsets of X_8 corresponding to the circles in Figure 3, i.e., they are the sets $\mathcal{O}_1 = \{1, 2, 3\}$, $\mathcal{O}_2 = \{5, 6, 7, 8\}$, and $\mathcal{O}_3 = \{4\}$, (a trivial orbit). These three orbits, of course, partition X_8. But, more importantly, the non-trivial orbits $\mathcal{O}_1$ and $\mathcal{O}_2$ partition the set $X_8 \backslash \{4\}$ consisting of the elements of X_8 *not* mapped to themselves under σ.

For $\sigma \in S_n$, we shall refer to the sets

$$F_\sigma = \{a \in X_n | \sigma(a) = a\}$$

and

$$\bar{F}_\sigma = \{a \in X_n | \sigma(a) \neq a\}$$

as the *fixed set* and the *changed set* of σ. Clearly F_σ and $\bar{F}_\sigma$ partition X_n, and the non-trivial orbits of σ, in turn, partition $\bar{F}_\sigma$, the changed set of σ.

Examining the two non-trivial orbits in our example, we observe that

$$\mathcal{O}_1 = \{1, \sigma(1), \sigma^2(1)\} = \{2, \sigma(2), \sigma^2(2)\} = \{3, \sigma(3), \sigma^2(3)\}$$

and

$$\begin{aligned} \mathcal{O}_2 &= \{5, \sigma(5), \sigma^2(5), \sigma^3(5)\} \\ &= \{6, \sigma(6), \sigma^2(6), \sigma^3(6)\} \\ &= \{7, \sigma(7), \sigma^2(7), \sigma^3(7)\} \\ &= \{8, \sigma(8), \sigma^2(8), \sigma^3(8)\}. \end{aligned}$$

Lemma: For $n \geq 1$, let $\sigma \in S_n$, and let $\mathcal{O}$ be one of the non-trivial orbits of σ. If $\mathcal{O}$ consists of k elements, then, for each $a \in \mathcal{O}$,

$$\mathcal{O} = \{a, \sigma(a), \ldots, \sigma^{k-1}(a)\}.$$

Proof: Since $\mathcal{O}$ is finite, the elements of $\sigma^h a$ ($h \in \mathbb{Z}$) cannot all be distinct. Hence there are integers $i < j$ such that $\sigma^i a = \sigma^j a$, and $\sigma^{j-i} a = a$. It follows (how?) that there is a positive integer k_0 such that $\sigma^{k_0} a = a$ and $k_0 \leq m$ for all positive integers m such that $\sigma^m a = a$. As in the proof of Theorem 2.3.6, we find that $a, \sigma a, \ldots, \sigma^{k_0 - 1} a$ are distinct elements of $\mathcal{O}$. Again as in the proof of Theorem 2.3.6, using the Division Algorithm (see Exercise 2.6.11), it is easy to show that for every integer s, $\sigma^s a = \sigma^r a$ for some $r \in \{0, 1, \ldots, k_0 - 1\}$. But then

$$\mathcal{O} = \{\sigma^0 a, \sigma^1 a, \ldots, \sigma^{k_0 - 1} a\}.$$

Since $\mathcal{O}$ has k elements, we conclude that $k_0 = k$, and that

$$\mathcal{O} = \{a, \sigma a, \ldots, \sigma^{k-1} a\},$$

as promised. ∎

Definition 2.6.3: For $n > 1$, a (*non-trivial*) *cycle* in S_n is a permutation $\sigma \in S_n$ such that σ has exactly one non-trivial orbit. The identity permutation $\iota \in S_n$ is called the *trivial cycle*. The number of elements in the non-trivial orbit of a cycle is called its *length*. A cycle of length k is called a *k-cycle*. Two cycles whose non-trivial orbits are disjoint are called *disjoint cycles*.

Notation: If $\sigma \in S_n$ is the cycle with non-trivial orbit $\mathcal{O} = \{a_1, a_2, \ldots, a_k\}$, where $\sigma a_i = a_{i+1}$ for $i = 1, \ldots, k-1$, and $\sigma a_k = a_1$, we shall write $\sigma = (a_1\ a_2 \ldots a_k)$ (*cycle notation*). The trivial cycle may be denoted by (1).

In our example from S_8, the permutation

$$\sigma_1 = \begin{pmatrix} 1 & 2 & 3 & 4 & 5 & 6 & 7 & 8 \\ 3 & 1 & 2 & 4 & 5 & 6 & 7 & 8 \end{pmatrix}$$

and

$$\sigma_2 = \begin{pmatrix} 1 & 2 & 3 & 4 & 5 & 6 & 7 & 8 \\ 1 & 2 & 3 & 4 & 7 & 8 & 6 & 5 \end{pmatrix}$$

are disjoint cycles of length 3 and 4, respectively. We have already observed that $\sigma = \sigma_1 \sigma_2$. In cycle notation: $\sigma_1 = (1\ 3\ 2)$, $\sigma_2 = (5\ 7\ 6\ 8)$, and $\sigma = (1\ 3\ 2)(5\ 7\ 6\ 8)$.

Remark: Disjoint cycles commute. In S_n, If σ_1 has non-trivial orbit $\mathcal{O}_1$, and σ_2 has non-trivial orbit $\mathcal{O}_2$, with $\mathcal{O}_2 \cap \mathcal{O}_2 = \varnothing$, then

$$\sigma_1\sigma_2(x) = \sigma_1 x = \sigma_2\sigma_1(x) \quad \text{for each} \quad x \in \mathcal{O}_1$$

$$\sigma_1\sigma_2(y) = \sigma_2 y = \sigma_2\sigma_1(y) \quad \text{for each} \quad y \in \mathcal{O}_2$$

and

$$\sigma_1\sigma_2(z) = z = \sigma_2\sigma_1(z) \quad \text{for each} \quad z \in X_n \backslash (\mathcal{O}_1 \cup \mathcal{O}_2).$$

But then $\sigma_1\sigma_2 = \sigma_2\sigma_1$.

Theorem 2.6.2: For $n > 1$, every permutation $\sigma \in S_n$, $\sigma \neq \iota$, is equal to a product of disjoint cycles. This representation is unique up to the order of the cycles.

Proof: Let $\sigma \in S_n$, $\sigma \neq \iota$, and let $\mathcal{O}_1, \ldots, \mathcal{O}_s$ be the non-trivial orbits of σ. Since $\sigma \neq \iota$, $s \geq 1$. By the Lemma, for each $i = 1, \ldots, s$,

$$\mathcal{O}_i = \{a_i, \sigma(a_i), \ldots, \sigma^{k_i - 1}(a_i)\}$$

where a_i is any element of $\mathcal{O}_i$, and k_i is the length of $\mathcal{O}_i$.

For each $i = 1, \ldots, s$, let σ_i be the cycle defined by $\sigma_i x = \sigma x$ for each $x \in \mathcal{O}_i$; $\sigma_i x = x$ for $x \notin \mathcal{O}_i$. Then, for each $i = 1, \ldots, s$, the non-trivial orbit of σ_i is $\mathcal{O}_i$, and the cycles σ_i are pairwise disjoint. For each $x \in F_\sigma$ (the fixed set of σ), we have

$$\sigma_1 \ldots \sigma_s x = x = \sigma x.$$

If $x \in \bar{F}_\sigma$ (the changed set of σ), then $x \in \mathcal{O}_i$ for exactly one i $(1 \le i \le s)$ and

$$\sigma_1 \ldots \sigma_s x = \sigma_1 \ldots \sigma_i x = \sigma_i x = \sigma x.$$

(In Exercise 2.6.12, you are asked to explain each of these equalities in detail.)

Hence $\sigma_1 \ldots \sigma_s = \sigma$. We leave the proof of uniqueness as an exercise (see Exercise 2.6.5). ∎

Henceforth, we shall use cycle notation to express permutations. The computation of products is greatly facilitated by this notation. Recall that, since we write function symbols on the left, the composite $\alpha\beta$ of two permutations performs β before α.

Example 1:

Express each of the following permutations α, β in S_7 in cycle notation and form the product $\alpha\beta$.

$$\alpha = \begin{pmatrix} 1 & 2 & 3 & 4 & 5 & 6 & 7 \\ 1 & 7 & 6 & 2 & 5 & 4 & 3 \end{pmatrix}$$

$$\beta = \begin{pmatrix} 1 & 2 & 3 & 4 & 5 & 6 & 7 \\ 2 & 1 & 3 & 6 & 5 & 7 & 4 \end{pmatrix}.$$

We have

$$\alpha = (2\ 7\ 3\ 6\ 4) \quad \text{(a single cycle!)}$$
$$\beta = (1\ 2)(4\ 6\ 7)$$

hence

$$\alpha\beta = (2\ 7\ 3\ 6\ 4)(1\ 2)(4\ 6\ 7) = (1\ 7\ 2)(3\ 6).$$

This result is obtained as follows: start with any element of X_7, say, with 1. Note that, under $\alpha\beta$, $1 \to 2$, and $2 \to 7$, hence $1 \to 7$, $7 \to 4$ and $4 \to 2$; hence $7 \to 2$; and $2 \to 1$. Thus, as shown in Figure 4, one orbit of $\alpha\beta$ is

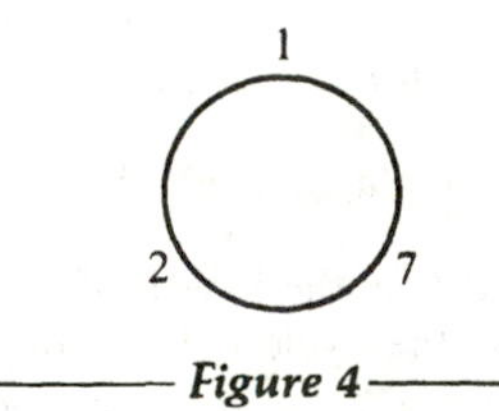

Figure 4

Next pick any element of X_7 not in the set $\{1, 7, 2\}$, say, 3. Note that, under $\alpha\beta$, $3 \to 6$; $6 \to 7$, and $7 \to 3$, hence $6 \to 3$. Thus, another orbit of $\alpha\beta$ shown in Figure 5 is

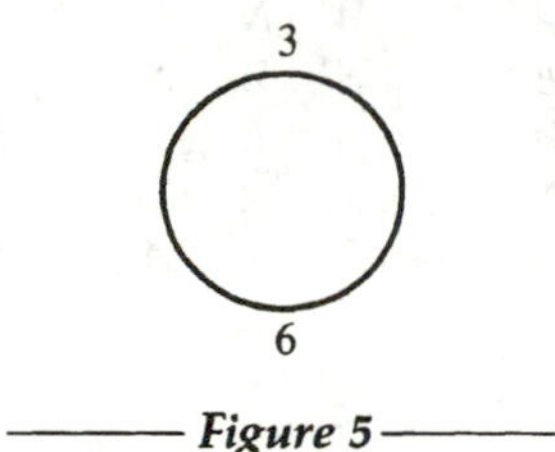

Figure 5

The remaining elements 4 and 5 of X_7 are mapped to themselves under $\alpha\beta$, hence their orbits are trivial. Thus $\alpha = (1\ 7\ 2)(3\ 6)$.

Example 2:

If $\sigma = (a_1 \dots a_n)$ is an n-cycle ($a_i \in X_n$), then, for each $i = 1, \dots, n$,

$$\begin{aligned} \sigma a_i &= a_{i +_n 1} \\ \sigma^2 a_i &= a_{i +_n 2} \\ &\vdots \\ \sigma^{n-1} a_i &= a_{i +_n (n-1)} \\ \sigma^n a_i &= a_{i +_n n} = a_i, \end{aligned} \tag{1}$$

where $+_n$ denotes addition modulo n. From this, it follows that n is the least positive integer such that $\sigma^n = \iota$, i.e., $o(\sigma) = n$ in S_n. The distinct powers of σ are $\iota, \sigma, \sigma^2, \dots, \sigma^{n-1}$. Thus for $s,t \in \mathbb{Z}$, $\sigma^s = \sigma^t$ if and only if $s \equiv t \bmod n$. In particular, $\sigma^s = \iota = \sigma^0$ if and only if $n \mid s$.

The powers of σ powers are easily computed, making use of (1). For example, in case $n = 5$, with $\sigma = (a_1 a_2 a_3 a_4 a_5)$, we obtain

$$\begin{aligned} \sigma^2 &= (a_1 a_3 a_5 a_2 a_4) \quad \text{(skip } \textit{one} \text{ index)} \\ \sigma^3 &= (a_1 a_4 a_2 a_5 a_3) \quad \text{(skip } \textit{two} \text{ indices)} \\ \sigma^4 &= (a_1 a_5 a_4 a_3 a_2) \quad \text{(skip } \textit{three} \text{ indices)} \\ \sigma^5 &= \iota. \end{aligned}$$

Note that $\sigma^{-1} = \sigma^4 = (a_1 a_5 a_4 a_3 a_2) = (a_5 a_4 a_3 a_2 a_1)$. In general, if $\sigma = (a_1 \dots a_n)$, then $\sigma^{-1} = (a_n \dots a_1)$.

Cycle notation is used next to list the elements of the symmetric groups S_3 and S_4, and is an indispensable tool for computing the entries in their Cayley tables.

Cayley Table for S_3

	A	E	J	N	S	W
A	A	E	J	N	S	W
E	E	J	A	W	N	S
J	J	A	E	S	W	N
N	N	S	W	A	E	J
S	S	W	N	J	A	E
W	W	N	S	E	J	A

$A = (1)$
$E = (123)$
$J = (132)$
$N = (12)$
$S = (23)$
$W = (13)$

Cayley Table for S_4

	A	B	C	D	E	F	G	H	J	K	L	M	N	O	P	R	S	T	U	V	W	X	Y	Z
A	A	B	C	D	E	F	G	H	J	K	L	M	N	O	P	R	S	T	U	V	W	X	Y	Z
B	B	A	D	C	G	H	E	F	M	L	K	J	O	N	R	P	U	V	S	T	Z	Y	X	W
C	C	D	A	B	H	G	F	E	K	J	M	L	R	P	O	N	T	S	V	U	Y	Z	W	X
D	D	C	B	A	F	E	G	H	L	M	J	K	P	R	N	O	V	U	T	S	X	W	Z	Y
E	E	F	G	H	J	K	L	M	A	B	C	D	W	X	Y	Z	N	O	P	R	S	T	U	V
F	F	E	H	G	L	M	J	K	D	C	B	A	X	W	Z	Y	P	R	N	O	V	U	T	S
G	G	H	E	F	M	L	K	J	B	A	D	C	Z	Y	X	W	O	N	R	P	U	V	S	T
H	H	G	F	E	K	J	M	L	C	D	A	B	Y	Z	W	X	R	P	O	N	T	S	V	U
J	J	K	L	M	A	B	C	D	E	F	G	H	S	T	U	V	W	X	Y	Z	N	O	P	R
K	K	J	M	L	C	D	A	B	H	G	F	E	T	S	V	U	Y	Z	W	X	R	P	O	N
L	L	M	J	K	D	C	B	A	F	E	H	G	V	U	T	S	X	W	Z	Y	P	R	N	O
M	M	L	K	J	B	A	D	C	G	H	E	F	U	V	S	T	Z	Y	X	W	O	N	R	P
N	N	O	P	R	S	T	U	V	W	X	Y	Z	A	B	C	D	E	F	G	H	J	K	L	M
O	O	N	R	P	U	V	S	T	Z	Y	X	W	B	A	D	C	G	H	E	F	M	L	K	J
P	P	R	N	O	V	U	T	S	X	W	Z	Y	D	C	B	A	F	E	H	G	L	M	J	K
R	R	P	O	N	T	S	V	U	Y	Z	W	X	C	D	A	B	H	G	F	E	K	J	M	L
S	S	T	U	V	W	X	Y	Z	N	O	P	R	J	K	L	M	A	B	C	D	E	F	G	H
T	T	S	V	U	Y	Z	W	X	R	P	O	N	K	J	M	L	C	D	A	B	H	G	F	E
U	U	V	S	T	Z	Y	X	W	O	N	R	P	M	L	K	J	B	A	D	C	G	H	E	F
V	V	U	T	S	X	W	Z	Y	P	R	N	O	L	M	J	K	D	C	B	A	F	E	H	G
W	W	X	Y	Z	N	O	P	R	S	T	U	V	E	F	G	H	J	K	L	M	A	B	C	D
X	X	W	Z	Y	P	R	N	O	V	U	T	S	F	E	H	G	L	M	J	K	D	C	B	A
Y	Y	Z	W	X	R	P	O	N	T	S	V	U	H	G	F	E	K	J	M	L	C	D	A	B
Z	Z	Y	X	W	O	N	R	P	U	V	S	T	G	H	E	F	M	L	K	J	B	A	D	C

$A = (1)$
$B = (12)(34)$
$C = (13)(24)$
$D = (14)(23)$

$E = (123)$
$F = (134)$
$G = (243)$
$H = (142)$

$J = (132)$
$K = (234)$
$L = (124)$
$M = (143)$

$N = (12)$
$O = (34)$
$P = (1324)$
$R = (1423)$

$S = (23)$
$T = (1342)$
$U = (1243)$
$V = (14)$

$W = (13)$
$X = (1234)$
$Y = (24)$
$Z = (1432)$

Definition 2.6.4: A cycle of length 2 is a *transposition.*

Starting with some arrangement of books on a shelf, can we obtain any desired rearrangement just by successively interchanging two books at a time? The answer is "yes," and is easily proved, making use of Theorem 2.6.3.

Theorem 2.6.3: For $n > 1$, every permutation in S_n is a product of transpositions.

Proof: The identity permutation, ι, is a product of 0 transpositions. (Also, for example, $\iota = (12)(12)$.) Any permutation different from the identity is a product of non-trivial cycles. Thus, it suffices to prove that every non-trivial cycle is a product of transpositions. Let $\gamma = (a_1a_2\ldots a_k)$ be a non-trivial cycle. Then $\gamma = (a_1a_k)(a_1a_{k-1})\ldots(a_1a_2)$, a product of transpositions. ∎

The representation of a permutation as a product of transpositions is far from unique. For example, $\sigma = (132)(45) = (12)(13)(45) = (31)(32)(45)(12)(12)$. Thus, not even the *number* of transpositions required is uniquely determined. However, in the example, observe that both representations of σ as a product of transpositions have an *odd* number of factors.

Theorem 2.6.4: For $n > 1$, let $\sigma \in S_n$. Then exactly one of the following statements is true:

(1) all representations of σ as a product of transpositions have an *even* number of factors;

or

(2) all representations of σ as a product of transpositions have an *odd* number of factors.

Proof: First note that any transposition is equal to its inverse. Now suppose that, for some $\sigma \in S_n$, there are transpositions α_i $(i = 1, \ldots, h)$ and β_j $(j = 1, \ldots, k)$ such that

$$\sigma = \alpha_1 \ldots \alpha_h = \beta_1 \ldots \beta_k,$$

where h is even and k is odd. Then

$$\iota = \sigma\ \sigma^{-1} = \alpha_1 \ldots \alpha_h\ \beta_k \ldots \beta_1,$$

the product of an odd number of transpositions. We shall prove that this leads to a contradiction.

Changing to simpler notation, suppose that

$$\iota = \tau_1 \ldots \tau_m$$

where the τ_i are transpositions and m is odd. We consider the action of elements of S_n on polynomials in n variables.

Let $p_n = \prod_{1 \le i < j \le n} (x_i - x_j)$ (a polynomial in $x_1, \ldots, x_n$). For $\sigma \in S_n$, define

$$\sigma(p_n) = \prod_{1 \le i < j \le n} (x_{\sigma i} - x_{\sigma j})$$

and $\sigma(-p_n) = -\sigma p_n$. We note that, for any $\sigma \in S_n$, either $\sigma(p_n) = p_n$ or $\sigma(p_n) = -p_n$, and that $\sigma\tau(p_n) = \sigma(\tau(p_n))$ for σ,τ in S_n.

For $\tau = (hk)$, $1 \leq h < k \leq n$, we determine $\tau(p_n)$ by examining the fate of each factor of p_n involving either h or k, as τ is applied to p_n. (See Figure 6.)

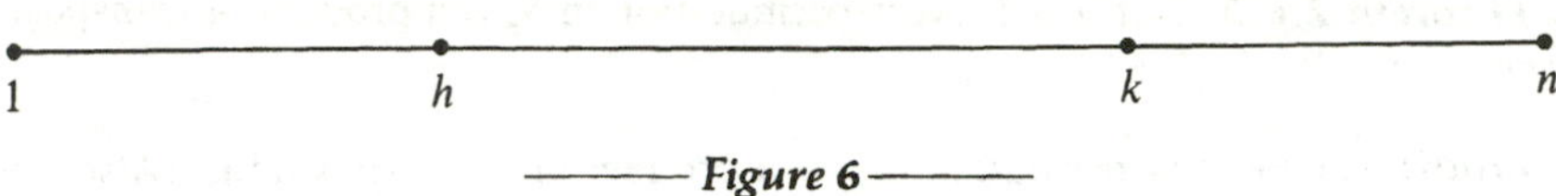

Figure 6

There are:

$h-1$ factors $x_i - x_h$ $(1 \leq i < h)$ that are transformed into factors $x_i - x_k$ of p_n;
$h-1$ factors $x_i - x_k$ $(1 \leq i < h)$ that are transformed into factors $x_i - x_h$ of p_n;
$n-k$ factors $x_h - x_j$ $(k < j \leq n)$ that are transformed into factors $x_h - x_j$ of p_n;
$n-k$ factors $x_k - x_j$ $(k < j \leq n)$ that are transformed into factors $x_h - x_j$ of p_n;
$k-h-1$ factors $x_h - x_j$ $(h < j < k)$ that are transformed into factors $x_k - x_j = -(x_j - x_k)$ (negatives of factors of p_n);
$k-h-1$ factors $x_i - x_k$ $(h < i < k)$ that are transformed into factors $x_i - x_h = -(x_h - x_i)$ (negatives of factors of p_n);

and one factor $x_h - x_k$ that is transformed into $x_k - x_h = -(x_h - x_k)$, the negative of a factor of p_n.

But then applying τ to p_n causes exactly $2(k-h-1)+1$ sign changes in factors of p_n, and so $\tau p_n = -p_n$.

Applying $\tau_1 \ldots \tau_m$ to p_n, we get

$$\tau_1 \ldots \tau_{m-1}(\tau_m p_n) = \tau_1 \ldots \tau_{m-1}(-p_n) = -\tau_1 \ldots (\tau_{m-1} p_m) = \tau_1 \ldots (\tau_{m-2} p_n), \text{ etc.,}$$

alternating signs at each step. Since m is odd, we end up with $\tau_1 \ldots \tau_m p_n = -p_n$. But this is impossible since $\tau_1 \ldots \tau_m = \iota$ and $\iota p_n = p_n$.

Thus, ι cannot be expressed as a product of an odd number of transpositions, and the theorem follows. ∎

Definition 2.6.5: For $n > 1$, a permutation σ in S_n is *even* if it can be expressed as the product of an even number of transpositions and *odd* if it can be expressed as the product of an odd number of transpositions.

Theorem 2.6.4 ensures that the terms *even* and *odd* are well-defined by Definition 2.6.5.

From Definition 2.6.5 and the additive properties of even and odd integers, it follows immediately that evenness and oddness of permutations obey the following crude "multiplication table":

·	even	odd
even	even	odd
odd	odd	even

Theorem 2.6.5: For $n > 1$, let A_n be the set of all even permutations in S_n. Then $A_n < S_n$, with $[S_n : A_n] = 2$.

Proof: As indicated in the preceding table, A_n is closed under composition. By the Corollary of Theorem 2.3.6, this suffices to ensure that $A_n < S_n$. Since $n \geq 2$, S_n contains an odd permutation (for example, any transposition is odd). Let β be an odd permutation in S_n, and consider the right coset $A_n\beta$. We prove that $A_n \cup A_n\beta = S_n$. Since all the even permutations are elements of A_n, it suffices to prove that any odd permutation is in $A_n\beta$. Thus, let α be an odd permutation in S_n. Then $\alpha = (\alpha\beta^{-1})\beta$. Since β is odd, so is β^{-1} (by Definition 2.6.5), hence $\alpha\beta^{-1}$ is even. It follows that $\alpha \in A_n\beta$. This implies that $A_n \cup A_n\beta = S_n$. Thus, A_n and $A_n\beta$ are the only two right cosets of A_n in S_n, and so $[S_n : A_n] = 2$. ∎

Definition 2.6.6: For $n > 1$, the group A_n consisting of all even permutations on the set $X_n = \{1, 2, \ldots, n\}$ is called the *alternating group of degree n.*

(We shall return to A_n in Section 2.10.)

Remark: The introduction of the polynomial p_n in the proof of Theorem 2.6.4 seems spurious, inasmuch as polynomials, even in a single indeterminate, have not yet been formally introduced. However, the polynomial p_n and its use as a tool for discriminating between even and odd permutations lie close to the beginnings of group theory and the connection between groups and polynomial equations. Indeed, if $r_1, \ldots, r_n$ are the roots of a polynomial equation $f(x) = 0$, of degree n, then

$$(p_n(r_1, \ldots, r_n))^2 = \prod_{1 \leq i < j \leq n} (r_i - r_j)^2$$

is the *discriminant* of f (see Definition 4.16.1), a useful device for determining the nature of the roots. For example, if $f = ax^2 + bx + c$, $a \neq 0$, $a, b, c \in \mathbb{R}$, then the roots of the equation $f(x) = 0$ are $r_1 = (-b + \sqrt{b^2 - 4ac})/2a$ and $r_2 = (-b - \sqrt{b^2 - 4ac})/2a$, and the well-known discriminant $b^2 - 4ac$ of f is equal to

$$(p_2(r_1, r_2))^2 = (r_1 - r_2)^2.$$

Exercises 2.6

True or False

1. The permutation $\begin{pmatrix} 1 & 2 & 3 & 4 & 5 \\ 3 & 4 & 5 & 1 & 2 \end{pmatrix}$ in S_5 is a cycle.
2. If two cycles commute, then they are disjoint.
3. The permutation (123456) is even.

4. The odd permutations in S_n $(n \geq 2)$ form a subgroup of S_n.
5. There is a non-abelian group of order 360.

2.6.1. Express the permutation in S_8 given by $\begin{pmatrix} 1 & 2 & 3 & 4 & 5 & 6 & 7 & 8 \\ 8 & 6 & 3 & 4 & 2 & 5 & 1 & 7 \end{pmatrix}$ as a product of disjoint cycles, and as a product of transpositions.

2.6.2. Express the product $(123)(24)(354)(25)(1325)$ as a product of disjoint cycles, and as a product of transpositions.

2.6.3. For $1 \leq n \leq 12$, find the maximal order of elements in S_n.

2.6.4. For $n \geq 1$, suppose $\sigma \in S_n$ is equal to $\sigma_1 \ldots \sigma_t$, where the σ_i are disjoint cycles, of lengths $k_1, \ldots, k_t$, respectively. Prove that $o(\sigma) = \operatorname{lcm}(k_1, \ldots, k_t)$.

2.6.5. Prove the uniqueness in Theorem 2.6.2.

2.6.6. Prove that S_4 has a subgroup isomorphic to S_3. How many isomorphic copies of S_3 can you find in S_4?

2.6.7. Prove that the group S_X of all permutations on any set X of n elements is isomorphic to S_n.

2.6.8. Prove that S_5 has no element of order 15. (Since it can be shown that every group of order 15 is cyclic, S_5 has, in fact, no subgroup of order 15.)

2.6.9. Prove that, for $n \geq 2$, the symmetric group S_n is generated by the transpositions $(12), (13), \ldots, (1n)$.

2.6.10. Prove that, for $n \geq 3$, the alternating group A_n is generated by the 3-cycles in S_n. In particular, prove that for each $a, b \in X_n = \{1, 2, \ldots, n\}$ the 3-cycles (abi), $i = 1, \ldots, n$, $i \neq a$, $i \neq b$, generate A_n. Hint: use the preceding exercise.

2.6.11. Supply the detail omitted from the proof of the Lemma preceding Definition 2.6.3: Let $\sigma \in S_n$ and let $a \in \mathcal{O}$, where $\mathcal{O}$ is one of the non-trivial orbits of σ. Let k_0 be the least positive integer t such that $\sigma^t a = a$. Prove that $\mathcal{O} = \{a, \sigma a, \ldots, \sigma^{k_0 - 1} a\}$.

2.6.12. Give a detailed explanation of each step of (1) in the proof of Theorem 2.6.4.

2.6.13. For $n \geq 2$, prove that S_n is generated by the n-cycle $(1\ 2\ \ldots\ n)$ and the transposition $(1\ 2)$.

2.6.14. From Exercise 2.6.13, conclude that, for p prime, S_p is generated by the p-cycle $(1\ 2\ \ldots\ p)$ and *any* transposition (hk), $1 \leq h < k \leq p$.

2.6.15. For $n \geq 2$, let $\alpha = \begin{pmatrix} 1 & 2 & \ldots & n \\ i_1 & i_2 & \ldots & i_n \end{pmatrix} \in S_n$. Each occurrence, in the n-tuple $(i_1, i_2, \ldots, i_n)$, of $i_h > i_k$ for $1 \leq h < k \leq n$ is called an *inversion* of α. Prove that the parity of α is equal to the parity of the number of inversions of α.

2.6.16. Let G be a group of order 6. Prove that either $G \cong \mathbb{Z}_6$ or $G \cong S_3$.

2.7

Cayley's Theorem; Geometric Groups

Historically, the first groups to be considered were permutation groups. In view of the variety of contexts in which groups occur within mathematics and its applications, it may be somewhat surprising that, abstractly, groups never strayed far from their origins: every group is isomorphic to a group of bijections; hence every finite group is isomorphic to a group of permutations.

Theorem 2.7.1. (Cayley's Theorem): Every group is isomorphic to a group of bijections.

Proof: Let $< G, \cdot >$ be a group. Visualize its Cayley table. For each $a \in G$, the column headed by a lists the elements xa $(x \in G)$, while the row "headed" by a lists the elements $ax(x \in G)$. By Theorem 2.2.6., every column of the table, and every row of the table, contains each element of G exactly once. Thus, each row or column of the table represents some particular arrangement of the elements of G, i.e., some bijection of the set G onto itself. Specifically, for each $a \in G$, define the left-multiplication $\alpha_a : G \to G$ by $\alpha_a(x) = ax$, for each $x \in G$. (The images under α_a are listed in the row belonging to a in the Cayley Table for G). Each α_a is a (well-defined) map of G into G. For $x, y \in G$, $\alpha_a(x) = \alpha_a(y) \Rightarrow ax = ay \Rightarrow x = y$ (by left-cancellation in G). Hence α_a is 1-1. For each $g \in G$, $g = a(a^{-1}g) = \alpha_a(a^{-1}g)$. Hence α_a is a bijection.

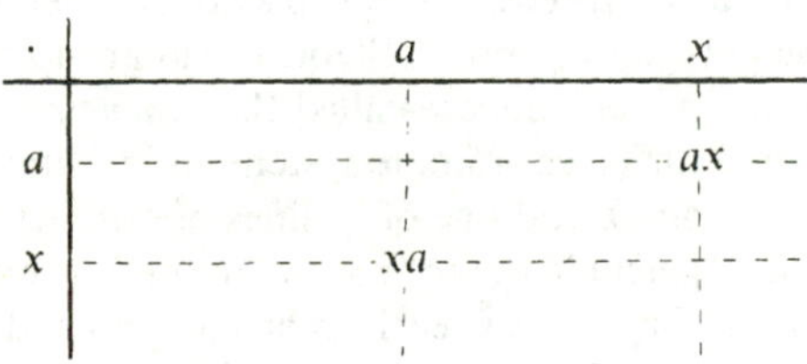

Let S_G be the group of all bijections on G. Define $\rho : G \to S_G$ by: $\rho(a) = \alpha_a$. Then ρ is a 1-1 map, for: if $\rho(a) = \rho(b)$ $(a, b \in G)$, then $ax = bx$ for each $x \in G$; hence, $a = b$. For $a, b \in G$, is $\rho(ab) = \rho(a)\rho(b)$? Let $x \in G$. Then $\alpha_{ab}(x) = (ab)x = a(bx) = \alpha_a(\alpha_b(x))$. It follows that $\alpha_{ab} = \alpha_a\alpha_b$ for each $a, b \in G$. But then $\rho(ab) = \rho(a)\,\rho(b)$. Thus, ρ is a 1-1 product-preserving mapping (a monomorphism of G into S_G. This implies (see Exercise 2.9.3) that Im ρ is a subgroup of S_G, isomorphic to G. ■

Corollary: If G is a group of finite order n, then G is isomorphic to a subgroup of the symmetric group S_n.

Proof: By the theorem, G is isomorphic to a subgroup of S_G which, in turn, is isomorphic to S_n (see Exercise 2.6.7). But then G is isomorphic to a subgroup of S_n. ■

Remarks: *(1)* The mapping $\rho : G \to S_G$ that associates with each $a \in G$ the left-multiplication α_a is called the *left-regular representation* of G.

(2) We could also define, for each $a \in G$, the "right-multiplication" $\beta_a : G \to G$ by: $\beta_a(x) = xa$, for each $x \in G$, and then define $\bar{\rho} : G \to S_G$ by: $\bar{\rho}(a) = \beta_a$ for each $a \in G$. Then $\bar{\rho}$ will be a 1-1 map that *reverses* multiplication — a so-called *anti-monomorphism* of G into S_G.

The mapping $\bar{\rho}$ is called the *right-regular representation* of G.

(3) Of course, for G a group of finite order n, S_n is not, in general, the smallest symmetric group in which G can be isomorphically embedded. (For example, for $m \geq 2$, A_m and S_m can be naturally embedded in S_m under the inclusion map.)

Geometric Groups

We now examine, rather informally, some groups of bijections that are of geometric interest.

Let A be the set of all points in a plane and let S_A be the group of all bijections on A. The elements of S_A are called *motions* of the plane. There are many interesting subgroups of S_A, among them the group R_A consisting of all distance preserving motions (also known as *rigid motions*, or *isometries* of the plane). The subgroups of R_A, in turn, include the group of all translations and the group of all rotations about any fixed point in the plane.

Now let P_n be a regular polygon of n sides, situated in the plane. Think of P_n as the set of all points that lie on its edges — a subset of A. Then the rigid motions of the plane that map the polygon P_n onto itself form a subgroup of R_n. This subgroup is the *dihedral group* D_n; its elements are called the *symmetries* of the polygon P_n.

In the plane, introduce an xy-coordinate system and position the polygon P_n so that its center is at the origin 0, and one of its lines of symmetry lies on the x-axis. Then the symmetries of P_n include the rotation ρ about 0 through an angle of $\frac{360^\circ}{n}$, and the reflection σ about the x-axis. Clearly, ρ has order n and σ has order 2 in the group S_A. It can be shown that D_n is equal to the subgroup of S_A which is generated by ρ and σ, i.e., $D_n = [\rho, \sigma]$. The subgroup $[\rho, \sigma]$ generated by ρ and σ obviously contains the $2n$ elements $\iota, \rho, \rho^2, \ldots, \rho^{n-1}, \sigma, \rho\sigma, \rho^2\sigma, \ldots, \rho^{n-1}\sigma$. It is not difficult to show that $\sigma\rho = \rho^{n-1}\sigma$ (see Exercise 2.7.3). Using this formula, one can convert every element of $[\rho, \sigma]$ (i.e., every product of powers of ρ and σ) into one of the $2n$ elements already listed. (For example, for $n = 5$, $\sigma\rho^3\sigma\rho = \sigma\rho^3\rho^4\sigma = \sigma\rho^7\sigma = \sigma\rho^2\sigma = \sigma\rho\,\rho\sigma = \rho^4\sigma\rho\sigma = \rho^4\rho^4\sigma\sigma = \rho^8\sigma^2 = \rho^3$.) Thus, $D_n = \{\iota, \rho, \ldots, \rho^{n-1}, \sigma, \rho\sigma, \ldots, \rho^{n-1}\sigma\}$.

In the tables for D_3 and D_4 that follow (Figures 7 and 8), products are computed, as usual, from right to left, i.e., $\rho\sigma$ means: first σ, then ρ.

With the vertices labeled $1, 2, \ldots, n$ (as above, for $n = 3$ and $n = 4$), there is an obvious correspondence between the elements of the dihedral group D_n and certain

D_3: Symmetries of an Equilateral Triangle

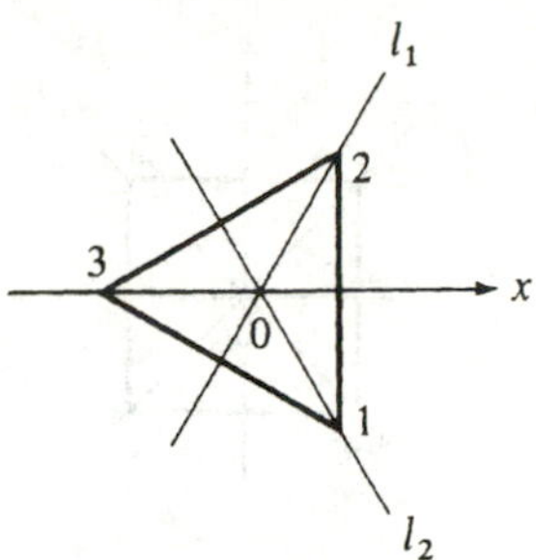

D_3

	ι	ρ	ρ^2	σ	$\rho\sigma$	$\rho^2\sigma$
ι	ι	ρ	ρ^2	σ	$\rho\sigma$	$\rho^2\sigma$
ρ	ρ	ρ^2	ι	$\rho\sigma$	$\rho^2\sigma$	σ
ρ^2	ρ^2	ι	ρ	$\rho^2\sigma$	σ	$\rho\sigma$
σ	σ	$\rho^2\sigma$	$\rho\sigma$	ι	ρ^2	ρ
$\rho\sigma$	$\rho\sigma$	σ	$\rho^2\sigma$	ρ	ι	ρ^2
$\rho^2\sigma$	$\rho^2\sigma$	$\rho\sigma$	σ	ρ^2	ρ	ι

ι identity map
ρ 120° rotation
ρ^2 240° rotation
σ reflection in x-axis
$\rho\sigma$ reflection in l_1
$\rho^2\sigma$ reflection in l_2

Figure 7

permutations in S_n: simply associate with each motion in D_n the permutation it induces on the vertices of the polygon P_n, or, equivalently, on the set $X_n = \{1, 2, \ldots, n\}$. This correspondence is clearly a monomorphism of D_n into S_n; hence D_n is isomorphic to a subgroup of S_n. For $n = 3$, since $|D_3| = 2 \cdot 3 = 3! = |S_3|$, D_3 is isomorphic to S_3. For $n > 3$, $2n < n!$; hence D_n is isomorphic to a proper subgroup of S_n.

D_4: Symmetries of a Square

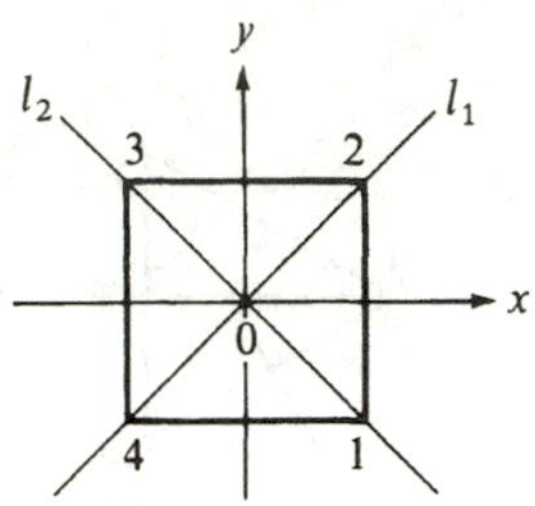

D_4

	ι	ρ	ρ^2	ρ^3	σ	$\rho\sigma$	$\rho^2\sigma$	$\rho^3\sigma$
ι	ι	ρ	ρ^2	ρ^3	σ	$\rho\sigma$	$\rho^2\sigma$	$\rho^3\sigma$
ρ	ρ	ρ^2	ρ^3	ι	$\rho\sigma$	$\rho^2\sigma$	$\rho^3\sigma$	σ
ρ^2	ρ^2	ρ^3	ι	ρ	$\rho^2\sigma$	$\rho^3\sigma$	σ	$\rho\sigma$
ρ^3	ρ^3	ι	ρ	ρ^2	$\rho^3\sigma$	σ	$\rho\sigma$	$\rho^2\sigma$
σ	σ	$\rho^3\sigma$	$\rho^2\sigma$	$\rho\sigma$	ι	ρ^3	ρ^2	ρ
$\rho\sigma$	$\rho\sigma$	σ	$\rho^3\sigma$	$\rho^2\sigma$	ρ	ι	ρ^3	ρ^2
$\rho^2\sigma$	$\rho^2\sigma$	$\rho\sigma$	σ	$\rho^3\sigma$	ρ^2	ρ	ι	ρ^3
$\rho^3\sigma$	$\rho^3\sigma$	$\rho^2\sigma$	$\rho\sigma$	σ	ρ^3	ρ^2	ρ	ι

ι	identity map	σ	reflection in x-axis
ρ	90° rotation	$\rho\sigma$	reflection in l_1
ρ^2	180° rotation	$\rho^2\sigma$	reflection in y-axis
ρ^3	270° rotation	$\rho^3\sigma$	reflection in l_2.

Figure 8

Exercises 2.7

True or False

1. Every finite group is isomorphic to one of the symmetric groups S_n.
2. A rectangle that is not a square has only two symmetries.
3. A regular hexagon has more than twelve symmetries.
4. Every permutation of the vertices of an equilateral triangle is induced by a symmetry of the triangle.

5. Every permutation of the vertices of a square is induced by a symmetry of the square.

2.7.1. Find the least positive integer n such that S_n contains an isomorphic copy of
(a) a cyclic group of order p, for p any prime;
(b) a cyclic group of order 6;
(c) a cyclic group of order 12.

2.7.2. In D_4, without consulting the table, verify that $\rho\sigma\rho = \sigma$. Conclude that $\sigma\rho = \rho^3\sigma$.

2.7.3. For $n \geq 3$, with the vertices of a regular n-gon, P_n, labeled consecutively $1, 2, \ldots, n$, express the rotation ρ through $\frac{360^\circ}{n}$ and a reflection σ about a line of symmetry as permutations on the vertices. Verify that $\rho\sigma\rho = \sigma$, and conclude that $\sigma\rho = \rho^{n-1}\sigma$.

2.7.4. In S_4, find three distinct subgroups isomorphic to D_4. (Hint: there is more than one way to label the vertices of a square!)
By the Sylow Theorems (Theorems 2.15.3 and 2.15.5), S_4 has exactly three subgroups of order 8.

2.7.5. Let H_1, H_2, H_3 be the subgroups of S_4, isomorphic to D_4, which you found in the preceding exercise. Express H_2 and H_3 as conjugates of H_1, i.e., find elements α and $\beta \in S_4$ such that $H_2 = \alpha H_1 \alpha^{-1}$, $H_3 = \beta H_1 \beta^{-1}$. (The labeling of the subgroups is immaterial.)

2.7.6. Make a Cayley table for the group, D_5, of symmetries of a regular pentagon.

2.7.7. Extend the notion of symmetry to three dimensions. Prove that the symmetries of a regular tetrahedron form a group isomorphic to the symmetric group S_4.

2.7.8. Find the order of the group of all symmetries of a cube.

2.7.9. In the group of all symmetries of a regular tetrahedron, identify those symmetries that can be demonstrated by moving a rigid tetrahedron physically in 3-space, and prove that they form a subgroup.

2.7.10. In the group of all symmetries of a cube, identify those symmetries that can be demonstrated by moving a rigid cube physically in 3-space, and prove that they form a subgroup.

2.7.11. Let X be a non-empty subset of a plane and let $\phi : X \to X$ be an isometry (i.e., a distance-preserving function). Prove that ϕ can be extended to an isometry of the plane (i.e., prove that there is an isometry ψ of the plane whose restriction to X is the given isometry ϕ).

2.8

Normal Subgroups

The fundamental importance of the class of subgroups which we are about to introduce will become increasingly apparent as we study homomorphisms, factor groups, and automorphisms of groups.

To begin with, consider, for example, the group S_3 (see the table on p. 72). Let $H_1 = [E]$, the subgroup generated by $E = (123)$, and let $H_2 = [N]$, the subgroup generated by $N = (12)$. From the table, it is readily apparent that the right cosets of H_1 are

$$H_1A = H_1E = H_1J = \{A, E, J\} = H_1$$

and

$$H_1N = H_1S = H_1W = \{N, S, W\},$$

and the left cosets are

$$AH_1 = EH_1 = JH_1 = \{A, E, J\} = H_1$$

and

$$NH_1 = SH_1 = WH_1 = \{N, S, W\}.$$

Thus, in this case, not only is the set of all right cosets equal to the set of all left cosets, but $H_1\sigma = \sigma H_1$ holds for each $\sigma \in S_3$.

A very different situation holds for the subgroup $H_2 = [N]$. Here, the right cosets of H_2 are

$$H_2A = H_2N = \{A, N\} = H_2$$
$$H_2E = H_2S = \{E, S\}$$
$$H_2J = H_2W = \{J, W\},$$

while the left cosets are

$$AH_2 = NH_2 = \{A, N\} = H_2$$
$$EH_2 = WH_2 = \{E, W\}$$
$$JH_2 = SH_2 = \{J, S\}.$$

In this case, the set of all right cosets is $\{\{A, N\}, \{E, S\}, \{J, W\}\}$, while the set of all left cosets is $\{\{A, N\}, \{E, W\}, \{J, S\}\}$. The right and left coset decompositions of S_3 with respect to H_2 yield two different partitions of G. The equality $H_2\sigma = \sigma H_2$ holds only for $\sigma = A$ or $\sigma = N$; for all other elements of S_3, it fails to hold.

It is for subgroups which, like H_1, yield only a single coset decomposition that a group structure can be defined on the set of all cosets (generalizing the construction of $\bar{\mathbb{Z}}_n$ from $\mathbb{Z}$). The fundamental importance of normal subgroups derives from this fact.

Definition 2.8.1: Let $\langle G, \cdot \rangle$ be a group, and let $H < G$. Then H is a *normal subgroup* of G (or: *H is normal in G*) if $Ha = aH$ for all $a \in G$.

Notation: We write "$H \lhd G$" to signify that H is a normal subgroup of G.

Warning: The condition "$Ha = aH$ for all $a \in G$" does NOT state that every $a \in G$ commutes individually with every $h \in H$. It merely requires that, for each $a \in G$ and each $h \in H$, $ha = ah'$ for some $h' \in H$, and $ah = h''a$ for some $h'' \in H$.

Of course, if for a particular $H < G$, every element $a \in G$ *does* commute with every element $h \in H$, then the condition $Ha = aH$ is trivially fulfilled. In particular, we have the following corollaries.

Corollary 1: Every subgroup of an abelian group is normal. Another simple, but important, consequence of Definition 2.8.1 is Corollary 2.

Corollary 2: Every subgroup of index 2 is normal.

Proof: Suppose that $H < G$, with $[G : H] = 2$. If $a \in H$, then $Ha = H = aH$. Let $a \in G \setminus H$. Then $H \neq Ha$, hence the right cosets of H in G are H and Ha; and $H \neq aH$, hence the left cosets of H in G are H and aH. Since the right cosets of H in G partition G, we have

$$G = H \cup Ha, \qquad H \cap Ha = \varnothing.$$

Similarly, since the left cosets of H in G partition G, we have

$$G = H \cup aH, \qquad H \cap aH = \varnothing.$$

But then $Ha = G \setminus H = aH$.

Thus, for all $a \in G$, $Ha = aH$, and so $H \lhd G$. ■

Corollary 3: For each positive integer n, $A_n \lhd S_n$.

(For mysterious psychological reasons, students often misread or misremember Corollary 2, and end up believing that every subgroup of *order* 2 is normal. This is certainly false—the example preceding Definition 2.8.1 serves as an easy counterexample.)

Like most important mathematical concepts, normality can be characterized in a variety of equivalent ways. In particular, the behavior of a subgroup H under the relation of conjugacy is crucial in determining whether or not H is a normal subgroup.

Definition 2.8.2: Let G be a group, and let $a, x, y \in G$. If $y = axa^{-1}$, then *y is conjugate to x via a.*

Theorem 2.8.1: Let $\langle G, \cdot \rangle$ be a group, and let $H < G$. Then the following statements are equivalent:

(1) $aH = Ha$ for each $a \in G$ (i.e., $H \lhd G$);

(2) $aha^{-1} \in H$ for each $h \in H$ and each $a \in G$;

(2a) $a^{-1}ha \in H$ for each $h \in H$ and each $a \in G$;

(3) $aHa^{-1} = H$ for each $a \in G$;

(3a) $a^{-1}Ha = H$ for each $a \in G$

(where $aHa^{-1} = \{aha^{-1} | h \in H\}$ and $a^{-1}Ha = \{a^{-1}ha | h \in H\}$).

Proof: Statements (2) and (2a) are clearly equivalent; so are statements (3) and (3a). (See Exercise 2.8.15.) Hence it suffices to prove the equivalence of (1), (2), and (3).

(1) $\Rightarrow$ (2). Suppose $aH = Ha$ for each $a \in G$. Then, for each $a \in G$ and each $h \in H$, $ah = h'a$ for some $h' \in H$. But then $aha^{-1} = h' \in H$, and so (2) holds.

(2) $\Rightarrow$ (3). Suppose $aha^{-1} \in H$ for each $h \in H$, $a \in G$. Then $aHa^{-1} \subset H$, for each $a \in G$. Since (2) $\Rightarrow$ (2a), we also have $a^{-1}ha \in H$ for each $h \in H$, $a \in G$, whence $h = ah'a^{-1}$ for some $h' \in H$, and so $aHa^{-1} \supset H$. But then $aHa^{-1} = H$, for each $a \in G$.

(3) $\Rightarrow$ (1). Suppose $aHa^{-1} = H$, for each $a \in G$. Since (3) $\Rightarrow$ (3a), we also have $a^{-1}Ha = H$, for each $a \in G$. Hence, if $a \in G$ and $h \in H$, then $aha^{-1} = h'$ for some $h' \in H$, and so $ah = h'a$. Thus, $aH \subset Ha$, for each $a \in G$. Also, $a^{-1}ha = h''$ for some $h'' \in H$; hence $ha = ah''$. Thus $Ha \subset aH$, and the equality follows: $aH = Ha$, for each $a \in G$. ∎

Example 1:

By Theorem 2.6.5 and Corollary 2 above, the alternating group A_n is a normal subgroup of the symmetric group S_n for each $n \geq 2$ (recall Definition 2.6.6).

For example, in S_3,

$$A_3 = [(123)] = \{(123), (132), (1)\} \lhd S_3.$$

Observe that $(123)(12) = (13)$, while $(12)(123) = (23)$. Thus, for example, (12) does not commute with each element of A_3—illustrating the Warning!

Example 2:

Let $G = GL_n(\mathbb{R})$, the group of all non-singular $n \times n$ matrices over $\mathbb{R}$, and let $H = SL_n(\mathbb{R})$, the subgroup of G consisting of all $n \times n$ matrices of determinant 1. Is $H \lhd G$?

It is often more convenient to settle the question of normality by using one of the conditions in Theorem 2.8.1 which is equivalent to the defining property (1). In this case, we use (2) of Theorem 2.8.1.

Let $M \in H = SL_n(\mathbb{R})$ and let $A \in G = GL_n(\mathbb{R})$. Then $\det(AMA^{-1}) = (\det A)(\det M)(\det A^{-1}) = (\det M)(\det A)(\det A)^{-1} = \det M = 1$. But then $AMA^{-1} \in H$. It follows that $H \lhd G$. (For a different approach, see Exercise 2.5.12.)

Example 3:

For $n > 2$, let $H = \{\sigma \in S_n | \sigma(1) = 1\}$. We have observed (Example 6, p. 45) that $H < S_n$. But $H \ntriangleleft S_n$, for (again using Condition 2 of Theorem 2.8.1), we have

$$(12) \in S_n, \quad (23) \in H,$$

but

$$(12)^{-1}(23)(12) = (12)(23)(12) = (13) \notin H.$$

Example 4:

Let G be a group, and let $C(G) = \{c \in G \mid cx = xc \text{ for each } x \in G\}$. ($C(G)$ is called the *center* of G.) Then every subgroup of $C(G)$ is normal in G. In particular, of course, $C(G) \lhd G$ (see Exercise 2.8.6).

(Note that, in this case, for any $H < C(G)$, it *is* true that each $x \in G$ commutes individually with every element of H.)

Example 5:

To illustrate Example 4, let $G = GL_n(\mathbb{R})$. The center of $GL_n(\mathbb{R})$ is the subgroup $C(G)$ consisting of all non-zero scalar matrices aI_n, $a \in \mathbb{R}^*$. $C(G)$ and all of its subgroups (e.g., the group of all scalar matrices bI_n, where $b \in \mathbb{R}^+$) are normal in G (see Exercise 2.8.9).

Exercises 2.8

True or False

1. If $H \lhd G$, then $ha = ah$ for each $a \in G$ and each $h \in H$.
2. Every subgroup of prime index is normal.
3. Every subgroup of order 2 is normal.
4. If $H \lhd G$, then $h^{-1}xh \in H$ for each $h \in H$, each $x \in G$.
5. If H is an abelian subgroup of a group G, then $H \lhd G$.

2.8.1. Find all subgroups of S_3 and classify them as normal or not normal.

2.8.2. Make a lattice diagram for the subgroups of the dihedral group D_4 and determine which subgroups are normal.

2.8.3. Let G be a group, $x, y \in G$. Then y is conjugate to x if $y = axa^{-1}$ for some $a \in G$. Prove that the relation "is conjugate to" is an equivalence relation on G. (The equivalence classes with respect to this relation are called *conjugate classes*.)

2.8.4. Let $\tau = (a_1\ a_2 \ldots a_k)$ be a cycle in S_n and let $\sigma \in S_n$ $(n \geq 1)$. Prove that $\sigma\tau\sigma^{-1}$ is the cycle $(\sigma a_1\ \sigma a_2\ \ldots\ \sigma a_k)$. Use this result to prove that two permutations in S_n are conjugate if and only if they have the same cycle structure. (The *cycle structure* of a permutation γ is the sequence of lengths of the disjoint cycles of γ, arranged in non-decreasing order.)

2.8.5. Use the preceding exercise to prove that the subgroup V_4 of S_4 consisting of (1), (12)(34), (13)(24), (14)(23) is normal in S_4. Find at least one non-normal subgroup of S_4 of each of the following orders: 2, 3, 4, 6, 8.

2.8.6. For any group G, let $C(G) = \{a \in G \mid xa = ax \text{ for all } x \in G\}$. Prove:

(1) $C(G) \lhd G$;

(2) Every subgroup H of $C(G)$ is normal in G.

(Note: $C(G)$ is called the *center* of G.)

2.8.7. Prove that, for $n \geq 3$, the center of S_n consists of the identity only.

2.8.8. Use the result of the preceding exercise to prove that, for $n \geq 3$, S_n has no normal subgroup of order 2.

2.8.9. Prove that the center of $GL_n(\mathbb{R})$ $(n \in \mathbb{Z}^+)$ consists of all scalar matrices cI_n, $c \in \mathbb{R}^*$.

2.8.10. *(1)* Prove that subgroup inclusion is transitive, while normal subgroup inclusion is not, i.e.,

(a) if $K < H$ and $H < G$, then $K < G$

but

(b) if $K \lhd H$ and $H \lhd G$, then K need not be normal in G. (Look in S_4 for a counterexample.)

(2) Prove: if $K \lhd G$, $H < G$, and $K \subset H$, then $K \lhd H$.

2.8.11. Let $n \geq 3$ be a positive integer. For each $i = 1, 2, \ldots, n$, let H_i be the subset of S_n defined by

$$H_i = \{\sigma \in S_n | \sigma(i) = i\}.$$

(1) Prove that $H_i < S_n$, but H_i is not normal in S_n.

(2) Prove that $H_i \cong S_{n-1}$.

(3) For $n \geq 3$, does S_n have a normal subgroup isomorphic to S_{n-1}? Prove that the answer is "no" for $n = 3$ and $n = 4$. For $n \geq 5$, see Exercise 2.8.16 or Exercise 2.10.2.

2.8.12. Prove that the group given by the following table is a non-abelian group, each of whose subgroups is normal. (This group is call the *quaternion group*. It is a subgroup of the multiplicative group of all non-zero real quaternions—see Chapter 3, p. 135)

	1	-1	i	$-i$	j	$-j$	k	$-k$
1	1	-1	i	$-i$	j	$-j$	k	$-k$
-1	-1	1	$-i$	i	$-j$	j	$-k$	k
i	i	$-i$	-1	1	k	$-k$	$-j$	j
$-i$	$-i$	i	1	-1	$-k$	k	j	$-j$
j	j	$-j$	$-k$	k	-1	1	i	$-i$
$-j$	$-j$	j	k	$-k$	1	-1	$-i$	i
k	k	$-k$	j	$-j$	$-i$	i	-1	1
$-k$	$-k$	k	$-j$	j	i	$-i$	1	-1

(Note: a group with the property that each of its subgroups is normal is called a *Hamiltonian group*.)

2.8.13. Let G be a group, $H < G$, and let $N_H = \{a \in G | aH = Ha\}$.
Prove:

(1) $N_H < G$;

(2) $H \lhd N_H$.

(N_H is called the *normalizer* of H in G.)

2.8.14. Let G be a group, $H < G$, and let $C_H = \{a \in G | ah = ha \text{ for all } h \in H\}$. Prove that $C_H < N_H$ (see the preceding exercise). Give examples to prove that, in general $H \not\subset C_H$, and $C_H \neq N_H$. (C_H is called the *centralizer* of H in G.)

2.8.15. Prove that, if $H < G$, then the statements

(2) $aha^{-1} \in H$ for each $h \in H$ and each $a \in G$

and

(2a) $a^{-1}ha \in H$ for each $h \in H$ and each $a \in G$
are equivalent (see Theorem 2.8.1). However, prove that the statements
(i) $aha^{-1} \in H$ for each $h \in H$
and
(ii) $a^{-1}ha \in H$ for each $h \in H$
are *not* equivalent for any particular $a \in G$. (Give an example where one of these statements holds and the other fails.) Conclude that if $aHa^{-1} \subset H$ for some particular $a \in G$, aHa^{-1} need not be equal to H.

2.8.16. For $n \geq 3$, prove: if $K \lhd A_n$ and K contains a 3-cycle, then $K = A_n$. (Use Exercises 2.6.10 and 2.8.4.)

2.9

Homomorphism, Factor Groups, and the Fundamental Theorem of Homomorphism for Groups

While a simple photograph will render a faithful likeness of a subject, an artistic photographer or painter may wish to blur some features of a subject in order to highlight others. Similarly, an isomorphic image of a group G will faithfully reflect the structure of G, while a homomorphic image may "blur" some features of G, highlighting others. The homomorphic images of a group may, in this way, provide important information about the group itself.

Definition 2.9.1: Let $\langle G, \cdot \rangle$ and $\langle G', \cdot' \rangle$ be groups and let $\phi : G \to G'$ be a mapping satisfying the condition

$$\phi(x \cdot y) = \phi(x) \cdot' \phi(y)$$

for each $x, y \in G$. Then ϕ is a *homomorphism* of G into G'.

(Various kinds of homomorphisms are designated by appropriate Greek prefixes. A homomorphism $\phi : G \to G'$ is

a *monomorphism* if ϕ is 1–1.
an *epimorphism* if ϕ is onto.
an *isomorphism* if ϕ is 1–1 onto (a bijection).

For the case where $\langle G', \cdot' \rangle = \langle G, \cdot \rangle$, we have: a homomorphism $\phi : G \to G$ is an *endomorphism* of G; an isomorphism $\phi : G \to G$ is an *automorphism* of G.)

A homomorphism $\phi : G \to G'$ will preserve some, but not necessarily all, structural properties. Any homomorphism does "preserve identities and inverses," in the sense of the following theorem.

Theorem 2.9.1: If $\phi : G \to G'$ is a homomorphism, then

(1) $\phi(e) = e'$;
(2) for each $a \in G$, $\phi(a^{-1}) = (\phi(a))^{-1}$.

Proof: *(1)* We have $\phi(e) \cdot' \phi(e) = \phi(e \cdot e) = \phi(e) = \phi(e) \cdot' e'$. By left cancellation in G', it follows that $\phi(e) = e'$.

(2) Let $a \in G$. Then $\phi(a) \cdot' \phi(a^{-1}) = \phi(a \cdot a^{-1}) = \phi(e) = e' = \phi(a) \cdot' [\phi(a)]^{-1}$. By left cancellation in G', it follows that $\phi(a^{-1}) = [\phi(a)]^{-1}$. ■

By Theorem 2.9.1, under a homomorphism $\phi : G \to G'$, at least one element of G maps to the identity of G'. We are now ready to introduce the heart of a homomorphism: its "kernel."

Notation: If A, B are sets, and $\phi : A \to B$ is a mapping, then for each $b \in \text{Im } \phi$, we write

$$\phi^{-1}(b) = \{a \in A \mid \phi(a) = b\}.$$

(This notation is in no way intended to suggest that ϕ has an inverse map, ϕ^{-1}.) The set $\phi^{-1}(b)$ is called the *inverse image of b* under ϕ.

Definition 2.9.2: If $\phi : G \to G'$ is a group homomorphism, then the *kernel of* ϕ is the set

$$\text{Ker } \phi = \{x \in G \mid \phi(x) = e'\} = \phi^{-1}(e').$$

Example:

Let $G = S_3$ and let $G' = \mathbb{Q}^*$. Define $\phi : G \to G'$ by: $\phi(\sigma) = 1$ if σ is an even permutation, and $\phi(\sigma) = -1$ if σ is an odd permutation. Then ϕ is a homomorphism. For: if σ, τ are both even, then $\sigma\tau$ is even, hence

$$\phi(\sigma\tau) = 1 = 1 \cdot 1 = \phi(\sigma)\phi(\tau).$$

If σ, τ are both odd, then $\sigma\tau$ is even, hence

$$\phi(\sigma\tau) = 1 = (-1)(-1) = \phi(\sigma)\phi(\tau).$$

If σ is even and τ is odd, then $\sigma\tau$ is odd, hence

$$\phi(\sigma\tau) = -1 = 1 \cdot (-1) = \phi(\sigma)\phi(\tau).$$

If σ is odd and τ is even, then $\sigma\tau$ is odd, hence

$$\phi(\sigma\tau) = -1 = (-1) \cdot 1 = \phi(\sigma)\phi(\tau).$$

Thus, ϕ is a homomorphism, with $\text{Ker } \phi = \phi^{-1}(1) = A_3$, the subgroup of S_3 consisting of all even permutations in S_3. It follows that $\text{Ker } \phi \lhd S_3$. Clearly, $\text{Im } \phi = \{1, -1\}$, a subgroup of $\mathbb{Q}^*$. What is $\phi^{-1}(-1)$? By definition of ϕ,

$\phi^{-1}(-1)$ is the set of all odd permutations in S_3. These form the second coset of A_3 in S_3, i.e.,

$$\phi^{-1}(-1) = A_3\tau,$$

where τ is any odd permutation in S_3. (See Figure 9.)

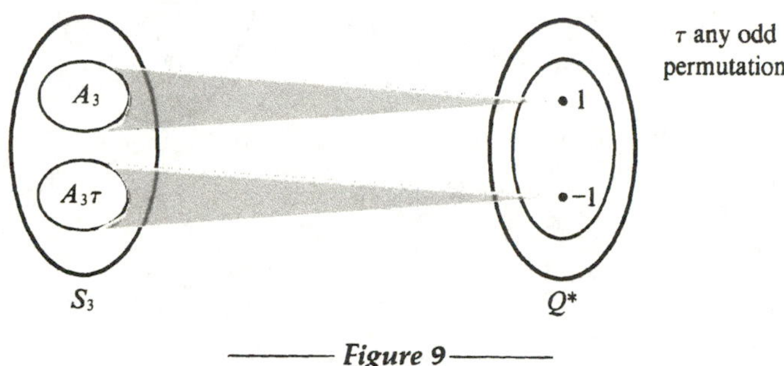

Figure 9

Theorem 2.9.2: Let $\phi : G \to G'$ be a group homomorphism. Then

(1) Ker $\phi \lhd G$.

(2) Im $\phi < G'$.

(3) If $a \in G$ and $b = \phi(a)$, then $\phi^{-1}(b) = (\text{Ker } \phi)a = a \text{ Ker } \phi$.

(4) $(\text{Ker } \phi)a \mapsto \phi(a)$ defines a bijection of the set of all cosets of Ker ϕ onto Im ϕ.

Proof: *(1)* Since $\phi(e) = e'$, we have $e \in \text{Ker } \phi$. If $a, b \in \text{Ker } \phi$, then $\phi(ab^{-1}) = \phi(a)[\phi(b)]^{-1} = e' \cdot e' = e'$; hence $ab^{-1} \in \text{Ker } \phi$. By the corollary of Theorem 2.3.1, it follows that Ker $\phi < G$. For $a \in G$, $x \in \text{Ker } \phi$, $\phi(axa^{-1}) = \phi(a)\phi(x)[\phi(a)]^{-1} = \phi(a)e'[\phi(a)]^{-1} = e'$; hence $axa^{-1} \in \text{Ker } \phi$. By Theorem 2.8.1, we conclude: Ker $\phi \lhd G$.

(2) Since $e' = \phi(e)$, $e' \in \text{Im } \phi$. If $a', b' \in \text{Im } \phi$, then $a' = \phi(a)$, $b' = \phi(b)$ for some $a, b \in G$, and $a'(b')^{-1} = \phi(a)[\phi(b)]^{-1} = \phi(ab^{-1}) \in \text{Im } \phi$. Thus Im $\phi < G'$.

(3) If $b = \phi(a)$, then $x \in \phi^{-1}(b) \Leftrightarrow \phi(x) = b \Leftrightarrow \phi(x) = \phi(a) \Leftrightarrow \phi(x)[\phi(a)]^{-1} = e' \Leftrightarrow \phi(xa^{-1}) = e' \Leftrightarrow xa^{-1} \in \text{Ker } \phi \Leftrightarrow x \in (\text{Ker } \phi)a$. Thus $\phi^{-1}(b) = (\text{Ker } \phi)a$. Since Ker $\phi \lhd G$, $(\text{Ker } \phi)a = a \text{ Ker } \phi$.

(4) For $a_1, a_2 \in G$, $a_1 \text{ Ker } \phi = a_2 \text{ Ker } \phi \Leftrightarrow a_1^{-1}a_2 \in \text{Ker } \phi \Leftrightarrow \phi(a_1^{-1}a_2) = e' \Leftrightarrow [\phi(a_1)]^{-1}\phi(a_2) = e' \Leftrightarrow \phi(a_1) = \phi(a_2)$. Thus, $a \text{ Ker } \phi \mapsto \phi(a)$ defines a 1-1 mapping, ψ, of the set of all cosets of Ker ϕ into Im ϕ. The mapping ψ is onto since each $a' \in \text{Im } \phi$ is equal to $\phi(a) = \psi(a \text{ Ker } \phi)$ for some $a \in G$. ■

Remark: Figure 10 shows how a homomorphism ϕ blurs the distinctions between elements of G belonging to the same coset of Ker ϕ, i.e., all elements belonging to the same coset of Ker ϕ are assigned the same image under ϕ.

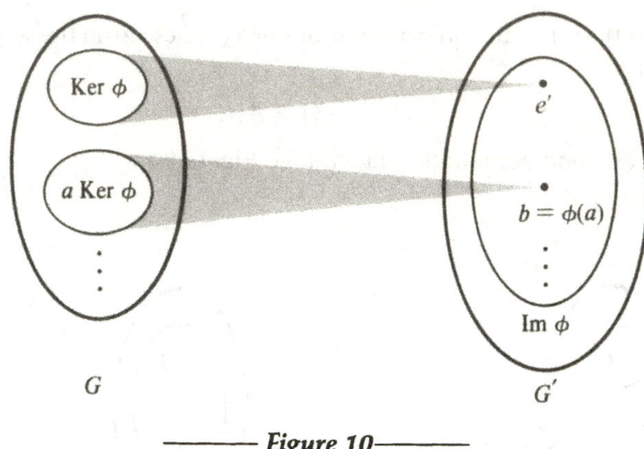

Figure 10

Example 1:

Define $\phi : \mathbb{R}^* \to \mathbb{R}^+$ by $\phi(x) = |x|$, for each $x \in \mathbb{R}^*$. Then $\phi(x \cdot y) = |x \cdot y| = |x| \cdot |y| = \phi(x) \cdot \phi(y)$, for each $x, y \in \mathbb{R}^*$. Hence ϕ is a homomorphism of $\langle \mathbb{R}^*, \cdot \rangle$ into $\langle \mathbb{R}^+, \cdot \rangle$, clearly onto, hence it is an epimorphism. Ker $\phi = \{x \in \mathbb{R}^* \mid |x| = 1\} = \{-1, 1\}$. (See Figure 11.)

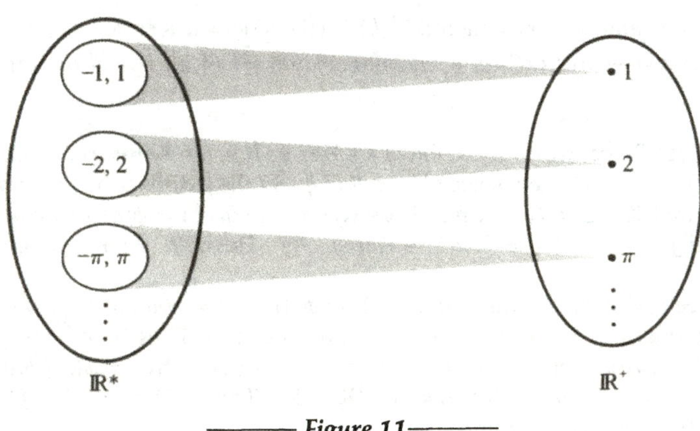

Figure 11

Example 2:

Define $\phi : \mathbb{C}^* \to \mathbb{R}^+$ by $\phi(z) = |z|$. Then $\phi(zw) = |zw| = |z|\,|w| = \phi(z)\phi(w)$ for each $z, w \in \mathbb{C}^*$. Hence ϕ is a homomorphism of $\langle \mathbb{C}^*, \cdot \rangle$ into $\langle \mathbb{R}^+, \cdot \rangle$, clearly onto; hence it is an epimorphism. Ker $\phi = \{z \in \mathbb{C}^* \mid |z| = 1\}$. (See Figure 12.)

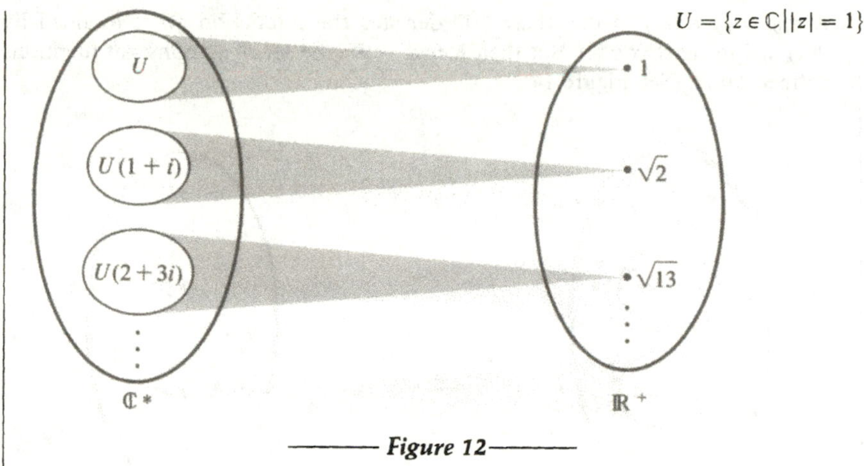

Figure 12

Example 3:

For $n \geq 2$, define $\phi : S_n \to \mathbb{Z}_2$ by: $\phi(\sigma) = \bar{0}$ if σ is even, $\phi(\sigma) = \bar{1}$ if σ is odd. If $\sigma, \tau \in S_n$ are both even, $\sigma\tau$ is even; hence $\phi(\sigma\tau) = \bar{0} = \bar{0} \oplus_2 \bar{0} = \phi(\sigma) \oplus_2 \phi(\tau)$; if σ is even, τ odd, $\sigma\tau$ is odd and $\phi(\sigma\tau) = \bar{1} = \bar{0} \oplus_2 \bar{1} = \phi(\sigma) \oplus_2 \phi(\tau)$; similarly, if σ is odd, τ even, $\phi(\sigma\tau) = \phi(\sigma) \oplus_2 \phi(\tau)$. Finally, if σ, τ are both odd, then $\sigma\tau$ is even; hence $\phi(\sigma\tau) = \bar{0} = \bar{1} \oplus_2 \bar{1} = \phi(\sigma) \oplus_2 \phi(\tau)$. Thus ϕ is a homomorphism, clearly onto; hence it is an epimorphism. Ker $\phi = \{\sigma \in S_n \mid \phi(\sigma) = \bar{0}\} = A_n$, the subgroup of S_n consisting of all even permutations. (See Figure 13.)

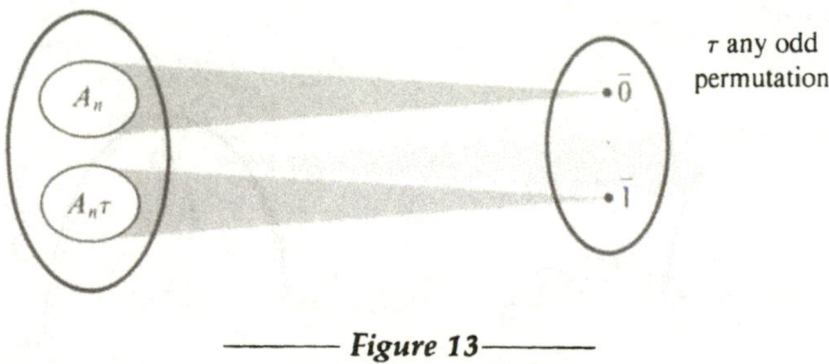

Figure 13

Example 4:

Let G be the additive group of all real-valued functions, defined and differentiable on $\mathbb{R}$, and let G' be the additive group of all real-valued functions defined on $\mathbb{R}$. Define $\phi : G \to G'$ by $\phi(f) = f'$ for each $f \in G$. Since $\phi(f + g) = (f + g)' = f' + g' = \phi(f) + \phi(g)$, for each $f, g \in G$, ϕ is a homomorphism of G into G'.

Ker $\phi = \{f \in G | f' = 0\}$, where "0" denotes the 0-function on $\mathbb{R}$ (defined by $0(x) = 0$ for each $x \in \mathbb{R}$). But then Ker $\phi = K$, the set of all constant functions defined on $\mathbb{R}$. (See Figure 14.)

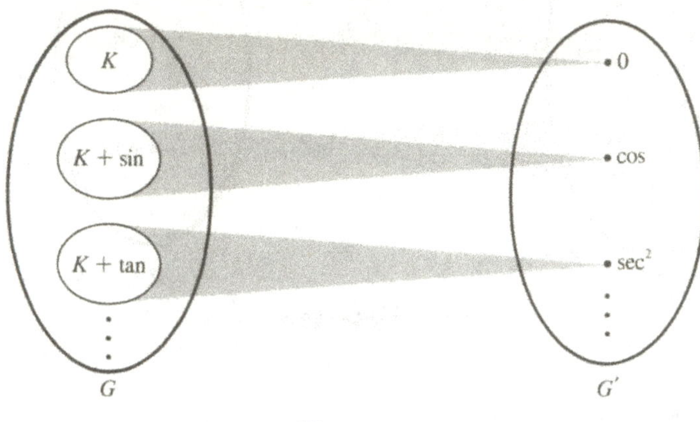

Figure 14

Example 5:

For $n \geq 1$, let $G = GL_n(\mathbb{R})$. Define $\phi : G \to \mathbb{R}^*$ by $\phi(A) = \det A$, for each $A \in G$. Since, for each $A, B \in G$, $\phi(AB) = \det(AB) = \det A \det B = \phi(A)\phi(B)$, ϕ is a homomorphism of G into $\mathbb{R}^*$. Ker $\phi = \{A \in G | \det A = 1\} = SL_n(\mathbb{R})$. (See Figure 15.) Since, for each $a \in \mathbb{R}^*$, we have $a = \phi\begin{pmatrix} a & 0 \\ 0 & 1 \end{pmatrix}$, ϕ is an epimorphism.

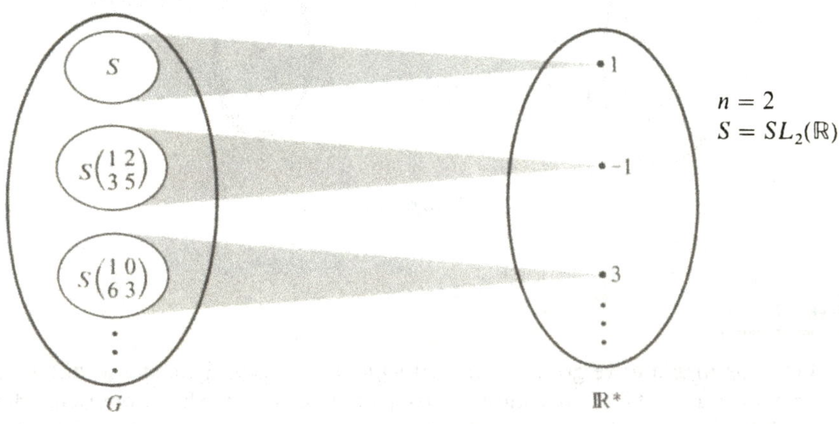

Figure 15

Example 6:

For $n \in \mathbb{Z}^+$, define $\phi : \mathbb{Z} \to \mathbb{Z}_n$ by: $\phi(h) = \bar{h}$ for each $h \in \mathbb{Z}$. Since $\phi(h + k) = \overline{h + k} = \bar{h} \oplus_n \bar{k} = \phi(h) \oplus_n \phi(k)$ for all $h, k \in \mathbb{Z}$, ϕ is a homomorphism, clearly onto, by the definition of $\mathbb{Z}_n$.

$$\text{Ker } \phi = \{h \in \mathbb{Z} \mid \phi(h) = \bar{0}\} = \{h \in \mathbb{Z} \mid \bar{h} = \bar{0}\} = \{h \in \mathbb{Z} \mid h \equiv 0 \text{ mod } n\} = n\mathbb{Z}.$$

But $n\mathbb{Z} = \bar{0}$, and the cosets of $n\mathbb{Z}$ in $\mathbb{Z}$ are the sets $n\mathbb{Z} + h = \bar{h}$ $(h \in \mathbb{Z})$. *Thus, in this case, the cosets of $n\mathbb{Z}$ themselves form a group, $\mathbb{Z}_n$, which is the image of $\mathbb{Z}$ under a homomorphism with kernel $n\mathbb{Z}$.* (See Figure 16.)

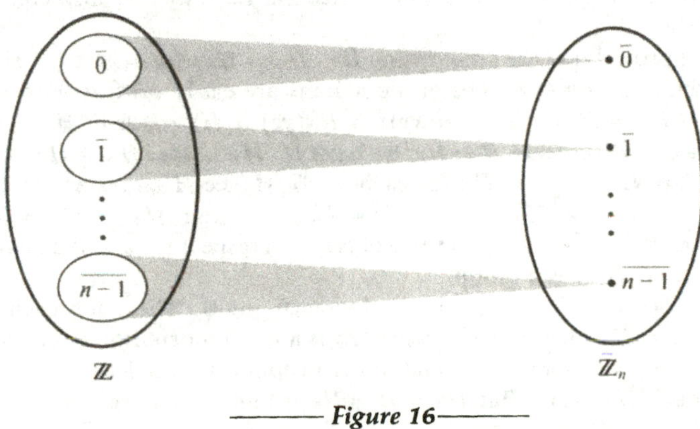

Figure 16

Our immediate aim is to explore the possibility of generalizing Example 6. We have seen that the kernel of every homomorphism with domain G is a normal subgroup of G. Is the converse true, i.e., is every normal subgroup of a group G the kernel of some homomorphism with domain G? The following theorem implies that the answer is "yes."

First recall how $\mathbb{Z}_n$, the set of all cosets of $n\mathbb{Z}$ in $\mathbb{Z}$, was made into a group. We simply defined *the sum of the cosets* $n\mathbb{Z} + x = \bar{x}$ and $n\mathbb{Z} + y = \bar{y}$ to be *the coset of the sum*, i.e., the coset $n\mathbb{Z} + (x + y) = \overline{x + y}$.

Translated into multiplicative language, for H a subgroup of a group $\langle G, \cdot \rangle$, this would suggest "defining" *the product of two cosets* Hx and Hy to be *the coset*, Hxy, *of the product* xy. We shall see that this "definition" works precisely when H is a *normal* subgroup of G.

Theorem 2.9.3: Let $\langle G, \cdot \rangle$ be a group, and let $H \lhd G$. Let G/H be the set of all cosets of H in G. For each $x, y \in G$, define $\cdot$ by: $Hx \cdot Hy = Hxy$. Then $\cdot$ is a binary operation on G/H such that $\langle G/H, \cdot \rangle$ is a group. The mapping ν defined by $\nu : G \to G/H$ such that $\nu(x) = Hx$ for each $x \in G$ is an epimorphism with kernel H.

Proof: First note that, since $H \lhd G$, we have $Hx = xH$ for each $x \in G$. Hence the term *cosets* in the statement of the theorem needs no modifying adjective *left* or *right*. However, we shall use right coset notation to designate the cosets.

For each $x, y \in G$, $Hxy \in G/H$; hence Hx and Hy *have* a "product" in G/H. It is incumbent upon us to show that the product of the two cosets Hx and Hy is independent of the particular representation of these cosets. Indeed, suppose $Hx = Hx'$ and $Hy = Hy'$ $(x, x', y, y' \in G)$. Is $Hxy = Hx'y'$?

From $Hx = Hx'$, $Hy = Hy'$, by Corollary 1, Theorem 2.5.1, we have $x(x')^{-1} \in H$ and $y(y')^{-1} \in H$. We need to determine whether $xy(x'y')^{-1} \in H$.

Now, $xy(x'y')^{-1} = xy(y')^{-1}(x')^{-1}$, and we know that $y(y')^{-1} = h$ for some $h \in H$; hence $xy(y')^{-1}(x')^{-1} = xh(x')^{-1}$. Since $H \lhd G$, $xH = Hx$. Hence there is some $\bar{h} \in H$ such that $xh = \bar{h}x$. But then $xy(x'y')^{-1} = xh(x')^{-1} = \bar{h}x(x')^{-1} \in H$ since $x(x')^{-1} \in H$. Again by Corollary 1, Theorem 2.5.1, we may now conclude that $Hxy = Hx'y'$.

We have proved that the prescription $Hx \cdot Hy = Hxy$ for each $x, y \in G$ defines a binary operation $\cdot$ on G/H. The group axioms are easily verified. For $x, y, z \in G$, $(Hx \cdot Hy) \cdot Hz = Hxy \cdot Hz = H(xy)z = Hx(yz) = Hx \cdot Hyz = Hx \cdot (Hy \cdot Hz)$. Thus, $\cdot$ is associative. Since $H = He$, we have $H \cdot Hx = He \cdot Hx = Hex = Hx$ and $Hx \cdot H = Hx \cdot He = Hxe = Hx$ for each $x \in G$. Hence H serves as identity for $\cdot$. For each $x \in G$, $Hx \cdot Hx^{-1} = Hxx^{-1} = He = H$, and $Hx^{-1} \cdot Hx = Hx^{-1}x = He = H$; hence Hx^{-1} serves as inverse of Hx with respect to $\cdot$ and the identity H. It follows that $\langle G/H, \cdot \rangle$ is a group.

Define $\nu : G \to G/H$ by: $\nu(x) = Hx$ for each $x \in G$. Then, for each $x, y \in G$, $\nu(xy) = Hxy = Hx \cdot Hy = \nu x \cdot \nu y$, whence ν is a homomorphism. Every element of G/H is $Hx = \nu(x)$ for some $x \in G$, and so ν is an epimorphism. Ker $\nu = \{x \in G \mid \nu x = e_{G/H}\} = \{x \in G \mid Hx = H\}$. But $Hx = H = He$ if and only if $xe^{-1} = x \in H$. Thus, Ker $\nu = H$. ■

As an immediate consequence of Theorem 2.9.3, we obtain one of the most fundamental characterizations of normality:

Corollary: *A subgroup H of a group G is normal if and only if it is the kernel of some homomorphism with domain G.*

Definition 2.9.3: If $H \lhd G$, then the group $\langle G/H, \cdot \rangle$ (where $G/H = \{Hx \mid x \in G\}$ and $\cdot$ is defined by: $Hx \cdot Hy = Hxy$ for each $x, y \in G$) is called the *factor group* (or *quotient group*) *of G modulo H*. The epimorphism $\nu : G \to G/H$ defined by $\nu(x) = Hx$ for each $x \in G$ is *the natural* (or *the canonical*) *homomorphism of G modulo H*.

Note: We have yet to make good on our earlier remark that the definition of coset multiplication "works" *precisely when* $H \lhd G$. Suppose $H < G$, but $H \ntriangleleft G$. Then there is some $x \in G$ and some $h \in H$ such that $x^{-1}hx \notin H$. Now, $Hx = Hhx$ (by Corollary 1, Theorem 2.5.1); hence, (if $\cdot$ were indeed a binary operation on the set of all right cosets), we should have $Hx^{-1} \cdot Hx = Hx^{-1} \cdot Hhx$, or, in other words, $Hx^{-1}x = Hx^{-1}hx$. But $Hx^{-1}x = H$, while $Hx^{-1}hx$ is *not* equal to H since its element $ex^{-1}hx = x^{-1}hx \notin H$, by hypothesis. Thus if H is a non-normal subgroup of G, the prescription $Hx \cdot Hy = Hxy$ for combining two right cosets does *not* define a binary operation on the set of all right cosets of H in G (nor does the analogous prescription $xH \cdot yH = xyH$ define a binary operation on the set of all left cosets of H in G, as you may easily verify).

We thus have yet another characterization of normality:

Corollary: If $H < G$, then $H \lhd G$ if and only if "$Hx \cdot Hy = Hxy$" defines a binary operation on the set of all right (or left) cosets of H in G.

Theorem 2.9.3 shows that the observations we made with regard to Example 6 may now be extended to arbitrary normal subgroups.

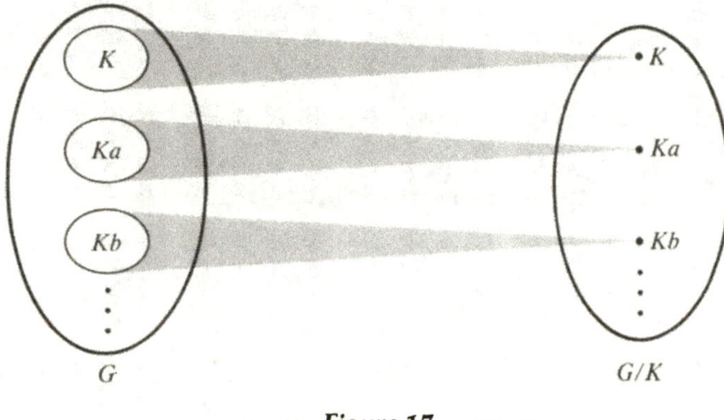

Figure 17

The canonical epimorphism $\nu : G \to G/K$ blurs the distinctions between elements belonging to the same coset of K; it maps all elements of a given coset X of K to X, now regarded as a single element of the factor group G/K. See Figure 17.

The visualization of factor groups may be aided also by the use of Cayley tables. Consider, for example, the subgroup A_3 of S_3. In the table for S_3, if we blur the distinctions between elements belonging to the same coset of A_3, we obtain the factor group S_3/A_3.

S_3

	A	E	J	N	S	W
A	A	E	J	N	S	W
E	E	J	A	W	N	S
J	J	A	E	S	W	N
N	N	S	W	A	E	J
S	S	W	N	J	A	E
W	W	N	S	E	J	A

$\longrightarrow$

S_3/A_3

	A_3	A_3N
A_3	A_3	A_3N
A_3	A_3N	A_3

$A_3 = \{A, E, J\}$

$A_3N = \{N, S, W\}$

In the table for S_4 as printed on p. 72, the elements of S_4 are arranged in batches of four: the cosets of $V_4 = \{\iota, (12)(34), (13)(24), (14)(23)\} = \{A, B, C, D\}$, a normal subgroup of S_4. Blurring of each of the cosets V_4A, V_4E, V_4J, V_4N, V_4S, V_4W yields the factor group S_4/V_4 (plainly isomorphic to S_3):

Cayley Table for S_4

	A	B	C	D	E	F	G	H	J	K	L	M	N	O	P	R	S	T	U	V	W	X	Y	Z
A	A	B	C	D	E	F	G	H	J	K	L	M	N	O	P	R	S	T	U	V	W	X	Y	Z
B	B	A	D	C	G	H	E	F	M	L	K	J	O	N	R	P	U	V	S	T	Z	Y	X	W
C	C	D	A	B	H	G	F	E	K	J	M	L	R	P	O	N	T	S	V	U	Y	Z	W	X
D	D	C	B	A	F	E	G	H	L	M	J	K	P	R	N	O	V	U	T	S	X	W	Z	Y
E	E	F	G	H	J	K	L	M	A	B	C	D	W	X	Y	Z	N	O	P	R	S	T	U	V
F	F	E	H	G	L	M	J	K	D	C	B	A	X	W	Z	Y	P	R	N	O	V	U	T	S
G	G	H	E	F	M	L	K	J	B	A	D	C	Z	Y	X	W	O	N	R	P	U	V	S	T
H	H	G	F	E	K	J	M	L	C	D	A	B	Y	Z	W	X	R	P	O	N	T	S	V	U
J	J	K	L	M	A	B	C	D	E	F	G	H	S	T	U	V	W	X	Y	Z	N	O	P	R
K	K	J	M	L	C	D	A	B	H	G	F	E	T	S	V	U	Y	Z	W	X	R	P	O	N
L	L	M	J	K	D	C	B	A	F	E	H	G	V	U	T	S	X	W	Z	Y	P	R	N	O
M	M	L	K	J	B	A	D	C	G	H	E	F	U	V	S	T	Z	Y	X	W	O	N	R	P
N	N	O	P	R	S	T	U	V	W	X	Y	Z	A	B	C	D	E	F	G	H	J	K	L	M
O	O	N	R	P	U	V	S	T	Z	Y	X	W	B	A	D	C	G	H	E	F	M	L	K	J
P	P	R	N	O	V	U	T	S	X	W	Z	Y	D	C	B	A	F	E	H	G	L	M	J	K
R	R	P	O	N	T	S	V	U	Y	Z	W	X	C	D	A	B	H	G	F	E	K	J	M	L
S	S	T	U	V	W	X	Y	Z	N	O	P	R	J	K	L	M	A	B	C	D	E	F	G	H
T	T	S	V	U	Y	Z	W	X	R	P	O	N	K	J	M	L	C	D	A	B	H	G	F	E
U	U	V	S	T	Z	Y	X	W	O	N	R	P	M	L	K	J	B	A	D	C	G	H	E	F
V	V	U	T	S	X	W	Z	Y	P	R	N	O	L	M	J	K	D	C	B	A	F	E	H	G
W	W	X	Y	Z	N	O	P	R	S	T	U	V	E	F	G	H	J	K	L	M	A	B	C	D
X	X	W	Z	Y	P	R	N	O	V	U	T	S	F	E	H	G	L	M	J	K	D	C	B	A
Y	Y	Z	W	X	R	P	O	N	T	S	V	U	H	G	F	E	K	J	M	L	C	D	A	B
Z	Z	Y	X	W	O	N	R	P	U	V	S	T	G	H	E	F	M	L	K	J	B	A	D	C

Cayley Table for S_4/V_4

	V_4A	V_4E	V_4J	V_4N	V_4S	V_4W
V_4A	V_4A	V_4E	V_4J	V_4N	V_4S	V_4W
V_4E	V_4E	V_4J	V_4A	V_4W	V_4N	V_4S
V_4J	V_4J	V_4A	V_4E	V_4S	V_4W	V_4N
V_4N	V_4N	V_4S	V_4W	V_4A	V_4E	V_4J
V_4S	V_4S	V_4W	V_4N	V_4J	V_4A	V_4E
V_4W	V_4W	V_4N	V_4S	V_4E	V_4J	V_4A

Now let $\phi : G \to G'$ be a group homomorphism. Since Ker $\phi \lhd G$, we can define the factor group $G/\text{Ker } \phi$ and the canonical epimorphism $\nu : G \to G/\text{Ker } \phi$. By Theorem 2.9.2, part (4), the mapping $\psi : G/\text{Ker } \phi \to \text{Im } \phi$ is a bijection. We claim that ψ is an isomorphism.

Theorem 2.9.4. Fundamental Theorem of Homomorphism for Groups: If $\phi : G \to G'$ is a group homomorphism, then $G/\text{Ker } \phi \cong \text{Im } \phi$.

Proof: By Theorem 2.9.2, part (4), if $\psi : G/\text{Ker } \phi \to \text{Im } \phi$ is defined by $\psi(a \text{ Ker } \phi) = \phi(a)$ for each $a \in G$, then ψ is a bijection.

For each $a, b \in G$, $\psi(KaKb) = \psi(Kab) = \phi(ab) = \phi(a)\phi(b) = \psi(Ka)\psi(Kb)$. Hence ψ is an isomorphism of G/K onto Im ϕ, and we have $\text{Im } \phi \cong G/\text{Ker } \phi$. ■

Corollary: If $\phi : G \to G'$ is a group epimorphism, with $K = \text{Ker } \phi$, and if $\nu : G \to G/K$ is the natural homomorphism of G modulo K, then there exists a unique isomorphism $\psi : G/K \to G'$ such that $\psi\nu = \phi$.

Proof: Here, Im $\phi = G'$, and so the isomorphism ψ of Theorem 2.8.4 maps G/K *onto* G'. For each $a \in G$, we have $(\psi\nu)(a) = \psi(\nu a) = \psi(Ka) = \phi(a)$, whence $\psi\nu = \phi$. If $\psi' : G/K \to G'$ is another isomorphism such that $\psi'\nu = \phi$, then, for each $a \in G$, $(\psi\nu)(a) = (\psi'\nu)(a)$; hence $\psi(\nu a) = \psi'(\nu a)$, for each $\nu a \in G/K$. But γ is onto, and so ψ and ψ' agree on their domain G/K, i.e., $\psi = \psi'$. ■

We represent this result by a "*commutative diagram*" in Figure 18,

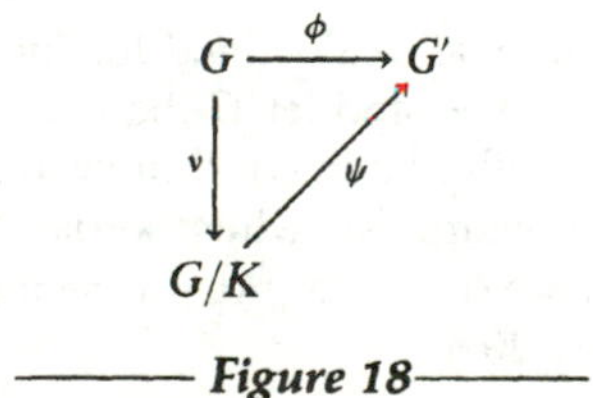

Figure 18

so called because it indicates two alternate routes to travel from G to G': directly via ϕ or indirectly via ν and ψ. We say that ϕ has been "factored through" the factor group G/K.

Notation: If $\alpha : X \to Y$ and $\alpha(x) = y (x \in X, y \in Y)$, we write $x \overset{\alpha}{\mapsto} y$. (Arrows between elements have "feathers"; arrows between sets have none.)

Example 1:

Let $G = [a]$ be a cyclic group of finite order n. Define $\phi : \mathbb{Z} \to G$ by $\phi(k) = a^k$, for each $k \in \mathbb{Z}$. Then ϕ is an epimorphism, with $\text{Ker } \phi = \{k \in \mathbb{Z} \mid a^k = e\} = n\mathbb{Z}$. Hence $G \cong \mathbb{Z}/n\mathbb{Z} = \bar{\mathbb{Z}}_n$. (This is the promised more elegant proof of Theorem 4.1.2.)

Example 2:

Let $\phi : \mathbb{R}^* \to \mathbb{R}^+$ be defined by $\phi(x) = |x|$, for each $x \in \mathbb{R}^*$. Then ϕ is an epimorphism with kernel $K = \{-1, 1\}$. Hence $\mathbb{R}^+ \cong \mathbb{R}^*/\{-1, 1\}$.

The elements of the factor group $\mathbb{R}^*/\{-1, 1\}$ are the sets $\{-x, x\}$ $(x \in \mathbb{R}^*)$, combined under the operation $\{-x, x\}\cdot\{-y, y\} = \{-xy, xy\}$ $(x, y \in \mathbb{R}^*)$. The isomorphism $\psi : \mathbb{R}^*/\{-1, 1\} \to \mathbb{R}^+$ sends each set $\{-x, x\}$ to $|x|$ $(x \in \mathbb{R}^*)$. Note that $\{-x, x\} = \phi^{-1}(|x|)$, consisting of all (i.e., both) real numbers which have the same absolute value as x.

Figure 19 indicates the fate of each element in the domain of the maps ϕ, ν and ψ. (See the Corollary of Theorem 2.9.4.)

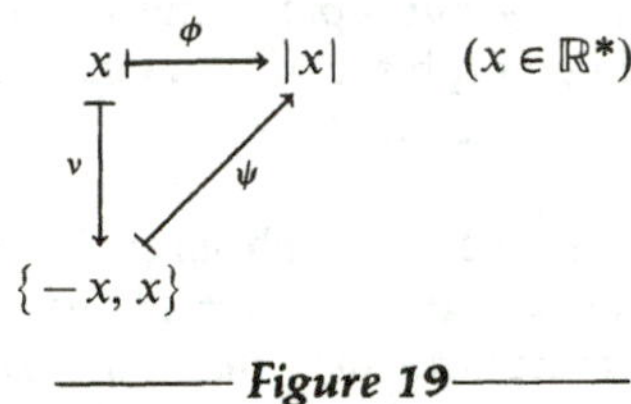

Figure 19

Example 3:

Let G be the additive group of all real-valued functions which are defined and continuously differentiable on $\mathbb{R}$, and let G' be the additive group of all real-valued functions which are defined and continuous on $\mathbb{R}$. Define $\phi : G \to G'$ by: $\phi(f) = f'$. Then ϕ is a homomorphism whose kernel K is the set of all constant functions defined on $\mathbb{R}$. Hence $G' \cong G/K$. Again, we indicate the fate of elements under ϕ, ν, and ψ in Figure 20.

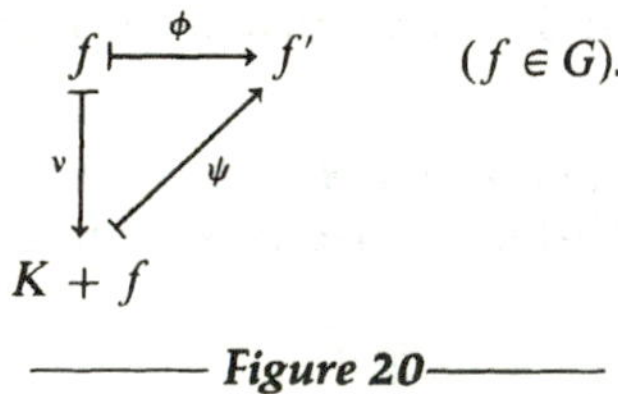

Figure 20

Note that $K + f$ is usually denoted by "$\int f'(x)\,dx$." Thus, an indefinite integral is (merely?) a coset of the subgroup of all constant functions.

Example 4:

For $n \geq 1$, define $\phi : GL_n(\mathbb{R}) \to \mathbb{R}^*$ by: $\phi(A) = \det A$, for each $A \in GL_n(\mathbb{R})$. Then ϕ is an epimorphism, with $\text{Ker } \phi = \{A \in GL_n(\mathbb{R}) | \det A = 1\} = SL_n(\mathbb{R})$. Hence $\mathbb{R}^* \cong GL_n(\mathbb{R})/SL_n(\mathbb{R})$. The elementwise diagram is shown in Figure 21.

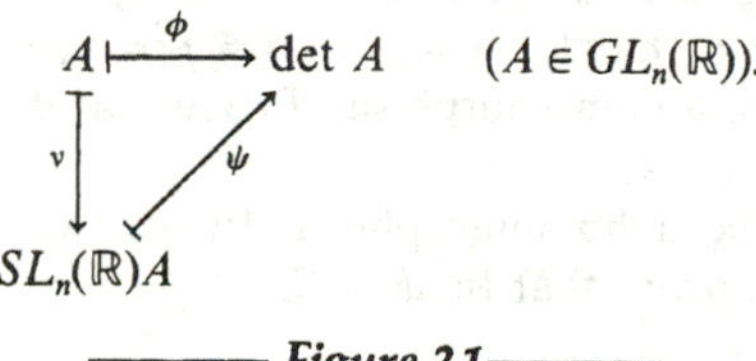

Figure 21

Note that, for each $A \in GL_n(\mathbb{R})$, the coset $SL_n(\mathbb{R})A$ is simply the set of all those matrices in $GL_n(\mathbb{R})$ which have the same determinant as A. For, by Theorem 2.9.2,

$$SL_n(\mathbb{R})A = \phi^{-1}(\phi(A)) = \phi^{-1}(\det A).$$

Remarks: This section has provided several major encounters with ideas of abstract algebra: with quotient structures, homomorphisms, and their kernels. We have generalized the act of abstraction involved in passing from $\mathbb{Z}$ to the groups $\bar{\mathbb{Z}}_n$ of residue classes modulo a positive integer n (see Theorem 2.2.2). For $H \lhd G$, we *abstract from* the differences between elements congruent to each other modulo H (i.e., between elements belonging to the same coset of H in G). *Each coset of H thus becomes a single element of the factor group G/H.* By defining the product of the cosets Ha and Hb of two elements $a, b \in G$ to be the coset, Hab, of the product ab, we ensure that the factor group G/H will be a homomorphic image of G under the homomorphism ν: $a \mapsto Ha$. Conversely, every homomorphic image of G under a homomorphism ϕ is isomorphic to the factor group $G/\text{Ker } \phi$.

Every homomorphism ϕ of a group G illuminates certain facets of the structure of G itself, since $\text{Im } \phi$ retains those properties of G that are independent of the differences between elements congruent to each other modulo $\text{Ker } \phi$.

Exercises 2.9

True or False

1. Every homomorphism is an isomorphism.
2. If $H \lhd G$, then the elements of the factor group G/H are elements of G.
3. If $\phi : G \to \bar{G}$ is a monomorphism, then $\text{Im } \phi$ is isomorphic to G.
4. A group of order 70 cannot have a homomorphic image of order 15.
5. For each $n \geq 2$, the group $\{-1, 1\} < \mathbb{Q}^*$ is a homomorphic image of S_n.

2.9.1. In each of the following examples, compute $Ha \circ Hb \in G/H$, where $a, b \in G$ and $\circ$ is the appropriate operation for the factor group. In each case, express the product coset Hc in a simple form by using a well-chosen representative element c:

(1) $G = \mathbb{R}^*, \quad H = \{-1, 1\}, \quad a = -2, \quad b = \pi;$

(2) $G = GL_2(\mathbb{R}), \quad H = SL_2(\mathbb{R}), \quad a = \begin{pmatrix} 2 & 3 \\ 1 & 4 \end{pmatrix}, \quad b = \begin{pmatrix} 1 & 0 \\ 2 & 1 \end{pmatrix};$

(3) $G = \mathbb{Z}, \quad H = 8\mathbb{Z}, \quad a = 13, \quad b = -185:$

(4) $G = D_4, \quad H = \{\iota, \rho^2\}, \quad a = \rho\sigma, \quad b = \rho^3\sigma$ (see the table on p. 80).

2.9.2. Let $\phi : G \to G'$ be a homomorphism. Prove that ϕ is a monomorphism if and only if $\text{Ker } \phi = \{e\}$.

2.9.3. Let $\phi : G \to G'$ be a homomorphism. Prove that $\text{Im } \phi < G'$. If ϕ is a monomorphism, prove that $\text{Im } \phi \cong G$.

2.9.4. On the vector space $\mathbb{R}^2$ consisting of all ordered pairs $\begin{pmatrix} x \\ y \end{pmatrix}$ of real numbers, define $\phi : \mathbb{R}^2 \to \mathbb{R}^2$ by: $\phi\begin{pmatrix} x \\ y \end{pmatrix} = \begin{pmatrix} x + 2y \\ 0 \end{pmatrix}$ $(x, y \in \mathbb{R})$. Prove that ϕ is a homomorphism of the additive group $\langle \mathbb{R}^2, + \rangle$ into itself. Find $\text{Ker } \phi$ and $\text{Im } \phi$. Define explicitly the isomorphism ψ guaranteed to exist by the Fundamental Theorem of Homomorphism.

2.9.5. Let $\phi : G \to G'$ be a homomorphism. Prove: if G is a finite group, then, for each $a \in G$, $o(\phi(a)) | o(a)$.

2.9.6. A group G is called a *torsion group* if each of its elements has finite order. A group G, all of whose non-identity elements have infinite order, is *torsion free*.

Let G be an abelian group, and let T be its torsion subgroup (recall Exercise 2.3.6).

(1) Prove that G/T is torsion free.

(2) If $\phi : G \to G'$ is a homomorphism such that $\text{Im } \phi$ is torsion free, prove that $T \subset \text{Ker } \phi$.

2.9.7. In a group G, let $Q = \{x^2 | x \in G\}$, and let $H = [Q]$, the subgroup of G generated by the subset Q. Prove:

(1) $H \lhd G$.

(2) Every non-trivial element G/H has order 2.

(3) If $\phi : G \to G'$ is a homomorphism such that every non-trivial element of $\text{Im } \phi$ has order 2, then $H \subset \text{Ker } \phi$.

2.9.8. For p any prime, G a group, let $Q_p = \{x^p | x \in G\}$ and let $H = [Q_p]$. Generalize the results of the preceding exercise to this case.

2.9.9. Let $\mathbb{R}$ be the additive group of all real numbers and let $M_2(\mathbb{R})$ be the additive group of all 2×2 matrices over $\mathbb{R}$. Define $\phi : M_2(\mathbb{R}) \to \mathbb{R}$ by $\phi\begin{pmatrix} a & b \\ c & d \end{pmatrix} = a + d$ $(a, b, c, d \in \mathbb{R})$. Prove that ϕ is an epimorphism. Find $\text{Ker } \phi$. (For $A \in M_2(\mathbb{R})$, $\phi(A)$ is called the *trace* of A.)

Prove that, for each $A, B \in M_2(\mathbb{R})$, $B \in \text{Ker } \phi + A$ if and only if A and B have the same trace. Prove that $\mathbb{R} \cong M_2(\mathbb{R})/K$, where K consists of all matrices in $M_2(\mathbb{R})$ whose trace is equal to zero.

2.9.10. Let G be a cyclic group of finite order n. Prove that, for each divisor m of n, G has exactly one subgroup H such that G/H is isomorphic to $\mathbb{Z}_m$.

2.9.11. In the additive group $\langle \mathbb{R}, + \rangle$, let $[\pi]$ be the subgroup generated by π. Give a graphic representation of the factor group $\mathbb{R}/[\pi]$. Prove: for $x, y \in \mathbb{R}$, $\tan x = \tan y$ if and only if $x \equiv y \bmod [\pi]$.

2.9.12. Let p be a prime, and let $\mathbb{Q}^*$ be the multiplicative group of all non-zero rational numbers. Let

$$G = \{x/y \mid p \nmid x, p \nmid y, \ x, y \in \mathbb{Z} \quad (\gcd(x, y) = 1)\}.$$

Prove that $G < \mathbb{Q}^*$. Define an epimorphism of $\mathbb{Q}^*$ onto G to prove that $G \cong \mathbb{Q}^*/[p]$.

2.9.13. Let A be an $n \times n$ matrix over $\mathbb{R}$, and let $\mathbb{R}^n$ be the vector space consisting of all ordered n-tuples of real numbers. Define $\phi_A : \mathbb{R}^n \to \mathbb{R}^n$ by $\phi_A(x) = Ax$ for each $x \in \mathbb{R}^n$.

(1) Prove that ϕ_A is an endomorphism of $\langle \mathbb{R}^n, + \rangle$.

(2) Prove that Ker ϕ_A is the nullspace of matrix A.

(Note: ϕ_A is *more* than a group homomorphism since it preserves not only addition but also scalar multiplication; this makes it a vector space homomorphism, usually referred to as a *linear transformation.*)

2.9.14. Let $A \in M_n(\mathbb{R})$ and let λ be an eigenvalue of A. Define an endomorphism ϕ of $\langle \mathbb{R}^n, + \rangle$ such that Ker ϕ is the eigenspace of A belonging to the eigenvalue λ.

2.10

Further Isomorphism Theorems; Simple Groups

The theorems that follow illustrate further the concept of factor groups and the Fundamental Theorem of Homomorphism, and are of interest in their own right.

The first theorem illustrates the quotient-like behavior of factor groups (also called "quotient groups").

Theorem 2.10.1: Let G be a group, $H \lhd G$ and $K \lhd G$, with $K \subset H$. Then $H/K \lhd G/K$, and $G/K/H/K \cong G/H$.

Proof: From $K \subset H$ and $K \lhd G$, we have $K \lhd H$ (see Exercise 2.8.10b); hence the factor group H/K referred to in the statement of the theorem makes sense.

"Define " $\phi : G/K \to G/H$ by $\phi(Ka) = Ha$, for each $a \in G$. Is ϕ a (well-defined) map? Clearly ϕ assigns an image in G/H to each element of G/K. We must show that the image under ϕ of a coset in G/K is independent of the representation of the coset. Thus, suppose for $a, a' \in G$, we have $Ka = Ka'$. By Corollary 1 of Theorem 2.5.1, we have $Ka = Ka' \Leftrightarrow a(a')^{-1} \in K$. Since $K \subset H$, this implies that $a(a')^{-1} \in H$, and so $Ha = Ha'$. We conclude that the prescription $\phi(Ka) = Ha$ $(a \in G)$ does define a map from G/K to G/H. The map ϕ is onto since each element of G/H is

$Ha = \phi(Ka)$ for some $a \in G$. If $a, b \in G$, then $\phi(Ka \cdot Kb) = \phi(Kab) = Hab = Ha \cdot Hb = \phi(Ka) \cdot \phi(Kb)$. Thus, ϕ is a homomorphism of G/K onto G/H. What is Ker ϕ?

$$\text{Ker}\ \phi = \{Ka \mid a \in G, \quad \phi(Ka) = e_{G/H}\} = \{Ka \mid a \in G, \quad Ha = H\} = \{Ka \mid a \in H\} = H/K.$$

But then, by the Fundamental Theorem of Homomorphism, with Im $\phi = G/H$ and Ker $\phi = H/K$, we have $H/K \lhd G/K$ and $G/K/H/K \cong G/H$. ■

Theorem 2.10.2: Let G be a group, $H < G$ and $K \lhd G$. Define HK by: $HK = \{hk \mid h \in H, \quad k \in K\}$. Then $HK < G$, $K \lhd HK$, $H \cap K \lhd H$, and $HK/K \cong H/H \cap K$.

Proof: The normality of K in G plays an essential role in proving that $HK = \{hk \mid h \in H, \quad k \in K\}$ is a subgroup of G. Of course, $e \in HK$ since $e = e \cdot e$, and $e \in H$, $e \in K$. Let $h_1, h_2 \in H$, $k_1, k_2 \in K$. Is $(h_1k_1)(h_2k_2) \in HK$? Since $K \lhd G$, $Kh_2 = h_2K$; hence $k_1h_2 = h_2\bar{k}$ for some $\bar{k} \in K$. Thus $h_1k_1h_2k_2 = h_1h_2\bar{k}k_2 \in HK$, and so HK is closed under the group operation.

For $h \in H, k \in K$, is $(hk)^{-1} \in HK$? We have $(hk)^{-1} = k^{-1}h^{-1}$. By the normality of K in G, $Kh^{-1} = h^{-1}K$; hence there is some $\bar{k} \in K$ such that $k^{-1}h^{-1} = h^{-1}\bar{k}$. But then $(hk)^{-1} = h^{-1}\bar{k} \in HK$. It follows that $HK < G$. Since $K = \{ek \mid k \in K\}$, $K \subset HK$; hence, being a normal subgroup of G, K is also a normal subgroup of HK (see Exercise 2.8.10).

Define $\phi : H \to HK/K$ by: $\phi(h) = Kh$, for each $h \in H$. ϕ is a (well-defined) map, and a homomorphism since, for $h_1, h_2 \in H$,

$$\phi(h_1h_2) = Kh_1h_2 = Kh_1Kh_2 = \phi(h_1)\phi(h_2).$$

Does ϕ map H *onto* HK/K? For each $h \in H$, $\phi(h) = Kh$ *is* indeed an element of HK/K, since $h \in KH$. We show that *every* element of HK/K is of this form. To start with, any element of HK/K is a coset Khk for some $h \in H$, $k \in K$. But $K \lhd G$, hence $(hk)h^{-1} = (h^{-1})^{-1}kh^{-1} \in K$ (by Theorem 2.8.1); by Corollary 1 of Theorem 2.5.1, we conclude that $Khk = Kh$. Thus, ϕ is an epimorphism.

Ker $\phi = \{h \in H \mid Kh = K\} = \{h \in H \mid h \in K\} = H \cap K$. Hence $H \cap K \lhd H$ and, by Theorem 2.9.4 (or its Corollary), we conclude that $HK/K \cong H/H \cap K$. ■

Theorems 2.9.4, 2.10.1 (or its variant, Theorem 2.14.1), and 2.10.2 are now often referred to as the First, Second, and Third Isomorphism Theorems. But the numbering of "Isomorphism Theorems" appears to be a function of age, and mathematicians brought up with a different numbering have not yet died out.

Definition 2.10.1: A group G is *simple* if its only normal subgroups are G and $\{e\}$.

By Theorem 2.9.4, the only homomorphic images of a simple group are either trivial or isomorphic to G. It is not surprising that such groups have turned out to be hard to describe. A mammoth effort, completed only recently, with the aid of high-speed computers, was required to classify all finite simple groups (see [4]).

Yet the importance of simple groups has been known since group theory began: for, it is the simplicity of the alternating group, A_n, for each $n \geq 5$, that underlies the unsolvability by radicals of the general polynomial equation of degree $n \geq 5$, (recall Section 1.1).

(An easy direct proof of the unsolvability of S_n, hence A_n, for $n \geq 5$, appears in Section 2.14, Theorem 2.14.2. Thus, the following theorem is not needed for purposes of Galois Theory.)

Theorem 2.10.3: For $n \geq 5$, the alternating group A_n is simple.

Proof: We make use of the result in Exercise 2.8.16: for $n \geq 3$, if $K \lhd A_n$ and K contains a 3-cycle, then $K = A_n$. Thus, to prove the theorem, it suffices to show that, for $n \geq 5$, every non-trivial normal subgroup of A_n does indeed contain a 3-cycle.

Let $K \lhd A_n$, $K \neq \{\iota\}$. Choose a permutation $\tau \in K$, $\tau \neq \iota$, such that the fixed set F_τ of τ is as large as possible (among the fixed sets of non-trivial elements of K). We want to show that τ *is* a 3-cycle.

Consider the changed set $\overline{F}_\tau$ of τ (consisting of all elements of $X_n = \{1, 2, \ldots, n\}$ which are *not* mapped to themselves by τ). Clearly, $\overline{F}_\tau$ contains more than one element of X_n (else τ is not 1-1) and, since τ cannot be a transposition (transpositions being odd), $\overline{F}_\tau$ consists of at least three elements.

Suppose $\overline{F}_\tau$ contains *more* than three elements.

Case 1: One of the disjoint cycles, γ_1, of τ has length ≥ 3. Then $\tau = (abc\ldots)\ \gamma_2 \ldots \gamma_s$ $(s \geq 1)$, where a, b, c are distinct elements of $\overline{F}_\tau$. Since $\overline{F}_\tau$ has at least four elements, it has at least five, since τ cannot be a 4-cycle (4-cycles being odd). Let d, e be two more elements in $\overline{F}_\tau$, and let $\sigma = (ced)$. Then

$$\tau_1 = \sigma\tau\sigma^{-1} = \sigma\gamma_1\sigma^{-1}\sigma\gamma_2\sigma^{-1} \ldots \sigma\gamma_s\sigma^{-1} = (abe\ldots)\ldots.$$

Since $\tau \in K$, $\sigma \in A_n$, and $K \lhd A_n$, we have $\tau_1 = \sigma\tau\sigma^{-1} \in K$; hence also $\tau^{-1}\tau_1 \in K$. Clearly, $\tau_1 \neq \tau$ (e.g., $\tau_1 b \neq \tau b$); hence, $\tau^{-1}\tau_1 \neq \iota$. Since σ involves only elements of the changed set of τ, $\tau_1 = \sigma\tau\sigma^{-1}$ fixes all the elements in the fixed set of τ; hence so does $\tau^{-1}\tau_1$. However, $\tau^{-1}\tau_1$ *also* fixes a! This violates the hypothesis that the fixed set of τ is maximal (among the fixed sets of non-trivial elements of K). Contradiction.

Case 2: The disjoint cycles of τ are all transpositions. Since τ is even, the number of disjoint transpositions is at least two; hence there are distinct elements $a, b, c, d \in \overline{F}_\tau$ such that $\tau = (ab)(cd)\ldots$. Let e be any element of X_n different from a, b, c, d, and let $\sigma = (ced)$. Then $\tau_1 = \sigma\tau\sigma^{-1} = (ab)(ce)\ldots \in K$. Again, we have $\tau^{-1}\tau_1 \neq \iota$, with $\tau^{-1}\tau_1$ fixing all, or all but one, of the elements in the fixed set of τ, and *also* fixing a and b, in violation of the maximality of the fixed set of τ. Contradiction.

We conclude that the changed set of τ consists of exactly three elements. But then τ is a 3-cycle. Thus, every non-trivial normal subgroup of A_n contains a 3-cycle, and is therefore equal to A_n. It follows that A_n is simple. ∎

Exercises 2.10

True or False

1. If $H < G$ and $K < G$, then $HK < G$.
2. If $K \lhd G$ and $H < G$, then $G/H \cong G/K/H/K$.
3. For $n \geq 5$, S_5 is a simple group.
4. The alternating groups A_n $(n \geq 5)$ are the only non-abelian simple groups.
5. The center of a non-abelian simple group is trivial.

2.10.1. Prove that the alternating group A_5 is a group of order 60 which has no subgroup of order 30.

2.10.2. For $n \geq 5$, prove that the alternating group A_n is the only normal subgroup of the symmetric group S_n.

2.10.3. Prove that a non-trivial abelian group is simple if and only if it is finite cyclic of prime order.

2.10.4. Let G be a group, $H \lhd G$, $K \lhd G$, with $K \subset H$. Prove: if G/K is abelian, then G/H is abelian.

2.10.5. Use Theorem 2.10.2 to prove that $2\mathbb{Z}/6\mathbb{Z} \cong \mathbb{Z}/3\mathbb{Z}$.

2.10.6. If $K \lhd G$, then K is a *maximal normal subgroup* of G if there is no normal subgroup H of G such that $K \subsetneq H \subsetneq G$. Prove: if K is a normal subgroup of G, then K is a maximal normal subgroup of G if and only if G/K is a simple group.

2.11

Automorphism and Invariant Subgroups

A symmetry is a 1–1 mapping of an object onto itself that preserves some essential features of the object. The symmetries of an object may provide important insights about its structure. In Section 2.7, we studied some groups of symmetries of geometric objects. We now turn our attention to the symmetries of groups. They are called *automorphisms*.

We have already defined an automorphism of a group G to be a 1–1 mapping of G onto itself that preserves the group operation. Trivially, every group *has* automorphisms, for: if G is a group, then the identity map ι_G is clearly an automorphism of G. In the following, we shall see that "most" abelian groups, and all non-abelian groups, have non-trivial automorphisms.

Example:

Let G be an *abelian* group. Define $\psi : G \to G$ by $\psi(a) = a^{-1}$, for each $a \in G$. Then, for each $a, b \in G$, $\psi(ab) = (ab)^{-1} = a^{-1}b^{-1}$ (since G is abelian). Thus, $\psi(ab) = \psi(a) \quad \psi(b)$, and so ψ is a homomorphism. Since each element of G is its own inverse's inverse, ψ is onto. For $a, b \in G$, $\psi(a) = \psi(b) \Rightarrow a^{-1} = b^{-1} \Rightarrow (a^{-1})^{-1} = (b^{-1})^{-1} \Rightarrow a = b$. Thus, ψ is 1–1; hence it is an automorphism of G.

In particular, for the group $\langle \mathbb{Z}, + \rangle$, the mapping $\psi : \mathbb{Z} \to \mathbb{Z}$ defined by $\psi(a) = -a$ for each $a \in \mathbb{Z}$ is an automorphism of $\mathbb{Z}$. It turns out that $\iota_{\mathbb{Z}}$ and ψ are, in fact the *only* automorphisms of $\langle \mathbb{Z}, + \rangle$.

In the special case where G is an abelian group, each of whose non-identity elements has order 2, every element of G is its own inverse, and the mapping ψ collapses to the identity map ι_G. For example, $\bar{\mathbb{Z}}_2$ is such a group. Its only automorphism is $\iota_{\mathbb{Z}_2}$.

A non-abelian group G always has at least one non-trivial automorphism since each element $a \in G$ determines its own "inner" automorphism: *conjugation by a*. If G is non-abelian, at least one of these automorphisms is different from the identity map:

Theorem 2.11.1: Let G be a group, $a \in G$. Define $\gamma_a : G \to G$ by $\gamma_a(x) = axa^{-1}$ for each $x \in G$. Then γ_a is an automorphism of G.

Proof: Let $a \in G$. Then γ_a defined by $\gamma_a(x) = axa^{-1}$ $(x \in G)$ is a (well-defined) map from G to G. If $x, y \in G$, then $\gamma_a(xy) = a\ xy\ a^{-1} = a\ xa^{-1}a\ ya^{-1} = \gamma_a(x)\ \gamma_a(y)$. If $\gamma_a(x) = \gamma_a(y)$, then $axa^{-1} = aya^{-1}$; hence $x = y$, by left and right cancellation in G. γ_a maps G *onto* G, for: if $x \in G$, then $x = a(a^{-1}xa)a^{-1} = \gamma_a(a^{-1}xa)$. Thus, γ_a is an automorphism of G. ■

Definition 2.11.1: If G is a group, $a \in G$, then the automorphism $\gamma_a : G \to G$ defined by $\gamma_a(x) = axa^{-1}$ for each $x \in G$ is called *the inner automorphism of G determined by a*.

The symmetries of a group themselves form groups of symmetries: if G is a group, then the automorphisms of G form a group under composition.

Notation: For G a group, we let A_G be the set of all automorphisms, I_G the set of all inner automorphisms, of G.

We relegate to the exercises (Exercise 2.11.3) the proofs of the following assertions.

For any group G,

(1) $A_G < S_G$ (the group of all bijections on G)

(2) $I_G \lhd A_G$.

(As a test of your abstract algebraic virtuosity, we most highly recommend the proof of the normality in 2.)

Recall (see Example 4, p. 85) the definition of the center $C(G)$ of a group G:

$$C(G) = \{c \in G \mid cx = xc \quad \text{for all} \quad x \in G\}.$$

Clearly, if $c \in C(G)$, then the inner automorphism γ_c collapses to ι_G. This causes all inner automorphisms determined by elements of the same coset of $C(G)$ to collapse to a single inner automorphism. Thus, there are only as many *distinct* inner automorphisms of G as $C(G)$ has cosets in G. All this, and more, is said more formally in the following theorem.

Theorem 2.11.2: Let G be a group with center $C(G)$, and let $\mathscr{I}_G$ be the group of all inner automorphisms of G. Then

$$\mathscr{I}_G \cong G/C(G).$$

Proof: Define $\phi: G \to \mathscr{I}_G$ by: $\phi(a) = \gamma_a$ for each $a \in G$. Then ϕ is a (well-defined) map, clearly onto. Is ϕ a homomorphism? For each $a, b \in G$, we have $\phi(ab) = \gamma_{ab}$. Now, for each $x \in G$,

$$\gamma_{ab}\ (x) = ab\ x(ab)^{-1} = a(bxb^{-1})a^{-1} = \gamma_a\gamma_b(x).$$

But then $\phi(ab) = \gamma_{ab} = \gamma_a\gamma_b = \phi(a)\phi(b)$, and so ϕ is a homomorphism.

$$\operatorname{Ker} \phi = \{a \in G | \gamma_a = \iota_G\} = \{a \in G | axa^{-1} = x \quad \text{for all} \quad x \in G\} = C(G).$$

But then, by the Fundamental Theorem of Homomorphism (Theorem 2.9.4), we have

$$\mathscr{I}_G \cong G/C(G). \qquad \blacksquare$$

Definition 2.11.2: If $\alpha: X \to X$ is a function and $Y \subset X$, then Y is *invariant under* α if $\alpha(y) \in Y$ for each $y \in Y$.

The following theorem provides one more characterization of normality.

Theorem 2.11.3: If $H < G$, then $H \lhd G$ if and only if H is invariant under every inner automorphism of G.

Proof: By Theorem 2.8.1, $H \lhd G \Leftrightarrow aha^{-1} \in H$ for each $a \in G$ and each $h \in H \Leftrightarrow \gamma_a(h) \in H$ for each $a \in G \Leftrightarrow H$ is invariant under each inner automorphism γ_a $(a \in G)$. ■

If a subgroup H of a group G satisfies the stronger condition of invariance under *every* automorphism of G, then H is a *characteristic* subgroup of G. (Of course, every characteristic subgroup is normal.)

An interesting class of such subgroups are the commutator subgroups.

Definition 2.11.3: Let $\langle G, \cdot\rangle$ be a group, and let $a, b \in G$. Then the element $a^{-1}b^{-1}ab$ is the commutator, $[a, b]$, of a and b.

The subgroup *generated* by the set of all commutators in a group G is the *commutator subgroup*, $[G, G]$, of G.

Remark 1: If a and b commute, then $a^{-1}b^{-1}ab = e$. Thus, if G is abelian, then $[G, G] = \{e\}$. In general, for two non-commuting elements $a, b \in G$, if $c = a^{-1}b^{-1}ab$, then $ab = bac$. In this sense, the commutator $c = [a, b]$ provides a kind of measure of the non-commutativity of a and b.

Remark 2: In general, the *set* of all commutators in a group G is not a subgroup of G. The commutator subgroup of a group G is the subgroup $[S]$ *generated* by the set S of all commutators in G, in the sense of Definition 2.3.2. By Theorem 2.3.3, $[G, G] = [S]$ is the set of all products of powers of commutators—or, equivalently, of all products of commutators and their inverses. However, since the inverse of a commutator $[a, b] = a^{-1}b^{-1}ab$ is the commutator $[b, a] = b^{-1}a^{-1}ba$, we can simplify our description of $[G, G]$ somewhat:

Corollary (of Definition 2.11.3): The commutator subgroup of a group G is equal to the set of all products of commutators in G.

Theorem 2.11.4: Let $\langle G, \cdot \rangle$ be a group, with commutator subgroup $[G, G]$. Then

(1) $[G, G]$ is a characteristic subgroup of G.

(2) $G/[G, G]$ is abelian.

(3) If $H \lhd G$, then G/H is abelian if and only if $[G, G] \subset H$.

Proof: (1) Let $\psi : G \to G$ be an automorphism of G, and let $x \in [G, G]$. Then there are commutators $c_1, \ldots, c_k$, not necessarily distinct, such that $x = c_1 \ldots c_k$. Hence $\psi(x) = \psi(c_1 \ldots c_k) = \psi(c_1) \ldots \psi(c_k)$. For each $i = 1, \ldots, k$, $c_i = a_i^{-1}b_i^{-1}a_ib_i$ for some $a_i, b_i \in G$. Hence $\psi(c_i) = \psi(a_i^{-1}b_i^{-1}a_ib_i) = (\psi a_i)^{-1}(\psi b_i)^{-1}\ \psi a_i\ \psi b_i$, itself a commutator, for each $i = 1, \ldots, k$. But then $\psi(x) \in [G, G]$, and so $[G, G]$ is a characteristic subgroup of G. (In fact, we nowhere used the hypothesis that ψ is a bijection. Thus, $[G, G]$ is invariant under arbitrary *endo*morphisms of G—a so-called *fully invariant* subgroup.)

(2) From (1), we infer that certainly $[G, G] \lhd G$. If $a, b \in G$, then the cosets $[G, G]ab$ and $[G, G]ba$ are equal, for: $(ba)^{-1}ab = a^{-1}b^{-1}ab \in [G, G]$. But then $[G, G]a \cdot [G, G]b = [G, G]b \cdot [G, G]a$ for each $a, b \in G$, and so $G/[G, G]$ is abelian.

(3) Suppose $H \lhd G$, with G/H abelian. Then, for each $a, b \in G$, $HaHb = HbHa$; hence $Hab = Hba$, and so $a^{-1}b^{-1}ab = (ba)^{-1}ab \in H$. Thus, H contains all commutators. But then all *products* of commutators are in H, and so $[G, G] \subset H$. Conversely, suppose $[G, G] \subset H$. Let $a, b \in G$. Then $(ba)^{-1}ab = a^{-1}b^{-1}ab \in H$. Hence, using Corollary 1 of Theorem 2.5.1, we have $Ha\ Hb = Hab = Hba = Hb\ Ha$, and so G/H is abelian. ∎

Corollary: If $\phi : G \to G'$ is a homomorphism such that Im ϕ is abelian, then Ker $\phi \supset [G, G]$.

Exercises 2.11

True or False

1. If $\phi : G \to G$ is an automorphism of a group G, then there is some $a \in G$ such that $\phi(x) = axa^{-1}$ for all $x \in G$.
2. The inner automorphisms of a group G are elements of G.
3. The commutators in any group G form a subgroup of G.
4. If the commutator subgroup of any group G consists of the identity only, then G is abelian.
5. If G is a non-abelian group of order n and if $[G, G] \neq G$, then G has a non-trivial abelian homomorphic image of order $< n$.

2.11.1. In S_5, find the image of $\sigma = (12)(345)$ under the inner automorphism γ_τ determined by $\tau = (451)$.

How many possible distinct images does σ have under inner automorphisms of S_5? (Recall Exercise 2.8.4.)

2.11.2. Find all automorphisms of a cyclic group
(a) of finite order n
(b) of infinite order.
(Hint: Recall Exercise 2.4.4.)

2.11.3. Let G be a group, $\mathscr{A}_G$ the set of all automorphisms of G and $\mathscr{I}_G$ the set of all inner automorphisms of G. Prove that $\mathscr{A}_G < S_G$ and $\mathscr{I}_G \lhd \mathscr{A}_G$.

2.11.4. Let G be a group, and let $a, b \in G$. Prove that $C(G)a = C(G)b$ if and only if $\gamma_a = \gamma_b$, where γ_a and γ_b are the inner automorphisms determined, respectively, by a and b.

2.11.5. Let G be a finite group of order n. Prove:
(1) The number of inner automorphisms of G is a divisor of n.
(2) The number of automorphisms of G is $\leq n!$
(3) If G is abelian and contains an element of odd order, then the number of automorphisms of G is even.

2.11.6. Prove that, for $n \geq 3$, S_n is isomorphic to its group of inner automorphisms.

2.11.7. *(1)* Prove that the commutator subgroup of S_3 is A_3. What is the commutator subgroup of A_3?
(2) Prove that the commutator subgroup of S_4 is A_4. What is the commutator subgroup of A_4?
(Hint: make use of Theorem 2.11.4.)

2.11.8. Find the center and the commutator subgroup of the dihedral group D_4. (Hint: make use of Theorem 2.11.4.)

2.11.9. Prove that $GL_n(\mathbb{R})/SL_n(\mathbb{R})$ is abelian. (Hint: What is the determinant of any commutator $[A, B]$, for $A, B \in GL_n(\mathbb{R})$?)

2.11.10. If $\phi : G \to G$ is an endomorphism of G *onto* G, prove that $\phi(C(G)) \subset C(G)$ (where $C(G)$ is the center of G).

2.11.11. For n a positive integer, find the center of $SL_n(\mathbb{R})$. (Recall Exercise 2.8.9.)

2.11.12. Let G be a group, $H < G$. Prove: if $[G, G] \subset H$, then $H \lhd G$.

2.11.13. Let G be a group, $H \lhd G$, and let K be a characteristic subgroup of H. Prove that $K \lhd G$. (Contrast this result with the result in Exercise 2.8.10, part (2)).

2.12

Direct Products of Groups

We now introduce a method of combining groups to create "products" of groups. The usefulness of this multiplication of groups will be two-fold: first, it will place in our hands a powerful tool for constructing examples of groups with special properties. Second, and even more important, applied "backwards," as it were, it will provide a way of factoring certain more complex groups into simpler components.

Theorem 2.12.1: For n a positive integer, let $\langle G_1, \cdot_1\rangle, \ldots, \langle G_n, \cdot_n\rangle$ be groups (not necessarily distinct). On the Cartesian product $G = G_1 \times \cdots \times G_n$ (see Section 1.2), define $\cdot$ by

$$(x_1, \ldots, x_n) \cdot (y_1, \ldots, y_n) = (x_1 \cdot_1 y_1, \ldots, x_n \cdot_n y_n).$$

Then $\cdot$ is a binary operation on G and $\langle G, \cdot\rangle$ is a group.

Proof: It is obvious that $\cdot$ is a binary operation on G, that $e = (e_1, \ldots, e_n)$ (where e_i is the identity of G_i for each $i = 1, \ldots, n$) serves as identity for $\cdot$, and that $(x_1^{-1}, \ldots, x_n^{-1}) \in G$ serves as $(x_1, x_2, \ldots, x_n)^{-1}$ with respect to $\cdot$ and the identity e. Just as easily, one sees that the associativity of $\cdot_i$, for each $i = 1, \ldots, n$, causes $\cdot$ to be associative. Thus, $\langle G, \cdot\rangle$ is a group. ■

Definition 2.12.1: The *direct product* of groups $\langle G_1, \cdot_1\rangle, \ldots, \langle G_n, \cdot_n\rangle$ is the group $\langle G, \cdot\rangle$, where $G = G_1 \times \cdots \times G_n$ and $\cdot$ is defined by

$$(x_1, \ldots, x_n) \cdot (y_1, \ldots, y_n) = (x_1 \cdot_1 y_1, \ldots, x_n \cdot_n y_n)$$

(for $(x_1, \ldots, x_n), (y_1, \ldots, y_n) \in G$).

(Sensibly, G could be called the (direct) sum when group operations are written additively. But while, for a finite set of groups, the distinction between *product* and *sum* is purely notational, there is an important conceptual distinction between the terms *direct product* and *direct sum* as used when the set of given groups $\langle G_i, \cdot_i\rangle$ is infinite. We spell out this distinction in the final Remark of this section.)

Example:

The direct product of the groups $\langle A_3, \cdot\rangle$ and $\langle \mathbb{Z}_2, \oplus_2\rangle$ is

$$G = \{((1), \bar{0}),\ \ ((1), \bar{1}),\ \ ((123), \bar{0}),\ \ ((123), \bar{1}),\ \ ((132), \bar{0}),\ \ ((132), \bar{1})\}.$$

In G, for example $((123), \bar{1}) \cdot ((132), \bar{1}) = ((1), \bar{0}) = e_G$.

Corollary (of Definition 2.12.1): If $G_1, \ldots, G_n$ are finite groups, then the order of their direct product G is $|G_1| \cdot \cdots \cdot |G_n|$.

The following theorem shows that if G is the direct product of the groups $G_1, \ldots, G_n$, then each of the groups G_i is isomorphic to a subgroup $\bar{G}_i$ of G_i, and that these subgroups $\bar{G}_i$ play a special role in G.

Theorem 2.12.2: Let $\langle G_1, \cdot_1\rangle, \ldots, \langle G_n, \cdot_n\rangle$ be groups, and let $\langle G, \cdot\rangle$ be their direct product. For each $i = 1, \ldots, n$, let

$$\bar{G}_i = \{(e_i, \ldots, e_{i-1}, a_i, e_{i+1}, \ldots, e_n) | a_i \in G_i\}.$$

Then, for each $i = 1, \ldots, n$, $\bar{G}_i < G$ and the groups G_i and $\bar{G}_i$ are isomorphic. Moreover:

(1) $xy = yx$ for each $x \in \bar{G}_i$, $y \in \bar{G}_j$, provided

$$i \neq j \quad (1 \leq i, j \leq n)$$

(2) Every element $x \in G$ is expressible in one and only one way as $x = x_1 \cdots x_n$, where $x_i \in \bar{G}_i$.

Proof: For each $i = 1, \ldots, n$, the mapping $\mu_i : G_i \to G$ defined by: $\mu_i(a_i) = (e_1, \ldots, e_{i-1}, a_i, e_{i+1}, \ldots, e_n)$ is easily seen to be a monomorphism with image $\bar{G}_i$; hence $\bar{G}_i < G$, and $\bar{G}_i \cong G_i$.

(1) If $i < j \quad (1 \leq i, j \leq n)$, then for $x \in \bar{G}_i$, and $y \in \bar{G}_j$, we have

$$x = (e_1, \ldots, e_{i-1}, a_i, e_{i+1}, \ldots, e_n),$$

$$y = (e_1, \ldots, e_{j-1}, b_j, e_{j+1}, \ldots, e_n),$$

for some $a_i \in G_i$, $b_j \in G_j$; hence

$$xy = (e_1, \ldots, e_{i-1}, a_i, e_{i+1}, \ldots, e_{j-1}, b_j, e_{j+1}, \ldots, e_n) = yx.$$

(2) For each $x \in G$, we have $x = (a_1, \ldots, a_n)$, $a_i \in G_i \quad (i = 1, \ldots, n)$. But then $x = (a_1, e_2, \ldots, e_n) \quad (e_1, a_2, e_3, \ldots, e_n) \cdots (e_1, \ldots, e_{n-1}, a_n)$. Denoting the factors by $x_1, \ldots, x_n$ we have the required representation $x = x_1 \cdots x_n$, with $x_i \in \bar{G}_i \quad (i = 1, \ldots, n)$.

If also $x = y_1 \cdots y_n$, $y_i \in \bar{G}_i \quad (i = 1, \ldots, n)$ then, for each $(i = 1, \ldots, n)$,

$$y_i = (e_1, \ldots, e_{i-1}, b_i, e_{i+1}, \ldots, e_n)$$

for some $b_i \in G_i$, and

$$x = y_1 \cdots y_n = (b_1, \ldots, b_n).$$

But then the n-tuples $(a_1, \ldots, a_n)$ and $(b_1, \ldots, b_n)$ are equal, whence $a_i = b_i$ for each $i = 1, \ldots, n$, and therefore $x_i = y_i$, for each $i = 1, \ldots, n$. ■

Definition 2.12.2: Let G be a group, and let $H_1, \ldots, H_n$ be subgroups of G such that

(1) $xy = yx$ for each $x \in H_i$, $y \in H_j$, provided $i \neq j \quad (1 \leq i, j \leq n)$

(2) every element $x \in G$ is expressible in one and only one way as $x = x_1 \cdots x_n$, where $x_i \in H_i$, for each $i = 1, \ldots, n$.

Then G is the *internal direct product of* the subgroups $H_1, \ldots, H_n$.

By Theorem 2.12.2, if G is the direct product of groups $G_1, \ldots, G_n$, then G is the internal direct product of its subgroups $\bar{G}_i$ defined as in Theorem 2.12.2. To emphasize the distinction, we may use the adjective *external* for the direct product of groups defined in Definition 2.12.1.

Conversely, it is not difficult to show that, if a group G is the internal direct product of its subgroups $H_1, \ldots, H_n$ (in the sense of Definition 2.12.2), then G is isomorphic to the (external) direct product $H_1 \times H_2 \times \cdots \times H_n$ of the groups H_i (in the sense of Definition 2.12.1). (See Exercise 2.12.16.)

Notations: Standard notations vary. Let us write $G = G_1 \times \cdots \times G_n$ to signify that G is the (external) direct product of the groups G_i and $G = H_1 \circ \cdots \circ H_n$ to signify that G is the internal direct product of its subgroups H_i.

Remark: If $\{G_i\}$ (where J is some indexing set, not necessarily finite) is a family of groups, we can generalize Definition 2.12.1 in two different ways. The first is to let P be the set of all sequences $\{x_i\}_{i \in J}$, where $x_i \in G_i$, for each $i \in J$. The second is to let S be the subset of P consisting of only those sequences $\{x_i\}_{i \in J}$ $(x_i \in G_i)$ which have *all but* finitely many components x_i equal to the identity e_i of the corresponding group G_i. Under componentwise multiplication $\cdot$, $\langle P, \cdot \rangle$ is a group and $S < P$.

Somewhat confusingly, P is then called the *direct product*, and S is called the *direct sum*, of the family of groups $\{G_i\}_{i \in J}$—this terminology being independent of the notation used for the group operation.

Exercises 2.12

True or False

1. If G_1 and G_2 are groups, then the direct product $G = G_1 \times G_2$ has a subgroup isomorphic to G_1.
2. If one of the groups $G_1, \ldots, G_n$ is non-abelian, then the direct product $G = G_1 \times \cdots \times G_n$ is non-abelian.
3. The direct product of two cyclic groups cannot be cyclic.
4. The direct product of two cyclic groups must be cyclic.
5. The direct product $A_4 \times \bar{\mathbb{Z}}_2$ is isomorphic to S_4.

2.12.1. Compute
(1) $((123), \bar{2}) \times ((23), \bar{3})$ in $S_3 \times \bar{\mathbb{Z}}_5$,
(2) $(2, 3) \times (-1, 5)$ in $\mathbb{Z} \times \mathbb{Z}$,
(3) $(2, 3) \times (-1, 5)$ in $\mathbb{Q} \times \mathbb{Q}^*$,
(4) $(\sigma\rho, (1253))^{-1}$ in $D_4 \times S_5$,
(5) $\left(\begin{pmatrix} 1 & 2 \\ 3 & 4 \end{pmatrix}, 5\right)^{-1}$ in $GL_2(\mathbb{R}) \times \mathbb{R}^*$,
(6) $(2 + 3i, 1 - i)^{-1}$ in $\mathbb{C}^* \times \mathbb{C}$.

2.12.2. Let G be a Klein 4-group, i.e., any group isomorphic to $V_4 < S_4$. Prove that G is the internal direct product of two groups of order 2. Conclude that G is isomorphic to $\bar{\mathbb{Z}}_2 \times \bar{\mathbb{Z}}_2$.

2.12.3. Give an example of a non-abelian group G that has an infinite cyclic homomorphic image.

2.12.4. Prove that each of the subgroups $\bar{G}_i$ in Theorem 2.12.2 is normal in G. Prove that $G/\bar{G}_i \cong G_1 \times \cdots \times G_{i-1} \times G_{i+1} \times \cdots \times G_n$, for each $i = 1, \ldots, n$. Find a subgroup $\bar{K}_i$ of G that is isomorphic to $G_1 \times \cdots \times G_{i-1} \times G_{i+1} \times \cdots \times G_n$, prove that it is normal, and that $G/\bar{K}_i \cong G_i$.

2.12.5. Give an example of a non-abelian group that has a normal subgroup of order 47, and a normal subgroup of index 47.

2.12.6. Give an example of a non-abelian group that has a normal subgroup of order m, and a normal subgroup of index m, for each $m = 1, \ldots, 100$.

2.12.7. Prove that the additive group of the vector space $\mathbb{R}^2$ is isomorphic to $\mathbb{R} \times \mathbb{R}$, where $\mathbb{R}$ is the additive group of all real numbers.

2.12.8. Prove that the additive group $M_2(\mathbb{R})$ of all 2×2 matrices over $\mathbb{R}$ is isomorphic to $\mathbb{R} \times \mathbb{R} \times \mathbb{R} \times \mathbb{R}$, where $\mathbb{R}$ is the additive group of all real numbers.

2.12.9. Prove that the additive group of the vector space $\mathbb{R}^4$ is isomorphic to the additive group $M_2(\mathbb{R})$.

2.12.10. Prove that the direct product of a set of groups $G_i \quad (i \in J)$ is abelian if and only if each G_i is abelian.

2.12.11. A group G is *indecomposable* if it is not equal to the internal direct product of proper subgroups. Prove that a non-abelian group of order 8 is indecomposable. Conclude that the dihedral group D_4 and the quaternion group (see Exercise 2.8.12) are indecomposable.

2.12.12. Prove that a cyclic group of prime power order is indecomposable.

2.12.13. Given two finite cyclic groups G_1 and G_2, find a necessary and sufficient condition for $G = G_1 \times G_2$ to be cyclic. Generalize your result to an arbitrary finite number of finite cyclic groups.

2.12.14. Give an example of an abelian group G that has infinitely many automorphisms.

2.12.15. Prove that the multiplicative group $\mathbb{R}^*$ of all non-zero real numbers is isomorphic to $\mathbb{Z}_2 \times \mathbb{R}^+$ (where $\mathbb{R}^+$ is the multiplicative group of all positive real numbers).

2.12.16. Prove: if G is the internal direct product of subgroups $H_1, \ldots, H_m$, then G is isomorphic to the (external) direct product $H_1 \times \cdots \times H_m$.

2.12.17. Let $G_1, \ldots, G_n$ be groups such that, for each $i = 1, \ldots, n$,

$$G_i = H_{i1} \times \cdots \times H_{ik_i},$$

i.e., each G_i is itself a direct product of groups $H_{ij}, j = 1, \ldots, k_i$. Prove that $G_1 \times \cdots \times G_n$ is isomorphic to $H_{11} \times \cdots \times H_{1k_1} \times H_{21} \times \cdots \times H_{2k_2} \times \cdots \times H_{n1} \times \cdots \times H_{nk_n}$.

2.12.18. Let J be an infinite set, and let $G = \bigcup_{j \in J} G_j$, where $\{G_j\}_{j \in J}$ is a family of groups, with each G_j of finite order. It can be proved in set theory that G and J have the same cardinality, i.e., there exists a bijection from J onto G. Use this result to prove that it is possible to define a group structure on any infinite set. (See Exercise 2.4.9.)

2.12.19. Prove that, for finite groups $G_1, \ldots, G_m$, $\quad |G_1 \times \cdots \times G_m| = \prod_{i=1}^{m} |G_i|$.

2.12.20. Prove that the additive group $\langle \mathbb{R}^n, + \rangle$ of the vector space $\mathbb{R}^n$ is isomorphic to a direct product of n groups, each isomorphic to $\langle \mathbb{R}, + \rangle$.

2.13

The Structure of Finite Abelian Groups

We have seen (Corollary 3, Theorem 2.5.2) that, if p is prime, then every group of order p is cyclic, hence isomorphic to $\mathbb{Z}_p$. There are other classes of positive integers n such that all groups of order n are completely determined (i.e., determined up to isomorphism). For example, if p, q are distinct primes, with $p < q$ and $q \not\equiv 1 \bmod p$, then all groups of order pq are cyclic, and hence isomorphic to $\mathbb{Z}_{pq}$. (See Exercise 2.15.9.) However, in general, given a positive integer n, there is no known classification of all "abstract groups" of order n (i.e., of all isomorphism classes of groups of order n). It is thus very gratifying that all abstract *abelian* groups of any given finite order are known.

We begin by stating and illustrating the main structure theorem for finite abelian groups. This theorem is a special case of the main structure theorem for *finitely generated* abelian groups (known as *The Fundamental Theorem of Abelian Groups*). That theorem, in turn, is a special case of a theorem on modules over a principal ideal domain which has, as another special case, one of the most important theorems in linear algebra. (cf [3].)

The proof of the Fundamental Theorem of Abelian Groups is beyond the scope of most introductory courses in abstract algebra. However, the structure theorem for *finite* abelian groups can be proved without going beyond the scope of the theory we have developed thus far, and we include such a proof at the end of this section. (See Theorem 2.13.1 and its corollary.) We start with a preliminary statement of the theorem (using only the concept of external direct product) which should suffice for readers not intending to proceed to the proof.

Structure Theorem for Finite Abelian Groups (Primary Version)

Every finite abelian group is isomorphic to a direct product of cyclic groups of prime power order. In particular, if G has order $n = p_1^{\alpha_1} \cdots p_m^{\alpha_m}$, where the p_i are distinct primes listed in increasing order ($m \geq 1$, $\alpha_i \geq 1$ for $i = 1, \ldots, m$), then G is isomorphic to exactly one direct product

$$\mathbb{Z}_{p_1^{\beta_{11}}} \times \cdots \times \mathbb{Z}_{p_1^{\beta_{1t_1}}} \times \cdots \times \mathbb{Z}_{p_m^{\beta_{m1}}} \times \cdots \times \mathbb{Z}_{p_m^{\beta_{mt_m}}},$$

where $\beta_{i1} \geq \beta_{i2} \geq \cdots \geq \beta_{it_i}$ for each $i = 1, \ldots, m$.

Remark: Since the order of G is equal to the product of the orders of the direct factors (see Exercise 2.12.19), we have $\sum_{j=1}^{t_i} \beta_{ij} = \alpha_i$, for each $i = 1, \ldots, m$.

Example 1:

Let us find all abstract abelian groups of order 12. We have $12 = 2^2 \cdot 3$. Thus, the possible sequences of prime power orders of the direct factors are 2, 2, 3 and 2^2, 3. Hence every abelian group of order 12 is isomorphic either to $\mathbb{Z}_2 \times \mathbb{Z}_2 \times \mathbb{Z}_3$ or to $\mathbb{Z}_4 \times \mathbb{Z}_3$. The uniqueness condition of the theorem implies that these two direct products are non-isomorphic. Thus, there are exactly two *abstract abelian groups* of order 12.

Example 2:

To find all abstract abelian groups of order 72, we write $72 = 2^3 \cdot 3^2$, and conclude that every abelian group of order 72 is isomorphic to exactly one of the following six direct products:

$$\begin{array}{ll} \mathbb{Z}_2 \times \mathbb{Z}_2 \times \mathbb{Z}_2 \times \mathbb{Z}_9 & \mathbb{Z}_2 \times \mathbb{Z}_2 \times \mathbb{Z}_2 \times \mathbb{Z}_3 \times \mathbb{Z}_3 \\ \mathbb{Z}_4 \times \mathbb{Z}_2 \times \mathbb{Z}_9 & \mathbb{Z}_4 \times \mathbb{Z}_2 \times \mathbb{Z}_3 \times \mathbb{Z}_3 \\ \mathbb{Z}_8 \times \mathbb{Z}_9 & \mathbb{Z}_8 \times \mathbb{Z}_3 \times \mathbb{Z}_3 \end{array}$$

Remark: The (more general) Fundamental Theorem of Abelian Groups (primary version) states that every *finitely generated* abelian group is isomorphic to a direct product of finitely many cyclic groups: some finite of prime power order, some infinite. Each infinite cyclic direct factor is, of course, isomorphic to $\mathbb{Z}$. For example, a typical finitely generated abelian group may have the following structure:

$$\mathbb{Z}_{2^3} \times \mathbb{Z}_2 \times \mathbb{Z}_5 \times \mathbb{Z}_{7^2} \times \mathbb{Z}_7 \times \mathbb{Z} \times \mathbb{Z}.$$

Our proof of the Structure Theorem for Finite Abelian Groups requires some preparation.

Definition 2.13.1: A group G is *p-primary* (or a *p-group*) for some prime p if the order of every element of G is a power of p. If a group G is *p-primary* for some prime p, then G is called a *primary group*.

Not all primary groups are finite. For example, for a given prime p, the complex numbers z such that $z^{p^k} = 1$ for some integer k form an infinite p-primary group.

Clearly, if G is a finite group of prime power order p^α, then G is a p-primary group (by Corollary 1 of La Grange's Theorem). Conversely, suppose G is a p-primary finite group. Is the order of G a power of p? The answer is "yes," as we shall see in Section 2.15. For finite *abelian* groups, we have the following lemmas.

Lemma 1: Let G be an abelian group of finite order n, and let p be a prime dividing n. Then G has an element of order p.

Proof: Suppose the theorem is false. Then there is a least positive integer n_o and a prime $p_o | n_o$ such that some abelian group G of order of n_o has *no* element of order p_o. If n_o is prime, then $p_o = n_o$ and G is cyclic, hence has an element of order $p_o = |G|$, contrary to hypothesis. Thus, n_o is composite, and G has a proper subgroup $H \neq \{e\}$, $H \neq G$. (See Exercise 2.3.17.) Let $m = |H|$. By Euclid's Lemma, either $p_o | m$ or $p_o \mid \frac{n_o}{m}$. In the first case, since $1 < m < n_o$, H has an element of order p_o, hence so has G. Contradiction! In the second case, since $\frac{n_o}{m} < n_o$, the factor group G/H $\left(\text{whose order is } \frac{n_o}{m}\right)$ has an element of order p_o. Suppose, for $x \in G$, Hx has order p_o in G/H. Let s be the order of x in G. Then $(Hx)^s = Hx^s = He = H$;

hence the order, p_o, of Hx divides s (by Theorem 2.3.7). But then the cyclic group $[x]$ contains an element of order p_o (see Exercise 2.3.10), and hence so does G. Contradiction! ■

Lemma 2: If G is a finite abelian p-primary group for some prime p, then the order of G is a power of p.

Proof: If $|G|$ is not a power of p, then there is some prime $q \neq p$ such that $q \mid |G|$. But then, by Lemma 1, G has an element of order q, contrary to the hypothesis that G is p-primary.

It follows that $|G|$ is a power of p. ■

Lemma 3: Let G be a finite abelian group of order $n = p_1^{\alpha_1} \cdots p_m^{\alpha_m}$ (p_i distinct primes, $m \geq 1$, $\alpha_i \geq 1$ for each $i = 1, \ldots, m$). For each $i = 1, \ldots, m$, let $H_i = \{x \in G \mid o(x) \text{ is a power of } p_i\}$. Then $H_i < G$ $(i = 1, \ldots, m)$.

Proof: For each $i = 1, \ldots, m$, $e \in H_i$ since e has order $1 = p_i^0$. If $x, y \in H_i$, then $o(x) = p_i^s$, $o(y) = p_i^t$ for some integers $s \geq 0$, $t \geq 0$. Let $r = \max\{s, t\}$. Since G is abelian, $(xy)^{p_i^r} = x^{p_i^r} y^{p_i^r} = e$; hence $o(xy)$ is a divisor of p_i^r. But then $o(xy)$ is a power of p_i, and so $xy \in H_i$. Finally, if $x \in H_i$, then since $o(x) = o(x^{-1})$ (see Exercise 2.3.5), we have $x^{-1} \in H_i$. By Theorem 2.8.1, $H_i < G$, for each $i = 1, \ldots, m$. ■

Lemma 4: If G is an abelian group of finite order $n = p_1^{\alpha_1} \cdots p_m^{\alpha_m}$ ($m \geq 1$, p_i primes, with $p_1 < p_2 < \cdots < p_m$, $\alpha_i \geq 1$, for each $i = 1, \ldots, m$), then G is the internal direct product of the subgroups $H_i = \{x \in G \mid o(x) \text{ is a power of } p_i\}$, $i = 1, \ldots, m$. The H_i are the only primary subgroups of which G is the internal direct product.

Proof: Since G is abelian, it suffices to prove that the subgroups H_i satisfy condition (2) of Definition 2.12.2. Let $a \in G$ and let $k = o(a)$. Since $k \mid |G|$, $k = p_1^{\beta_1} \cdots p_m^{\beta_m}$, where $0 \leq \beta_i \leq \alpha_i$ for each $i = 1, \ldots, m$. Define $k_i = \dfrac{k}{p_i^{\beta_i}}$ $(i = 1, \ldots, m)$. Then the integers $k_1, \ldots, k_m$ are relatively prime.

By Exercise 1.3.11, there are integers $t_1, \ldots, t_m$ such that

$$1 = t_1 k_1 + \cdots + t_m k_m.$$

But then $a = a^1 = a^{t_1 k_1 + \cdots + t_m k_m} = a^{t_1 k_1} \cdot \cdots \cdot a^{t_m k_m}$. Since, for each $i = 1, \ldots, m$,

$$(a^{t_i k_i})^{p_i^{\beta_i}} = (a^k)^{t_i} = e,$$

the order of $a^{t_i k_i}$ is a power of p_i, and so $a^{t_i k_i} \in H_i$. Thus, we have expressed a as a product of elements of the H_i:

$$a = a_1 \cdot \cdots \cdot a_m$$

where $a_i = a^{t_i k_i}$ $(i = 1, \ldots, m)$.

To establish the uniqueness of this factorization, suppose there are elements $b_i \in H_i \quad (i = 1, \ldots, m)$ such that

$$a = a_1 \cdots a_m = b_1 \cdots b_m.$$

Since G is abelian, we have

$$e = a_1 b_1^{-1} \quad a_2 b_2^{-1} \cdots a_m b_m^{-1} = c_1 \cdots c_m, \quad \text{where}$$
$$c_i = a_i b_i^{-1} \in H_i, \quad (i = 1, \ldots, m).$$

But then, for each $i = 1, \ldots, m$,

$$c_i^{-1} = c_1 \cdots c_{i-1} \quad c_{i+1} \cdots c_m.$$

Now, the order of each c_j on the right-hand side is $p_j^{\gamma_j}$ for some integer γ_j, $1 \leq \gamma_j \leq \alpha_j$. Let $s = \prod_{j \neq i} p_j^{\gamma_j}$. Then

$$c_i^{-s} = c_1^s \cdots c_{i-1}^s \quad c_{i+1}^s \cdots c_m^s = e,$$

and so the order of c_i divides s. But the order of c_i is a power of p_i, and p_i and s are relatively prime. This is possible only if $o(c_i) = p_i^0 = 1$, and $c_i = e$. Thus, $a_i b_i^{-1} = e$ and $a_i = b_i$, for each $i = 1, \ldots, m$.

We conclude that G is the internal direct product of the subgroups H_i.

To prove uniqueness of the decomposition, suppose that $\bar{H}_1, \ldots, \bar{H}_{\bar{m}}$ are $\bar{p}_i$-primary subgroups of G, with $|\bar{H}_i| = \bar{p}_i^{\bar{\alpha}_i} \quad (i = 1, \ldots, \bar{m})$, $\bar{p}_1 < \bar{p}_2 < \cdots < \bar{p}_{\bar{m}}$, and that G is the internal direct product of the $\bar{H}_i \quad (i = 1, \ldots, \bar{m})$. Then (by Exercises 2.12.16 and 2.12.19), $|G| = p_1^{\alpha_1} \cdots p_m^{\alpha_m} = \bar{p}_1^{\bar{\alpha}_1} \cdots \bar{p}_{\bar{m}}^{\bar{\alpha}_{\bar{m}}}$. By the Fundamental Theorem of Arithmetic, we conclude that $\bar{m} = m$, $\bar{p}_i = p_i$, $\bar{\alpha}_i = \alpha_i$, for each $i = 1, \ldots, m$. By the definition of the H_i, we have $\bar{H}_i \subset H_i \quad (i = 1, \ldots, m)$. But, since $|\bar{H}_i| = |H_i|$, this implies $\bar{H}_i = H_i$, $i = 1, \ldots, m$. ∎

The subgroups H_i are called the *primary components* of G.

To prove the structure theorem for finite abelian groups, it suffices (see Exercise 2.12.17) to prove that every primary component of a finite abelian group is a direct product of cyclic subgroups.

Lemma 5: Let p be a prime, and let H be a finite p-primary abelian group. Then H is the internal direct product of cyclic subgroups.

Proof: Since H is finite, H contains an element, a_1, of maximal order p^{k_1}, and H has a subgroup T_1 which is maximal with respect to the property $[a_1] \cap T_1 = \{e\}$ (i.e., T_1 is not properly contained in a larger subgroup with this property). We wish to prove that H is the internal direct product of $[a_1]$ and T_1.

We already know that $[a_1] \cap T_1 = \{e\}$; hence if $h \in H$ is expressible as $h = xy$ where $x \in [a_1]$, $y \in T_1$, then this representation is unique. (For, if $x_1 y_1 = x_2 y_2$ $(x_1, x_2 \in [a_1], \quad y_1, y_2 \in T_1)$, then $x_1 x_2^{-1} = y_1 y_2^{-1} \in [a_1] \cap T_1$, whence $x_1 x_2^{-1} = e$, $y_1 y_2^{-1} = e$, and so $x_1 = x_2$, $y_1 = y_2$). Thus we need show only that *every* element of H *is* expressible in this form.

Let $H^* = [a_1, T_1]$ (the subgroup of H generated by the set $\{a_1\} \cup T_1$). We need to show that $H^* = H$. Suppose $H^* \neq H$. Then there is some $x \in H \backslash H^*$, such that

$x^p \in H^*$. (For, given any $y \in H \backslash H^*$, $y^{p^s} = e \in H^*$ if $o(y) = p^s$; hence there is a least positive integer r, $1 \le r \le s$, such that $y^{p^r} \in H^*$. But then, with $x = y^{p^{r-1}}$, we have $x \in H \backslash H^*$ and $x^p \in H^*$). By Theorem 2.3.3, $x^p = ta_1^q$ for some $t \in T_1$ and some $q \in \mathbb{Z}$. Since p^{k_1} is maximal along the orders of elements of H, we have $o(x) = p^h$, $h \le k_1$; hence

$$(ta_1^q)^{p^{k_1-1}} = x^{p^{k_1}} = e,$$

and so $t^{p^{k_1-1}} = a_1^{-1p^{k_1-1}} \in [a_1] \cap T_1 = \{e\}$. Thus, $a_1^{qp^{k_1-1}} = e$; but then $p^{k_1} | qp^{k_1-1}$, and $p | q$. Write $q = pj \quad (j \in \mathbb{Z})$. Then

$$t = x^p a_1^{-q} = x^p a_1^{-pj} = (xa_1^{-j})^p \in T_1.$$

But $xa_1^{-j} \notin T_1$ since $x \notin [a_1, T_1] = H^*$. Thus, $[T_1, xa_1^{-j}] \supsetneq T_1$. Since T_1 is maximal with respect to the property of having trivial intersection with $[a_1]$, $[T_1, xa_1^{-j}] \cap [a_1] \ne \{e\}$. Hence there is some integer r such that $a_1^r \ne e$ and $a_1^r \in [T_1, xa_1^{-j}]$. Then $a_1^r = t'(xa_1^{-j})^s$ for some $t' \in T_1$, $s \in \mathbb{Z}$. If $p | s$ then, since $(xa_1^{-j})^p \in T_1$, we have $a_1^r = t'(xa_1^{-j})^s \in T_1$, contrary to the hypothesis that $[a_1] \cap T_1 = \{e\}$. Thus, $\gcd(p, s) = 1$, and there are integers u, v such that $1 = up + vs$. Since x^p and $x^s = (t')^{-1}a_1^{r+js}$ are both in H^*, $x = x^{up+vs} = (x^p)^u \cdot (x^s)^v \in H^*$. But x was chosen to be in $H \backslash H^*$. Contradiction.

We conclude that $H = [a_1] \circ T_1$. Now, T_1 is again a finite abelian p-group, and we can find an element $a_2 \in T_1$ of maximal order in T_1 and a subgroup T_2 of T_1 such that $T_1 = [a_2] \circ T_2$, and therefore (see Exercises 2.12.16 and 2.12.17), $H = [a_1] \circ [a_2] \circ T_2$. Continuing in this manner, since H is finite, we ultimately arrive at an integer k such that $|T_k| = 1$, and

$$H = [a_1] \circ \cdots \circ [a_k],$$

an internal direct product of (p-primary) cyclic groups. (Note that the sequence of orders of the cyclic subgroups $[a_1]$ is non-increasing.) ■

Remark: It can be shown (see Exercise 2.13.5) that the sequence of orders of the cyclic direct factors $[a_i]$ is uniquely determined, for any finite abelian p-group H.

The main theorem of this section is now essentially proved.

Theorem 2.13.1. Structure Theorem for Finite Abelian Groups: Let G be an abelian group of finite order $|G| = p_1^{\alpha_1} \cdots p_m^{\alpha_m}$, with $p_1 < p_2 < \cdots < p_m$ ($m \ge 1$, $\alpha_i \ge 1$ p_i primes). Then G is the internal direct product of cyclic subgroups of orders

$$p_1^{\beta_{11}}, \ldots, p_1^{\beta_{1t_1}}; \ldots; p_m^{\beta_{m1}}, \ldots, p_m^{\beta_{mt_m}} \quad (\beta_{ij} \ge 1). \tag{1}$$

If the direct factors of G are arranged so that the exponents β_{ij} belonging to each prime p_i are non-increasing, then the sequence (1) of their orders is uniquely determined by G.

Proof: By Lemma 4, if $|G| = p_1^{\alpha_1} \cdots p_m^{\alpha_m}$, with $p_1 < p_2 < \cdots < p_m$, and $H_i = \{x \in G | o(x) \text{ is a power of } p_i\}$ $i = 1, \ldots, m$, then G is the internal direct product of the subgroups H_i—each p_i-primary. By Lemma 5, each subgroup H_i is the internal direct product of p_i-primary cyclic groups. But then G is the internal direct product of the cyclic direct factors of the p_i-primary components H_i of G (see Exercise 2.12.17). If the orders are arranged as prescribed, then the uniqueness of the sequence of orders follows from the uniqueness conditions of Lemmas 4 and 5. ∎

Corollary: If G is an abelian group of finite order $n = p_1^{\alpha_1} \cdots p_m^{\alpha_m}$ ($m \geq 1$, p_i primes with $p_1 < p_2 < \cdots < p_m$, $\alpha_i \geq 1$, for each $i = 1, \ldots, m$), then G is isomorphic to exactly one external direct product

$$\mathbb{Z}_{p_1{}^{\beta_{11}}} \times \cdots \times \mathbb{Z}_{p_1{}^{\beta_{1t_1}}} \times \cdots \times \mathbb{Z}_{p_m{}^{\beta_{m1}}} \times \cdots \times \mathbb{Z}_{p_m{}^{\beta_{mt_m}}}$$

where $\beta_{i1} \geq \cdots \geq \beta_{it_i}$ for each $i = 1, \ldots, m$.

Proof: The Corollary follows immediately from Theorem 2.13.1, in view of Theorem 2.4.1 and Exercise 2.12.16. ∎

Exercises 2.13

True or False

1. If n is a product of distinct primes, then every abelian group of order n is isomorphic to $\mathbb{Z}_n$.
2. Every abelian group of order 60 has a subgroup of order 30.
3. There are exactly 2 abstract abelian groups of order 20.
4. There are exactly two abstract groups of order 12.
5. There are exactly six abstract groups of order 72.

2.13.1. Up to isomorphism, find all abelian groups of order ≤ 30.

2.13.2. Up to isomorphism, find all abelian groups of order 100.

2.13.3. Up to isomorphism, find all abelian groups of order 60.

2.13.4. Let G be an abelian group of finite order n, and let $m|n$. Prove that G has at least one subgroup of order m.

2.13.5. To prove the uniqueness of the sequence of orders in the decomposition of a p-primary group G into cyclic subgroups of non-increasing order (Lemma 5), proceed as follows: Suppose

$$G = H_1 \circ \cdots \circ H_s = K_1 \circ \cdots \circ K_t$$

where $|H_i| = p^{\alpha_i}$, $\alpha_1 \geq \cdots \geq \alpha_s$; $|K_j| = p^{\beta_j}$, $\beta_1 \geq \cdots \geq \beta_t$.

First let $P = \{x \in G | x^p = e\}$. Prove that $P < G$, with

$$|P| = p^s = p^t, \text{ whence } s = t.$$

Next suppose that, for some i $(1 \leq i \leq s)$, $\alpha_i < \beta_i$. Let h be the least such index i. Compute the order of G^{p^h} using, in turn, the decompositions $H_1 \circ \cdots \circ H_s$ and $K_1 \circ \cdots \circ K_s$ of G to obtain a contradiction.

2.13.6. Prove that there is a non-abelian group of order n for every *even* integer $n > 4$.

2.14

Solvable Groups

(The material in Sections 2.14 and 2.15 is included primarily for use in conjunction with Galois Theory. In a course which includes Galois Theory, these sections may be deferred until just prior to Section 4.12.)

In the preceding section, we observed the role of direct products in characterising the structure of finite abelian groups. Another important type of tool for studying group structure is *normal series* of groups (Definition 2.14.1). Some interesting classes of groups can be characterized in terms of properties of their normal series. In particular, the *solvable groups* (Definition 2.14.1) play a fundamental role in Galois Theory.

Definition 2.4.1: Let G be a finite group, and let $G_0, G_1, \ldots, G_n$ be subgroups of G such that

$$G = G_n \rhd G_{n-1} \rhd \ldots \rhd G_1 \rhd G_0 = \{e\}.$$

Then the sequence $\mathcal{N} : \{G_i\}_{i=0}^{n}$ is a *normal series* for G; the subgroups G_i $(i = 0, \ldots, n)$ are the *terms* of $\mathcal{N}'$, and the factor groups G_i/G_{i-1} $(i = 1, \ldots, n)$ are the *factors* of $\mathcal{N}$.

If $\mathcal{N}'$ and $\mathcal{N}$ are normal series of G such that every term of $\mathcal{N}$ is also a term of $\mathcal{N}'$, then $\mathcal{N}'$ is a *refinement* of $\mathcal{N}$.

A normal series, $\mathcal{N}$, of G is *solvable* if each of its factors is abelian.

G is a *solvable group* if it has a solvable normal series.

Obviously, every abelian group is solvable. Obviously, too, every non-abelian simple group fails to be solvable. Two major breakthroughs of recent times, related to solvability, are:

(1) the theorem of J. G. Thompson and W. Feit (1962) which states that every finite group of odd order is solvable ([14]); and

(2) the completion in 1983, making heavy use of computers, of the mammoth task of classifying all finite simple groups ([116]).

We begin with an isomorphism theorem that is a somewhat more detailed variant of Theorem 2.10.1. This theorem, which will be useful in its own right, is interesting also because of its structural resemblance to the Fundamental Theorem of Galois Theory (Theorem 4.10.1).

Theorem 2.14.1. (Correspondence Theorem): Let G, $\bar{G}$ be groups and let $\nu : G \to \bar{G}$ be an epimorphism with Ker $\nu = K$. Let S_K be the set of all subgroups of G containing K, and let $\bar{S}$ be the set of *all* subgroups of $\bar{G}$,

(1) If $\phi : S_K \to \bar{S}$ is defined by: $\phi(H) = \nu H \quad (H \in S_K)$ and $\psi : \bar{S} \to S_K$ is defined by $\psi(\bar{H}) = \nu^{-1}\bar{H} \quad (\bar{H} \in \bar{S})$, then ϕ and ψ are mutually inverse inclusion-preserving bijections.

(2) For H_1, H_2 subgroups of G, $H_1 \lhd H_2$ implies $vH_1 \lhd vH_2$, with vH_2/vH_1 a homomorphic image of H_2/H_1. In particular, if $H_1, H_2 \in S_K$, with $H_1 \lhd H_2$, then $vH_2/vH_1 \cong H_2/H_1$.

(3) For $\bar{H}_1, \bar{H}_2 \in \bar{S}$, if $\bar{H}_1 \lhd \bar{H}_2$, then $v^{-1}\bar{H}_1 \lhd v^{-1}\bar{H}_2$, and $v^{-1}\bar{H}_2/v^{-1}\bar{H}_1 \cong \bar{H}_2/\bar{H}_1$.

Proof: *(1)* Clearly, ϕ is a well-defined mapping of S_K into $\bar{S}$. Also, ψ is a well-defined mapping of $\bar{S}$ into S_K since, for each $\bar{H} \in \bar{S}$, $v^{-1}(\bar{H}) \supset v^{-1}\{\bar{e}\} = K$.

For each $H \in S_K$, $(\psi\phi)(H) = \psi(\phi(H)) = v^{-1}(vH)$. If $x \in v^{-1}(vH)$, then $vx = vh$ for some $h \in H$, whence $v(xh^{-1}) = \bar{e}$; hence $xh^{-1} \in K \subset H$ and so $x \in H$. Thus, $v^{-1}(vH) \subset H$. If $h \in H$, then obviously, $h \in v^{-1}(vH)$. It follows that $v^{-1}(vH) = H$, for each $H \in S_K$. Hence $\psi\phi = \iota_{S_K}$.

For each $\bar{H} \in \bar{S}$, $(\phi\psi)(\bar{H}) = \phi(\psi(\bar{H})) = v(v^{-1}\bar{H})$. If $\bar{x} \in v(v^{-1}\bar{H})$, then $\bar{x} = vx$ for some $x \in v^{-1}\bar{H}$; hence $vx = \bar{x} \in \bar{H}$. Thus, $v(v^{-1}\bar{H}) \subset \bar{H}$. On the other hand, if $\bar{h} \in \bar{H}$, then $\bar{h} = vx$ for some $x \in v^{-1}\bar{H}$, whence $\bar{h} \in v(v^{-1}\bar{H})$. It follows that $\bar{H} = v(v^{-1}\bar{H})$, for each $\bar{H} \in \bar{S}$. Hence $\phi\psi = \iota_{\bar{S}}$. But then ϕ and ψ are inverse bijections. Obviously, both ϕ and ψ preserve set inclusion.

(2) Suppose $H_1 \lhd H_2$ (H_1, H_2 subgroups of G). If $\bar{x} \in vH_2$ and $\bar{y} \in vH_1$, then $\bar{x} = vx$ and $\bar{y} = vy$, where $x \in H_2$, $y \in H_1$; hence $\overline{xyx}^{-1} = vxvy(vx)^{-1} = v(xyx^{-1}) \in vH_1$, and so $vH_1 \lhd vH_2$. The mapping $\mu : H_2 \to vH_2/vH_1$ defined by: $\mu(x) = vH_1vx$ $(x \in H_2)$ is easily seen to be an epimorphism with $H_1 \subset \text{Ker } \mu$. Since $H_1 \lhd H_2$, we have $H_1 \lhd \text{Ker } \mu$. By Theorem 2.10.1, $H_2/H_1/\text{Ker } \mu/H_1 \cong H_2/\text{Ker } \mu \cong \text{Im } \mu = vH_2/vH_1$. Thus, vH_2/vH_1 is a homomorphic image of H_2/H_1. In particular, if $H_1, H_2 \in S_K$, with $H_1 \lhd H_2$, then $\text{Ker } \mu = \{x \in H_2 \mid vH_1vx = vH_1\} = \{x \in H_2 \mid vx \in vH_1\} = \{x \in H_2 \mid x \in v^{-1}vH_1\} = H_1$. Hence, in this case,

$$H_2/H_1 = H_2/\text{Ker } \mu \cong vH_2/vH_1.$$

(3) Suppose $\bar{H}_1 \lhd \bar{H}_2$ $(\bar{H}_1, \bar{H}_2 \in \bar{S})$. If $x \in v^{-1}\bar{H}_2$ and $y \in v^{-1}\bar{H}_1$, then $vx \in vv^{-1}\bar{H}_2 = \bar{H}_2$ and $vy \in vv^{-1}\bar{H}_1 = \bar{H}_1$; hence $v(xyx^{-1}) = vxvy(vx)^{-1} \in \bar{H}_1$, and so $xyx^{-1} \in v^{-1}\bar{H}_1$. It follows that $v^{-1}\bar{H}_1 \lhd v^{-1}\bar{H}_2$.

Since $K = v^{-1}\{\bar{e}\} \subset v^{-1}\bar{H}_1$, we have $v^{-1}\bar{H}_1$, $v^{-1}\bar{H}_2 \in S_K$; hence $v^{-1}\bar{H}_2/v^{-1}\bar{H}_1 \cong v(v^{-1}\bar{H}_2)/v(v^{-1}\bar{H}_1) = \bar{H}_2/\bar{H}_1$, by part (2). (See Figure 22.) ∎

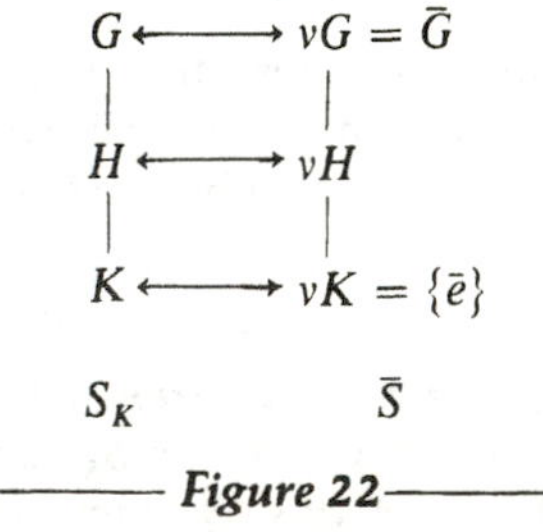

Figure 22

The following lemma provides a useful characterization of solvability for finite groups.

Lemma: A finite group G is solvable if and only if it has a normal series, all of whose factors are cyclic of prime order.

Proof: Obviously, if G has a normal series all of whose factors are cyclic, then G is solvable.

Conversely, suppose G is solvable. Let

$$G = G_n \rhd G_{n-1} \rhd \dots \rhd G_0 = \{e\} \tag{1}$$

be a solvable normal series. Each G_i $(i = 1, \dots, n)$ is finite; hence, between G_i and G_{i-1}, there are subgroups $H_{i1}, \dots, H_{is_i}$ such that

$$G_i = H_{is_i} \rhd H_{is_i-1} \rhd \dots \rhd H_{i0} = G_{i-1},$$

where each of the H_{ij-1} is *maximal* normal in H_{ij} $(j = 1, \dots, s_i)$. By the Correspondence Theorem, it follows that, for each $j = 1, \dots, s_i$, H_{ij}/H_{ij-1} is a simple group. Since $G_{i-1} \lhd G_i$, we have $G_{j-1} \lhd H_{ij}$ for each $j = 1, \dots, s_i$. By Theorem 2.10.1, $H_{ij}/H_{ij-1} \cong H_{ij}/G_{i-1}/H_{ij-1}/G_{i-1}$. But then H_{ij}/H_{ij-1} is a homomorphic image of H_{ij}/G_{i-1}, which is a subgroup of G_i/G_{i-1}, hence abelian. By Exercise 2.10.3, a simple abelian group must be cyclic of prime order. Hence the normal series obtained from (1) by splicing in the H_{ij} $(j = 1, \dots, s_i)$ for each $i = 1, \dots, n$, is of the required form. ■

Remark: A normal series, all of whose factors are simple, is called a *composition series*, and its factors are called composition factors. The preceding proof shows that for a finite solvable group, every solvable normal series can be refined to a composition series, all of whose factors have prime order. These primes, and their multiplicities, are uniquely determined for a given group G since their product is equal to the order of G. Hence the structure of the composition factors and their "multiplicities" (i.e., the frequencies with which the abstract cyclic groups occur as composition factors) are uniquely determined by G. (The *sequence* of the composition factors, however, is *not* determined by G (see Exercise 2.14.3).

These results represent special cases of much more general theorems.

(1) If G is a group which has a composition series, then every normal series of G can be refined to a composition series (Schreier Refinement Theorem). (cf. [3], [5], [7], [8].)

(2) If $\mathcal{N}$ and $\mathcal{N}'$ are two composition series of a group G, then there is a 1-1 correspondence between the *factors* of $\mathcal{N}$ and the *factors* of $\mathcal{N}'$ such that corresponding factors are isomorphic (Jordan-Hölder Theorem). (cf. [3], [5], [7], [8].)

In Theorem 2.10.3, we proved the simplicity of the alternating group A_n, for $n \geq 5$. From this, it easily follows that the symmetric group, S_n, is not solvable for $n \geq 5$. (See Exercise 2.14.4.)

The following simpler proof establishes the unsolvability of S_n $(n \geq 5)$ directly.

Theorem 2.14.2: The symmetric group S_n is solvable if and only if $n \leq 4$.

Proof: For $n = 1, 2$, S_n is abelian, hence solvable. For $n = 3$, $S_3 \rhd A_3 \rhd \{\iota\}$ is a solvable normal series. For $n = 4$,

$$S_4 \rhd A_4 \rhd V_4 \rhd \{\iota\},$$

where $V_4 = \{(1), (12)(34), (13)(24), (14)(23)\}$, is a solvable normal series.

Suppose $n \geq 5$. We first prove that, if $H \lhd K < S_n$ such that K/H is abelian and K contains every 3-cycle in S_n, then H also contains every 3-cycle. If K/H is abelian, then (by Theorem 2.11.4) all commutators of elements of K are in H. Let (abc) be a 3-cycle ($a, b, c \in I_n$), and let d, e be two additional elements of I_n. (Such elements exist since $n \geq 5$.) Then

$$(abc) = \sigma\tau\sigma^{-1}\tau^{-1}$$

where $\sigma = (abe)$, $\tau = (cda)$ in K. Thus, H contains every 3-cycle. Now, if S_n is solvable, then there is a solvable normal series

$$S_n = H_t \rhd H_{t-1} \rhd \dots \rhd H_0 = \{\iota\},$$

with H_i/H_{i-1} abelian for each $i = 1, \dots, t$. Obviously, $S_n = H_t$ contains every one of its 3-cycles, hence so does H_{t-1}, so does H_{t-2}, etc., and finally, so does $H_0 = \{\iota\}$. But this is absurd, and so S_n is not a solvable group. ∎

(This elegant proof, due to A. N. Milgram, appears in [10].)

Theorem 2.14.3: *(1)* Every subgroup of a solvable group is solvable.

(2) Every factor group (hence every homomorphic image) of a solvable group is solvable.

(3) If G is a group, $H \lhd G$ such that H and G/H are solvable, then G is solvable.

Proof: *(1)* Let

$$G = G_n \rhd G_{n-1} \rhd \dots \rhd G_0 = \{e\} \tag{1}$$

be a solvable normal series for G. If $H < G$, then

$$H = G_n \cap H \supset G_{n-1} \cap H \supset \cdots \supset G_0 \cap H = \{e\}. \tag{2}$$

The inclusions in (2) are normal inclusions. For: if $y \in G_{i-1} \cap H$ and $x \in G_i \cap H$ $(1 \leq i \leq n)$, then $xyx^{-1} \in G_{i-1}$ since $G_{i-1} \lhd G_i$, and $xyx^{-1} \in H$, by closure of H. Thus, $xyx^{-1} \in G_{i-1} \cap H$. It follows that $G_{i-1} \cap H \lhd G_i \cap H$ for each $i = 1, \dots, n$.

For $x, y \in G_i \cap H$, since G_i/G_{i-1} is abelian, the commutator $xyx^{-1}y^{-1}$ is an element of G_{i-1}, hence of $G_{i-1} \cap H$. But then $(G_i \cap H)/(G_{i-1} \cap H)$ is abelian, for each $i = 1, \dots, n$, and so (2) is a solvable normal series for H.

(2) Again let (1) be a normal series for G. If $H \lhd G$, let $\nu : G \to G/H$ be the natural homomorphism. By the Correspondence Theorem (Theorem 2.14.1(2),

$$G/H = \nu G_n \rhd \nu G_{n-1} \rhd \cdots \rhd \nu G_0 = \{\nu e\}, \tag{3}$$

with $\nu G_i/\nu G_{i-1}$ a homomorphic image of G_i/G_{i-1}, hence abelian, for each $i = 1, \dots, n$. Thus, (3) is a solvable normal series for G/H, and so G/H is solvable.

(3) Suppose $H \lhd G$, with H and G/H both solvable. Let

$$H = H_s \rhd H_{s-1} \rhd \cdots \rhd H_0 = \{e\} \tag{4}$$

be a solvable normal series for H, and let

$$G/H = \bar{G}_t \rhd \bar{G}_{t-1} \rhd \cdots \rhd \bar{G}_0 = \{\bar{e}\} \tag{5}$$

be a solvable normal series for G/H. Let $\nu : G \to G/H$ be the natural homomorphism. Then, by the Correspondence Theorem,

$$G = \nu^{-1}\bar{G}_t \rhd \nu^{-1}\bar{G}_{t-1} \rhd \cdots \rhd \nu^{-1}\bar{G}_0 = H \tag{6}$$

with $\nu^{-1}\bar{G}_i/\nu^{-1}\bar{G}_{i-1}$ isomorphic to $\bar{G}_i/\bar{G}_{i-1}$, hence abelian, for each $i = 1, \ldots, t$. But then

$$G = \nu^{-1}\bar{G}_t \rhd \cdots \rhd \nu^{-1}G_0 = H \rhd H_{s-1} \rhd \cdots \rhd H_0 = \{e\} \tag{7}$$

is a solvable normal series for G. ■

Exercises 2.14

True or False

1. Every finite group of even order is solvable.
2. If a simple group is solvable, then it is cyclic of prime order.
3. If G is a group which has a solvable normal subgroup, then G is solvable.
4. If G is a group which has a solvable homomorphic image, then G is solvable.
5. G_2 is one of the factors of the normal series $G = G_0 \rhd G_1 \rhd G_2 \rhd G_3 \rhd G_4 = \{e\}$.

2.14.1. Find all composition series for the dihedral group D_4.

2.14.2. Find all composition series for the symmetric group S_4.

2.14.3. Give an example to illustrate that the *sequence* of composition factors is not uniquely determined for a given finite group. (Hint: use direct products.)

2.14.4. For $n \geq 5$, use the simplicity of A_n (Theorem 2.10.3) to prove that S_n is not solvable.

2.14.5. If $G_1, \ldots, G_n$ are solvable groups, prove that the direct product $G = G_1 \times \ldots \times G_n$ is a solvable group.

2.14.6. Let G be a group with center $C(G)$. Prove: if $G/C(G)$ is solvable, then so is G.

2.14.7. Let G be a group such that, for some $a \in G$, $[a] \lhd G$ and $G/[a]$ is solvable. Prove that G is solvable.

2.14.8. Prove that a group G is solvable if and only if its commutator subgroup $[G, G]$ is solvable. (Do *not* use Exercise 2.14.9.)

2.14.9. Let G be a group. Prove that G is solvable if and only if G has a normal series of the form

$$G = G_n \rhd G_{n-1} \rhd \cdots \rhd G_0 = \{e\},$$

where $G_{i-1} = [G_i, G_i]$ for each $i = 1, \ldots, n$.

2.14.10. Deduce the result of Exercise 2.14.8 from the result of Exercise 2.14.9.

2.14.11. Let m be a positive integer divisible by $n!$ for some $n \geq 5$. Prove that there exists a non-solvable group of order m. Hint: use direct products.

2.14.12. Let m be a positive integer such that $2^3 | m$ and $3 | m$. Prove that there exists a non-abelian solvable group of order m.

2.14.13. Let L and L' be lattices with partial order relations R and R', respectively. A bijection $\beta : L \to L'$ such that $aRb \Rightarrow \beta(a)\ R'\beta(b)$ for each $a, b \in L$ is a *lattice isomorphism* of L onto L'.

In Theorem 2.14.1, interpret S_K and $\bar{S}$ as lattices and define a lattice isomorphism of S_K onto $\bar{S}$.

2.15

Primary Groups and the Sylow Theorems

(*Note*: Only the first Sylow Theorem (Theorem 2.15.3) and Theorem 2.15.6 are required as preparation for Galois Theory.)

By Lagrange's Theorem (Theorem 2.5.2), the order of every subgroup, hence of every element, of a finite group, G, is a divisor of the order of G. Conversely, in the very special case where G is a finite cyclic group of order n, there is a unique subgroup of order m for each positive divisor m of n; hence there are $\varphi(m)$ elements of order m. (See Exercise 2.3.10.) For G an arbitrary finite abelian group of order n, Theorem 2.13.1 implies that, for each divisor m of n, there is a subgroup, not necessarily unique, of order m (see Exercise 2.13.4), but there need not be an element of order m (e.g., if G is non-cyclic of order n, there is no element of order n). For a non-abelian finite group G of order n, we have seen that, for some divisors m of n, there may be no subgroup (hence certainly no element) of order m. (See Exercise 2.10.1.) The question thus arises: are there *special* divisors m of n for which the existence of subgroups, or even the existence of elements, of order m *can* be guaranteed? We shall see that the existence of a subgroup of order m can be guaranteed whenever m is a prime power divisor of n; hence the existence of *elements* of order m can be guaranteed whenever m is a prime divisor of n.

For the special case where G is an *abelian* group of finite order n, we have already proved (*without* employing the Fundamental Theorem of Abelian Groups) that, for every prime divisor p of n, G has an element of order p. (Lemma 1, p. 114). This lemma will once more be useful to us here.

We have seen (Exercise 2.8.3) that the relation R of *conjugacy* defined on a group G by:

$$yRx \quad \text{if} \quad y = axa^{-1} \quad \text{for some} \quad a \in G \quad (x, y \in G)$$

is an equivalence relation on G. The equivalence classes with respect to R are the *conjugate classes* of G. We write $\bar{x}$ for the conjugate class of $x \in G$.

For $x, a \in G$, $axa^{-1} = x$ if and only if $ax = xa$. Thus, the number of distinct elements in $\bar{x}$ will depend on the number of elements $a \in G$ such that a commutes with x.

Definition 2.15.1: Let G be a group, $x \in G$. Then the *centralizer*, $C(x)$, *of* x *in* G is the set

$$C(x) = \{a \in G \mid ax = xa\}.$$

It is easily verified that, for each $x \in G$, $C(x) < G$. (See Exercise 2.15.8).

Theorem 2.15.1: Let G be a group, and let $x \in G$. Then the cardinality of the conjugate class, $\bar{x}$, of x is equal to the index, $[G : C(x)]$, of the centralizer $C(x)$ in G.

Proof: It suffices to exhibit a 1-1 correspondence between the elements of $\bar{x}$ and the left cosets of $C(x)$ in G. For $a, b \in G$, we have

$$aC(x) = bC(x) \Leftrightarrow a^{-1}b \in C(x) \Leftrightarrow a^{-1}bx = xa^{-1}b \Leftrightarrow bxb^{-1} = axa^{-1}.$$

This implies that $aC(x) \mapsto axa^{-1}$ is a (well-defined) 1-1 map of the set of all left cosets of $C(x)$ in G onto the conjugate class $\bar{x}$. It follows that the cardinality of $\bar{x}$ is equal to $[G : C(x)]$. ■

In the particular case where $x \in C(G)$, the center of G, we obviously have $C(x) = G$ and $\bar{x} = \{x\}$. (We shall refer to a conjugate class of more than one element as a *non-trivial* conjugate class, and to a conjugate class of one element as a trivial one.) This simple remark, together with Theorem 2.15.1, leads to the following important theorem.

Theorem 2.15.2. (Class Equation): Let G be a finite group, and let $x_1, \ldots, x_t \in G$ be a set of representative elements, one chosen from each non-trivial conjugate class of G. Then

$$|G| = \sum_{i=1}^{t} [G : C(x_i)] + |C(G)|. \tag{1}$$

Proof: The total number of elements belonging to trivial conjugate classes is equal to $|C(G)|$. The total number of elements belonging to non-trivial conjugate classes is

$$\sum_{i=1}^{t} |\bar{x}_i| = \sum_{i=1}^{t} [G : C(x_i)],$$

where $|\bar{x}_i|$ is the number of elements in $\bar{x}_i$. From this, (1) follows immediately. ■

The following theorem is the first one of the Sylow Theorems (named for the Norwegian mathematician Peter Ludvig Mejdell Sylow, 1832-1918).

Theorem 2.15.3 (Sylow 1): Let G be a group of finite order n, and let p be a prime such that $p \mid |G|$. If k is a positive integer such that $p^k \mid n$, then G has a subgroup of order p^k.

Proof: We proceed by induction on n. If $n = 1$, the theorem holds vacuously. For $n > 1$, suppose the theorem holds for groups of order $< n$. We distinguish two cases.

Case 1: $p \nmid [G:H]$ for some $H \subsetneqq G$. Then $p^k \mid |H|$. By the induction hypothesis, since $|H| < n$, H has a subgroup of order p^k; hence so has G.

Case 2: $p \mid [G:H]$ for each $H \subsetneqq G$. In particular, for each $x \in G \setminus C(G)$, $C(x) \subsetneqq G$ and so $p \mid [G:C(x)]$. If $x_1, \ldots, x_t$ is a representative set of elements of the non-trivial conjugate classes of G, then

$$|G| = \sum_{i=1}^{t} [G:C(x_i)] + |C(G)|.$$

Since $p \mid |G|$ and $p \mid \sum_{i=1}^{t} [G:C(x_i)]$, we have $p \mid |C(G)|$. But $C(G)$ is abelian; hence (by Lemma 1, p. 114), $C(G)$ has an element, t, hence a subgroup $T = [t]$, of order p. From $T \subset C(G)$, we have $T \lhd G$ (see Exercise 2.8.6), and $|G/T| = \frac{n}{p} < n$. But then $p^{k-1} \mid |G/T|$, and the induction hypothesis guarantees that $|G/T|$ has a subgroup, Q, of order p^{k-1}. If $v: G \to G/T$ is the natural homomorphism, then (by the Correspondence Theorem) $Q = vS = S/T$ for some subgroup, S, of G. But then $|S| = [S:T] \cdot |T| = p^{k-1} \cdot p = p^k$, and so S is a subgroup of G with the required property. ■

Corollary 1: If G is a finite group and p is a prime such that $p \mid |G|$, then G has an element of order p. (This generalizes Lemma 1, p. 114, to arbitrary finite groups).

Corollary 2: If p is a prime and G is a finite group, then G is a p-group if and only if $|G|$ is a power of p.

Proof: By Corollary 1, Theorem 2.5.2, if $|G|$ is a power of p, then $o(x)$ is a power of p for each $x \in G$, hence G is a p-group. Conversely, suppose G is a p-group. If $q \mid |G|$ for some prime $q \neq p$, then G has an element of order q, contrary to Definition 2.12.1 of a p-group. It follows that $|G|$ is a power of p. ■

Definition 2.15.2: Let G be a finite group, p a prime, and let p^s be the highest power of p that divides $|G|$. Then any subgroup of G of order p^s is a *Sylow p-subgroup of G*.

By Theorem 2.15.3, for every prime p, every finite group G has a Sylow p-subgroup (trivial in case $s = 0$).

The remaining Sylow Theorems concern the class of all Sylow p-subgroups of a given group G for a given prime p.

We first extend the notion of conjugacy from elements to subsets.

Definition 2.15.3: Let G be a finite group and let X, Y be subsets of G. Then Y *is conjugate to* X if $Y = aXa^{-1}$ for some $a \in G$. More generally, if $K < G$, then Y *is* K*-conjugate to* X if $Y = kXk^{-1}$ for some $k \in K$.

Note that conjugacy of subsets is a special case of K-conjugacy, with $K = G$.

It is easy to check that, for $K < G$, K-conjugacy of subsets is an equivalence relation on the power set of G. The equivalence classes with respect to K-conjugacy of subsets are called *K-conjugacy classes of subsets*, or simply *conjugacy classes of subsets* in case $K = G$. If, in particular, X is a sub*group* of G, then the class of all subsets (K-)conjugate to X consists entirely of sub*groups*.

Definition 2.15.4: Let G be a group, X a subset of G. Then the *normalizer of X in G* is the set

$$N(X) = \{a \in G \mid aXa^{-1} = X\}.$$

It is easy to verify (see Exercise 2.15.8) that $N(X)$ is a sub*group* of G for any subset X of G.

While, clearly, $aX = Xa$ for each $a \in N(X)$, the elements of $N(X)$ do not, in general, commute individually with the elements of X. The set $C(X) = \{a \in G \mid ax = xa \text{ for all } x \in X\}$ is called the *centralizer* of X in G. (See Exercise 2.15.8). For any subset X of G, $C(X) \subset N(X)$. In the special case where X consists of a single element, x, we have $N(X) = C(X)$ equal to $C(x)$, the centralizer of the element x (as previously defined).

If $H < G$, then, clearly, $H \lhd N(H)$, and $H \lhd G$ if and only if $N(H) = G$.

Theorem 2.15.1 can now be generalized to the following.

Theorem 2.15.4: Let G be a group, $K < G$ and $X \subset G$. Then the cardinality of the class of all subsets of G that are K-conjugate to X is equal to $[K:K \cap N(X)]$.

Proof: It suffices to exhibit a 1–1 correspondence between the elements of the K-conjugacy class, $\bar{X}$, of X and the left cosets of $K \cap N(X)$ in K. For $a, b \in K$, we have

$$\begin{aligned} a(K \cap N(X)) = b(K \cap N(X)) &\Leftrightarrow a^{-1}b \in K \cap N(X) \Leftrightarrow a^{-1}b \in N(X) \\ &\Leftrightarrow (a^{-1}b)^{-1}Xa^{-1}b = X \Leftrightarrow b^{-1}aXa^{-1}b = X \Leftrightarrow aXa^{-1} \\ &= bXb^{-1}. \end{aligned}$$

This implies that $a(K \cap N(X)) \mapsto aXa^{-1}$ is a (well-defined) 1–1 map of the set of all left cosets of $K \cap N(X)$ in K onto the class, $\bar{X}$, of all subsets of G which are K-conjugate to X. It follows that the cardinality of $\bar{X}$ is equal to $[K:K \cap N(X)]$. ■

Corollary: If G is a finite group, $K < G$, then the number of elements in any K-conjugacy class of subsets of G is a divisor of the order of G.

Proof: For $X \subset G$, $[K:K \cap N(X)]$ is a divisor of $|K|$ which, in turn, is a divisor of $|G|$. ■

To prove the remaining Sylow Theorems, we require the following lemma.

Lemma: Let G be a finite group, p a prime, and S a Sylow p-subgroup of G. Let H be another p-subgroup of G. Then $H \cap N(S) = H \cap S$.

Proof: Since $S \lhd N(S)$ and $H \cap N(S) < N(S)$, Theorem 2.10.2 implies that

$$(H \cap N(S))S/S \cong H \cap N(S)/(H \cap N(S)) \cap S = H \cap N(S)/H \cap S.$$

Since H is a p-group, so are $H \cap N(S)$ and its homomorphic image $H \cap N(S)/H \cap S$. Hence $(H \cap N(S))S/S$ is a p-group. Since S is a p-group, this implies that $(H \cap N(S))S$ is a p-group. But S is a p-subgroup of maximal order in G, and $S \subset (H \cap N(S))S$. Hence $(H \cap N(S))S = S$, and so

$$|(H \cap N(S))S/S| = |H \cap N(S)/H \cap S| = 1.$$

Thus, $H \cap N(S) = H \cap S$. ■

Remark: If $K < G$, then any two K-conjugate subsets of G are, clearly, (G-)conjugate. Thus, every (G-)conjugacy class of subsets of G may be partitioned into K-conjugacy classes. (Put another way: the partitioning of the power set P_G into K-conjugacy classes is a refinement of its partitioning into (G-)conjugacy classes.) We make use of this fact in the proof of the following theorem.

Theorem 2.15.5 (Sylow 2 and 3): Let G be a finite group, p a prime, and S a Sylow p-subgroup of G. Then:

(1) The conjugacy class of S is the set of all Sylow p-subgroups of G. (Sylow 2.)

(2) The number of Sylow p-subgroups of G is equal to $[G:N(S)]$, which divides $|G|$ and is $\equiv 1 \bmod p$. (Sylow 3.)

Proof: *(1)* Clearly, every conjugate of S is a subgroup of the same maximal p-power order, hence a Sylow p-subgroup. We need to prove that *every* Sylow p-subgroup of G *is* conjugate to S. Let K be a Sylow p-subgroup of G. Then the conjugacy class, $\bar{S}$, of S is equal to the disjoint union of K-conjugacy classes $\bar{S}_i$ $(i = 1, \ldots, r)$, where $S_1 = S, S_2, \ldots, S_r$ are subgroups of G conjugate to S. By Theorem 2.15.4 and the Lemma, we have

$$[G:N(S)] = \sum_{i=1}^{r} [K:K \cap N(S_i)] = \sum_{i=1}^{r} [K:K \cap S_i]. \tag{1}$$

Since S is a subgroup of maximum p-power order, and $S \subset N(S) \subset G$, p does not divide $[G:N(S)]$. Since all the terms on the right-hand side of (1) are powers of p, it follows that there is some i $(1 \leq i \leq r)$ such that $[K:K \cap S_i] = 1$, hence $K \subset S_i$. Since K is a Sylow p-subgroup, we have $|K| = |S_i| = |S|$, and so $K = S_i$, a conjugate of S.

(2) By (1) and Theorem 2.15.4, it follows that the number of Sylow p-subgroups of G is equal to $[G:N(S)]$, a divisor of $|G|$. As we have seen, *at least* one of the terms on the right-hand side of (1) (where K is any Sylow p-subgroup of G) is equal to $1 = p^0$. Indeed, *only* one term *can* be equal to 1, for: if $[K:K \cap S_j] = 1$ $(1 \leq j \leq r)$, then $S_j = K$. But only *one* of the S_i can be equal to K since the S_i belong to distinct K-conjugacy classes. Hence all but one of the terms on the right-hand side of (1) is a non-trivial power of p, and we therefore have

$$[G:N(S)] \equiv 1 \bmod p.$$

■

Corollary: Let G be a finite group, p a prime, and S a Sylow p-subgroup of G. Then $S \lhd G$ if and only if S is the *only* Sylow p-subgroup of G.

Proof: Using Theorem 2.15.5, we have $S \lhd G \Leftrightarrow N(S) = G \Leftrightarrow [G:N(S)] = 1 \Leftrightarrow S$ is the only conjugate of S in $G \Leftrightarrow S$ is the only Sylow p-subgroup of G. ■

Example 1:

Let G be a group of order 60. Since $60 = 2^2 \cdot 3 \cdot 5$, G has Sylow 2-subgroups of order 4, Sylow 3-subgroups of order 3, and Sylow 5-subgroups of order 5.

The number of Sylow 2-subgroups of G must be a divisor of 60 that is $\equiv 1 \bmod 2$. Hence there are either 1, 3, 5, or 15 Sylow 2-subgroups of G.

The number of Sylow 3-subgroups of G must be a divisor of 60 that is $\equiv 1 \bmod 3$. Hence there are either 1, 4, or 10 Sylow 3-subgroups of G.

The number of Sylow 5-subgroups of G must be a divisor of 60 that is $\equiv 1 \bmod 5$. Hence there are either 1 or 6 Sylow 5-subgroups of G.

Example 2:

Let G be a group of order 28. Since $28 = 2^2 \cdot 7$, G has Sylow 2-subgroups of order 4 and Sylow 7-subgroups of order 7. The number of Sylow 7-subgroups must be a divisor of 28 that is $\equiv 1 \bmod 7$. Since 1 is the only such divisor, there is only one Sylow 7-subgroup, necessarily normal, by the preceding corollary. Thus, a group of order 28 cannot be a simple group. (For a generalization of this result, see Exercise 2.15.11.)

We conclude this section by examining further the structure of finite p-groups.

Lemma: If G is a finite non-trivial p-group (p prime), then the center, $C(G)$, of G is non-trivial.

Proof: If G is abelian, then $C(G) = G$, non-trivial by hypothesis. If G is non-abelian, of order p^n $(n > 1)$, then (by Theorem 2.15.2)

$$p^n = \sum_{i=1}^{t} [G:C(x_i)] + |C(G)|, \tag{1}$$

where $\{x_1, \ldots, x_t\}$ is a complete set of representatives of the non-trivial conjugate classes of G. For each $i = 1, \ldots, t$, $[G:C(x_i)]$ is a divisor of $|G| = p^n$, hence a power of p (different from 1). It follows that $p \mid |C(G)|$, and so $C(G)$ is non-trivial. ■

Theorem 2.15.6: Every finite p-group is solvable. If $|G| = p^n$, then G has a solvable normal series

$$G = G_n \rhd G_{n-1} \rhd \cdots \rhd G_0 = \{e\},$$

where $|G_i| = p^i$ for each $i = 0, \ldots, n$.

Proof: First note that, if H is an *abelian* p-group of order p^k, then Theorem 2.15.3 guarantees the existence of subgroups H_i $(i = 0, \ldots, k)$ such that

$$H = H_k \supset H_{k-1} \supset \cdots \supset H_0 = \{e\},$$

with $|H_i| = p^i$ for each $i = 0, \ldots, k$, and all inclusions obviously normal.

For G an arbitrary finite p-group, of order p^n, we proceed by induction on n. If $n = 0$, the theorem holds trivially. For $n > 0$, suppose the theorem holds for p-groups of order p^k, $k < n$. By the Lemma, $C(G)$ is non-trivial; hence $|C(G)| = p^k$, $1 \leq k \leq n$. If $k = n$, then G is abelian; hence G is certainly solvable, with a normal series of the required form. If $k < n$, then $|G/C(G)| = p^h$, where $h = n - k$, with $1 \leq h < n$. By the induction hypothesis, $G/C(G)$ has a solvable normal series

$$G/C(G) = \bar{G}_h \rhd \bar{G}_{h-1} \rhd \cdots \rhd \bar{G}_0 = \{e\}, \tag{1}$$

with $|\bar{G}_i| = p^i$ for each $i = 0, \ldots, h$. Let $v: G \to G/C(G)$ be the natural homomorphism. By the Correspondence Theorem, for each $i = 0, \ldots, h$, $v^{-1}(\bar{G}_i)/C(G) \cong \bar{G}_i$; hence $|v^{-1}\bar{G}_i| = p^{i+k}$ for each $i = 0, \ldots, h$, and

$$G = v^{-1}\bar{G}_h \rhd \cdots \rhd v^{-1}\bar{G}_0 = C(G), \tag{2}$$

with $v^{-1}\bar{G}_i/v^{-1}\bar{G}_{i-1} \cong \bar{G}_i/\bar{G}_{i-1}$, hence abelian $(i = 1, \ldots, h)$.

Since $C(G)$ is abelian, there are subgroups H_j of $C(G)$ such that

$$C(G) = H_k \rhd H_{k-1} \rhd \cdots \rhd H_0 = \{e\} \tag{3}$$

with $|H_j| = p^j$ for each $j = 0, \ldots, k$, clearly a solvable normal series for $C(G)$.

Splicing together (2) and (3), we obtain a solvable normal series for G, of the required form:

$$G = v^{-1}\bar{G}_h \rhd \cdots \rhd v^{-1}\bar{G}_0 = C(G) \rhd H_{k-1} \rhd \cdots \rhd H_0 = \{e\}. \quad ■$$

Exercises 2.15

True or False

1. If G is a group such that $360 \mid |G|$, then G has a subgroup of order 360.
2. If G is a group such that 4913 divides $|G|$, then G has a subgroup of order 289.
3. A group of order 250 has a homomorphic image of order 2.
4. A group of order 72 may have exactly two subgroups of order 8.
5. A group of order 128677176 must have an element of order 2.

2.15.1. Find the Sylow subgroups of S_4.

2.15.2. Prove that all subgroups of order 8 in S_4 are isomorphic to D_4.

2.15.3. Let G be a group of order 45. Prove that G is not a simple group.

2.15.4. Let G be a group of order 56. Prove that G is not a simple group.

2.15.5. Modify the proof of Theorem 2.15.5(1) to obtain the result: if S is a Sylow p-subgroup of a finite group G (p prime), then every p-subgroup of G is contained in some conjugate of S.

2.15.6. Let G be a finite group, p a prime, and S a Sylow p-subgroup of G. Prove: if $S \lhd G$, then S contains all elements $x \in G$ such that $o(x)$ is a power of p.

2.15.7. Construct a non-abelian 2-group of order 64 and find a composition series for it.

2.15.8. Let G be a group and let X be a subset of G. Prove:

(1) $C(X) < G$, $N(X) < G$, with $C(X) \subset N(X)$.

(2) If $X < G$, then $X \lhd N(X)$; $X \lhd G \Leftrightarrow N(X) = G$.

2.15.9. (1) Prove: if p, q are primes, with $p < q$ and $q \not\equiv 1 \bmod p$, then any group of order pq is cyclic.

(2) For any odd prime q, prove that there is a non-abelian group of order $2q$. Why does this not contradict part (1)?

2.15.10. Prove that every finite abelian group is isomorphic to the direct product of its Sylow subgroups.

2.15.11. Let p, q be distinct primes and let G be a group of order p^2q. Prove that G is not a simple group.

2.15.12. Without using the simplicity of the alternating group A_5, prove that A_5 has no normal Sylow subgroups.

3

Rings, Modules, and Vector Spaces

3.1

Rings and Subrings

The most widely used algebraic systems have *two* binary operations — one of them a group operation. We encounter such systems quite early, as we learn to add and multiply integers, rational numbers, real numbers, complex numbers, polynomials, and, eventually, matrices. All of these systems form rings, in the sense of the following definition.

Definition 3.1.1: Let A be a set, and let $+, \cdot$ be binary operations on A. Then the ordered triple $\langle A, +, \cdot \rangle$ is a *ring* if:

(1) $\langle A, + \rangle$ is an abelian group.

(2) $\cdot$ is associative.

(3) $\cdot$ is left and right distributive over $+$, i.e.,

$$a \cdot (b + c) = a \cdot b + a \cdot c$$

$$(b + c) \cdot a = b \cdot a + c \cdot a$$

for all $a, b, c \in A$.

(When no confusion is likely regarding the operations, we shall say simply "A is a ring.")

Note that, while $\langle A, + \rangle$ is required to be an abelian group, $\cdot$ is not necessarily commutative, nor is the existence of an identity for $\cdot$ a requirement of our definition of a ring. (However, in some of the contemporary mathematical literature, the existence of a multiplicative identity *is* assumed in all rings under discussion.)

Notation: If $\langle A, +, \cdot \rangle$ is a ring, we denote by "0" the identity of $\langle A, + \rangle$. We write "$-a$" for the additive inverse of $a \in A$, and "$a - b$" for the element $a + (-b)$ $(a, b \in A)$.

Caution: Because of their occurrence in familiar examples of rings, it is customary to use the symbols " + " and " · " to denote the operations in a ring. Beginners are sometimes tempted to assume that some intrinsic meaning adheres to these symbols, and they attribute to the operations represented by " + " and " · " *all* of the familiar properties of addition and multiplication of numbers. Don't be fooled by labels! Such symbols as "∇" and "□" for the ring operations would do just as well — provided we attribute to them *no more and no less* than Definition 3.1.1 (*mutatis mutandis*) requires.

Theorem 3.1.1: Let $\langle A, +, \cdot\rangle$ be a ring. Then:

(1) $a \cdot 0 = 0 = 0 \cdot a$, for each $a \in A$;

(2) $a \cdot (-b) = -ab = (-a) \cdot b$, for each $a, b \in A$;

(3) $(-a) \cdot (-b) = ab$, for each $a, b \in A$;

(4) $a \cdot (b - c) = a \cdot b - a \cdot c$, for each $a, b, c \in A$;
$(b - c) \cdot a = b \cdot a - c \cdot a$, for each $a, b, c \in A$.

Proof: *(1)* For $a \in A$, we have $a \cdot 0 = a \cdot (0 + 0) = a \cdot 0 + a \cdot 0$. Thus, $a \cdot 0 + 0 = a \cdot 0 + a \cdot 0$. By (left) cancellation in the group $\langle A, +\rangle$ (see Theorem 2.2.4), we conclude that $0 = a \cdot 0$. Similarly, $0 \cdot a = 0$ (see Exercise 3.1.1).

(2) Because of the uniqueness of inverses in the abelian group $\langle A, +\rangle$, the first half of (2) will be proved if we can show that $a(-b) + ab = 0$. But this is certainly true since $a(-b) + ab = a[-b + b] = a \cdot 0 = 0 = ab + a(-b)$ (the final equality by the commutativity of $+$). It follows that $a(-b)$ is the additive inverse of ab, i.e., $a(-b) = -ab$.

We leave the proofs of the second half of (2), and of (3) and (4) as exercises (Exercise 3.1.1). ■

Before discussing specific examples of rings, we introduce names for special kinds of rings.

Definition 3.1.2: Let $\langle A, +, \cdot\rangle$ be a ring.

(1) If A contains an identity, 1, for $\cdot$, then $\langle A, +, \cdot\rangle$ is a *ring with identity*. (In the literature, the multiplicative identity is sometimes called a *unity* and rings with identity are referred to as "rings with unity.")

(2) If $\cdot$ is commutative, then $\langle A, +, \cdot\rangle$ is a *commutative ring*.

(3) If the product of two non-zero elements of A cannot be equal to zero, then A is a *ring without zero divisors*.

(4) A commutative ring with identity $1 \neq 0$, and without zero divisors, is an *integral domain*.

(5) A ring with identity $1 \neq 0$ is a *division ring* (or *skew field*) provided that each of its non-zero elements has a multiplicative inverse with respect to the identity 1.

(6) A commutative division ring is a *field*. (Thus, *a field is a commutative ring with identity* $1 \neq 0$, *each of whose non-zero elements has a multiplicative inverse*).

Why the emphasis on "$1 \neq 0$" in Definition 3.1.2? Because a ring with identity $1 = 0$ consists of 0 only and is thus very uninteresting. (See Exercise 3.1.2.)

Note that, by Theorems 2.1.1 and 2.1.2, if $\langle A, +, \cdot \rangle$ is a ring, then A has exactly one additive identity relative to which every element of A has exactly one additive inverse; A has at most one multiplicative identity relative to which any element of A can have at most one multiplicative inverse. ("At most one," of course, means: either none or one.)

Definition 3.1.3: In any ring A with identity, the elements of A which have multiplicative inverses are called the *units* of A.

The units in a ring A with identity form a multiplicative group, A^*, called the *unit group* of A. (See Exercise 3.1.3.)

Examples of Rings:

Example 1:

The familiar number systems. $\langle \mathbb{Q}, +, \cdot \rangle$, $\langle \mathbb{R}, +, \cdot \rangle$, $\langle \mathbb{C}, +, \cdot \rangle$ are very special rings. They are, in fact, fields.

$\langle \mathbb{Z}, +, \cdot \rangle$ is a commutative ring with identity $1 \neq 0$, and without zero divisors. It is thus an integral domain, *but it is not a field.* (Which integers have multiplicative inverses?)

Example 2:

Some finite rings: the residue class rings $\mathbb{Z}_n$ of $\mathbb{Z}$. Let n be a positive integer. On the set $\mathbb{Z}_n = \{\bar{0}, \bar{1}, \ldots, \overline{n-1}\}$ of all residue classes modulo n, define $\oplus$ and $\circ$ by $\bar{a} \oplus \bar{b} = \overline{a+b}$, $\bar{a} \circ \bar{b} = \overline{ab}$ (see Theorem 2.2.2 and Exercise 2.5.14(1)). Then $\langle \mathbb{Z}_n, \oplus, \circ \rangle$ is a commutative ring with identity $\bar{1}$, consisting of n elements. We shall see in the Corollary of Theorem 3.4.1 that $\langle \mathbb{Z}_n, \oplus, \circ \rangle$ is a field if and only if n is prime.

Example 3:

Some rings without identity. Let $2\mathbb{Z}$ be the set of all even integers, and let $+, \cdot$ be addition and multiplication of integers, restricted to $2\mathbb{Z}$. Then $\langle 2\mathbb{Z}, +, \cdot \rangle$ is a ring without zero divisors, with *no* identity. More generally, for each integer $n > 1$, the set, $n\mathbb{Z}$, of all multiples of n forms a ring with *no* identity. (See Exercise 3.1.4.)

Example 4:

Some non-commutative rings: matrix rings. First, consider the set $M_n(\mathbb{R})$ of all $n \times n$ matrices over $\mathbb{R}$, for $n \geq 1$. Under the usual operations of matrix addition and matrix multiplication, $M_n(\mathbb{R})$ forms a ring which, for $n \geq 2$, is non-commutative and has zero divisors Exercise 3.1.5.)

More generally, if A is any ring, not necessarily commutative, and n is a positive integer, we can form the set $M_n(A)$ of all $n \times n$ matrices with entries in A. As in $M_n(\mathbb{R})$, define $+$ and $\cdot$ on $M_n(A)$ by:

$$(a_{ij})_{n \times n} + (b_{ij})_{n \times n} = (c_{ij})_{n \times n}$$

where $c_{ij} = a_{ij} + b_{ij}$ for each (i, j), $1 \leq i, j \leq n$;

$$(a_{ij})_{n \times n} \cdot (b_{ij})_{n \times n} = (c_{ij})_{n \times n}$$

where $c_{ij} = \sum_{h=1}^{n} a_{ih}b_{hj}$ for each (i, j), $1 \leq i, j \leq n$. Then $\langle M_n(A), +, \cdot \rangle$ is a ring.

If A is a ring with identity, then so is $M_n(A)$. If A is not a "zero ring," i.e., if there are elements $a, b \in A$ such that $ab \neq 0$ then, for $n \geq 2$, $M_n(A)$ is a non-commutative ring with zero divisors. (See Exercise 3.1.5.)

Example 5:

A non-commutative division ring: the ring of real quaternions. Let $\mathbb{H}$ be the set of all *formal expressions* $a + bi + cj + dk$ $(a, b, c, d \in \mathbb{R})$, added and multiplied like polynomials in i, j, and k, but subject to the conditions $i^2 = j^2 = k^2 = -1$, $ij = k$, $jk = i$, $ki = j$. Then $\langle \mathbb{H}, +, \cdot \rangle$ is a non-commutative division ring (see Exercise 3.1.6). It is called the ring of *real quaternions*.

Another important class of examples, the polynomial rings, will be discussed in Section 3.5. The classes of rings defined in Definition 3.1.2 are not unrelated. One important relationship is expressed in the following theorem.

Theorem 3.1.2: Every division ring is a ring without zero divisors.

Proof: Let A be a division ring. Suppose $a, b \in A$ and $ab = 0$. We show that either $a = 0$ or $b = 0$. This is logically equivalent to either one of the statements:

(1) if $a \neq 0$, then $b = 0$

or

(2) if $b \neq 0$, then $a = 0$.

We prove statement (1). Suppose $a \neq 0$. Then a has a multiplicative inverse, a^{-1}, in A. From $ab = 0$, we have $a^{-1}(ab) = a^{-1}0$; hence $(a^{-1}a)b = 0$, and so $b = 0$. It follows that A has no zero divisors. ■

Corollary: Every field is an integral domain.

Proof: If A is a field, then A is a commutative division ring; hence it has no zero divisors. But then A is a commutative ring with identity $1 \neq 0$, without zero divisors, hence an integral domain. ■

Warning: We have already observed that the converse of Theorem 3.1.2, and of its corollary, is false; e.g., $\langle \mathbb{Z}, +, \cdot \rangle$ is an integral domain, but not a field.

Rings without zero divisors have another interesting property: they are *cancellation rings*. In fact, we have the following theorem.

Theorem 3.1.3: Let $\langle A, +, \cdot \rangle$ be a ring. Then the following statements are equivalent:

(1) If $a, b \in A$ and $ab = 0$, then $a = 0$ or $b = 0$ (i.e., A has no zero divisors).

(2) If $a, b \in A$ and $x \neq 0$ in A, then $ax = bx$ implies $a = b$ (right cancellation of non-zero factors holds in A).

(3) If $a, b \in A$ and $x \neq 0$ in A, then $xa = xb$ implies $a = b$ (left cancellation of non-zero factors holds in A).

Proof: We prove that (1) and (2) are equivalent. By a symmetric argument, the equivalence of (1) and (3) then follows.

(1) $\Rightarrow$ (2): Suppose A has no zero divisors. For $a, b \in A$, $x \neq 0$ in A, suppose $ax = bx$. Then $0 = ax - bx = (a - b)x$. Since $x \neq 0$, we conclude that $a - b = 0$; hence $a = b$. Thus, right cancellation of non-zero factors holds in A.

(2) $\Rightarrow$ (1): Suppose right cancellation of non-zero factors holds in A. For $a, b \in A$, suppose $ab = 0$. If $b \neq 0$, then, from $ab = 0 = 0b$, we have $a = 0$, by right cancellation. Thus, A has no zero divisors. ∎

Definition 3.1.4: Let A be a ring, and let B be a subset of A. Then B is a *subring* of A if B forms a ring under the operations of A (restricted to B).

As in the case of groups, we need a convenient operational test for determining whether a given subset of a ring *is* a subring.

Theorem 3.1.4: Let $\langle A, +, \cdot \rangle$ be a ring, $B \subset A$. Then B is a subring of A if and only if

(1) $B \neq 0$.

(2) $x - y \in B$, for each $x, y \in B$.

(3) $xy \in B$, for each $x, y \in B$.

Proof: If B is a subring of A, then B forms a group under $+$, restricted to B, hence forms a subgroup of $\langle A, + \rangle$. By the Corollary of Theorem 2.3.1, (1) and (2) hold. Since B forms a ring under $+$ and $\cdot$, restricted to B, (3) holds.

Conversely, suppose (1), (2), and (3) hold. Again by the Corollary of Theorem 2.3.1, B is a subgroup of $\langle A, + \rangle$; hence it is clearly abelian. By (3), $\cdot$ (restricted to B) serves as a binary operation on B. The associativity and distributivity conditions of Definition 3.1.1 are clearly inherited from A, and so B is a subring of A. ∎

Example 1:

$\mathbb{Z}$ is a subring of $\mathbb{Q}$, $\mathbb{Q}$ is a subring of $\mathbb{R}$, and $\mathbb{R}$ is a subring of $\mathbb{C}$.

Example 2:

$2\mathbb{Z}$ is a subring of $\mathbb{Z}$. (Thus, a ring without identity can be a subring of a ring with identity.)

Example 3:

The matrices $\begin{pmatrix} a & b \\ 0 & 0 \end{pmatrix}$, $a, b \in \mathbb{R}$, form a subring, A, of the ring $M_2(\mathbb{R})$ of all 2×2 matrices over $\mathbb{R}$. This ring, $A = \left\{\begin{pmatrix} a & b \\ 0 & 0 \end{pmatrix} \middle| a, b \in \mathbb{R}\right\}$, has no identity (see Exercise 3.1.7). Now consider $B = \left\{\begin{pmatrix} a & 0 \\ 0 & 0 \end{pmatrix} \middle| a \in \mathbb{R}\right\}$. Then B is a subring of A, with identity $E = \begin{pmatrix} 1 & 0 \\ 0 & 0 \end{pmatrix}$. (See Exercise 3.1.7.) Thus, a ring without identity can have a subring with identity. Indeed, since B is also a subring of $M_2(\mathbb{R})$ $\left(\text{which has identity } I_2 = \begin{pmatrix} 1 & 0 \\ 0 & 1 \end{pmatrix}\right)$, we see that a subring of a ring with identity may have an identity which is different from the identity of the whole ring.

Exercises 3.1

True or False

1. The positive integers form a ring under addition and multiplication of integers.
2. The power set of a set A forms a ring under the operations $\cup$ and $\cap$.
3. The nonzero elements of any ring $\langle A, +, \cdot \rangle$ form a group under multiplication.
4. If $\langle A, +, \cdot \rangle$ is a ring, then so is $\langle A, \cdot, + \rangle$.
5. Every group is the additive group of some ring.

3.1.1. Complete the proof of Theorem 3.1.1, i.e., prove: if $\langle A, +, \cdot \rangle$ is a ring then, for $a, b, c \in A$,

$$0 \cdot a = 0$$

$$(-a)b = -ab$$

$$(-a)(-b) = ab$$

$$a(b - c) = ab - ac$$

$$(b - c)a = ba - ca.$$

3.1.2. Prove that, if A is a ring with identity 1, and if $1 = 0$, then $A = \{0\}$.

3.1.3. Let $\langle A, +, \cdot \rangle$ be a ring with identity 1, and let A^* be the set of all units in A (see Definition 3.1.3). Prove that $\langle A^*, \cdot \rangle$ is a group (where " $\cdot$ " represents the restriction to A^* of the multiplication operation of A. (A^* is called the *unit group* of the ring A.)

3.1.4. Prove that, for every integer $n > 1$, the set $n\mathbb{Z}$ of all multiples of n in $\mathbb{Z}$ forms a ring without identity under the operations $+, \cdot$ of $\mathbb{Z}$, restricted to $n\mathbb{Z}$.

3.1.5. *(1)* Prove that, for $n \geq 2$, the set $M_n(\mathbb{R})$ of all $n \times n$ matrices over $\mathbb{R}$ forms a ring with respect to matrix addition and multiplication. Prove that the ring $M_n(\mathbb{R})$ is a non-commutative ring with identity, and that it has zero divisors.

(2) A ring in which *all* products are zero is called a "zero-ring." Generalize (1) to $M_n(A)$ $(n \geq 2)$ where A is any ring (not necessarily commutative). Show that

(a) $M_n(A)$ has zero divisors if $A \neq \{0\}$;

(b) $M_n(A)$ has an identity if A has an identity;

(c) $M_n(A)$ is non-commutative if and only if A is not a zero ring.

3.1.6. Verify that the set A of all real quaternions forms a ring under the operations described in Example 5. Prove that A is a non-commutative division ring, with $ji = -k$, $kj = -i$, $ik = -j$. Verify that A^* has a subgroup isomorphic to the quaternion group (see Exercise 2.8.12).

3.1.7. In the ring $M_2(\mathbb{R})$, let $A = \left\{\begin{pmatrix} a & b \\ 0 & 0 \end{pmatrix} | a, b \in \mathbb{R}\right\}$ and let $B = \left\{\begin{pmatrix} a & 0 \\ 0 & 0 \end{pmatrix} | a \in \mathbb{R}\right\}$. Prove that A is a subring, without identity, of the ring $M_2(\mathbb{R})$, and that B is a subring of A, hence of $M_2(\mathbb{R})$, with identity different from the identity of $M_2(\mathbb{R})$.

3.1.8. Prove: if F is a field and A is a subring of F with identity $1 \neq 0$, then A is an integral domain, and $1 = 1_F$.

3.1.9. Prove that every finite integral domain is a field. (Compare Exercise 2.2.13.) (Note: a much more powerful theorem regarding the effect of finiteness on a ring is the famous theorem of Wedderburn which asserts that every finite division ring is commutative, hence a field.)

3.1.10. Find the unit group of each of those examples on pages 134 and 135, which are rings with identity.

3.1.11. Let A be a ring. An element $e \in A$ is *idempotent* if $e^2 = e$; an element $k \in A$ is *square nilpotent* if $k^2 = 0$; if A has identity 1, then an element $v \in A$ is *involutory* if $v^2 = 1$.

Let A be a ring with identity, and let $e \in A$ be idempotent. Prove:

(1) $1 - e$ is idempotent;

(2) for each $x \in A$, $ex(1 - e)$ is square nilpotent;

(3) for each $x \in A$, $e + ex(1 - e)$ is idempotent;

(4) for each $x \in A$, $1 + ex(1 - e)$ is a unit in A;

(5) $2e - 1$ is involutory.

3.1.12. Give examples of non-trivial idempotents, square nilpotent elements, and involutions in $M_2(\mathbb{R})$ (see the preceding exercise).

3.1.13. An element a of a ring A is *nilpotent* if there is some positive integer $n \geq 1$ such that $a^n = 0$. Prove that, in any commutative ring A, the nilpotent elements form a subring.

3.1.14. Let A be a ring, and let $a \in A$ be a nilpotent element. If $n > 1$ is a positive integer such that $a^{n-1} \neq 0$ and $a^n = 0$, prove that a^{n-1} is square nilpotent, i.e., $(a^{n-1})^2 = 0$.

3.1.15. Let $F_{\mathbb{R}}$ be the set of all real-valued functions defined on $\mathbb{R}$. Define $+, \cdot$ by $f + g = h$ where $h(x) = f(x) + g(x)$ for each $x \in \mathbb{R}$; $fg = k$ where $k(x) = f(x) \cdot g(x)$ for each $x \in \mathbb{R}$. Let $C(-\infty, \infty)$ be the set of all real-valued continuous functions defined on $\mathbb{R}$. Prove that $\langle F_{\mathbb{R}}, +, \cdot \rangle$ is a commutative ring with identity, and that $C(-\infty, \infty)$ is a subring of $F_{\mathbb{R}}$. Identify all idempotent, nilpotent, and involutory elements of $F_{\mathbb{R}}$, and determine which ones are in $C(-\infty, \infty)$.

3.1.16. Let $\langle G, + \rangle$ be an abelian group, and let R be the set of all group endomorphisms of G (i.e., of all group homomorphisms of G into G). On R, define $+$ by: $\alpha + \beta = \gamma$ if $\gamma(x) = \alpha(x) + \beta(x)$ for each $x \in G$. Define $\cdot$ by: $\alpha\beta = \gamma$ if $\gamma(x) = \alpha(\beta(x))$ for each $x \in G$. Prove that $\langle R, +, \cdot \rangle$ is a ring. If G contains an element a such that $2a \neq 0_G$, prove that R contains an involution different from ι.

3.1.17. Let R, S be subrings of a ring A.

(1) Prove that $R \cap S$ is a subring of A.

(2) Prove that $R \cup S$ is a subring of A if and only if either $R \subset S$ or $S \subset R$.

3.1.18. Find all units and all zero divisors in $\mathbb{Z}_{12}$.

3.1.19. Let A be a ring with identity. Let (I) be the condition: $ab = 0 \Rightarrow ba = 0$ for $a, b \in A$. Let (II) be the condition: $ab = 1 \Rightarrow ba = 1$, for $a, b \in A$.

(1) Prove that (I) implies (II).

(2) Give an example to prove that (II) does not imply (I). (Hint: try $A = M_2(\mathbb{R})$.)

3.1.20. Let A be a ring with identity. Prove: if a is a zero divisor of A, then a is not a unit of A.

3.1.21. Let A be a ring with identity, and let U_A be the subset of A consisting of all elements of A that are expressible as sums of units of A. Prove that U_A forms a subring of A. (Even in a ring with very few units, it is possible for U_A to be equal to A. Can you give some examples of rings A such that $U_A = A$?)

3.1.22. Prove that every abelian group is the additive group of some ring. (Hint: form a "zero ring" — see Exercise 3.1.5(2).)

3.2

Ring Homomorphism, Ideals, Residue Class Rings, and Simple Rings

The concepts of isomorphism and, more generally, homomorphism, that we developed for groups in Chapter 2, extend in the obvious way to rings: we now have two operations to preserve, one of them $(+)$ a group operation.

Definition 3.2.1: Let $\langle A, +, \cdot \rangle$, $\langle A', +', \cdot' \rangle$ be rings, and let $\phi : A \to A'$ be a map such that

(1) $\phi(x + y) = \phi(x) +' \phi(y)$

and

(2) $\phi(x \cdot y) = \phi(x) \cdot' \phi(y)$.

Then ϕ is a (ring) homomorphism of A into A'.

If $\phi : A \to A'$ is a (ring) homomorphism, then ϕ is:

a (ring) monomorphism if ϕ is 1–1;
a (ring) epimorphism if ϕ is onto;
a (ring) isomorphism if ϕ is bijective.

A (ring) homomorphism of a ring A into itself is a (ring) endomorphism; a (ring) isomorphism of a ring A onto itself is a (ring) automorphism. (The designation "(ring)" will be omitted if it is clear from the context what is intended.)

Note that a homomorphism ϕ of the ring $\langle A, +, \cdot \rangle$ into the ring $\langle A', +', \cdot' \rangle$ is, in particular, a homomorphism of the group $\langle A, + \rangle$ into the group $\langle A', +' \rangle$ since it satisfies condition (1) of Definition 3.2.1. Thus, we already know a good deal about it. For example, we know that $\phi(0) = 0'$, and that $\phi(-a) = -\phi(a)$, for each $a \in A$ (see Theorem 2.9.1). We also know that $\phi^{-1}\{0'\} = \{a \in A \mid \phi(a) = 0'\}$ is a (normal) subgroup of $\langle A, + \rangle$ (see Theorem 2.9.2). We shall call it the kernel, Ker ϕ, of the ring homomorphism ϕ. However, the normality of Ker ϕ as a subgroup of $\langle A, + \rangle$ can be of no special interest here since $\langle A, + \rangle$ is an abelian group (*all* of whose subgroups are normal).

We thus ask: what special kind of subgroup of the additive group of a ring will serve as the kernel of a ring homomorphism with domain A?

Lemma: Let $\langle A, +, \cdot \rangle, \langle A', +', \cdot' \rangle$ be rings, and let $\phi : A \to A'$ be a ring homomorphism. Then

$$\operatorname{Ker} \phi = \phi^{-1}\{0'\} = \{a \in A \mid \phi(a) = 0'\}$$

has the following properties:

(1) if $k \in \operatorname{Ker} \phi$ and $a \in A$, then $ak \in \operatorname{Ker} \phi$;
(2) if $k \in \operatorname{Ker} \phi$ and $a \in A$, then $ka \in \operatorname{Ker} \phi$.

Proof: If $k \in K$ and $a \in A$, then $\phi(ka) = \phi(k) \cdot' \phi(a) = 0' \cdot' \phi(a) = 0'$ and $\phi(ak) = \phi(a) \cdot' \phi(k) = \phi(a) \cdot' 0' = 0'$. Thus, ka and ak are in Ker ϕ. ■

Definition 3.2.2: Let $\langle A, +, \cdot \rangle$ be a ring, and let K be a subgroup of the additive group of A. Then K is a (*2-sided*) *ideal* of A if, for each $k \in K$ and each $a \in A$, ka and ak are both elements of K. (One-sided ideals may be defined by omitting one or the other of these conditions. Thus, a subgroup K of $\langle A, + \rangle$ is a *right ideal* of A if $ka \in K$ for each $k \in K$, $a \in A$, and a *left ideal* if $ak \in K$ for each $k \in K$, $a \in A$.)

While one-sided ideals play an important role in the structure theory of rings, our interest here lies mainly in 2-sided ideals. In the following, we shall often use the term *ideal* (without a modifying adjective) to mean "2-sided ideal."

Caution: The multiplicative conditions in Definition 3.2.2 do not suffice to guarantee that a subset K of a ring A is an ideal of A. For example, let $K = \mathbb{Z}\backslash\{-1, 1\}$. For each $k \in K$ and each $a \in \mathbb{Z}$, we have $ak = ka \in K$. However, K is not a subgroup of $\langle \mathbb{Z}, +\rangle$, hence not an ideal of $\mathbb{Z}$. (This simple example is due to Kathy Toussaint, who, as an undergraduate, created it during a final exam.) For another example, see Exercise 3.5.11.

Note that, in any ring A, $\{0\}$ and A itself are ideals of A.

Corollary: If $\phi : A \to A'$ is a ring homomorphism, then Ker ϕ is an ideal of A.

With ideals thus established as possible ring-theoretic analogues of normal subgroups in the theory of groups, we proceed to investigate if the analogy is complete: is every ideal of a ring A the kernel of some ring homomorphism with domain A?

We first introduce some language and notation: if K is an ideal of A, we refer to the (additive) cosets of K in the group $\langle A, +\rangle$ as the *residue classes, in A, modulo the ideal K* (this language stems from number theoretic usage, as exemplified in Example 1 on p. 142). We denote by A/K the set of all residue classes, in A, modulo K.

$$\text{Thus, } A/K = \{K + a \mid a \in A\}.$$

Our criterion for equality of cosets (Corollary 1, Theorem 2.5.1), translated, of course, into additive notation, now provides us immediately with a *criterion for equality of residue classes*: if K is an ideal in A, then, for $a, b \in A$,

$$K + a = K + b$$

if and only if $a - b \in K$.

Theorem 3.2.1: Let A be a ring, and let K be an ideal in A. On the set A/K of all residue classes in A, modulo K, define $+$ and $\cdot$ by

$$(K + a) + (K + b) = K + (a + b)$$
$$(K + a) \cdot (K + b) = K + ab.$$

Then $\langle A/K, +, \cdot\rangle$ is a ring.

If $v : A \to A/K$ is defined by: $v(a) = K + a$, for each $a \in A$, then v is a ring epimorphism, and

$$\text{Ker } v = K.$$

Proof: $\langle A/K, +\rangle$ is merely the factor group of $\langle A, +\rangle$ with respect to its (*a fortiori*) normal subgroup K; hence $\langle A/K, +\rangle$ is clearly abelian.

We need to show that the $\cdot$ operation defined by $(K + a) \cdot (K + b) = K + ab$ is, indeed, well-defined, i.e., if, for $a, a', b, b' \in A$,

$$K + a = K + a' \quad \text{and} \quad K + b = K + b',$$

then $K + ab = K + a'b'$. This is where the multiplicative properties of the ideal K will play an important role.

If $K + a = K + a'$ and $K + b = K + b'$, $(a, a', b, b' \in A)$, then $a - a' \in K$ and $b - b' \in K$. We need to show that $ab - a'b' \in K$.

Now $ab - a'b' = ab - ab' + ab' - a'b' = a(b - b') + (a - a')b'$. Since K is an ideal, $a(b - b')$ and $(a - a')b'$ are both in K; hence so is their sum. We conclude that $ab - a'b' \in K$, whence

$$K + ab = K + a'b',$$

as required.

The remainder of the proof is routine. We need show only that the binary operation $\cdot$ on A/K is associative, and that it is right and left distributive over the operation $+$ on A/K. We leave the details to the reader (in Exercise 3.2.1). Thus, $\langle A/K, +, \cdot\rangle$ is a ring.

The canonical group epimorphism $\nu: A \to A/K$ defined by $\nu a = K + a$ has kernel K (see Definition 2.9.3) and is thus a good candidate for the ring homomorphism we seek. It suffices to prove that this group homomorphism ν also preserves multiplication. For $a, b \in A$, we have

$$\nu(ab) = K + ab = (K + a)\cdot(K + b) = \nu(a)\cdot\nu(b).$$

Thus, ν is a ring epimorphism of A onto A/K, with kernel K. ■

By Theorem 3.2.1, the answer to our question is "yes"; every ideal of a ring A *is* the kernel of a ring homomorphism with domain A. Thus, in complete analogy to the corollary of Theorem 2.9.3, we now have the following corollary.

Corollary: A subgroup K of the additive group $\langle A, +\rangle$ of a ring A is an ideal of A if and only if K is the kernel of a ring homomorphism with domain A.

With the analogy to normal subgroups thus complete, we accord to ideals in the theory of rings the same distinguished role as that played by normal subgroups in the theory of groups.

Notation: If K is an ideal of a ring A, we sometimes write $K \lhd A$.

Remark 1: A ring homomorphism ϕ is a monomorphism if and only if Ker $\phi = \{0\}$. (Compare Exercise 2.9.2.)

Remark 2: Every right, left, or 2-sided ideal of a ring A is a subring of A. (See Exercise 3.2.2.)

Example 1:

In the ring $\langle \mathbb{Z}, +, \cdot\rangle$ of all integers, *every* subgroup of $\langle \mathbb{Z}, +\rangle$ is an ideal! (This is a very rare property of rings.) For: the subgroups of the cyclic group $\langle \mathbb{Z}, +\rangle$ are the cyclic groups $[m]$, $m \in \mathbb{Z}$. Since m and $-m$ are both in $[m]$, every subgroup of $\langle \mathbb{Z}, +\rangle$ is of the form $[n]$, for $n \geq 0$. Since $[n] = \{nk \mid k \in \mathbb{Z}\}$, we frequently write $n\mathbb{Z}$ for the subgroup of $\langle \mathbb{Z}, +\rangle$ generated by n. Now, if $n \geq 0$, and if $k \in \mathbb{Z}$ and $a \in \mathbb{Z}$, then $a(nk) = (nk)a = n(ak) \in n\mathbb{Z}$. Thus, $n\mathbb{Z}$ has the strong closure properties of Definition 3.2.2, i.e., $n\mathbb{Z}$ is an ideal.

For $n \geq 0$ in $\mathbb{Z}$, each residue class $n\mathbb{Z} + a = \{nk + a \mid k \in \mathbb{Z}\}$ is an element of $\mathbb{Z}_n$ (see Theorems 2.2.2 and 2.5.3). In fact $\mathbb{Z}_n$ and $\mathbb{Z}/n\mathbb{Z}$ are precisely the same set. The operations $\oplus$ and $\circ$ on $\mathbb{Z}_n$, which we introduced in Chapter 2, are precisely the operations on residue classes which we defined on A/K in Theorem 3.2.1, with $A = \mathbb{Z}$ and $K = n\mathbb{Z}$. Thus $\langle \mathbb{Z}_n, \oplus, \circ \rangle$ *is* the residue class ring of $\mathbb{Z}$ with respect to the ideal $n\mathbb{Z}$.

The canonical ring homomorphism $\nu : \mathbb{Z} \to \mathbb{Z}_n = \mathbb{Z}/n\mathbb{Z}$ sends each integer a to its residue class $\bar{a} = n\mathbb{Z} + a$, modulo n.

Example 2:

In the ring $M_n(\mathbb{Z})$, let $K = M_n(k\mathbb{Z})$, where k is a non-negative integer. Then K is an ideal in $M_n(\mathbb{Z})$. In fact, every ideal in $M_n(\mathbb{Z})$ is of this form (see Exercise 3.2.4). The elements of $M_n(\mathbb{Z})/M_n(k\mathbb{Z})$ are the additive cosets $M_n(k\mathbb{Z}) + Q$, where Q is a matrix in $M_n(\mathbb{Z})$. The canonical ring epimorphism

$$\nu : M_n(\mathbb{Z}) \to M_n(\mathbb{Z})/M_n(k\mathbb{Z})$$

sends each matrix Q to the residue class $M_n(k\mathbb{Z}) + Q$. For example, if $n = 2$, $k = 3$, and $Q = \begin{pmatrix} 7 & 5 \\ -8 & -10 \end{pmatrix}$, then

$$\nu Q = M_2(3\mathbb{Z}) + \begin{pmatrix} 7 & 5 \\ -8 & -10 \end{pmatrix} = M_2(3\mathbb{Z}) + \begin{pmatrix} 1 & -1 \\ 1 & -1 \end{pmatrix}$$

(the residue class of the matrix obtained from Q by reducing each entry modulo 3). This should make it plausible that $M_2(\mathbb{Z})/M_2(3\mathbb{Z})$ is isomorphic to $M_2(\mathbb{Z}/3\mathbb{Z})$. In general, for n a positive integer, we shall show in Section 3.3 that

$$M_n(\mathbb{Z})/M_n(k\mathbb{Z}) \cong M_n(\mathbb{Z}/k\mathbb{Z})$$

for each non-negative integer k.

You may wonder why we do not use $\mathbb{R}$ or $M_n(\mathbb{R})$ as illustrations here. The reason is that these rings contain no interesting ideals—none, in fact, except the trivial subring and the whole ring.

Definition 3.2.3: A ring A is *simple* if the only (2-sided) ideals of A are $\{0\}$ and A.

(Usage varies: in the literature, the condition "A is not a zero-ring," in the sense that not all products in A are equal to 0 (see Exercise 3.2.3), is an additional requirement imposed on a simple ring A.)

Theorem 3.2.2: Every division ring is a simple ring. In fact, if A is a division ring, then A has no right or left ideals other than $\{0\}$ and A.

Proof: Let K be a left ideal in A. If $K \neq \{0\}$, there is some $k \in K$, $k \neq 0$. Hence $k^{-1} \in A$ and $k^{-1}k \in K$. Thus, $1 \in K$. But then, for each $a \in A$, $a = a \cdot 1 \in K$. It follows that $K = A$. Thus, A has no left ideals other than $\{0\}$ and A. (Similarly, A has no right ideals other than $\{0\}$ and A.) The non-existence of left ideals other than

$\{0\}$ and A suffices to allow us to conclude: the only (2-sided) ideals of A are $\{0\}$ and A. Thus, A is simple. ■

More surprising is the simplicity of the rings $M_n(D)$ where D is a division ring, n a positive integer. For $n > 1$, these rings have interesting one-sided ideals, but no proper 2-sided ideals.

Theorem 3.2.3: Let D be a division ring, n a positive integer, and let $A = M_n(D)$. Then A is simple.

Proof: If $n = 1$, then the ring A of all 1×1 matrices over D is clearly isomorphic to D, hence simple, by Theorem 3.2.2. For $n > 1$, let K be an ideal in A. If $K \neq \{0\}$, then K contains a matrix $Q = (q_{ij}) \neq 0_n$. Hence, for some h, k, $1 \leq h, k \leq n$, we have $q_{hk} \neq 0$. Write $q = q_{hk}$. For each i, j $(0 \leq i, j \leq n)$, define an $n \times n$ matrix E_{ij} having its i, j-entry equal to 1, all other entries equal to 0. (These matrices are sometimes called *matrix units.*) Then

$$\sum_{i=1}^{n} E_{ih}QE_{ki} = qI_n \in K. \tag{1}$$

But then $q^{-1}(qI_n) = (q^{-1}q)I_n = I_n \in K$, and so, for each $L \in A$, $L = LI_n \in K$. It follows that $K = A$. Thus, A is simple. ■

Note that, in (1), full use is made of the fact that K is a *2-sided* ideal. We illustrate the computations for the case where $n = 2$ and $Q = \begin{pmatrix} a & b \\ q & c \end{pmatrix}$, $a, b, c, q \in \mathbb{R}$, $q \neq 0$. Note that q is the $(2, 1)$-element of Q. Thus, $(h, k) = (2, 1)$ here. We define $E_{11} = \begin{pmatrix} 1 & 0 \\ 0 & 0 \end{pmatrix}$, $E_{12} = \begin{pmatrix} 0 & 1 \\ 0 & 0 \end{pmatrix}$, $E_{21} = \begin{pmatrix} 0 & 0 \\ 1 & 0 \end{pmatrix}$, $E_{22} = \begin{pmatrix} 0 & 0 \\ 0 & 1 \end{pmatrix}$. Then

$$E_{12}QE_{11} = \begin{pmatrix} 0 & 1 \\ 0 & 0 \end{pmatrix}\begin{pmatrix} a & b \\ q & c \end{pmatrix}\begin{pmatrix} 1 & 0 \\ 0 & 0 \end{pmatrix} = \begin{pmatrix} q & c \\ 0 & 0 \end{pmatrix}\begin{pmatrix} 1 & 0 \\ 0 & 0 \end{pmatrix} = \begin{pmatrix} q & 0 \\ 0 & 0 \end{pmatrix}$$

$$E_{22}QE_{12} = \begin{pmatrix} 0 & 0 \\ 0 & 1 \end{pmatrix}\begin{pmatrix} a & b \\ q & c \end{pmatrix}\begin{pmatrix} 0 & 1 \\ 0 & 0 \end{pmatrix} = \begin{pmatrix} 0 & 0 \\ q & c \end{pmatrix}\begin{pmatrix} 0 & 1 \\ 0 & 0 \end{pmatrix} = \begin{pmatrix} 0 & 0 \\ 0 & q \end{pmatrix}$$

whence $E_{12}QE_{11} + E_{22}QE_{12} = \begin{pmatrix} q & 0 \\ 0 & q \end{pmatrix} = qI_2$. (If Q is an element of a 2-sided ideal K, then so are $E_{12}QE_{11}$, $E_{22}QE_{12}$; hence so is their sum, qI_2.)

Even more striking, and a great deal harder to prove, is the following "converse" of Theorem 3.2.3.

Theorem 3.2.4 (Wedderburn Structure Theorem for Simple Rings): Let A be a simple ring satisfying the conditions:

(1) $A^2 \neq \{0\}$, i.e., there is some $a, b \in A$ such that $ab \neq 0$.

(2) There is no infinite properly descending chain $K_1 \supsetneq K_2 \supsetneq K_3 \supsetneq \ldots$ of right ideals in A.

Then A is isomorphic to $M_n(D)$ for some division ring D, and some positive integer n. (cf. [3], [39].)

Exercises 3.2

True or False

1. Every subgroup of the group $\langle \mathbb{Q}, + \rangle$ is an ideal in the ring $\langle \mathbb{Q}, +, \cdot \rangle$.
2. If K is an ideal in the ring A, then the residue class ring A/K is a subring of A.
3. If A is a simple ring, then every non-zero homomorphic image of A is isomorphic to A.
4. If K is a left but not right ideal of a ring A, then

$$(K + x) \cdot (K + y) = K + xy$$

defines a binary operation on A/K.

5. The mapping $\phi : \mathbb{Q} \to \mathbb{Q}$ defined by: $\phi(x) = |x|$ for all $x \in \mathbb{Q}$ is a ring homomorphism of $\langle \mathbb{Q}, +, \cdot \rangle$.

3.2.1. Supply the missing details in the proof of Theorem 3.2.1. In particular, identify the zero element and the additive inverse of each element of A/K, and prove that $\cdot$ is associative and distributive (right and left) over $+$.

Prove that, if A is commutative, then so is A/K; and if A is a ring with identity, then so is A/K.

3.2.2. Prove that any right, left, or 2-sided ideal of a ring A is a subring of A. Give examples of subrings that are *not* ideals!

3.2.3. Prove that, in a zero ring A (i.e., a ring A in which all products are equal to zero), every subgroup of $\langle A, + \rangle$ is an ideal of A. (Do you know a non-zero ring with this property?) Conclude that a zero ring A is simple if and only if $\langle A, + \rangle$ is of prime order.

3.2.4. Let $R = M_2(\mathbb{Z})$, the ring of all 2×2 matrices over $\mathbb{Z}$.

(1) Prove that, for each non-negative integer k, the set of all 2×2 matrices over $k\mathbb{Z}$ forms an ideal of R.

(2) Prove that every ideal of R is of the form $M_2(k\mathbb{Z})$ for some $k \geq 0$ in $\mathbb{Z}$.

(3) Generalize the results of this exercise to the ring $M_n(\mathbb{Z})$ for any $n \geq 2$ in $\mathbb{Z}$.

3.2.5. Let A be a ring with identity, and let K be an ideal of A. Prove that the following statements are equivalent:

(1) $K = A$

(2) $1 \in K$

(3) $u \in K$ for some unit $u \in A$.

3.2.6. (1) In $M_2(\mathbb{R})$, prove that $K = \left\{ \begin{pmatrix} a & b \\ 0 & 0 \end{pmatrix} \middle| a, b \in \mathbb{R} \right\}$ is a right ideal, but not a left ideal.

(2) Give another example in $M_2(\mathbb{R})$ of a right, but not left, ideal.

(3) Give examples in $M_2(\mathbb{R})$ of left, but not right, ideals.

3.2.7. For $k > 1$ in $\mathbb{Z}$, prove that the ring $\mathbb{Z}/k\mathbb{Z}$ has no zero divisors if and only if k is prime.

3.2.8. *(1)* Prove that $M_2(\mathbb{Z})/M_2(k\mathbb{Z})$ has zero divisors for each $k > 1$ in $\mathbb{Z}$.
(2) Is it true that, if A has zero divisors, then A/K has zero divisors for each ideal $K \neq A$? Prove or disprove.

3.2.9. Let A be a ring, and let L be a left ideal of A. Prove that the set $K = \{a \in A \mid al = 0 \text{ for each } l \in L\}$ forms a 2-sided ideal of a.

3.2.10. Let A be a ring with identity 1 and let $\phi : A \to A'$ be a ring epimorphism. Prove:
(1) $\phi(1)$ serves as identity for A'.
(2) If $u \in A$ is a unit, then $\phi(u)$ is a unit in A'.

3.2.11. If A is a ring with identity 1 and $\phi : A \to A'$ is a ring homomorphism, not necessarily onto, then $\phi(1)$ need not be an identity for A'. For example, let A be the set of all diagonal matrices in $M_2(\mathbb{R})$, and let

$$\phi : A \to A$$

be defined by: $\phi\begin{pmatrix} a & 0 \\ 0 & b \end{pmatrix} = \begin{pmatrix} a & 0 \\ 0 & 0 \end{pmatrix}$ (for each $a, b \in \mathbb{R}$). Find $\phi(I_2)$ and prove that it is not an identity for A. (Note that $\phi(I_2)$ *is* an identity for the image ring $\phi(A)$.)

3.2.12. Let A be a commutative ring, and let N be the set of all nilpotent elements of A. (See Exercise 3.1.13.)
(1) Prove that N is an ideal of A.
(2) Prove that A/N has no non-zero nilpotent elements.

3.2.13. *(1)* Let H, K be ideals of a ring A. Prove that $H \cap K$ is an ideal of A.
(2) Let J be a non-empty set of ideals of a ring A. Prove that $\bigcap_{K \in J} K$ is an ideal of A.

3.2.14. Let A be a ring, and let S be a subset of A. Prove that $(S) = \bigcap_{\substack{K \lhd A \\ S \subset K}} K$ is an ideal of A. ((S) is called the *ideal of A generated by the subset* S. If $S = \{a\}$, $a \in A$, we call (S) the *ideal*, (a), *generated by* a.)

3.2.15. *(1)* In any ring A, what is the ideal generated by $\varnothing$? What is the ideal generated by $\{0\}$? (See Exercise 3.2.14.)
(2) In any ring A, prove that the ideal, (a), generated by an element $a \in A$ is the set of all sums of elements of the form ax, ya, cad, and na $(x, y, c, d \in A, n \in \mathbb{Z})$.
(3) In a ring A with identity, prove that the ideal, (a), generated by $a \in A$ is the set of all sums of elements of the form $xay \quad (x, y \in A)$.
(4) In a commutative ring A, prove that the ideal, (a), generated by $a \in A$ is the set $aA + \mathbb{Z}a = \{ax + na \mid x \in A,\ n \in \mathbb{Z}\}$.
(5) In a commutative ring A with identity, prove that the ideal, (a), generated by $a \in A$ is the set

$$aA = \{ax \mid x \in A\}.$$

3.2.16. Let A be a commutative ring with identity, and let $a, b \in A$. Prove that the ideal of A generated by the subset $\{a, b\}$ is equal to the set $aA + bA = \{ax + by \mid x, y \in A\}$.

3.2.17. Let a, b be relatively prime integers. Prove that $a\mathbb{Z} \cap b\mathbb{Z} = ab\mathbb{Z}$ and $a\mathbb{Z} + b\mathbb{Z} = (1) = \mathbb{Z}$.

3.2.18. Prove that the ideal of $M_2(\mathbb{R})$ generated by any non-zero matrix is the whole ring, $M_2(\mathbb{R})$.

3.2.19. Let A be a commutative ring, and let K be an ideal of A. Denote by $\sqrt{K}$ the set of all $x \in A$ such that $x^n \in K$ for some positive integer n (depending on x). Prove that $\sqrt{K}$ is an ideal of A. ($\sqrt{K}$ is called the *radical* of the ideal K.) What is the radical of the ideal $K = \{0\}$? (Compare Exercise 3.1.13 and Exercise 3.2.12.)

3.2.20. In $\mathbb{Z}$, what is the radical of each of the following ideals? (See Exercise 3.2.19.)

(1) $3\mathbb{Z}$
(2) $6\mathbb{Z}$
(3) $12\mathbb{Z}$
(4) $30\mathbb{Z}$
(5) $36\mathbb{Z}$.

3.2.21. If $n = p_1^{\alpha_1} \cdots p_k^{\alpha_k}$, p_i primes in $\mathbb{Z}$, $\alpha_i \geq 1$, $k \geq 1$, what is the radical of the ideal $n\mathbb{Z}$?

3.2.22. Let F be the ring of all real-valued functions defined on $\mathbb{R}$ (see Exercise 3.1.15) and let $p \in \mathbb{R}$. Let $K = \{f \in F \mid f(p) = 0\}$. Prove that K is an ideal in F.

3.2.23. Determine whether ideal inclusion in an arbitrary ring A is transitive, i.e., if $K \lhd H$ and $H \lhd A$, need $K \lhd A$?

3.2.24. Let $\langle A, +, \cdot \rangle$ be a commutative ring with identity and let $<$ be an order relation on A (see Definition 1.2.2) such that, for $a, b, x \in A$,

(1) $a < b \Rightarrow a + x < b + x$
(2) $a < b,\ 0 < x \Rightarrow ax < bx$

Prove that $\langle A, +, \cdot \rangle$ is an integral domain. (The quadruple $\langle A, +, \cdot, < \rangle$ is called an *ordered integral domain.*)

3.2.25. Let $\langle A, +, \cdot, < \rangle$ be an ordered integral domain and let

$$A^+ = \{a \in A \mid 0 < a\}.$$

Prove: *(1)* A^+ is closed under addition.
(2) A^+ is closed under multiplication.
(3) If $a \in A$, then exactly one of the following holds:

$$a \in A^+$$
$$a = 0$$
$$-a \in A^+$$

(4) $1 \in A^+$.

(A^+ is called the *positive cone* of A.)

3.2.26. Let $\langle A, +, \cdot, < \rangle$ be an ordered integral domain satisfying the additional condition: If $S \subset A^+$, $S \neq \varnothing$, then S has a least element, i.e., there is some $s_0 \in S$ such that $s_0 \leq s$ for all $s \in S$. (Put another way: the positive cone A^+ of A is well-ordered with respect to $<$.)

Now let $\langle A', +', \cdot', <' \rangle$ be another ordered integral domain, satisfying the condition that *its* positive cone, $(A')^+$, is well-ordered with respect to $<'$.

Prove that there is a ring isomorphism $\phi : A \to A'$ such that

$$a < b \Rightarrow \phi(a) <' \phi(b).$$

(Thus, A and A' are isomorphic, as ordered integral domains.)

As a consequence of this result, we may characterize $\mathbb{Z}$ by the following set of axioms:

(1) $\langle \mathbb{Z}, +, \cdot, < \rangle$ is an ordered integral domain.

(2) The positive cone, $\mathbb{Z}^+$, of $\langle \mathbb{Z}, +, \cdot, < \rangle$ is well-ordered with respect to $<$.

Since these conditions determine an ordered integral domain up to isomorphism, we say that (1) and (2) form a *categorical set of axioms* for $\mathbb{Z}$. (cf. [42].)

3.3

Fundamental Theorem of Homomorphism for Rings

In analogy to Theorem 2.9.4, we have the following theorem.

Theorem 3.3.1: Let A, A' be rings, and let $\phi : A \to A'$ be a ring epimorphism with kernel K. Then Im $\phi \cong A/\text{Ker}\,\phi$, i.e., $A' \cong A/K$ (where the symbol "$\cong$" signifies the existence of a ring isomorphism). In fact, if $v : A \to A/K$ is the canonical ring homomorphism modulo the ideal K, then there is a unique ring isomorphism $\psi : A/K \to A'$ such that $\psi v = \phi$.

Proof: Since ϕ is a group epimorphism of A onto A', with kernel K, and v is the canonical group epimorphism of A onto A/K, sending each $a \in K$ to its additive coset (residue class) $K + a \in A/K$, Theorem 2.9.4 implies the existence of a unique group isomorphism $\psi : A/K \to A'$ such that $\psi v = \phi$. (Thus, $\psi(K + a) = \psi(va) = (\psi v)(a) = \phi(a)$, for each $a \in A$.) To complete the proof of the present theorem, it suffices to prove that ψ is, in fact, a ring homomorphism. Let $a, b \in A$. Then

$$\begin{aligned}\psi((K + a)(K + b)) &= \psi(K + ab) = \psi(v(ab)) = (\psi v)(ab) = \phi(ab) = \phi(a)\phi(b)\\ &= (\psi v)(a)(\psi v)(b) = \psi(va)\psi(vb) = \psi(K + a)\psi(K + b).\end{aligned}$$

We conclude that ψ is the required ring isomorphism. ■

Example:

We prove that the ring $M_2(\mathbb{Z}/k\mathbb{Z})$ is isomorphic to the residue class ring $M_2(\mathbb{Z})/M_2(k\mathbb{Z})$. By Theorem 3.3.1, it suffices to find an epimorphism $\phi : M_2(\mathbb{Z}) \to M_2(\mathbb{Z}/k\mathbb{Z})$, with Ker $\phi = M_2(k\mathbb{Z})$.

Define $\phi : M_2(\mathbb{Z}) \to M_2(\mathbb{Z}/k\mathbb{Z})$ by: $\phi(q_{ij})_{2\times 2} = (\bar{q}_{ij})_{2\times 2}$, where $\bar{q}_{ij}$ is the residue class mod k of q_{ij}, in $\mathbb{Z}$, for each $i, j (1 \le i, j \le 2)$. One verifies easily that

ϕ is a ring homomorphism. Explicitly, if $P = (p_{ij})_{2\times 2}$, $Q = (q_{ij})_{2\times 2}$, then

$$\begin{aligned}\phi(PQ) &= \phi\left(\begin{pmatrix} p_{11} & p_{12} \\ p_{21} & p_{22}\end{pmatrix}\begin{pmatrix} q_{11} & q_{12} \\ q_{21} & q_{22}\end{pmatrix}\right)\\ &= \phi\begin{pmatrix} p_{11}q_{11} + p_{12}q_{21} & p_{11}q_{12} + p_{12}q_{22} \\ p_{21}q_{11} + p_{22}q_{21} & p_{21}q_{12} + p_{22}q_{22}\end{pmatrix}\\ &= \begin{pmatrix} \overline{p_{11}q_{11} + p_{12}q_{21}} & \overline{p_{11}q_{12} + p_{12}q_{22}} \\ \overline{p_{21}q_{11} + p_{22}q_{21}} & \overline{p_{21}q_{12} + p_{22}q_{22}}\end{pmatrix}\\ &= \begin{pmatrix} \bar{p}_{11}\bar{q}_{11} + \bar{p}_{12}\bar{q}_{21} & \bar{p}_{11}\bar{q}_{12} + \bar{p}_{12}\bar{q}_{22} \\ \bar{p}_{21}\bar{q}_{11} + \bar{p}_{22}\bar{q}_{21} & \bar{p}_{21}\bar{q}_{12} + \bar{p}_{22}\bar{q}_{22}\end{pmatrix}\\ &= \begin{pmatrix} \bar{p}_{11} & \bar{p}_{12} \\ \bar{p}_{21} & \bar{p}_{22}\end{pmatrix}\begin{pmatrix} \bar{q}_{11} & \bar{q}_{12} \\ \bar{q}_{21} & \bar{q}_{22}\end{pmatrix} = \phi(P)\cdot\phi(Q).\end{aligned}$$

Thus, ϕ preserves multiplication. We leave it to the reader to verify that ϕ perserves addition.

ϕ is, clearly, an epimorphism, for: if

$$\bar{Q} = \begin{pmatrix} \bar{q}_{11} & \bar{q}_{12} \\ \bar{q}_{21} & \bar{q}_{22}\end{pmatrix} \in M_2(\mathbb{Z}/k\mathbb{Z}), \quad \text{then} \quad \bar{Q} = \phi(Q), \quad \text{where}$$

$$Q = \begin{pmatrix} q_{11} & q_{12} \\ q_{21} & q_{22}\end{pmatrix} \in M_2(\mathbb{Z}).$$

Finally, $\operatorname{Ker}\phi = \{Q \in M_2(\mathbb{Z}) | \phi(Q) = \begin{pmatrix} \bar{0} & \bar{0} \\ \bar{0} & \bar{0}\end{pmatrix}$ in $M_2(\mathbb{Z}/k\mathbb{Z})\}$. But then, for $Q = (q_{ij})_{2\times 2}$, we have $Q \in \operatorname{Ker}\phi \Leftrightarrow \bar{q}_{ij} = \bar{0}$ for each i, j $(1 \le i, j \le 2) \Leftrightarrow q_{ij} \in k\mathbb{Z}$ for each i, j $(1 \le i, j \le 2) \Leftrightarrow Q \in M_2(k\mathbb{Z})$.

Thus, $\operatorname{Ker}\phi = M_2(k\mathbb{Z})$. But then, by Theorem 3.3.1, we may conclude that $\operatorname{Im}\phi = M_2(\mathbb{Z}/k\mathbb{Z})$ is isomorphic to $M_2(\mathbb{Z})/\operatorname{Ker}\phi = M_2(\mathbb{Z})/M_2(k\mathbb{Z})$.

(For simplicity, we considered 2×2 matrices in this example. Our result extends readily to $n \times n$ matrices, for each positive integer n—see Exercise 3.3.10.)

Exercises 3.3

True or False

1. If $\phi : A \to A'$ is a ring epimorphism with kernel K, then the ring A' is isomorphic to the ring A/K.
2. If A/K is a commutative ring for each 2-sided ideal $K \neq \{0\}$ of A, then every homomorphic image of A is either commutative or isomorphic to A.
3. A commutative ring may have a non-commutative homomorphic image.
4. If $\phi : A \to A'$ is a ring homomorphism such that $\phi(1_A) = 0_{A'}$, then $\operatorname{Ker}\phi = A$.
5. If $\phi : A \to A'$ is a ring homomorphism, and $x \in A$ is such that $\phi(x^2) = 0_{A'}$, then $\operatorname{Ker}\phi + x^2 = \operatorname{Ker}\phi$.

3.3.1. Let $\phi : A \to A'$ be a ring epimorphism with Ker $\phi = K$. If A' is a ring with zero divisors, prove that there are elements $a, b \in A$ such that $ab \in K$, but $a \notin K$ and $b \notin K$.

3.3.2. Let $\phi : A \to A'$ be a ring epimorphism such that A' is a commutative ring. Prove that $ab - ba \in \text{Ker } \phi$, for each $a, b \in A$.

3.3.3. In a ring A, let C be the ideal generated by the elements $ab - ba$ $(a, b \in A)$. (These elements are called (ring theoretic) *commutators*. Compare Definition 2.11.3 for group theoretic commutators.) Thus,

$$C = \bigcap_{\substack{K \lhd A \\ S \subset K}} K \text{ where } S = \{ab - ba/a, b \in A\}.$$

Prove: if $\phi : A \to A'$ is a ring epimorphism, then A' is commutative if and only if $C \subset \text{Ker } \phi$.

3.3.4. Let $\phi : A \to A'$ be a ring epimorphism such that A' contains no nonzero nilpotent elements (see Exercise 3.1.13).

(1) Prove: if $a^n \in \text{Ker } \phi$ for some $a \in A$, $n \in \mathbb{Z}^+$, then $a \in \text{Ker } \phi$.

(2) Conclude that, if A is commutative, then $\sqrt{0} \subset \sqrt{\text{Ker } \phi} \subset \text{Ker } \phi$ (see Exercise 3.2.19).

3.3.5. Let $\phi : A \to A'$ be a ring homomorphism. If A is a ring with identity 1, prove that $\phi(1)$ is an idempotent element of A'. (See Exercise 3.1.11.) If ϕ is an epimorphism, then $\phi(1)$ serves as identity for A'.

3.3.6. A ring A with identity each of whose elements is idempotent (see Exercise 3.1.11) is called a *Boolean ring*. Let $\phi : A \to A'$ be a ring epimorphism. Prove that A' is a Boolean ring if and only if $a^2 - a \in \text{Ker } \phi$, for each $a \in A$.

3.3.7. Let A be a ring with identity, and let $\phi : A \to A'$ be a ring epimorphism. Prove that every element of A' is involutory (see Exercise 3.1.11) if and only if $x^2 - 1 \in \text{Ker } \phi$ for each $x \in A$.

3.3.8. Let $\phi : A \to A'$ and $\psi : A' \to A''$ be ring homomorphisms. Prove

(1) Ker $\phi \subset$ Ker $\psi\phi$.

(2) Ker $\psi\phi = \phi^{-1}$ (Im $\phi \cap$ Ker ψ).

3.3.9. Use the Fundamental Theorem of Homomorphism for Rings to prove that $3\mathbb{Z}/6\mathbb{Z}$ is isomorphic to $\mathbb{Z}/2\mathbb{Z}$ (and is thus a field). Hint: Define an epimorphism of $\mathbb{Z}$ onto $3\mathbb{Z}/6\mathbb{Z}$.

3.1.10. Generalize the Example on pp. 148–149, i.e., prove that for k, n positive integers greater than 1,

$$M_n(\mathbb{Z}/k\mathbb{Z}) \cong M_n(\mathbb{Z})/M_n(k\mathbb{Z}).$$

3.4

Maximal and Prime Ideals

Let p be a prime in $\mathbb{Z}$. Then the ideal $p\mathbb{Z}$ has several interesting properties. For example, if n is a positive integer such that $p\mathbb{Z} \subset n\mathbb{Z}$, then, since $p = p1 \in p\mathbb{Z}$, we have $p \in n\mathbb{Z}$, and so $n|p$. Since p is prime, this implies that either $n = 1$ or $n = p$. If $n = 1$, then $n\mathbb{Z} = 1\mathbb{Z} = \mathbb{Z}$, and if $n = p$, then $n\mathbb{Z} = p\mathbb{Z}$. We have just proved that, if p is prime in $\mathbb{Z}$, then $p\mathbb{Z}$ is a maximal ideal, in the sense of the following definition.

Definition 3.4.1: Let A be a ring, and let $K \lhd A$. Then K is a *maximal ideal* of A provided

(1) $K \neq A$

and

(2) if $H \lhd A$ such that $K \subset H \subset A$, then either $H = K$ or $H = A$.

Example 1:

As we have already observed, if p is a prime in $\mathbb{Z}$, then $p\mathbb{Z}$ is a maximal ideal of $\mathbb{Z}$. Conversely, suppose K is a maximal ideal of $\mathbb{Z}$. Since $\mathbb{Z}$ has non-zero ideals, $K \neq \{0\}$. Hence $K = n\mathbb{Z}$ for some $n > 1$ in $\mathbb{Z}$. If n is composite, then $n = hk$ for some $h, k \in \mathbb{Z}$, $1 < h < n$, $1 < k < n$. But then $n\mathbb{Z} \subsetneq h\mathbb{Z} \subsetneq \mathbb{Z}$. (Note that $n\mathbb{Z} = h\mathbb{Z}$ cannot hold since $n \nmid h$, and $h\mathbb{Z} = \mathbb{Z}$ cannot hold since $h \nmid 1$.) This contradicts the hypothesis that $n\mathbb{Z}$ is a maximal ideal of $\mathbb{Z}$. It follows that n is prime.

Example 2:

Again, let p be a prime in $\mathbb{Z}$; let $A = M_2(\mathbb{Z})$, and let $K = M_2(p\mathbb{Z})$. Then K is a maximal ideal of A. For, suppose $H \lhd A$, $K \subset H \subset A$. For each i, j $(1 \leq i, j \leq 2)$, let H_{ij} be the set of all integers that occur in the (i, j)-position of some matrix Q in H. It is easy to verify that each of the H_{ij} is an ideal in $\mathbb{Z}$ (see Exercise 3.4.8). Hence there are non-negative integers t_{ij} such that $H_{ij} = t_{ij}\mathbb{Z}$, for each i, j $(1 \leq i, j \leq 2)$. Then, for each i, j $(1 \leq i, j \leq 2)$, there is some matrix $T_{ij} \in H$ having t_{ij} in the (i, j)-position. Now, using matrix units E_{ij} (such as we defined in the proof of Theorem 3.2.3) $(1 \leq i, j \leq 2)$, we can show that each of the t_{ij} occurs in *any* arbitrary position as an entry of some matrix belonging to H. (For example, $E_{21}T_{11}E_{12}$ has t_{11} in the (2,2)-position.) This implies that each of the non-negative integers t_{ij} divides every one of the others — from which it follows that $t_{11} = t_{12} = t_{21} = t_{22} = t \in \mathbb{Z}$. But then each of the ideals H_{ij} of $\mathbb{Z}$ is equal to the ideal $t\mathbb{Z}$, and so $H = M_2(t\mathbb{Z})$. Now, since $K = M_2(p\mathbb{Z}) \subset H = M_2(t\mathbb{Z})$, we have $p\mathbb{Z} \subset t\mathbb{Z}$. But $p\mathbb{Z}$ is a maximal ideal of $\mathbb{Z}$; hence $t\mathbb{Z} = p\mathbb{Z}$ or $t\mathbb{Z} = \mathbb{Z}$, whence either $H = K$ or $H = A$. We have proved that K is a maximal ideal of A.

This example generalizes readily to the rings $M_n(\mathbb{Z})$, for n any positive integer (see Exercise 3.4.9).

Let us now examine the residue class rings modulo the maximal ideals in Examples 1 and 2. In Example 1, $\mathbb{Z}/p\mathbb{Z}$ is a field, for: it is a commutative ring with identity $p\mathbb{Z} + 1 \neq p\mathbb{Z} + 0$; and, if $p\mathbb{Z} + k \neq p\mathbb{Z} + 0$, then $k \notin p\mathbb{Z}$, hence $\gcd(p, k) = 1$ and $kx + py = 1$ for some $x, y \in \mathbb{Z}$ (see the Corollary of Definition 1.3.4 and Theorem 1.3.5). But then $(p\mathbb{Z} + k)(p\mathbb{Z} + x) = p\mathbb{Z} + kx = p\mathbb{Z} + 1$ (since $kx - 1 \in p\mathbb{Z}$) — see the group-theoretic criterion for two cosets to be equal (Corollary 1, Theorem 2.5.1). Thus, every non-zero element of $\mathbb{Z}/p\mathbb{Z}$ has a multiplicative inverse in $\mathbb{Z}/p\mathbb{Z}$, and so $\mathbb{Z}/p\mathbb{Z}$ is a field.

On the other hand, in Example 2, $A/K = M_2(\mathbb{Z})/M_2(p\mathbb{Z})$ is isomorphic to $M_2(\mathbb{Z}/p\mathbb{Z})$ (as we showed in Section 3.3); hence it has zero divisors and is therefore not even a division ring.

Thus, the residue class ring A/K of a ring A with respect to a maximal ideal K of A may be a more or less "special" kind of ring, depending on the nature of A.

For A itself sufficiently special, we can guarantee that A/K will be a field if K is a maximal ideal of A.

Theorem 3.4.1: Let A be a commutative ring with identity, and let K be an ideal in A. Then K is a maximal ideal of A if and only if A/K is a field.

Proof: $\Rightarrow$: Suppose A is a commutative ring with identity, and K is a maximal ideal of A. Then A/K is a commutative ring (being a homomorphic image of the commutative ring A), and A/K has identity $K + 1$. Since $K \neq A$, we have $1 \notin K$ (see Exercise 3.2.5); hence $K + 1 \neq K + 0$, i.e., $1_{A/K} \neq 0_{A/K}$. To complete the proof that A/K is a field, it thus suffices to prove that every non-zero element of A/K has a multiplicative inverse in A/K.

For $a \in A$, suppose $K + a \neq K + 0$. Then $a = a - 0 \notin K$, and so the set

$$H = \{k + ax \mid k \in K, \quad x \in A\}$$

contains K properly: $K \subsetneq H$. We prove that H is an ideal of A. Of course, $H \neq \phi$ (e.g., $a = 0 + a \cdot 1 \in H$). If $k_1, k_2 \in K$ and $x_1, x_2 \in A$, then $(k_1 + ax_1) - (k_2 + ax_2) = k_1 - k_2 + a(x_1 - x_2) \in H$. Thus, H is a subgroup of $\langle A, + \rangle$. Let $y \in A$. Then, for each $k \in K$, $x \in A$, $y(k + ax) = yk + y(ax) = yk + a(yx) \in H$. (In the last equality, observe the use of commutativity, as well as the fact that K is an ideal!) Since A is commutative, we conclude that H is an ideal in A. But K is a maximal ideal in A, and

$$K \subsetneq H \subset A.$$

Hence $H = A$. Thus,

$$A = \{k + ax \mid k \in K, \quad x \in A\},$$

i.e., every element of A is of the form $k + ax$, for some $k \in K$, $x \in A$. In particular, there is some $k_0 \in K$ and some $x_0 \in A$ such that

$$1 = k_0 + ax_0,$$

whence $1 - ax_0 \in K$, and so

$$K + 1 = K + ax_0 = (K + a)(K + x_0).$$

But then $K + a$ has inverse $K + x_0$ in A/K. We conclude that A/K is a field.

$\Leftarrow$: Conversely, suppose that A/K is a field. Then $1_{A/K} \neq 0_{A/K}$, i.e., $K + 1 \neq K + 0$, whence $1 \notin K$, and so $K \neq A$. Suppose H is an ideal in A such that $K \subset H \subset A$. If $H \neq K$, then there is some $h \in H$ such that $h \notin K$, whence $K + h \neq K + 0$. Since A/K is a field, there is some $y \in A$ such that $K + y$ serves as the multiplicative inverse of $K + h$ in A/K. Thus, $(K + h) \cdot (K + y) = K + hy =$

$K + 1$, and so $1 - hy \in K \subset H$. Since H is an ideal in A, we have $hy \in H$; hence $1 \in H$. But then $H = A$. Thus, K is a maximal ideal of A. ∎

Remark: Note that, while commutativity played an essential role in the first part ($\Rightarrow$) of the proof of the preceding theorem, it played no essential role in the second part ($\Leftarrow$). In fact, if A is a ring, not necessarily commutative, K an ideal of A such that A/K is a division ring, then a similar argument shows that K is a maximal ideal of A. (See Exercise 3.4.2.)

Corollary: $\mathbb{Z}/n\mathbb{Z}$ is a field if and only if n is prime.

Proof: By Theorem 3.4.1, it suffices to recall (see Example 2 of this section) that $n\mathbb{Z}$ is a maximal ideal of $\mathbb{Z}$ if and only if n is prime. ∎

We began this section by stating that the ideals $p\mathbb{Z}$ in $\mathbb{Z}$, for p prime, have *several* interesting properties. The maximality of $p\mathbb{Z}$ is one such property. Another interesting property of the ideals $p\mathbb{Z}$ (p prime) stems from the Euclid property of primes: if p is a prime and a, b are integers such that $p|ab$, then $p|a$ or $p|b$. (p. 22). This implies that, if p is a prime and $ab \in p\mathbb{Z}$, then $a \in p\mathbb{Z}$ or $b \in p\mathbb{Z}$.

Definition 3.4.2: Let A be a commutative ring. Then an ideal K of A is a *prime ideal* of A if K has the properties:

(1) $K \neq A$

and

(2) $ab \in K \quad (a, b \in A) \Rightarrow a \in K \quad \text{or} \quad b \in K.$

Example:

If p is a prime in $\mathbb{Z}$, then $p\mathbb{Z}$ is a prime ideal of $\mathbb{Z}$. For, if $ab \in p\mathbb{Z}$, then $p|ab$, hence $p|a$ or $p|b$, and so either $a \in p\mathbb{Z}$ or $b \in p\mathbb{Z}$.

Conversely, suppose K is a prime ideal of $\mathbb{Z}$. Note that $\{0\}$ is a prime ideal of $\mathbb{Z}$ since $ab \in \{0\}$ $(a, b, \in \mathbb{Z})$ implies $ab = 0$; hence $a = 0$ or $b = 0$, and hence $a \in \{0\}$ or $b \in \{0\}$. If k is a *non-zero* prime ideal of $\mathbb{Z}$, then $K = n\mathbb{Z}$ for some $n > 1$. If n is composite, then $n = hk$, $1 < h, k < n$, hence $hk \in n\mathbb{Z}$. But then either $h \in n\mathbb{Z}$ or $k \in n\mathbb{Z}$. Thus, either $n|h$ or $n|k$. On the other hand, we have $n = hk$, hence $h|n$ and $k|n$, and so either $h = n$ or $k = n$.
Contradiction! It follows that n is prime.

Summarizing our results, we have: K is a prime ideal of $\mathbb{Z}$ if and only if $K = \{0\}$ or $K = p\mathbb{Z}$ for some prime p.

We already know that, for p prime, $\mathbb{Z}/p\mathbb{Z}$ is a field, hence it is certainly an integral domain. More generally, we have the following theorem.

Theorem 3.4.2: If A is a commutative ring with identity and K is an ideal of A, then K is a prime ideal of A if and only if A/K is an integral domain.

Proof: Suppose K is a prime ideal of A. Since $K \neq A$, we have $K + 1 \neq K + 0$, i.e., $1_{A/K} \neq 0_{A/K}$. Since A is commutative, so is its homomorphic image, A/K. For $a, b \in A$, suppose that

$$(K + a) \cdot (K + b) = K + 0 = 0_{A/K}.$$

Then $K + ab = K + 0$; hence $ab = ab - 0 \in K$. Since K is a prime ideal of A, either $a \in K$ or $b \in K$. But then either $K + a = K + 0 = 0_{A/K}$ or $K + b = K + 0 = 0_{A/K}$. It follows that A/K has no zero divisors and is, thus, an integral domain.

Conversely, suppose A/K is an integral domain. Then $0_{A/K} \neq 1_{A/K}$, i.e., $K + 0 \neq K + 1$, and so $1 \notin K$. Thus, $K \neq A$. If $a, b \in A$ such that $ab \in K$, then $K + ab = K + 0$, so $(K + a)(K + b) = 0_{A/K}$. Since A/K is an integral domain, it follows that either $K + a = 0_{A/K}$ or $K + b = 0_{A/K}$, i.e., either $K + a = K + 0$ or $K + b = K + 0$, whence either $a \in K$ or $b \in K$. But then K is a prime ideal. ■

Corollary 1: If A is a commutative ring with identity, then every maximal ideal of A is a prime ideal.

Proof: If K is a maximal ideal of A, then A/K is a field (by Theorem 3.4.1). But then A/K is an integral domain and, by Theorem 3.4.2, it follows that K is a prime ideal. ■

Corollary 2: If $n \geq 0$ in $\mathbb{Z}$, then $\mathbb{Z}/n\mathbb{Z}$ is an integral domain if and only if $n = 0$ or n is prime.

Proof: This follows immediately from the preceding theorem and our characterization of prime ideals in $\mathbb{Z}$ (Example 1, p. 151). ■

Corollary 3: For n a positive integer, the following statements are equivalent:

(1) n is prime;

(2) $\mathbb{Z}/n\mathbb{Z}$ is a field;

(3) $\mathbb{Z}/n\mathbb{Z}$ is an integral domain.

Exercises 3.4

True or False

1. A maximal ideal in a ring A may be equal to the whole ring.
2. The zero ideal in $M_2(\mathbb{Q})$ is a maximal ideal.
3. In any ring A, the whole ring A is a prime ideal.
4. In any integral domain, the zero ideal is a prime ideal.
5. In any division ring, the zero ideal is a maximal ideal.

3.4.1. Prove that, if A is an integral domain, but not a field, then the zero ideal of A is a prime ideal, but not a maximal ideal.

3.4.2. Let A be a ring and let K be an ideal of A such that A/K is a division ring. Prove that K is a maximal ideal of A.

3.4.3. Prove that a zero ring $A \neq \{0\}$ (i.e., a ring in which all products are equal to 0) has no prime ideals.

3.4.4. In the field $\mathbb{Z}/7\mathbb{Z}$, find the (multiplicative) inverse of the residue class $7\mathbb{Z} - 237$.

3.4.5. In the ring $M_2(\mathbb{Z})/M_2(7\mathbb{Z})$, determine whether the residue class $M_2(7\mathbb{Z}) + \begin{pmatrix} 2 & 5 \\ 6 & 8 \end{pmatrix}$ is a unit. (See Exercise 3.1.20.)

3.4.6. Polynomials are treated formally in the next section. This exercise draws on your informal experience with polynomials.

(1) Prove that the polynomials in x, with integer coefficients, form an integral domain under the usual operations $+$, $\cdot$ on polynomials. Denote this domain by $\mathbb{Z}[x]$.

(2) Let $K \subset \mathbb{Z}[x]$ be the set of all polynomials with zero constant term. Prove that K is a prime ideal, but not a maximal ideal, of $\mathbb{Z}[x]$. (Hint: consider the set of all polynomials with even constant term, in $\mathbb{Z}[x]$.)

3.4.7. Let $\mathbb{Q}_{(2)} = \{x \in \mathbb{Q} \mid x = a/b \text{ for some } a, b \in \mathbb{Z}, b \text{ odd}\}$.

(1) Prove that $\mathbb{Q}_{(2)}$ is a subdomain of $\mathbb{Q}$.

(2) Let $M = \{y \in \mathbb{Q}_{(2)} \mid y = c/d \text{ for some } c, d \in \mathbb{Z}, c \text{ even}, d \text{ odd}\}$. Prove that M is an ideal of $\mathbb{Q}_{(2)}$.

(3) Prove that M is a maximal ideal (hence a prime ideal) of $\mathbb{Q}_{(2)}$.

(4) Prove that the units of $\mathbb{Q}_{(2)}$ are precisely the elements of $\mathbb{Q}_{(2)} \backslash M$; hence $(\mathbb{Q}_{(2)})^* = \mathbb{Q}_{(2)} \backslash M$.

(5) Conclude that every ideal K of $\mathbb{Q}_{(2)}$, $K \neq \mathbb{Q}_{(2)}$, is contained in M; hence M is the unique maximal ideal of $\mathbb{Q}_{(2)}$.

(6) Prove that every ideal $K \neq \mathbb{Q}_{(2)}$ of $\mathbb{Q}_{(2)}$ is of the form $M_s = \{y \in \mathbb{Q} \mid y = c/d, c, d \in \mathbb{Z}, d \text{ odd, with } 2^s \mid c\}$, for some positive integer s. Note that, if "M^s" denotes the set $\{y_1 \dots y_s \mid y_i \in M, i = 1, \dots, s\}$, then $M_s = M^s$.

(7) Prove that, for each $s \in \mathbb{Z}^+$, the ring $\mathbb{Q}_{(2)}/M_s$ is isomorphic to the ring $\mathbb{Z}/2^s\mathbb{Z}$. ($\mathbb{Q}_{(2)}$ is called the *local ring* of $\mathbb{Q}$ at 2 — see Exercise 3.4.11.)

3.4.8. Prove that if H is an ideal in $M_2(n\mathbb{Z})$, where n is a positive integer, then, for each i, j $(1 \leq i, j \leq 2)$, the set $H_{ij} = \{q_{ij} \mid Q = (q_{hk})_{2\times 2} \in H\}$ is an ideal in $M_2(n\mathbb{Z})$. (See the proof in Example 2 of this section.)

3.4.9. For n a positive integer greater than 1, prove that, if K is a maximal ideal in the ring $M_n(\mathbb{Z})$, then $K = M_n(p\mathbb{Z})$, where p is a prime. (This generalizes Example 2 of this section.)

3.4.10. Zorn's Lemma in set theory (cf. [44]) implies that, if $\mathscr{X}$ is a non-empty set of subsets of a set A such that the union of every ascending chain $K_1 \subset K_2 \subset \cdots$ of sets in $\mathscr{X}$ is, itself, in $\mathscr{X}$, then $\mathscr{X}$ contains a maximal element (i.e., there is a set $K \in \mathscr{X}$ such that if $H \supset K$, $H \in \mathscr{X}$, then $H = K$). Use Zorn's Lemma to prove: if A is a ring with identity, then every non-unit of A is contained in a maximal ideal of A.

(Hint: For $a \in A$, let $\mathscr{X}$ be the set of all ideals of A which contain a.)

3.4.11. Let A be a commutative ring with identity, and let M be an ideal of A. Prove that M is the unique maximal ideal of A if and only if $A \backslash M$ is the unit group, A^*, of A. (A commutative ring with 1 and with a unique maximal ideal is a *local ring*.)

(Hint: You may find Exercise 3.4.10 useful.)

3.4.12. Let $M_1 \neq A$ and $M_2 \neq A$ be ideals of a ring A. If $M_1 \cap M_2$ is a maximal ideal of A, prove that $M_1 = M_2$.

3.4.13. Let A be a commutative ring with identity, and let F be a field. If $\phi : A \to F$ is a ring homomorphism such that $\phi a \neq 0$ for some $a \in A$, prove that Ker ϕ is a prime ideal of A. Need Ker ϕ be a maximal ideal?

3.4.14. Let F be a field. Define $\phi : \mathbb{Z} \to F$ by: $\phi(k) = k1$ for each $k \in \mathbb{Z}$. (Note: $k1$ in F is defined by: $01 = 0$, $(k + 1)1 = k1 + 1$ for each $k > 0$ in $\mathbb{Z}$; $(-k)1 = -k1$ for each $k > 0$ in $\mathbb{Z}$.)

(1) Prove that ϕ is a ring homomorphism.

(2) Prove that Ker ϕ is a prime ideal of $\mathbb{Z}$.

(3) Conclude that either Ker $\phi = \{0\}$, or Ker $\phi = p\mathbb{Z}$ for some prime p.

(If Ker $\phi = \{0\}$, then F has *characteristic* 0; if Ker $\phi = p\mathbb{Z}$ (p prime), then F has *characteristic* p.)

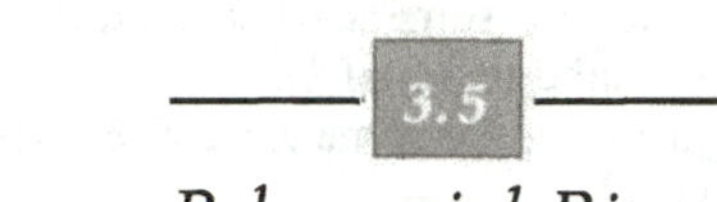

Polynomial Rings

The preceding section suffered somewhat from a dearth of examples. We partially remedy this situation now by considering an important class of rings.

For R any commutative ring with identity, consider the set of all "polynomials in an indeterminate x, with coefficients in R." Naively, we recognize these "polynomials" as familiar-looking objects $a_0 + a_1x + \cdots + a_nx^n$, $a_i \in R$. But what is an "indeterminate," and how do you multiply an indeterminate by a ring element?

A more formal approach recognizes the fact that, for all practical (i.e., computational) purposes, the powers x^i are merely place keepers, and the polynomial $a_0 + a_1x + \cdots + a_nx^n$ may be thought of as a sequence $(a_0, a_1, \ldots, a_n, 0, 0, \ldots)$ of ring elements from R, with all terms equal to zero beyond the $(n + 1)$st term a_n. We take this point of view in our formal description of polynomial rings.

Notation: We write $(a_i)_{i=0}^{\infty}$ for the sequence $(a_0, a_1, \ldots)$.

Theorem 3.5.1: Let R be a commutative ring with identity, and let $\bar{R}$ be the set of all sequences $(a_i)_{i=0}^{\infty}$, where each $a_i \in R$, and all but finitely many of the a_i are equal to 0. On $\bar{R}$, define $+$, $\cdot$ by:

$$(a_i)_{i=0}^{\infty} + (b_i)_{i=0}^{\infty} = (a_i + b_i)_{i=0}^{\infty}$$

$$(a_i)_{i=0}^{\infty} \cdot (b_i)_{i=0}^{\infty} = (c_i)_{i=0}^{\infty},$$

$$\text{where } c_i = \sum_{\substack{h+k=i \\ 0 \le h,\, 0 \le k}} a_h b_k.$$

Then $\langle \bar{R}, +, \cdot \rangle$ is a commutative ring with identity. Let $x = (0, 1, 0, \ldots)$. For $a \in R$, denote by "a" the sequence $(a, 0, 0, \ldots)$. Then, for $a_0, a_1, \ldots, a_n \in R$,

$$(a_0, a_1, \ldots, a_n, 0, 0, \ldots) = a_0 + a_1x + \cdots + a_nx^n.$$

Proof: For the operation +, note first that, if $(a_i)_{i=0}^{\infty}$, $(b_i)_{i=0}^{\infty}$ are sequences in $\bar{R}$, each has all but finitely many terms equal to 0; hence $(a_i + b_i)_{i=0}^{\infty}$ has all but finitely many terms equal to 0, and is therefore in $\bar{R}$. Commutativity and associativity of + follow immediately from the corresponding properties of addition in R. The sequence (0), all of whose terms are zero, serves as additive identity. For a sequence $(a_i)_{i=0}^{\infty}$, the sequence $(-a_i)_{i=0}^{\infty}$ serves as additive inverse. Thus, $\langle \bar{R}, + \rangle$ is an abelian group.

Before examining the $\cdot$ operation, let us illustrate how it works. For example, if

$$(a_i)_{i=0}^{\infty} = (1, 2, 0, 0, \ldots) \quad \text{and} \quad (b_i)_{i=0}^{\infty} = (3, 5, 7, 0, 0, \ldots),$$

$$\text{then} \quad (a_i)_{i=0}^{\infty} \cdot (b_i)_{i=0}^{\infty} = (c_i)_{i=0}^{\infty}$$

$$\text{where } c_i = \sum_{\substack{h+k=i \\ 0 \le h,\, 0 \le k}} a_h b_k, \quad \text{for each} \quad i = 0, 1, 2, \ldots.$$

Thus,

$$\begin{aligned}
c_0 &= a_0 b_0 = 1 \cdot 3 = 3 \\
c_1 &= a_0 b_1 + a_1 b_0 = 1 \cdot 5 + 2 \cdot 3 = 11 \\
c_2 &= a_0 b_2 + a_1 b_1 + a_2 b_0 = 1 \cdot 7 + 2 \cdot 5 + 0 \cdot 3 = 17 \\
c_3 &= a_0 b_3 + a_1 b_2 + a_2 b_1 + a_3 b_0 \\
&= 0 + 2 \cdot 7 + 0 \cdot 5 + 0 \cdot 3 = 14 \\
c_4 &= a_0 b_4 + a_1 b_3 + a_2 b_2 + a_3 b_1 + a_4 b_0 \\
&= 0 + 0 + 0 + 0 + 0 = 0.
\end{aligned}$$

Clearly, $c_i = 0$ for all $i \ge 4$. Thus,

$$(1, 2, 0, 0, \ldots) \cdot (3, 5, 7, 0, 0, \ldots) = (3, 11, 17, 14, 0, 0, \ldots).$$

It should now be clear that, for $(a_i)_{i=0}^{\infty}, (b_i)_{i=0}^{\infty} \in \bar{R}$, the sequence $(c_i)_{i=0}^{\infty}$ defined by

$$c_i = \sum_{\substack{h+k=i \\ 0 \le h,\, 0 \le k}} a_h b_k \quad (i = 0, 1, \ldots)$$

will be an element of the set $\bar{R}$.

To prove associativity of $\cdot$, let $(a_i)_{i=0}^{\infty}, (b_i)_{i=0}^{\infty}, (c_i)_{i=0}^{\infty} \in \bar{R}$. For each $i = 0, 1, 2, \ldots$, the ith term of

$$\begin{aligned}
((a_i)_{i=0}^{\infty}(b_i)_{i=0}^{\infty})(c_i)_{i=0}^{\infty} \quad \text{is} \quad & \sum_{j+l=i} \left(\sum_{h+k=j} a_h b_k \right) c_l \\
&= \sum_{h+k+l=i} (a_h b_k) c_l = \sum_{h+k+l=i} a_h (b_k c_l) \\
&= \sum_{j+h=i} a_h \left(\sum_{k+l=j} b_k c_l \right),
\end{aligned}$$

equal to the ith term of

$$(a_i)_{i=0}^{\infty}\left((b_i)_{i=0}^{\infty}(c_i)_{i=0}^{\infty} \right).$$

We leave as an exercise the proof that $\cdot$ distributes over $+$, and that $\cdot$ is commutative (Exercise 3.5.1). Thus, $\langle \bar{R}, +, \cdot \rangle$ is a commutative ring. The element $(1, 0, 0, \ldots)$ serves as identity for $\bar{R}$.

Let $x = (0, 1, 0, \ldots)$. Writing $(a_i)_{i=0}^{\infty} = (0, 1, 0, \ldots) = x$, we obtain the coefficients of $x^2 = (b_i)_{i=0}^{\infty}$:

$$b_0 = a_0^2 = 0$$

$$b_1 = a_0a_1 + a_1a_0 = 0$$

$$b_2 = a_0a_2 + a_1a_1 + a_2a_0 = 1$$

$$b_3 = a_0a_3 + a_1b_2 + a_2a_1 + a_3a_0 = 0$$

and

$$b_i = 0 \quad \text{for all} \quad i \geq 3.$$

Thus, $x^2 = (0, 0, 1, 0, \ldots)$. It is now easy to see that, for each positive integer i, $x^i = (0, 0, \ldots, 0, 1, 0, \ldots)$, the i-index term of the sequence being 1. If we write "a" for the sequence $(a, 0, 0, \ldots)$, $(a \in R)$, then any sequence $(a_0, a_1, \ldots, a_n, 0, 0, \ldots)$ emerges as—a *polynomial*! For,

$$\begin{aligned}(a_0, a_1, \ldots, a_n, 0, 0, \ldots) &= \\ &(a_0, 0, 0, \ldots) + (a_1, 0, 0, \ldots)(0, 1, 0, \ldots) \\ &+ (a_2, 0, 0, \ldots)(0, 0, 1, 0, \ldots) + \cdots \\ &+ (a_n, 0, 0, \ldots)(\underbrace{0, 0, \ldots, 0}_{n \text{ zeros}}, 1, 0, \ldots) \\ &= a_0 + a_1x + a_2x^2 + \cdots + a_nx^n.\end{aligned}$$

■

Having thus removed the mystery from "the indeterminate x" by unmasking it as the sequence $x = (0, 1, 0, 0, \ldots)$, we now revert to the customary notation for polynomials, and denote by $R[x]$ the ring $\bar{R}$ constructed in Theorem 3.5.1. Informally,

$$R[x] = \{a_0 + a_1x + \cdots + a_nx^n \mid n \geq 0, a_i \in R \text{ for each } i = 0, \ldots, n\},$$

to be known as *the ring of all polynomials in x, with coefficients in the ring R.*

Definition 3.5.1: Let R be a commutative ring, $R[x]$ the ring of all polynomials in x, with coefficients in R. If $f \in R[x]$, $f \neq 0$, then the *degree of f* is defined by: $\deg f = n$ if $f = a_0 + a_1x + \cdots + a_nx^n$, $a_n \neq 0$. (The 0-polynomial has no degree.)

We now examine the properties of polynomial rings over rings satisfying special conditions.

Theorem 3.5.2: If R is an integral domain, then so is $R[x]$. For each $f, g \in R[x], f \neq 0, g \neq 0$,

$$\deg fg = \deg f + \deg g.$$

Proof: We know that $R[x]$ is a commutative ring. Since R has identity $1 \neq 0$, the constant polynomial $1 \in R[x]$ is different from the 0-polynomial. If

$$f = a_0 + a_1x + \cdots + a_mx^m, a_m \neq 0,$$

and

$$g = b_0 + b_1x + \cdots + b_nx^n, b_n \neq 0,$$

then

$$fg = a_0b_0 + (a_0b_1 + a_1b_0)x + \cdots + a_mb_nx^{m+n}.$$

Since R is an integral domain, $a_mb_n \neq 0$, and so $fg \neq 0$. Thus, $F[x]$ is an integral domain and, if $\deg f = m$ and $\deg g = n$, then $\deg fg = m + n = \deg f + \deg g$. ∎

Corollary: If F is a field, then $F[x]$ is an integral domain.

Remark: Even for F a field, $F[x]$ is far from being a field. For, if $f \neq 0$ in $F[x]$ has a multiplicative inverse g in $F[x]$, then $fg = 1$; hence $\deg f = \deg g = 0$, and f and g are constant polynomials. Thus, no polynomial of degree > 0 can have an inverse in $F[x]$, and so $F[x]$ is *not* a field.

We now examine more closely the polynomial rings $F[x]$, where F is a field. It will turn out that these rings behave in many ways much like the ring of integers.

Theorem 3.5.3. (Division Algorithm for Polynomials): Let F be a field, and let $f, g \in F[x]$, $g \neq 0$. Then there are polynomials q and r in $F[x]$ such that

$$f = gq + r,$$

where either $r = 0$ or $\deg r < \deg g$. The polynomials q and r are uniquely determined for each f and each $g \neq 0$ in $F[x]$. (As usual, f, g, q, and r are referred to as *dividend*, *divisor*, *quotient*, and *remainder*.)

Proof: *Existence*. In the special case where f is the zero polynomial, we have

$$f = 0 = g \cdot 0 + 0.$$

Thus $q = 0$ and $r = 0$ will serve as quotient and remainder. For $f \neq 0$, we proceed by induction on $\deg f$.

If $\deg f = 0$, then f is a constant polynomial—say, $f = a$, for some $a \neq 0$ in F. If g is also a constant polynomial—say, $g = b$, $b \neq 0$ in F, then

$$f = a = b(b^{-1}a) + 0 = g(g^{-1}f) + 0,$$

and so $g^{-1}f$ and 0 serve as quotient and remainder. If g is not a constant polynomial, and $\deg f = 0$, we have

$$f = g \cdot 0 + f.$$

Since $\deg f < \deg g$, 0 and f will serve as quotient and remainder. We have thus anchored the induction at 0.

Now suppose that, for some $n \geq 1$, the theorem holds for dividends of degree $< n$. (Induction hypothesis.)

Let

$$f = a_0 + a_1x + \cdots + a_nx^n, a_n \neq 0 \quad \text{in } F[x],$$

and

$$g = b_0 + b_1x + \cdots + b_mx^m, b_m \neq 0 \quad \text{in } F[x].$$

If $n < m$, i.e., if $\deg f < \deg g$, we do not need to use the induction hypothesis. For, in this case, since

$$f = g \cdot 0 + f,$$

we can use $q = 0$, $r = f$ as quotient and remainder.

Suppose $n \geq m$. We now manufacture a polynomial to which we can apply the induction hypothesis. Let $f_1 = f - a_nb_m^{-1}x^{n-m}g$. Note that the highest power term in $a_nb_m^{-1}x^{n-m}g$ is a_nx^n. Thus, one of two things happens when $a_nb_m^{-1}x^{n-m}g$ is subtracted from f: the difference, f_1, is either equal to 0, or it is a polynomial of degree $< n$.

If $f_1 = f - a_nb_m^{-1}x^{n-m}g = 0$, then $f = ga_nb_m^{-1}x^{n-m} + 0$, and so $q = a_nb_m^{-1}x^{n-m}$ and $r = 0$ will serve as quotient and remainder.

If $f_1 = f - a_nb_m^{-1}x^{n-m}g \neq 0$, then since $\deg f_1 < \deg f = n$, the induction hypothesis applies to f_1 and we have

$$f_1 = g\bar{q} + \bar{r}$$

for some $\bar{q}, \bar{r} \in F[x]$, $\bar{r} = 0$, or $\deg \bar{r} < \deg g$. But then

$$f = f_1 + ga_nb_m^{-1}x^{n-m} = g(\bar{q} + a_nb_m^{-1}x^{n-m}) + \bar{r}.$$

Letting $q = \bar{q} + a_nb_m^{-1}x^{n-m}$ and $r = \bar{r}$, we have the required quotient and remainder for dividend f and divisor g.

Uniqueness. For $f \in F[x]$ and $g \neq 0$ in $F[x]$, suppose

$$f = gq_1 + r_1 = gq_2 + r_2,$$

where $q_1, q_2, r_1, r_2 \in F[x]$,

$$r_1 = 0 \quad \text{or} \quad \deg r_1 < \deg g; r_2 = 0 \quad \text{or} \quad \deg r_2 < \deg g.$$

Then $g(q_1 - q_2) = r_2 - r_1$. If $r_2 \neq r_1$, then

$$\deg(r_2 - r_1) \leq \max\{\deg r_1, \deg r_2\} < \deg g,$$

(see Exercise 3.5.2), while

$$\deg g(q_1 - q_2) = \deg g + \deg(q_1 - q_2) \geq \deg g.$$

(Note that $q_1 - q_2 \neq 0$ since $r_2 - r_1 \neq 0$.)
Contradiction! It follows that $r_1 = r_2$. But then

$$g(q_1 - q_2) = r_2 - r_1 = 0.$$

Since $g \neq 0$ and $F[x]$ is an integral domain, it follows that $q_1 - q_2 = 0$, and so $q_1 = q_2$. ∎

What other properties do the polynomial domains $F[x]$ (F a field) share with the ring of integers? In $\mathbb{Z}$, every ideal consists of the multiples of a single element.

Do the rings $F[x]$ have this property? The answer is "yes"—though these rings certainly do not share the stronger property of $\mathbb{Z}$: that every subgroup of the additive group is an ideal.

Definition 3.5.2: Let A be a commutative ring with identity. An ideal K of A is *principal* if there is some element $k \in K$ such that $K = kA = \{kx | x \in A\}$.

Theorem 3.5.4: Let F be a field. Then every ideal in $F[x]$ is principal.

Proof: Let K be an ideal in $A = F[x]$. If $K = \{0\}$, then $K = A0$, and is thus principal. If $K \neq \{0\}$, then there is some polynomial $g \neq 0$ in K. Let D_K be the set of all degrees of polynomials in K. Then D_K is a non-empty subset of $\mathbb{Z}^+ \cup \{0\}$. By the Well-Ordering Axiom for $\mathbb{Z}^+$, D_K contains a least element, i.e., K contains a polynomial of least degree—call it k. Since K is an ideal, $kA = \{kf \mid f \in A\} \subset K$. Let $h \in K$. By Theorem 3.5.3, there are polynomials q and r in $F[x]$ such that $h = kq + r$, where $r = 0$ or $\deg r < \deg k$. Now, from $h \in K$ and $k \in K$, we have $r = h - kq \in K$. Since k was chosen to be a non-zero polynomial *of least degree* in K, r cannot be a non-zero polynomial (else $r \in K$ and $\deg r < \deg k$). Hence $r = 0$. But then $h = kq + 0 = kq \in kA$. It follows that $K \subset kA$. But $kA \subset K$, and so $K = kA$, a principal ideal. ■

Definition 3.5.3: An integral domain A is a *principal ideal domain* if every ideal in A is a principal ideal.

Corollary: The ring $\mathbb{Z}$ of integers, and the polynomial domains $F[x]$, for F a field, are principal ideal domains.

On the other hand, for B an integral domain, the polynomial domain $B[x]$ need not be a principal ideal domain. For example, $\mathbb{Z}[x]$ is *not* a principal ideal domain. For, consider the subset K of $\mathbb{Z}[x]$ consisting of all polynomials in $\mathbb{Z}[x]$ that have even constant term. K is easily seen to be an ideal in $\mathbb{Z}[x]$. Suppose K is a principal ideal. Then there is some polynomial $k \in K$ such that $K = k\mathbb{Z}[x]$. Since the polynomials x and 2 are in K, both are multiples of k. But $k|2$ implies $k = \pm 2$ (note that $1 \notin K$), and $2 \nmid x$ in $\mathbb{Z}[x]$. Contradiction. Thus, K is not a principal ideal, and so $\mathbb{Z}[x]$ is not a principal ideal domain.

Exercises 3.5

True or False

1. The zero polynomial in $\mathbb{Q}[x]$ has degree zero.
2. The sum of any two polynomials of degree 5 is a polynomial of degree 5.
3. When a polynomial of degree 15 in $\mathbb{Q}[x]$ is divided by a polynomial of degree 3 in $\mathbb{Q}[x]$, the remainder may be a polynomial of degree 5 in $\mathbb{Q}[x]$.
4. The Division Algorithm (Theorem 3.5.3) applies when F is replaced by $\mathbb{Z}$.
5. The polynomial $2x^4 + x^3 - 5x^2 - 2x + 10$ in $\mathbb{Q}[x]$ is contained in the ideal generated by $x^2 - 2$. (See Exercise 3.2.15(5).)

3.5.1. Complete the proof of Theorem 3.5.1, i.e., prove that $\cdot$ is left and right distributive over $+$, and that $\cdot$ is commutative.

3.5.2. Prove: if R is a commutative ring with identity, then for f, g and $f+g$ non-zero polynomials in $R[x]$, $\deg(f+g) \leq \max\{\deg f, \deg g\}$. (Equality holds if $\deg f \neq \deg g$.)

3.5.3. Let R be a commutative ring with identity. Prove that the mapping $\alpha: R \to R[x]$ defined by $\alpha(a) = (a, 0, 0, \ldots)$ is a ring monomorphism ("isomorphic embedding").

In the following, we identify each $a \in R$ with its image $\alpha(a)$ and consider R to be a subring of $R[x]$.

3.5.4. If R is an integral domain, prove that the units in $R[x]$ are precisely the units of R. Conclude that, if F is a field, then the units in $F[x]$ are the non-zero elements of F (i.e., the non-zero constant polynomials).

3.5.5. In the complex number field $\mathbb{C}$, let $\mathbb{Z}[i] = \{a + bi \mid a, b \in \mathbb{Z}\}$. Prove that $\mathbb{Z}[i]$ is an integral domain. Find the units in $\mathbb{Z}[i]$ and in the polynomial domain $\mathbb{Z}[i][x]$. ($\mathbb{Z}[i]$ is the ring of *Gaussian integers*. We shall discuss it further in Section 3.7.)

3.5.6. Let R be a commutative ring with identity 1, and let B be a commutative ring with identity 1, containing R as a subring. Let $\beta \in B$. Define ϕ_β: $R[x] \to B$ by $\phi_\beta(f) = f(\beta)$ for each $f \in R[x]$ (i.e., if $f = \sum_{i=1}^{n} a_i x^i$, $a_i \in R$, then $\phi_\beta f = \sum_{i=1}^{n} a_i \beta^i$).

(1) Prove that ϕ_β is a ring homomorphism.

(2) Prove that $\operatorname{Ker} \phi_\beta = \{f \in R[x] \mid f(\beta) = 0\}$. ($\phi_\beta$ is called the *substitution homomorphism* of R into B, with respect to $\beta \in B$. If $f(\beta) = 0$, then β is called a *zero* of f. Thus, $\operatorname{Ker} \phi_\beta$ is the set of all polynomials in $R[x]$ having β as a zero, or: all polynomials in $R[x]$ that vanish at β.)

3.5.7. If B is an integral domain and R is a subring of B such that R is itself an integral domain, with identity equal to the identity of B, then R is called a *subdomain* of B.

(1) In the preceding exercise, if B is an integral domain, R a subdomain of B, prove that $\operatorname{Ker} \phi_\beta$ is a *prime* ideal, for each $\beta \in B$.

(2) Conclude that $R[x]/\operatorname{Ker} \phi_\beta$ is an integral domain.

3.5.8. Let R be a commutative ring with identity, B a ring having R as a subring. Let S be a subset of B, and let $K_S = \{f \in R[x] \mid f(\alpha) = 0 \text{ for all } \alpha \in S\}$. Prove that K_S is an ideal of $R[x]$.

3.5.9. Let R be a commutative ring with identity, and let $f \in R[x]$. Define $\psi_f: R \to R$ by: $\psi_f(\alpha) = f(\alpha) = f(\alpha)$ for each $\alpha \in R$. (ψ_f is the polynomial *function* on R associated with the polynomial f.) Is the correspondence $f \mapsto \psi_f$ a 1-1 correspondence for every ring R?
(Hint: Let $R = \mathbb{Z}_p$, p prime. Consider the polynomial $f = x^p - x$ in $\mathbb{Z}_p[x]$. What is $\psi_f(\alpha)$, for each $\alpha \in \mathbb{Z}_p$?)

3.5.10. For $f \in \mathbb{Q}[x]$, define ψ_f as in the preceding exercise. Assuming as known that a polynomial of degree n with coefficients in $\mathbb{Q}$ can have at most n distinct zeros in $\mathbb{Q}$, prove that the correspondence $f \mapsto \psi_f$ of polynomials in $\mathbb{Q}[x]$ to polynomial functions from $\mathbb{Q}$ to $\mathbb{Q}$ *is* 1-1. (This result generalizes to any infinite field.)

3.5.11. Let F be a field, and let n be a positive integer. Let K be the subset of $F[x]$ consisting of the zero polynomial and all polynomials of degree $\geq n$ in $F[x]$. Prove that, for each $k \in K$ and each $f \in F[x]$, the product $kf = fk$ is an element of K. Is K an ideal of $F[x]$?

3.6

Principal Ideal Domains

In the preceding section, we found that, like the ring of integers, the polynomial rings $F[x]$, F a field, have the property: every ideal consists of all multiples of a single element, i.e., every ideal is principal. What is remarkable about this property is the fact that it implies unique factorization, as exemplified in $\mathbb{Z}$ by the Fundamental Theorem of Arithmetic (Theorem 1.3.7).

Before we can discuss unique factorization in an integral domain, we must consider divisibility, and decide how to generalize the notion of a prime.

Definition 3.6.1: Let A be an integral domain, and let $a, b \in A$. Then $b|a$ (b *divides* a) if there is some $x \in A$ such that $a = bx$. In particular, $b \sim a$ (b *is an associate of* a) if $a = bu$, where u is a unit in A.

(Note this definition allows $0|0$, which does no harm, and will, on occasion, keep us from having to discuss exceptional cases.)

Theorem 3.6.1: Let A be an integral domain. Then

(1) $a \sim a$ for each $a \in A$.

(2) If $a \sim b$, then $b \sim a$ $(a, b \in A)$.

(3) If $a \sim b$ and $b \sim c$, then $a \sim c$ $(a, b, c \in A)$. (Thus, $\sim$ is an equivalence relation on A.)

(4) $a|a$ for each $a \in A$.

(5) For $a, b \in A$, $a|b$ and $b|a$ if and only if $a \sim b$.

(6) If $a|b$ and $b|c$, then $a|c$ $(a, b, c \in A)$.

(Thus, divisibility falls just short of being a partial order relation on A because it is not anti-symmetric (see Definition 1.2.2). However, it may be a partial order on a well-chosen subset of A. For example, if $A = \mathbb{Z}$, then $|$ serves as a partial order on $\mathbb{Z}^+$.)

Proof: *(1)* For each $a \in A$, $a = a \cdot 1$. Since 1 is a unit in A, $a \sim a$.

(2) If $a \sim b$, then $b = au$ for some unit u in A; hence $a = bu^{-1}$, and so $b \sim a$.

(3) If $a \sim b$, and $b \sim c$, then $b = au$, $c = bv$, where u, v are units in A. Hence $c = bv = (au)v = a(uv)$. Since uv is a unit in A, $a \sim c$.

(4) For each $a \in A$, $a = a \cdot 1$, hence $a|a$.

(5) Let $a, b \in A$. If $a|b$ and $b|a$, then $b = ay$, $a = bx$, for some $x, y \in A$. Hence $b = ay = (bx)y = b(xy)$. If $b \neq 0$, then (since A is an integral domain, hence a cancellation ring) we have $1 = xy$, and so x and y are both units in A. But then $a \sim b$. If $b = 0$, then from $a = bx$, we have $a = 0$ and so, certainly, $a \sim b$.

Conversely, if $a \sim b$, then $b = au$ for some unit $u \in A$, whence $a = bu^{-1}$, and we thus have $a|b$ and $b|a$.

(6) If $a|b$ and $b|c$, then $b = ax$, $c = by$, for some $x, y \in A$. Hence $c = by = (ax)y = a(xy)$, and so $a|c$. ∎

Example 1:

In $\mathbb{Z}$, the units are ± 1; hence the associates of an integer a are $\pm a$.

Example 2:

In the polynomial domain $F[x]$, F a field, the units are the non-zero constant polynomials k $(k \neq 0$ in $F)$ (see Exercise 3.5.4). Hence the associates of a polynomial f are the polynomials kf $(k \neq 0$ in $F)$. For example, in $\mathbb{Q}[x]$, the associates of $x^2 - 1$ include $\frac{2}{3}x^2 - \frac{2}{3}$, $-5x^2 + 5$, etc.

Example 3:

In the polynomial domain $A[x]$, A an integral domain, the units in $A[x]$ are the constant polynomials $k \neq 0$, where k is a unit in A, i.e., $k \in A^*$. Thus, for example, in $\mathbb{Z}[x]$, the only associates of $x^2 - 1$ are $x^2 - 1$ and $-x^2 + 1$.

Remark: In any integral domain A, u is a unit if and only if u is an associate of 1.

For: $u \sim 1$ if and only if $u|1$ and $1|u$, if and only if there is some $v \in A$ such that $uv = 1$, if and only if u is a unit in A.

Definition 3.6.2: Let A be an integral domain and let $a, b \in A$. Then b is a *proper divisor* of a if $b|a$ and b is neither a unit nor an associate of a. An element $q \in A$ is *irreducible* if $q \neq 0$, q is not a unit, and q has no proper divisors in A.

Example 1:

The irreducible elements in $\mathbb{Z}$ are the integers $q = \pm p$, where p is a prime.

Example 2:

Let F be a field. To begin with, to be irreducible, a polynomial $q \in F[x]$ must be non-zero and not a unit; hence it must be of positive degree. The proper divisors of a polynomial q of positive degree n are polynomials of positive degree less than n. (For, if $q = gh$, with $\deg g = \deg q$, then $\deg h = 0$, whence h is a unit and g an associate of q.) Thus, the irreducible elements in $F[x]$ are the polynomials of positive degree having no divisors of lower positive degree in $F[x]$.

For example, in $\mathbb{Q}[x]$, the polynomial $2x^2 - 4$ is irreducible since it has no factors of degree 1 in $\mathbb{Q}[x]$ (the factorization $2x^2 - 4 = 2(x^2 - 2)$ is not a *proper* factorization of $2x^2 - 4$ since 2 is a unit and $x^2 - 2$ is an associate of $2x^2 - 4$).

Example 3:

In $\mathbb{Z}[x]$, on the other hand, the polynomial $2x^2 - 4$ is *not* an irreducible element since it has 2 and $x^2 - 2$ as proper factors.

The irreducible elements in $\mathbb{Z}[x]$ are the constant polynomials q, where q is plus or minus a prime in $\mathbb{Z}$, and the polynomials $q = a_0 + a_1x + \cdots + a_nx^n$ $(a_n \neq 0, n > 0)$ satisfying the two conditions

(1) $\gcd(a_0, a_1, \ldots, a_n) = 1$

and

(2) q has no factors of positive degree less than n. Thus, the polynomials $-3, x^2 - 3, 5x^2 - x - 7$ are irreducible elements of $\mathbb{Z}[x]$, while the polynomials $10, 2x^2 - 4, 15x^2 - 3x - 21$ are reducible elements of $\mathbb{Z}[x]$. (Polynomials with property (1) will be featured in Section 3.9. They are called *primitive polynomials* — see Definition 3.9.1.)

We now proceed to work toward the goal of proving that the Fundamental Theorem of Arithmetic (Theorem 1.3.7) generalizes to principal ideal domains.

The following lemma provides us with a kind of dictionary that will enable us to translate freely back and forth between divisibility properties and ideal theoretic properties.

Lemma (Dictionary): Let A be an integral domain, and let $a, b \in A$. Then

(1) $b|a \Leftrightarrow aA \subset bA$.

(2) $a \sim b \Leftrightarrow aA = bA$.

(3) b is a proper divisor of $a \Leftrightarrow bA \neq A$ and $aA \subsetneq bA$.

(4) $u \in A$ is a unit $\Leftrightarrow uA = A$.

Proof: *(1)* If $b|a$, then $a = bx$ for some $x \in A$. For each $y \in A$, $ay = (bx)y = b(xy) \in bA$. Thus, $aA \subset bA$. Conversely, suppose $aA \subset bA$. Then $a = a1 \in bA$. But then $a = bx$ for some $x \in A$, and so $b|a$.

(2) By Theorem 3.6.1 (5), $a \sim b$ if and only if $a|b$ and $b|a$. But then, by part (1) of this Lemma, $a \sim b \Leftrightarrow bA \subset aA$ and $aA \subset bA \Leftrightarrow aA = bA$.

(3) b is a proper divisor of $A \Leftrightarrow b|a$ and b is neither a unit nor an associate of $a \Leftrightarrow aA \subset bA$, $bA \neq A$, and $aA \neq bA \Leftrightarrow bA \neq A$ and $aA \subsetneq bA$.

(4) $u \in A$ is a unit $\Leftrightarrow u \sim 1 \Leftrightarrow uA = 1A = A$. ■

We have seen (Section 3.4, Example 1) that an ideal $n\mathbb{Z}$ in $\mathbb{Z}$ $(n > 0)$ is maximal if and only if n is prime. Dropping the restriction "$n > 0$" and noting that $n\mathbb{Z} = (-n)\mathbb{Z}$ for any integer n, we can restate this result thus: for $m \in \mathbb{Z}\backslash\{0\}$, $m\mathbb{Z}$ is a maximal ideal of $\mathbb{Z}$ if and only if m is an irreducible element of $\mathbb{Z}$, i.e., $m = \pm p$, where p is a prime.

This result generalizes immediately to arbitrary principal ideal domains.

Theorem 3.6.2: Let A be a principal ideal domain, and let $K \neq \{0\}$ be an ideal in A. Then K is a maximal ideal of A if and only if $K = qA$, where q is an irreducible element of A.

Proof: Suppose $K = qA$, where q is an irreducible element of A. Then $q \neq 0$ and q is not a unit in A, whence $qA \neq \{0\}$ and $qA \neq A$. Suppose H is an ideal of A such that $qA \subset H \subset A$. Since A is a principal ideal domain, there is some $a \in A$ such that $H = aA$. From $qA \subset aA \subset A$, we have $a|q$. Since q is an irreducible element of A, it follows that either a is a unit or a is an associate of q. But then, by the Dictionary Lemma, either $aA = A$ or $aA = qA$, i.e., either $H = A$ or $H = qA$, and so qA is a maximal ideal of A.

Conversely, suppose that $K \neq \{0\}$ is a maximal ideal of A. Since A is a principal ideal domain, there is some $q \in A$ such that $K = qA$. By hypothesis, $K \neq \{0\}$; hence $q \neq 0$. By Definition 3.4.1, $K \neq A$; hence q is not a unit in A. Suppose $a|q$, for some $a \in A$. By the Dictionary Lemma, $qA \subset aA \subset A$. But $K = qA$ is a maximal ideal of A; hence either $aA = qA$ or $aA = A$. But then, by the Dictionary Lemma, either $a \sim q$ or a is a unit. It follows that q is an irreducible element of A. ■

Recall that, for A a commutative ring with identity, an ideal K of A is maximal if and only if the residue class ring A/K is a field. (Theorem 3.4.1.) Combined with the preceding theorem on principal ideal domains, this gives the following important corollary.

Corollary: Let A be a principal ideal domain, and let K be a non-zero ideal in A. Then A/K is a field if and only if $K = qA$, where q is an irreducible element of A.

Remark: If A is a principal ideal domain, then the 0-ideal is a maximal ideal of A if and only if A is itself a field. (See Exercise 3.6.12.)

Example 1:

We have already noted that $\mathbb{Z}/n\mathbb{Z}$ $(n \geq 1)$ is a field if and only if n is prime.

Example 2:

For F a field, $F[x]/pF[x]$ is a field if and only if p is an irreducible polynomial in $F[x]$. Thus, for example $\mathbb{Q}[x]/(x^2 - 2)\mathbb{Q}[x]$ is a field, while $\mathbb{Q}[x]/(x^2 - 4)\mathbb{Q}[x]$ has zero divisors, hence is certainly not a field. (For example,

$$[(x^2 - 4)\mathbb{Q}[x] + (x - 2)] \cdot [(x^2 - 4)\mathbb{Q}[x] + (x + 2)] = (x^2 - 4)\mathbb{Q}[x],$$

the 0-element of $\mathbb{Q}[x]/(x^2 - 4)\mathbb{Q}[x]$.)

One of the important properties of primes in $\mathbb{Z}$ is expressed by Euclid's Lemma (p. 22): *if p is prime and $p|ab$ $(a, b \in \mathbb{Z})$ then $p|a$ or $p|b$.*

Definition 3.6.3: Let A be a commutative ring with identity, and let $q \in A$, $q \neq 0$, q not a unit. Then q is a *prime element* of A if q has the property:

$$q|ab \quad \text{implies} \quad q|a \quad \text{or} \quad q|b \quad (a, b \in A).$$

Note: if q is a prime element of A, then an easy induction shows that $q|a_1 \ldots a_k$ $(a_i \in A)$ implies $q|a_i$ for some $i = 1, \ldots, k$.

Clearly, every irreducible element of $\mathbb{Z}$, i.e., every integer $n = \pm p$ (p prime) is a prime element of $\mathbb{Z}$. In fact, in any principal ideal domain, the set of all irreducible elements coincides with the set of all prime elements.

Theorem 3.6.3: Let A be a principal ideal domain, and let $p \in A$, $p \neq 0$, p not a unit. Then p is an irreducible element of A if and only if p is a prime element of A.

Proof: Suppose p is an irreducible element of A. If $p|ab$ for some $a, b \in A$, then $ab \in pA$. Since A is a principal ideal domain, Theorem 3.6.2 implies that pA is a maximal ideal. Hence, by Corollary 1 of Theorem 3.4.2, pA is a prime ideal. But then either $a \in pA$ or $b \in pA$, whence either $p|a$ or $p|b$.

Conversely, suppose that p is a prime element of A. If $p = hk$ for some $h, k \in A$, then $p|hk$; hence $p|h$ or $p|k$. On the other hand, $h|p$ *and* $k|p$. Hence we have either $p|h$ and $h|p$, or $p|k$ and $k|p$. But then either $h \sim p$ or $k \sim p$. Hence p is an irreducible element of A. ∎

Note that the first part of the proof of Theorem 3.6.3 made use of the hypothesis that A is a principal ideal domain. The second part works in any integral domain, i.e., in *any* integral domain, any prime element is irreducible. However, not every irreducible element in an arbitrary integral domain need be a prime element (see Exercise 3.6.11).

In any commutative ring A with identity, the prime elements are simply the generators of the non-zero principal prime ideals of A.

Theorem 3.6.4: Let A be a commutative ring with identity. Then $p \in A$ is a prime element of A if and only if pA is a non-zero prime ideal in A.

Proof: Suppose p is a prime element of A. By Definition 3.6.3, $p \neq 0$ and p is not a unit in A; hence $pA \neq \{0\}$ and $pA \neq A$. If $ab \in pA$ for some $a, b \in A$, then $p|ab$; hence $p|a$ or $p|b$. But then either $a \in pA$ or $b \in pA$, and so pA is a non-zero prime ideal of A.

Conversely, suppose pA is a non-zero prime ideal of A. Then $p \neq 0$ and p is not a unit in A. If $p|ab$ for some $a, b \in A$, then $ab \in pA$. Since pA is a prime ideal, it follows that either $a \in pA$ or $b \in pA$, whence either $p|a$ or $p|b$. Thus, p is a prime element of A. ∎

In view of Theorems 3.6.2, 3.6.3, and 3.6.4, we now have the following corollary.

Corollary: If A is a principal ideal domain, then a non-zero ideal K of A is a maximal ideal if and only if it is a prime ideal.

Proof: For K a non-zero ideal of A, there exists an element $q \neq 0$ in A such that $K = qA$. From Theorems 3.6.2, 3.6.3, and 3.6.4, we have

$$qA \text{ is maximal} \Leftrightarrow q \text{ is irreducible} \Leftrightarrow q \text{ is a prime element} \Leftrightarrow qA \text{ is a (non-zero) prime ideal.}$$ ∎

Note: The zero ideal in a commutative ring A with identity is a prime ideal if and only if A has no zero divisors. Thus, in any integral domain, $\{0\}$ is a prime ideal.

We are now ready to prove that unique factorization holds in any principal ideal domain.

Our aim is to show that every "reasonable" element a of a principal ideal domain A (i.e., every non-zero element a that is not a unit) can be expressed as a product of irreducible elements, and that, in some "reasonable" sense, this factorization is unique. Clearly, in view of commutativity of multiplication, the order of the factors should be of no consequence; e.g., $3 \cdot 5$ and $5 \cdot 3$ are equivalent factorizations of 15 in $\mathbb{Z}$. It is also reasonable to consider $3 \cdot 5$ and $(-3) \cdot (-5)$ to be equivalent factorizations of 15 as a product of irreducible elements of $\mathbb{Z}$. More generally, it seems reasonable to consider two factorizations equivalent if some of the irreducible factors in one of the factorizations are replaced by associates in the other: thus, $(x - 2) \cdot (x + 2)$ and $(3x - 6) \cdot (\frac{1}{3}x + \frac{2}{3})$ may reasonably be regarded as equivalent factorizations of $x^2 - 4$ in $\mathbb{Q}[x]$.

We lay down our ground rules for uniqueness in the following definitions:

Definition 3.6.4: Let A be an integral domain, $a \in A$, $a \neq 0$, a not a unit. Two factorizations $a = p_1 \dots p_h = q_1 \dots q_k$, where p_i $(i = 1, \dots, h)$ and q_j $(j = 1, \dots, k)$ are irreducible elements of A, are *equivalent* if $h = k$ and there is a 1–1 correspondence between the factors p_i and the factors q_j such that corresponding factors are associates.

We now define a unique factorization domain using this notion of equivalence.

Definition 3.6.5: An integral domain A is a *unique factorization domain* if every non-zero, non-unit element $a \in A$ may be represented as a product of irreducible elements of A, and if any two such representations of an element $a \in A$ are equivalent.

In order to prove the existence of such factorizations in principal ideal domains, we require an ideal-theoretic lemma that arises very naturally from considerations of divisibility.

Consider, for example, an integer such as 60. We can form chains of successive divisors of 60 in various ways, such as 60, 30, 15, 5; or 60, 20, 10, 2; or 60, 12, 6, 3. Whatever chain we decide upon, it seems that, after finitely many steps, we hit a prime. By our Dictionary Lemma, these divisor chains, going down, correspond to ideal chains, going up (with proper divisibility corresponding to proper inclusion): $60\mathbb{Z} \subsetneq 30\mathbb{Z} \subsetneq 15\mathbb{Z} \subsetneq 5\mathbb{Z}$; or $60\mathbb{Z} \subsetneq 20\mathbb{Z} \subsetneq 10\mathbb{Z} \subsetneq 2\mathbb{Z}$; or $60\mathbb{Z} \subsetneq 12\mathbb{Z} \subsetneq 6\mathbb{Z} \subsetneq 3\mathbb{Z}$. Each of these ideal chains terminates in a maximal ideal, which is properly contained in no ideal other than the whole ring $\mathbb{Z}$.

Our existence proof will move in the opposite direction. We shall show first that, in a principal ideal domain, there are no infinite, properly ascending ideal chains; hence there are no infinite properly descending divisor chains. From this, it will follow that every non-zero non-unit $a \in A$ has an irreducible divisor and is, in fact, a product of irreducible elements.

Lemma 1: Let A be a principal ideal domain, and let $(K_i)_{i \in \mathbb{Z}^+}$ be a sequence of ideals in A such that $K_i \subset K_{i+1}$ for each $i \in \mathbb{Z}^+$. Then there is some positive integer m such that $K_i = K_m$ for all $i \geq m$. (We can say more simply: every ascending chain of ideals of A terminates; or, equivalently: there are no infinite *properly* ascending chains of ideals in A.)

Proof: Let $(K_i)_{i \in \mathbb{Z}^+}$ be a sequence of ideals of A, with $K_i \subset K_{i+1}$ for each $i \in \mathbb{Z}^+$, and let $K = \bigcup_{i \in \mathbb{Z}^+} K_i$. Then K is an ideal in A, for:

(1) $K \neq \phi$ since $0 \in K_1$, hence $0 \in K$;

(2) If $x, y \in K$, then there are $i, j \in \mathbb{Z}^+$ such that $x \in K_i$ and $y \in K_j$. Without loss of generality, suppose $i \leq j$. Then $K_i \subset K_j$, and therefore $x, y \in K_j$. But then $x - y \in K_j$, and so $x - y \in K$.

(3) If $k \in K$ and $x \in A$, then $k \in K_i$, for some $i \in \mathbb{Z}^+$; hence, $kx \in K_i$, and therefore $kx \in K$.

Since A is a principal ideal domain, there is some $\bar{k} \in K$ such that $K = \bar{k}A$. But $K = \bigcup_{i \in \mathbb{Z}^+} K_i$, and so $\bar{k} \in K_m$ for some $m \in \mathbb{Z}^+$. It follows that $K = \bar{k}A \subset K_m$. On the other hand, $K_m \subset K$, and so $K = K_m$. But then, for any $i \geq m$, we have $K = K_m \subset K_i \subset K$, whence $K_i = K_m$. ■

Lemma 2: Let A be a principal ideal domain, and let $(a_i)_{i \in \mathbb{Z}^+}$ be a sequence of elements of A such that $a_{i+1} | a_i$ for each $i \in \mathbb{Z}^+$. Then there is some positive integer m such that $a_i \sim a_m$ for each $i \geq m$. (Equivalently: there are no infinite sequences $(a_i)_{i \in \mathbb{Z}^+}$ in A such that a_{i+1} is a *proper* divisor of a_i for each $i \in \mathbb{Z}^+$.)

Proof: If $a_1, a_2, \ldots$ is a sequence of non-zero non-units of A such that $a_{i+1} | a_i$ for each $i \in \mathbb{Z}$, then $a_iA \subset a_{i+1}A$ for each $i \in \mathbb{Z}^+$. By Lemma 1, there is some $m \in \mathbb{Z}^+$ such that $a_i\mathbb{Z} = a_m\mathbb{Z}$ for each $i \geq m$, whence $a_i \sim a_m$ for each $i \geq m$. ■

Theorem 3.6.5: Every principal ideal domain is a unique factorization domain.

Proof: Let A be a principal ideal domain, and let a be a non-zero non-unit in A.

Existence. We prove first that a has an irreducible divisor. If a is itself irreducible, stop: then a is an irreducible divisor of a. If a is not irreducible, then a has a proper divisor a_1. If a_1 is irreducible, stop: a_1 is an irreducible divisor of a. If a_1 is not irreducible, then a_1 has a proper divisor a_2. If a_2 is itself irreducible, stop: a_2 is an irreducible divisor of a_1, hence of a. Continuing in this way, we build a sequence of elements $a, a_1, a_2, \ldots$ such that each term of the sequence is a proper divisor of its predecessor. By Lemma 2, there can be no infinite sequence with this property. Hence, after finitely many steps, the process stops, i.e., there is some positive integer m such that a_m is irreducible, $a_m | a_{m-1}, a_{m-1} | a_{m-2}, \ldots, a_2 | a_1, a_1 | a$, all divisibility relations being proper. By the transitivity of divisibility, a_m is an irreducible divisor of a.

To prove that a is a product of irreducible elements, let p_1 be an irreducible divisor of a. Then $a = p_1k_1$ for some $k_1 \in A$. If k_1 is a unit, stop: then a is irreducible (see Exercise 3.6.5), hence a "product of one irreducible factor." If k_1 is not a unit, then k_1 has an irreducible divisor p_2 and $k_1 = p_2k_2$ for some $k_2 \in A$, whence $a = p_1p_2k_2$. If k_2 is a unit, stop: then p_2k_2 is irreducible, and $a = p_1(p_2k_2)$ is a product of two irreducible elements. If k_2 is not a unit, then k_2 has an irreducible divisor p_3 and $k_2 = p_3k_3$ for some $k_3 \in A$, whence $a = p_1p_2p_3k_3$. Continuing in this way, we build a sequence $k_1, k_2, k_3, \ldots$ such that each term of the sequence is either a unit or a proper divisor of its predecessor (proper since the p_i are non-units). By Lemma 2, there can be no infinite sequence of elements of A, where each term is a proper divisor of its predecessor. Hence there is some positive integer n such that k_n is a unit, and so $a = p_1 \ldots p_nk_n = p_1 \ldots p_{n-1}\bar{p}_n$, with $\bar{p}_n = p_nk_n$, and $p_1, \ldots, p_{n-1}, \bar{p}_n$ all irreducible.

Uniqueness. Let $a \in A$ be a non-zero non-unit, and suppose that

$$a = p_1 \ldots p_n = q_1 \ldots q_m,$$

where the p_i and the q_j are irreducible elements of A ($i = 1, \ldots, n; j = 1, \ldots, m$). To prove that the products $p_1 \ldots p_n$ and $q_1 \ldots q_m$ are equivalent (see Definition 3.6.5), we proceed by induction on n.

For $n = 1$, suppose $a = p_1 = q_1 \ldots q_m$. Since p_1 is irreducible, we conclude that $m = 1$ and $p_1 = q_1$.

For $n > 1$, suppose that any two products $p_1 \ldots p_{n-1} = q_l \ldots q_1$ (p_i, q_j irreducible) are equivalent. (Induction hypothesis.)

If $a = p_1 \ldots p_n = q_1 \ldots q_m$ (p_i, q_j irreducible), then since $p_n | q_1 \ldots q_m$ and p_n is a prime element of A (Theorem 3.6.3), we conclude that $p_n | q_j$ for some $j (1 \leq j \leq m)$. Because multiplication in A is commutative, we lose no generality by relabeling the elements $q_1, \ldots, q_m$ so that $p_n | q_m$. Since q_m is irreducible, $p_n \sim q_m$, and so $q_m = p_nu$ where u is a unit in A. Since A is an integral domain, we can cancel the factor p_n from

$$p_1 \cdots p_n = q_1 \cdots q_{m-1}p_nu,$$

obtaining $p_1 \ldots p_{n-1} = q_1 \ldots q_{m-2}\bar{q}_{m-1}$ where $\bar{q}_{m-1} = q_{m-1}u$, itself an irreducible element of A. By the Induction Hypothesis, the products $p_1 \ldots p_{n-1}$ and $q_1 \cdots q_{m-2}\bar{q}_{m-1}$ are equivalent: thus, $n - 1 = m - 1$ (hence $m = n$), and there is a 1-1 correspondence between the factors $p_1, \ldots, p_{n-1}$ and the factors $q_1, \ldots, q_{m-2}, \bar{q}_{m-1}$ such that corresponding factors are associates. Note that, if $p_i \sim \bar{q}_{m-1}$ then since $\bar{q}_{m-1} \sim q_{m-1}$, we have $p_i \sim q_{m-1}$. But then there is a 1-1 correspondence between the original factors $p_1, \ldots, p_n$ and the factors $q_1, \ldots, q_m$ such that corresponding factors are associates, i.e., the two products $p_1 \ldots p_n$ and $q_1 \ldots q_m$ representing a are equivalent. By the Principle of Induction, we conclude that the two products $p_1 \ldots p_n$ and $q_1 \ldots q_m$ representing a are equivalent, for all positive integers n. ■

Corollary: If F is a field, then the polynomial domain $F[x]$ is a unique factorization domain.

In any integral domain A, greatest common divisors and least common multiples can be defined much as they are in $\mathbb{Z}$ (see Definition 1.3.3 and Exercise 1.3.12).

Definition 3.6.6: Let A be an integral domain, and let $a, b \in A$. Then:

(1) $d \in A$ is a *greatest common divisor* of a and b if

(a) $d|a$ and $d|b$

and

(b) if $c|a$ and $c|b$ $(c \in A)$, then $c|d$;

(2) $m \in A$ is a *least common multiple* of a and b if

(a) $a|m$ and $b|m$

and

(b) if $a|n$ and $b|n$ $(n \in A)$, then $m|n$.

From Definition 3.6.6, it follows immediately that, if $d_1 \in A$ is a greatest common divisor of a and b, then $d_2 \in A$ is also a greatest common divisor of a and b $\Leftrightarrow d_1|d_2$ and $d_2|d_1 \Leftrightarrow d_1 \sim d_2$; if $m_1 \in A$ is a least common multiple of a and b, then $m_2 \in A$ is also a least common multiple of a and $b \Leftrightarrow m_1|m_2$ and $m_2|m_1 \Leftrightarrow m_1 \sim m_2$.

In integral domains with more units, the associate relation is less restrictive than in $\mathbb{Z}$ (where the only units are ± 1). Thus, in the polynomial domain $F[x]$ (F a field), if d is a greatest common divisor of f and g, so is ud, for each $u \neq 0$ in F. For example, all polynomials of the form

$$kx^2 - 2k, \quad k \neq 0 \quad \text{in} \quad \mathbb{R}$$

are greatest common divisors of $f = x^4 - 4$ and $g = x^4 - 4x^2 + 4$. Just as, in $\mathbb{Z}$, one generally works with the (unique) positive greatest common divisor of two integers, in $F[x]$ one generally selects the (unique) greatest common divisor with leading coefficient 1.

Theorem 3.6.6: If A is a unique factorization domain, then any two elements of A have a greatest common divisor and a least common multiple, each determined uniquely, up to replacement by associates.

Proof: See Exercise 3.6.8. ∎

Corollary: If A is a principal ideal domain, then any two elements of A have a greatest common divisor and a least common multiple in A, each determined uniquely up to replacement by associates.

While the Corollary follows immediately from Theorems 3.6.5 and 3.6.6, it can be proved directly, without the use of unique factorization. (See Exercise 3.6.2.)

Remark (a word about zero): In Definition 3.6.1, we allowed $0|0$. By Definition 3.6.6, it follows that 0 is a least common multiple, hence the only least common multiple, of two elements a and b if either $a = 0$ or $b = 0$, or $a = b = 0$. For greatest common divisors, we have: for $a \neq 0$, a is a greatest common divisor of a and 0, and 0 is a greatest common divisor (the only one) of 0 and 0.

Our seemingly eccentric inclusion of $0|0$ has as a consequence the assertions (without exception!) in Exercise 3.6.1: if A is a principal ideal domain, $a, b \in A$, then (1) $d \in A$ is a greatest common divisor of a and b if and only if $aA + bA = dA$; (2) $m \in A$ is a least common multiple of a and b if and only if $aA \cap bA = mA$. This leads to a very elegant existence proof for greatest common divisors and least common multiples in a principal ideal domain. (See Exercise 3.6.2.)

Exercises 3.6

True or False

1. $\mathbb{Z}$ is a principal ideal domain.

2. $\mathbb{Z}[x]$ is a principal ideal domain.

3. $\mathbb{C}[x]$ is a principal ideal domain.

4. $2x^2 - 4$ is irreducible in $\mathbb{Q}[x]$.

5. $2x^2 - 4$ is irreducible in $\mathbb{Z}[x]$.

3.6.1. Let A be a principal ideal domain, and let $a, b, d, m \in A$. Prove:
(1) d is a greatest common divisor of a and b if and only if $aA + bA = dA$;
(2) m is a least common multiple of a and b if and only if $aA \cap bA = mA$.

3.6.2. Use the result of the preceding exercise to prove that, if A is a principal ideal domain, then any two elements of A have a greatest common divisor and a least common multiple, each unique up to associates.

3.6.3. Let A be a principal ideal domain. Define $a, b \in A$ to be relatively prime if the only common divisors of a and b are units in A. Prove that the following statements are equivalent:
(1) a and b are relatively prime.
(2) There are elements $x, y \in A$ such that $ax + by = 1$.
(3) $aA + bA = 1A = A$.

3.6.4. Let A be a principal ideal domain, and let $a, b \in A$ be relatively prime (see the preceding exercise).
(1) If $c \in A$ such that $a | bc$, prove that $a | c$.
(2) If $y \in A$ such that $a | y$ and $b | y$, prove that $ab | y$.

3.6.5. In an integral domain A, prove that $p \in A$ is irreducible if and only if up is irreducible for each unit $u \in A$.

3.6.6. Let A be a unique factorization domain, and let p be an irreducible element of A. Prove that p is a prime element of A. Conclude that, in any unique factorization domain A, an element p is irreducible if and only if it is prime. (This generalizes Theorem 3.6.3 from principal ideal domains to the larger class of rings: unique factorization domains. However, Theorem 3.6.2 and the Corollary of Theorem 3.6.4 do *not* generalize from principal ideal domains to arbitrary unique factorization domains. Exercise 3.4.6 shows that a unique factorization domain may have prime ideals that are not maximal.)

3.6.7. Let A be a unique factorization domain, and let p be an irreducible element of A. For α, β nonnegative integers, prove that $p^\alpha | p^\beta$ if and only if $\alpha \leq \beta$.

3.6.8. Let A be a unique factorization domain. In order to prove that every pair of elements of A has a greatest common divisor and a least common multiple in A, it suffices (why?) to prove the following. Let $a \neq 0$, $b \neq 0$ be elements of A and let $p_1, \ldots, p_s$ be irreducible elements of A such that $a = p_1^{\alpha_1} \ldots p_s^{\alpha_s}$, $b = p_1^{\beta_1} \ldots p_s^{\beta_s}$, where $0 \leq \alpha_i$, $0 \leq \beta_i$, for each $i = 1, \ldots, s$. Let $\gamma_i = \min\{\alpha_i, \beta_i\}$, $\delta_i = \max\{\alpha_i, \beta_i\}$ for each $i = 1, \ldots, s$. Then $\prod_{i=1}^{s} p_i^{\gamma_i}$ is a greatest common divisor, and $\prod_{i=1}^{s} p_i^{\delta_i}$ is a least common multiple, of a and b. (Compare Exercise 1.3.13 for the case where $A = \mathbb{Z}$.)

3.6.9. Prove that, if A is a unique factorization domain, a polynomial $f = \sum_{i=1}^{n} a_i x^i$, $a_i \neq 0$ in A, $n > 0$, is irreducible if and only if

(a) $a_0, \ldots, a_n$ have no common divisors in A other than units

and

(b) f is not the product of two polynomials g and $h \in A[x]$, both of degree less than n.

3.6.10. Let $A = \{a + b\sqrt{-5} \,|\, a, b \in \mathbb{Z}\}$.

(1) Prove that A is an integral domain.

(2) On A, define $N : A \to \mathbb{Z}$ by: $N(a + b\sqrt{-5}) = a^2 + 5b^2 \quad (a, b \in \mathbb{Z})$. Prove that $N(zw) = N(z)N(w)$ for each $z, w \in A$.

(3) Find all units in A.

3.6.11. Let $A = \{a + b\sqrt{-5} \,|\, a, b \in \mathbb{Z})$ (see the preceding exercise). Note that $9 = (2 + \sqrt{-5})(2 - \sqrt{-5}) = 3 \cdot 3$. Use the results of Exercise 6.10 to prove:

(1) $3, 2 + \sqrt{-5}, 2 - \sqrt{-5}$ are irreducible, but not prime, elements of A.

(2) 3 is not an associate of either $2 + \sqrt{-5}$ or $2 - \sqrt{-5}$.

(3) A is not a unique factorization domain.

Conclude that A is not a principal ideal domain.

3.6.12. Prove that, if A is a principal ideal domain, then the 0-ideal is a maximal ideal of A if and only if A is a field.

3.6.13. Our definition of *principal ideal* was tailored for use in commutative rings with identity. In general, an ideal K in a ring A is principal if there is some $a \in A$ such that $K = (a) = \bigcap_{\substack{H \lhd A \\ a \in H}} H$ (see Exercises 3.2.14 and 3.2.15).

Define an ideal $L \neq \{0\}$ in a ring A to be *minimal* if for each ideal H of A such that $H \subset L$, either $H = \{0\}$ or $H = L$.

Prove that every minimal ideal in a ring A is principal.

3.6.14. Find all minimal ideals and all maximal ideals in the ring $\mathbb{Z}_{12} = \mathbb{Z}/12\mathbb{Z}$.

3.6.15. In the ring $M_2(\mathbb{Z}_{12})$, prove that the matrices whose entries are either $\bar{0}$ or $\bar{6}$ form a minimal ideal, K. Demonstrate that K is principal by finding a generator for it.

3.6.16. Let F be a field, and let f be a polynomial of positive degree in $F[x]$. Let $K = f \ \ F[x]$. If $f = p_1^{\alpha_1} \ldots p_k^{\alpha_k}$, where p_i is an irreducible polynomial in $F[x]$ for each $i = 1, \ldots, k \quad (\alpha_i \geq 1)$, what is the radical, $\sqrt{K}$, of the ideal K? (See Exercise 3.2.19).

3.6.17. Prove the *Factor Theorem*: Let F be a field, and let f be a polynomial of positive degree in $F[x]$. If $f(\alpha) = 0$ for some $\alpha \in F$, prove that $f = (x - \alpha)g$ for some $g \in F[x]$.

3.6.18. Let f be a polynomial of positive degree in $\mathbb{C}[x]$ and let $K = f\mathbb{C}[x]$. Let $S_K = \{\alpha \in \mathbb{C} \mid g(\alpha) = 0 \text{ for all } g \in K\}$.

(1) Prove that S_K is equal to the set of all zeros of f.

(2) Let $H = \{h \in \mathbb{C}[x] \mid h(\alpha) = 0 \text{ for each } \alpha \in S_K\}$. Prove that H is an ideal in $\mathbb{C}[x]$.

(3) Prove that $H = \sqrt{K}$. (See Exercise 3.2.19.)
Hint: Use the Fundamental Theorem of Algebra, which states that every polynomial in $\mathbb{C}[x]$ factors into linear factors in $\mathbb{C}[x]$. Also: use Exercises 3.6.4 and 3.6.17. (Note: this exercise represents an easy special case of the Hilbert Nullstellensatz (cf. [8]) which makes an analogous assertion regarding polynomials in any finite number of indeterminates.)

3.7

Euclidean Domains

A property common to the principal ideal domains we have thus far examined, i.e., to $\mathbb{Z}$ and the polynomial domains $F[x]$ (F a field) is the existence of a division algorithm (see Theorem 1.3.4 and Theorem 3.5.3). The condition relating remainder and divisor in these algorithms employed absolute values in $\mathbb{Z}$ and degrees in the polynomial domains $F[x]$.

In order to arrive at a class of "division algorithm domains" that includes the preceding examples and many other rings, we introduce the concept of a Euclidean norm.

Definition 3.7.1: Let A be an integral domain, and let $\delta : A\backslash\{0\} \to \mathbb{Z}^+ \cup \{0\}$ be a function such that:

(1) if $b \mid a$ $(a \neq 0, b \neq 0 \text{ in } A)$, then $\delta b \leq \delta a$

and

(2) for each $b \neq 0$ in A and $a \in A$,

there are elements $q, r \in A$ such that $a = bq + r$, where either $r = 0$ or $\delta r < \delta b$. Then δ is a *Euclidean norm* on A. The integral domain A forms a *Euclidean domain* with respect to the Euclidean norm δ.

Example 1:

Define $\delta : \mathbb{Z}\backslash\{0\} \to \mathbb{Z}^+ \cup \{0\}$ by: $\delta a = |a|$ for each $a \in \mathbb{Z}$. Then δ is a Euclidean norm, and $\mathbb{Z}$ forms a Euclidean domain with respect to δ, by Theorem 1.3.4.

Example 2:

For F a field, define

$$\delta : F[x]\backslash\{0\} \to \mathbb{Z}^+ \cup \{0\}$$

by: $\delta f = \deg f$. Then δ is a Euclidean norm and $F[x]$ forms a Euclidean domain with respect to δ, by Theorems 3.5.2 and 3.5.3.

Before giving further examples of Euclidean domains, we prove that Euclidean domains are necessarily principal ideal domains, hence necessarily unique factorization domains.

Theorem 3.7.1: Every Euclidean domain is a principal ideal domain.

Proof: Let A be a Euclidean domain, with Euclidean norm δ, and let K be an ideal in A. If $K = \{0\}$, then $K = 0A$, hence principal. Suppose $K \neq \{0\}$. Since $\mathbb{Z}^+ \cup \{0\}$ is well-ordered, there is a least non-negative integer s_0 that serves as the value of the Euclidean norm for some nonzero element of K. Let $k_0 \in K$ be an element of K such that $\delta k_0 = s_0$. We prove that $K = k_0 A$.

Since K is an ideal in A, clearly $k_0 A \subset K$. Let $k \in K$. Then

$$k = k_0 q + r$$

for some $q, r \in A$, where either $r = 0$, or $\delta r < \delta k_0 = s_0$. Since $r = k - k_0 q$, we conclude that $r \in K$. By the choice of k_0 as an element of least norm, s_0, in K, it follows that $r = 0$. Thus, $k_0 | k$, and we have $k \in k_0 A$. But then $K = k_0 A$, and so A is a principal ideal domain. ■

Corollary: Every Euclidean domain is a unique factorization domain.

The converse of Theorem 3.6.5 and the converse of Theorem 3.7.1 are both false: there exist unique factorization domains that are not principal ideal domains, and there exist principal ideal domains that are not Euclidean. Moreover, there exist integral domains that are not unique factorization domains. We illustrate these assertions in the following examples.

Example 1:

We have already seen that $\mathbb{Z}[x]$ is not a principal ideal domain. In Theorem 3.9.1, we will show that, for A any unique factorization domain, $A[x]$ is a unique factorization domain. Thus, $\mathbb{Z}[x]$ is a unique factorization domain, but not a principal ideal domain.

Example 2:

For d any squarefree integer, the set $\mathbb{Q}(\sqrt{d}) = \{a + b\sqrt{d} | a, b \in \mathbb{Q}\}$ forms a field (see Exercise 3.7.1). The elements $u \in \mathbb{Q}(\sqrt{d})$ such that $f(u) = 0$ for some $f \in \mathbb{Z}[x]$ with leading coefficient 1 form a subring I_d of $\mathbb{Q}(\sqrt{d})$. (See Exercise

3.7.5.) For each squarefree integer d, the field $\mathbb{Q}(\sqrt{d})$ is called a *quadratic number field*, and the ring I_d (necessarily an integral domain — why?) is referred to as the *ring of integers* of $\mathbb{Q}(\sqrt{d})$. It can be shown that

$$I_d = \{a + b\sqrt{d} \mid a, b \in \mathbb{Z}\}$$

if $d \equiv 2$ or $3 \bmod 4$, and

$$I_d = \{a + b\sqrt{d} \mid 2a, 2b \in \mathbb{Z}\}$$

if $d \equiv 1 \bmod 4$.

The integral domains I_d (see Example 2) have been the subject of extensive research in algebraic number theory. For $d < 0$, there exists a Euclidean norm for I_d if and only if d is one of the numbers $-1, -2, -3, -7, -11$. Thus, $I_{-1}, I_{-2}, I_{-3}, I_{-7}, I_{-11}$ are all Euclidean, hence principal ideal, unique factorization domains. There are only four other negative values of d for which I_d is a unique factorization domain: $-19, -43, -67$, and -163. The four domains $I_{-19}, I_{-43}, I_{-67}$, and I_{-163} are, in fact, principal ideal domains, all of them non-Euclidean.

For $d > 0$, it has been proved that the statements "I_d is Euclidean" and "I_d is a principal ideal domain" would be equivalent, provided certain generalizations of a famous conjecture, known as the Riemann Hypothesis, were valid. The validity of the Riemann Hypothesis, however, remains one of the great unsolved problems of number theory and, indeed, of mathematics. (cf. [33], [35], [36], [38], [41]).

For $d = -1$, since $-1 \equiv 3 \bmod 4$, we have $I_{-1} = \{a + b\sqrt{-1} \mid a, b \in \mathbb{Z}\}$. This ring, known as the ring of all *Gaussian integers*, is usually denoted by $\mathbb{Z}[i]$. It is not difficult to prove that $\mathbb{Z}[i]$ is Euclidean, and we do so in the following theorem.

Theorem 3.7.2: The ring $\mathbb{Z}[i]$ of all Gaussian integers $a + bi$ $(a, b \in \mathbb{Z})$ is a Euclidean domain.

Proof: Being a subring of the field $\mathbb{C}$, $\mathbb{Z}[i]$ is an integral domain (see Exercise 3.1.8 or 3.5.5). Define $N(a + bi) = a^2 + b^2$ (the square of the absolute value of $a + bi$). Then it is easy to verify that $N(uv) = N(u)N(v)$ for each $u, v \in \mathbb{Z}[i]$. We show that N serves as a Euclidean norm on $\mathbb{Z}[i]$. If $u \neq 0$, $v \neq 0$, and $u \mid v$, then $v = uz$ for some $z \in \mathbb{Z}[i]$; hence $N(v) = N(u)N(z)$. Since $N(v), N(u), N(z)$ are all ≥ 1, we have $N(u) \leq N(v)$.

Now suppose $v \neq 0$ in $\mathbb{Z}[i]$ and $u \in \mathbb{Z}[i]$, with

$$u = a + bi, \qquad v = c + di.$$

Then, in $\mathbb{C}$, we have

$$\frac{u}{v} = \frac{a + bi}{c + di} = \frac{(a + bi)(c - di)}{c^2 + d^2}$$

$$= \frac{(ac + bd) + (bc - ad)i}{c^2 + d^2} = e + fi,$$

where

$$e = \frac{ac + bd}{c^2 + d^2} \qquad f = \frac{bc - ad}{c^2 + d^2}.$$

Now, e and f are rational numbers, and every rational number differs by no more than $\frac{1}{2}$ from the nearest integer. Thus, there are integers $s, t \in \mathbb{Z}$ such that $|e - s| \leq \frac{1}{2}$ and $|f - t| \leq \frac{1}{2}$. But then we have

$$\frac{u}{v} = e + fi = s + ti + (e - s) + (f - t)i,$$

and so

$$u = v(s + ti) + v[(e - s) + (f - t)i].$$

We need to determine whether $s + ti$ can serve as the quotient, $v[(e - s) + (f - t)i]$ as the remainder, in the division of u by v in $\mathbb{Z}[i]$. Clearly, $s + ti$ and $v[(e - s) + (f - t)i]$ are elements of $\mathbb{Z}[i]$, and

$$\begin{aligned} N(v[(e - s) + (f - t)i]) &= N(v)[(e - s)^2 + (f - t)^2] \\ &\leq N(v)(\tfrac{1}{4} + \tfrac{1}{4}) = \tfrac{1}{2}N(v) < N(v). \end{aligned}$$

Thus, with $q = s + ti$, $r = v[(e - s) + (f - t)i]$, we have $u = vq + r$, where $N(r) < N(v)$. It follows that $\mathbb{Z}[i]$ forms a Euclidean domain, with Euclidean norm $\delta = N$ (or, strictly speaking, by our definition of Euclidean domain, $\delta = N$, restricted to $\mathbb{Z}[i] - \{0\}$). ■

Exercises 3.7

True or False

1. Every unique factorization domain is Euclidean.
2. Every principal ideal domain is Euclidean.
3. In a Euclidean domain, an element p is prime if and only if it is irreducible.
4. In a Euclidean domain, every ascending chain of ideals terminates.
5. In a Euclidean domain, any two elements have a least common multiple and a greatest common divisor.

3.7.1. Let d be a squarefree integer, and let $\mathbb{Q}(\sqrt{d}) = \{p + q\sqrt{d} \mid p, q \in \mathbb{Q}\}$ (where "$\sqrt{d}$" represents either one of the square roots of d in case $d < 0$). Prove that $\mathbb{Q}(\sqrt{d})$ forms a subfield of $\mathbb{C}$.

3.7.2. Let d be a squarefree integer. On the field $\mathbb{Q}(\sqrt{d}) = (p + q\sqrt{d} \mid p, q \in \mathbb{Q}\}$, define $N : (\mathbb{Q}(\sqrt{d}) \to \mathbb{Q}$ by:

$$N(p + q\sqrt{d}) = p^2 - dq^2.$$

Prove that $N(wz) = N(w)N(z)$ for each $w, z \in \mathbb{Q}(\sqrt{d})$.

3.7.3. Imitate the proof of Theorem 3.7.2 to prove that $I_{-2} = \{a + b\sqrt{-2} \mid a, b \in \mathbb{Z}\}$ forms a Euclidean domain, with respect to the Euclidean norm N given by: $N(a + b\sqrt{-2}) = a^2 + 2b^2$.

3.7.4. Attempt to prove that $N: I_{-5} \to \mathbb{Z}$ defined by $N(a + b\sqrt{-5}) = a^2 + 5b^2$ $(a, b \in \mathbb{Z})$ serves as a Euclidean norm on I_{-5}. Observe where the proof breaks down. (Note: I_{-5} does not have unique factorization (see Exercise 3.6.11); hence it *cannot* be a Euclidean domain.)

3.7.5. With $\mathbb{Q}(\sqrt{d})$ defined as in Exercise 3.7.1, prove that the set $I_d = \{u \in \mathbb{Q}(\sqrt{d}) \mid f(u) = 0$ for some $f \in \mathbb{Z}[x]$ with leading coefficient $1\}$ forms a subring of $\mathbb{Q}(\sqrt{d})$.

3.8

Fields of Quotients of Integral Domains

Let A be an integral domain. We wish to construct a field containing A that is as small as possible. Our construction will generalize the relationship between $\mathbb{Z}$ and $\mathbb{Q}$.

Every rational number may be represented by any one of infinitely many fractions, e.g., $\frac{2}{3}, \frac{-2}{-3}, \frac{4}{6}, \frac{6}{9}, \frac{40}{30}, \ldots$, all represent a single rational number. A symbol such as $\frac{2}{3}$ is merely a way of writing an ordered pair, in this case, the ordered pair (2, 3). Thus, we observe that, to each rational number, there corresponds an infinite class of ordered pairs (a, b), $b \neq 0$. Two fractions $\dfrac{a}{b}$ and $\dfrac{c}{d}$ represent the same rational number if and only if $ad = cb$. Put another way, two ordered pairs (a, b) and (c, d) $(b \neq 0, d \neq 0)$ belong to the same class if and only if $ad = cb$. Our construction will simply identify each rational number with the class of ordered pairs that corresponds to it.

Theorem 3.8.1: *(1)* Let A be an integral domain, and let X be the set of all ordered pairs (a, b), $a \in A$, $b \neq 0$ in A. (Thus, $X = A \times A - \{0\}$.) On X, define a binary relation $\sim$ by:

$$(a, b) \sim (c, d) \quad \text{if} \quad ad = cb \quad ((a, b), (c, d) \in X).$$

Then $\sim$ is an equivalence relation on X.

(2) Denote by $\dfrac{a}{b}$ the equivalence class of (a, b) $((a, b) \in X)$, and let $F = \left\{\dfrac{a}{b} \,\middle|\, a, b \neq 0 \text{ in } A\right\}$. On F, define $+, \cdot$ by:

$$\frac{a}{b} + \frac{c}{d} = \frac{ad + cb}{bd}$$

$$\frac{a}{b} \cdot \frac{c}{d} = \frac{ac}{bd} \qquad (a, c \in A, b \neq 0, d \neq 0 \quad \text{in} \quad A).$$

Then $\langle F, +, \cdot \rangle$ is a field.

(3) Define $\phi: A \to F$ by: $\phi(a) = \dfrac{a}{1}$. Then ϕ is a ring monomorphism.

(4) If E is a field containing A as a subring, then E contains an isomorphic copy of F.

Proof: *(1)* $\sim$ is an equivalence relation on X, for: clearly, $(a, b) \sim (a, b)$, since $ab = ab$ (for each $a \in A$, $b \neq 0$ in A).

If $(a, b) \sim (c, d)$, then $ad = cb$; hence $cb = ad$, and so $(c, d) \sim (a, b)$ $(a, c \in A, b \neq 0, d \neq 0$ in $A)$.

If $(a, b) \sim (c, d)$ and $(c, d) \sim (e, f)$ $(a, c, e \in A, b \neq 0, d \neq 0, f \neq 0$ in $A)$, then from $ad = cb$ and $cf = ed$, we have $(ad)f = (cb)f = (cf)b = (ed)b$; hence, $(af)d = (eb)d$, and so $af = eb$. But then $(a, b) \sim (e, f)$. Hence $\sim$ is reflexive, symmetric, and transitive, and is thus an equivalence relation on X.

(2) Write "$\frac{a}{b}$" for the equivalence class $(a, b) \in X$. Let $F = \left\{\frac{a}{b} \,\middle|\, a \in A, b \neq 0 \text{ in } A\right\}$. We need to show that the proposed binary operations $+, \cdot$ are well-defined, i.e., every pair of elements of F has a unique sum and a unique product in F.

Let $(a, b), (c, d) \in X$. Then $(ad + cb, bd)$ and (ac, bd) are elements of X since $bd \neq 0$ if $b \neq 0$ and $d \neq 0$ in the integral domain A. Thus, for each $\frac{a}{b}, \frac{c}{d} \in F$, we have

$$\frac{ad + cb}{bd} \quad \text{and} \quad \frac{ac}{bd}$$

in F. Now suppose

$$\frac{a}{b} = \frac{a'}{b'}, \quad \frac{c}{d} = \frac{c'}{d'}$$

$(a, c, a', c' \in A,\ b, b', d, d'$ non-zero elements of $A)$. Then

$$ab' = a'b \quad \text{and} \quad cd' = c'd.$$

Hence, $(ac)(b'd') = ab'\,cd' = a'\,bc'\,d = (a'\,c')bd$, and so

$$\frac{ac}{bd} = \frac{a'\,c'}{b'\,d'}.$$

Thus, the proposed operation $\cdot$ is a (well-defined) binary operation on F. Furthermore, we have

$$\begin{aligned}(ad + cb)b'\,d' &= adb'\,d' + cbb'\,d' \\ &= a'\,bdd' + c'\,dbb' = (a'\,d' + c'\,b')bd,\end{aligned}$$

and so the proposed operation $+$ is a (well-defined) binary operation on F. It is a simple exercise to verify that $\langle F, +, \cdot\rangle$ is a commutative ring with identity $\frac{1}{1}$, different from the 0-element $\frac{0}{1}$, hence a good candidate for being a field. (See Exercise 3.8.1.) For each $\frac{a}{b} \in F$, $\frac{a}{b} \neq \frac{0}{1}$, we have $a \cdot 1 \neq 0 \cdot b$, thus $a \neq 0$, and so $\frac{b}{a} \in F$. Thus, F is indeed a field.

(3) Define $\phi : A \to F$ by

$$\phi(a) = \frac{a}{1}$$

for each $a \in A$. Since, for $a, b \in A$, $\phi(a + b) = \dfrac{a+b}{1} = \dfrac{a}{1} + \dfrac{b}{1} = \phi(a) + \phi(b)$, and $\phi(a \cdot b) = \dfrac{ab}{1} = \dfrac{a}{1} \cdot \dfrac{b}{1} = \phi(a) \cdot \phi(b)$, ϕ is a ring homomorphism. But $\phi(a) = \dfrac{0}{1} \Leftrightarrow \dfrac{a}{1} = \dfrac{0}{1} \Leftrightarrow a = 0$; hence ϕ is a monomorphism.

(4) Let E be a field containing A as a subring, and let $\bar{F} = \{ab^{-1} | a \in A, b \neq 0 \text{ in } A\}$. Define $\psi : F \to E$ by: $\psi\left(\dfrac{a}{b}\right) = ab^{-1}$ for each $a \in A$, $b \neq 0$ in A. ψ is well-defined, for: if $\dfrac{a}{b} = \dfrac{a'}{b'}$ $(a, a' \in A,\ b, b'$ non-zero in $A)$, then $ab' = a'b$; hence $ab^{-1} = a'(b')^{-1}$. ψ is a homomorphism, for: if $a, c \in A$, b, d non-zero in A, then

$$\psi\left(\frac{a}{b} + \frac{c}{d}\right) = \psi\left(\frac{ad + cb}{bd}\right) = (ad + cb)(bd)^{-1} = (ad + cb)d^{-1}b^{-1}$$
$$= ab^{-1} + cd^{-1} = \psi\left(\frac{a}{b}\right) + \psi\left(\frac{c}{d}\right),$$

and

$$\psi\left(\frac{a}{b} \cdot \frac{c}{d}\right) = \psi\left(\frac{ac}{bd}\right) = ac(bd)^{-1} = acd^{-1}b^{-1} = (ab^{-1})(cd^{-1}) = \psi\left(\frac{a}{b}\right) \cdot \psi\left(\frac{c}{d}\right).$$

Finally, ψ is a monomorphism, for: if $\psi\left(\dfrac{a}{b}\right) = ab^{-1} = 0$, then $a = 0b = 0$; hence $\dfrac{a}{b} = \dfrac{0}{b} = \dfrac{0}{1}$. Thus, $\text{Im}\psi = \bar{F}$ is a field isomorphic to F. ∎

The field F constructed from the integral domain A in Theorem 3.8.1 is called the *field of quotients of A*. A monomorphism such as $\phi : A \to F$ is often referred to as an *isomorphic embedding*. It is customary to identify the elements of A with their images $\phi(a) = \dfrac{a}{1}$ in F. Each element of F is then simply a quotient of elements of A: we have $\dfrac{a}{b} = \dfrac{a}{1} \cdot \left(\dfrac{b}{1}\right)^{-1}$; hence, after identification, simply $\dfrac{a}{b} = ab^{-1}$.

We shall refer to the subfield $\bar{F}$ of E in Theorem 3.8.1(4) as the *field of quotients of A in E*.

Example 1:

$\mathbb{Q}$ is the field of quotients of $\mathbb{Z}$.

Example 2:

For F a field, the field of quotients of the polynomial domain $F[x]$ is the set of all quotients $\frac{f}{g}$, $f \in F[x]$, $g \neq 0$ in $F[x]$. This field of quotients is denoted by $F(x)$.

Example 3:

For A an integral domain with field of quotients F, the field of quotients of $A[x]$ is $F(x)$. (Thus, for example, $\mathbb{Q}(x)$ is the field of quotients of $\mathbb{Z}[x]$, as well as of $\mathbb{Q}[x]$.)

Example 4:

The field of quotients of $\mathbb{Z}[i]$ (the domain of Gaussian integers) is the field, $\mathbb{Q}(i)$, consisting of all complex numbers $p + qi$ $(p, q \in \mathbb{Q})$.

Exercises 3.8

True or False

1. The rational number $\frac{2}{3}$ is an equivalence class consisting of ordered pairs of integers.
2. The binary relation R defined on $\mathbb{Z} \times \mathbb{Z}$ by: $(a, b)\ R\ (c, d)$ if $ad = cb$ is an equivalence relation.
3. $\mathbb{Z}[x]$ and $\mathbb{Q}[x]$ have the same field of quotients.
4. $\mathbb{Q}$ is isomorphic to the field of quotients of every integral domain contained in $\mathbb{Q}$.
5. Every integral domain is a subring of a field.

3.8.1. Let A be a commutative ring with identity, and let S be a subset of A such that

(1) $1 \in S$

and

(2) if $a, b \in S$, then $ab \in S$.

(A subset, S, of a ring A with identity satisfying conditions (1) and (2) is called a *multiplicative set* in A.) On the set $A \times S = \{(a, s) | a \in A, s \in S\}$, define a binary relation $\sim$ by:

$$(a, s) \sim (b, t) \quad (a, b \in A,\ s, t \in S)$$

if $s'(ta - sb) = 0$ for some $s' \in S$. Prove that $\sim$ is an equivalence relation on $A \times S$.

3.8.2. Let $\sim$ be the equivalence relation defined in the preceding exercise. For $a \in A$, $s \in S$, denote by $\frac{a}{s}$ the equivalence class of (a, s), and let $S^{-1}A$ be the set of all equivalence classes with respect to $\sim$. On $S^{-1}A$, define $+, \cdot$ by

$$\frac{a}{s} + \frac{b}{t} = \frac{ta + sb}{st}$$

$$\frac{a}{s} \cdot \frac{b}{t} = \frac{ab}{st}.$$

(1) Prove that $+, \cdot$ are binary operations on $S^{-1}A$ and that $\langle S^{-1}A, +, \cdot \rangle$ is a ring with identity. ($S^{-1}A$ is called the *ring of quotients* of A with respect to S.)

(2) Prove that $\phi : A \to S^{-1}A$ defined by $\phi(a) = \frac{a}{1}$ $(a \in A)$ is a ring homomorphism.

(3) If $0 \notin S$ and A has no zero divisors, prove that ϕ is a monomorphism.

3.8.3. Interpret the field of quotients of an integral domain as a ring of quotients in the sense of Exercise 3.8.2.

3.8.4. Let A be a commutative ring with identity and let P be an ideal of A. Prove that P is a prime ideal of A if and only if $S = A \setminus P$ is a multiplicative set in A (see Exercise 3.8.1).

3.8.5. Let A be a commutative ring with identity and let $S = A \setminus P$, where P is a prime ideal of A. Prove that the ring of quotients $S^{-1}A$ has a unique maximal ideal M. (A ring having a unique maximal ideal is called a *local ring*. For P a prime ideal in A and $S = A \setminus P$, the ring $S^{-1}A$ is called the local ring of A at the prime ideal P.) (See Exercise 3.4.11.)

3.8.6. Interpret the ring $\mathbb{Q}_{(2)}$ of Exercise 3.4.7 as the local ring of $\mathbb{Z}$ at the prime ideal $2\mathbb{Z}$. Generalize the results of Exercise 3.4.7 to the local ring, $\mathbb{Q}_{(p)}$, of $\mathbb{Z}$ at the prime ideal $p\mathbb{Z}$, where p is an arbitrary prime in $\mathbb{Z}$.

3.8.7. Let A be an integral domain and let F be its field of quotients. Prove that for every subring B of F such that $A \subset B$, F is also the field of quotients of B.

3.8.8. Find infinitely many subrings of $\mathbb{Q}$ for which $\mathbb{Q}$ is the field of quotients.

3.8.9. Prove that, for p prime, the field $\mathbb{Z}_p$ is the field of quotients only of itself.

3.8.10. Find the field of quotients of $\mathbb{Z}[i]$.

3.8.11. Find the field of quotients $\mathbb{Z}[i][x]$.

3.8.12. Find the field of quotients of $I_{-3} = \mathbb{Z}[\sqrt{-3}]$.

3.8.13. Find the field of quotients of $I_{-3}[x]$.

3.8.14. Find an integral domain, not a field, of which $\mathbb{R}$ is the field of quotients.

3.8.15. Let D_1 and D_2 be integral domains, with fields of quotients F_1 and F_2, respectively. If $\phi : D_1 \to D_2$ is an isomorphism of D_1 onto D_2, prove that there is a unique isomorphism $\bar{\phi} : F_1 \to F_2$ such that $\bar{\phi}|_{D_1} = \phi$.

3.9

Polynomials over Unique Factorization Domains

The polynomial $2x^2 - 4$ is irreducible in $\mathbb{Q}[x]$, but reducible in $\mathbb{Z}[x]$, since it has the constant factor 2 which is not a unit in $\mathbb{Z}$. A necessary, but far from sufficient, condition that a polynomial be irreducible in $\mathbb{Z}[x]$ is that its coefficients have no common factors in $\mathbb{Z}$ other than ± 1. More generally, such polynomials play an important role in any ring $A[x]$, where A is a unique factorization domain.

Definition 3.9.1: Let A be a unique factorization domain, and let $p = a_0 + a_1x + \cdots + a_nx^n$ $(a_n \neq 0,\ n > 0)$ be a polynomial in $A[x]$. Then p is a *primitive* polynomial if the only common divisors, in A, of all of the coefficients a_i $(i = 0, \ldots, n)$ of p are units in A.

Example:

$x^2 - 2$ in $\mathbb{Z}[x]$ is primitive, $2x^2 - 4$ is not.

Our chief aim in this section is to show that, if A is a unique factorization domain, then so is $A[x]$. We require several lemmas, all regarding primitive polynomials.

Lemma 1: Let A be a unique factorization domain, and let $f \in A[x]$ be a polynomial of positive degree. Then $f = ap$ where $a \in A$ and $p \in A[x]$ is a primitive polynomial. If $f = a_1p_1 = a_2p_2$, where $a_1, a_2 \in A$ and p_1, p_2 are primitive polynomials in $A[x]$, then $a_1 \sim a_2$ in A and $p_1 \sim p_2$ in $A[x]$.

Proof: Let $f = a_0 + a_1x + \cdots + a_nx^n$, $a_n \neq 0$, $a_i \in A$ $(i = 0, \ldots, n)$, and let $d \in A$ be a greatest common divisor of the a_i. Then, clearly, $f = dp$, where $p = \bar{a}_0 + \bar{a}_1x + \cdots + \bar{a}_nx^n$, with $a_i = \bar{a}_id$ $(i = 0, \ldots, n)$, is a primitive polynomial in $A[x]$.

Conversely, if $f = ap$, where $a \in A$ and p is a primitive polynomial in $A[x]$, then a is a greatest common divisor of the coefficients of f. (See Exercise 3.9.1(1).) But then, if $f = a_1p_1 = a_2p_2$ $(a_1, a_2 \in A;\ p_1, p_2$ primitive polynomials in $A[x])$, we have $a_1 \sim a_2$ in A. If $a_1 = ua_2$, u a unit in A (hence a unit in $A[x]$), then from $a_1p_1 = ua_2p_1 = a_2p_2$ follows $up_1 = p_2$; hence $p_1 \sim p_2$ in $A[x]$. ■

Lemma 2 (Gauss): If A is a unique factorization domain, then the product of any two primitive polynomials in $A[x]$ is a primitive polynomial.

Proof: Let $p, q \in A[x]$ be primitive polynomials, with $p = a_0 + a_1x + \cdots + a_mx^m$, $a_m \neq 0$, $a_i \in A$ $(i = 0, \ldots, m)$, and $q = b_0 + b_1x + \cdots + b_nx^n$, $b_n \neq 0$, $b_j \in A$ $(j = 0, \ldots, n)$. Let $f = pq$. Then $f = c_0 + c_1x + \cdots + c_{m+n}x^{m+n}$ for some

$c_l \in A$ $(l = 0, \ldots, m + n)$. Suppose f is *not* primitive. Then there is some irreducible element $\alpha \in A$ such that $\alpha | c_i$ for each $i = 0, \ldots, m + n$. Since p is a primitive polynomial, α does not divide all of the coefficients of p. Hence there is a least non-negative integer s such that $\alpha \nmid a_s$. Similarly, there is a least non-negative integer t such that $\alpha \nmid b_t$. Now consider the coefficient c_{s+t} of f. We have

$$c_{s+t} = a_0 b_{s+t} + \cdots + a_{s-1} b_{t+1} + a_s b_t + a_{s+1} b_{t-1} + \cdots + a_{s+t} b_0. \tag{1}$$

By our choice of s and t, α divides all terms preceding and all terms following $a_s b_t$ in the sum on the right-hand side of (1). Since, by hypothesis, $\alpha | c_{s+t}$, we conclude that $\alpha | a_s b_t$. But then (see Exercise 3.6.6) either $\alpha | a_s$ or $\alpha | b_t$. Contradiction!

It follows that $f = pq$ is a primitive polynomial. ■

The next lemma is tacitly assumed in most high school discussions of factorization. Consider a polynomial such as $f = 2x^3 + 3x^2 + x + 1$. This polynomial is easily seen to have no proper factorization as a product of two polynomials with *integer* coefficients. Does this imply that f does not have a proper factorization as a product of two polynomials with *rational* coefficients?

Lemma 3: Let A be a unique factorization domain, and let F be its field of quotients. Let f be a primitive polynomial in $A[x]$. Then f is irreducible in $A[x]$ if and only if f is irreducible in $F[x]$.

Proof: Note that, since f is a primitive polynomial in $A[x]$, it has no proper divisor of degree 0 in $A[x]$. Hence the only possible proper divisors of f in $A[x]$ are polynomials of positive degree less than deg f.

If f is irreducible in $F[x]$, then f has no proper divisor in $F[x]$; hence certainly f has no proper divisor in $A[x]$.

Conversely, suppose f is irreducible in $A[x]$. Suppose there are polynomials $g, h \in F[x]$ such that $f = gh$, with $0 < \deg g < \deg f$, $0 < \deg h < \deg f$. By Lemma 1, (see Exercise 3.9.1(2)), there are elements $a, b, c, d \in A$ such that $g = \frac{a}{b}\bar{g}$, $h = \frac{c}{d}\bar{h}$, where $\bar{g}, \bar{h}$ are primitive polynomials in $A[x]$. Hence $f = \frac{ac}{bd}\bar{g}\bar{h}$, and so $bdf = ac\bar{g}\bar{h}$. By Lemma 2, $\bar{g}\bar{h}$ is itself a primitive polynomial. Hence, by Lemma 1, f is an associate of $\bar{g}\bar{h}$ in $A[x]$. Thus, $f = u(\bar{g}\bar{h}) = (u\bar{g})\bar{h}$, for u some unit in A. But $0 < \deg(u\bar{g}) < \deg f$, $0 < \deg \bar{h} < \deg f$, and so f is reducible in $A[x]$. Contradiction! It follows that f is irreducible in $F[x]$. ■

Lemma 4: Let p, q be primitive polynomials in $A[x]$. Then $p \sim q$ in $A[x]$ if and only if $p \sim q$ in $F[x]$.

Proof: Since the units in $A[x]$ are units also in $F[x]$, $p \sim q$ in $A[x]$ implies $p \sim q$ in $F[x]$. Conversely, suppose $p \sim q$ in $F[x]$. The units in $F[x]$ are the non-zero constant polynomials in $F[x]$, i.e., the non-zero elements of F. Thus, there are $a \in A$, $b \neq 0$ in A, such that $p = \frac{a}{b}q$. But then $bp = aq$ and, by Lemma 1, we have $p \sim q$ in $A[x]$. ■

We are now ready to prove the main theorem.

Theorem 3.9.1: If A is a unique factorization domain, then so is $A[x]$.

Proof: Let F be the field of quotients of A. We first prove that every *primitive* polynomial $p \in A[x]$ has a unique factorization (in the sense of Definition 3.6.4) as a product of irreducible polynomials in $A[x]$. If p is a primitive polynomial in $A[x]$, then p is a polynomial of positive degree in $F[x]$. Since $F[x]$ is a unique factorization domain (by the Corollary of Theorem 3.6.5), there are irreducible polynomials $p_1, \ldots, p_h \in F[x]$ such that $p = \prod_{i=1}^{h} p_i$. From Lemma 1, it follows that there are elements $a_1, \ldots, a_h; b_1, \ldots, b_h \in A$ such that, for each $i = 1, \ldots, h$, $p_i = \frac{a_i}{b_i} \bar{p}_i$ where $\bar{p}_i$ is a primitive polynomial in $A[x]$. By Lemma 3, each $\bar{p}_i$ is irreducible in $A[x]$. By Lemma 2, $\prod_{i=1}^{h} \bar{p}_i$ is a primitive polynomial in $A[x]$. Hence

$$\left(\prod_{i=1}^{h} b_i\right)p = \prod_{i=1}^{h} a_i \prod_{i=1}^{h} \bar{p}_i,$$

and, by Lemma 1, it follows that p is an associate in $A[x]$ of the product $\prod_{i=1}^{h} \bar{p}_i$, hence equal to a product $(u\bar{p}_1)\bar{p}_2 \ldots \bar{p}_h$ (u unit in A) of irreducible elements of $A[x]$. This establishes the existence of a factorization of p as a product of irreducible elements of $A[x]$.

To establish uniqueness, suppose

$$p = r_1 \ldots r_h = s_1 \ldots s_k$$

where the r_i and the s_j are irreducible polynomials of positive degree in $A[x]$. Then each factor is necessarily a primitive polynomial in $A[x]$, hence (by Lemma 3) irreducible also in $F[x]$. Since $F[x]$ is a unique factorization domain, it follows that $h = k$, and there is a 1-1 correspondence between the r_i and the s_j such that corresponding factors are associates in $F[x]$, hence associates in $A[x]$, by Lemma 4. It follows that the two factorizations of p are equivalent, in the sense of Definition 3.6.4.

Now suppose $f \in A[x]$. If $\deg f = 0$, then f has a unique factorization as a product of irreducible elements of the unique factorization domain A. (Clearly, no polynomials of positive degree can be factors of $a \in A$.)

If $\deg f > 0$, then, by Lemma 1, $f = ap$, $a \in A$, p primitive in $A[x]$. If a is a unit, then f is primitive, and we have already proved that it has a unique factorization as a product of irreducible elements. If a is not a unit, then $a = \alpha_1 \ldots \alpha_s$ (α_i irreducible in A, hence in $A[x]$) and $p = p_1 \ldots p_h$, p_i irreducible of positive degree in $A[x]$, whence f is a product of irreducible elements of $A[x]$.

Suppose

$$f = \alpha_1 \ldots \alpha_s p_1 \ldots p_h = \beta_1 \ldots \beta_t q_1 \ldots q_k \tag{1}$$

where $\alpha_1, \ldots, \alpha_s, \beta_1, \ldots, \beta_t$ are irreducible elements of A and $p_1, \ldots, p_h, q_1, \ldots, q_k$ are irreducible polynomials of positive degree in $A[x]$. The p_i and the q_j are (*a fortiori*) primitive. Hence, by Lemma 2, $\prod_{i=1}^{h} p_i$ and $\prod_{j=1}^{k} q_i$ are primitive polynomials in $A[x]$. But then, from (1), using Lemma 1, we have

$$\begin{aligned} \alpha_1 \ldots \alpha_s &\sim \beta_1 \ldots \beta_t \quad \text{in} \quad A \\ \text{and } p_1 \ldots p_h &\sim q_1 \ldots q_k \quad \text{in} \quad A[x]. \end{aligned} \tag{2}$$

Since A is a unique factorization domain, and since uniqueness of factorization holds for *primitive* polynomials in $A[x]$, it now follows from (2) that the products in (1) are equivalent. (See Exercise 3.9.3.)

We conclude that $A[x]$ is a unique factorization domain. ∎

From Theorem 3.9.1, it follows that, in particular, for any principal ideal domain A, and even more particularly, for any Euclidean domain A, the polynomial domain $A[x]$ is a unique factorization domain. Thus, for example, $\mathbb{Z}[x]$ and $\mathbb{Z}[i][x]$ are unique factorization domains. Moreover, if F is a field, the polynomial ring $F[x][y]$ is a unique factorization domain. Let us look more closely at $F[x][y]$. Its elements are polynomials in y, whose coefficients are polynomials in x, whose coefficients, in turn, are in F. Thus, for example, $f = (3x + 2)y^3 + 4x^2y^2 - (x^3 + 5)y + (x^2 - 10)$ is an element of $\mathbb{Q}[x][y]$. But, simplifying, we have

$$f = 3xy^3 + 2y^3 + 4x^2y^2 - x^3y - 5y + x^2 - 10,$$

manifestly a polynomial with coefficients in $\mathbb{Q}$, in the two indeterminates x and y.

More generally, for k any positive integer, the polynomial ring $A[x_1, \ldots, x_k]$ over a commutative ring A with identity, in the indeterminates $x_1, \ldots, x_k$, is, naively speaking, the set of all formal finite sums

$$\sum a_{i_1 \ldots i_k} x_1^{i_1} \ldots x_k^{i_k},$$

summed over some finite set of k-tuples $(i_1, \ldots, i_k)$ of non-negative integers, the $a_{i_1 \ldots i_k}$ being elements of A. (A more sophisticated description of $A[x_1, \ldots, x_n]$ generalizes the description of $A[x]$ in terms of sequences that we gave in Section 3.5. Each element of $A[x_1 \ldots x_k]$ can be thought of as simply an infinite sequence of terms $a_{i_1 \ldots i_k} \in A$, indexed on the set of all k-tuples of positive integers, with all but finitely many terms equal to 0. We spare you the details.) It is easy to see that any finite iteration of polynomial domains in a single indeterminate, such as A, $A[x_1]$, $A[x_1][x_2], \ldots, A[x_1][x_2] \ldots [x_k]$, leads to a ring isomorphic to the ring $A[x_1, \ldots, x_n]$ described previously.

Granting this, we now have the following Corollary of Theorem 3.9.1.

Corollary: If F is a field, k a positive integer, then the polynomial domain $F[x_1, \ldots, x_k]$ in k indeterminates over F is a unique factorization domain.

Proof: We need only consider the iterated polynomial domain

$$F[x_1][x_2] \ldots [x_k],$$

which is a unique factorization domain, by the Corollary of Theorem 3.6.5, and $(k-1)$-fold application of Theorem 3.9.1. ∎

There is, however, a catch! Just as, in $\mathbb{Z}$, it is a major problem to determine whether or not any given integer is prime, it is, likewise, non-trivial to determine whether, in a polynomial domain $A[x]$, a given polynomial is irreducible.

The following criterion characterizes a special class of irreducible polynomials in $A[x]$, where A is a unique factorization domain.

Theorem 3.9.2. (Eisenstein's Irreducibility Criterion): Let A be a unique factorization domain with field of quotients F, and let $f = a_0 + a_1x + \cdots + a_nx^n$, $a_n \neq 0$, be a primitive polynomial in $A[x]$. Suppose there exists a prime (i.e., irreducible) element $q \in A$ such that

(1) $q | a_i$ for each $i = 0, \ldots, n - 1$

(2) $q \nmid a_n$

(3) $q^2 \nmid a_0$.

Then f is irreducible over $F[x]$.

Proof: Suppose f is reducible over $A[x]$. Since f is primitive, there are polynomials g, h of degree $< n$ in $A[x]$ such that $f = gh$. If $g = b_0 + b_1x + \cdots + b_sx^s$, and $h = c_0 + c_1x + \cdots + c_tx^t$, then $s < n$, $t < n$, $a_n = b_sc_t$, and $a_0 = b_0c_0$. From condition (2), it follows that $q \nmid b_s$ and $q \nmid c_t$. Conditions (1) and (3) imply that either $q | b_0$ and $q \nmid c_0$, or $q \nmid b_0$ and $q | c_0$. Without loss of generality, suppose the first alternative holds: $q | b_0$ and $q \nmid c_0$. Since $q \nmid b_s$, there is a least positive integer h, $0 < h \leq s$, such that $q \nmid b_h$. Consider the coefficient a_h of f.

$$a_h = b_0c_h + b_1c_{h-1} + \cdots + b_{h-1}c_1 + b_hc_0.$$

Since $h \leq s < n$, we have $q | a_h$. Since $h > 0$, $q | b_0, \ldots, q | b_{h-1}$; hence q divides $b_0c_h + \cdots + b_{h-1}c_1$. But then $q | b_hc_0$, and so either $q | b_h$ or $q | c_0$. Contradiction!

It follows that p is irreducible in $A[x]$. But then, by Lemma 3, p is irreducible also in $F[x]$. ■

Corollary: Let A be a unique factorization domain, F its quotient field, and let $f = a_0 + a_1x + \cdots + a_nx^n$ be a polynomial of positive degree in $A[x]$. If there is a prime element $q \in A$ satisfying conditions (1), (2), (3) of Theorem 3.9.2, then f is not expressible as the product of two polynomials of degree $< n$, in $F[x]$.

We leave the proof as an exercise. (See Exercise 3.9.4.)

Polynomials satisfying the conditions of Theorem 3.9.2 are called *Eisenstein polynomials.* They are a joy to anyone in need of examples of irreducible polynomials. However, many irreducible polynomials are not "Eisenstein."

Some Eisenstein polynomials in $\mathbb{Z}[x]$: $x^5 + 9x^4 - 45x^3 + 24x^2 - 123x + 210$; $5x^3 + 10x^2 - 20x + 14$; $x^{100} - 1214$; $x^2 - 2$. (The last example provides a sledge hammer for proving $\sqrt{2}$ irrational—a task more easily accomplished simply by using the Fundamental Theorem of Arithmetic.)

Some irreducible polynomials in $\mathbb{Z}[x]$ that are not Eisenstein: $x^4 + x^3 + x^2 + x + 1$; $3x^3 + x^2 + 1$; $2x^5 + 3x^4 - 2x^3 + 5x^2 + 1$; $2x^7 + 1$. (See Exercise 3.9.5.)

Exercises 3.9

True or False

1. For each positive integer n, the ring $\mathbb{Z}[x_1, \ldots, x_n]$ is a unique factorization domain.
2. The polynomial
$$2x^4 + 14x^3 + 28x^2 + 42x + 70$$
is irreducible over the rational field.
3. The polynomial
$$x^4 + 3x^3 + 9x^2 + 9x + 18$$
is irreducible over the rational field.
4. If f is a primitive polynomial in $\mathbb{Z}[x]$ and $g = x^3 + 2x^2 + 3x + 5$, then fg is a primitive polynomial in $\mathbb{Z}[x]$.
5. If $f, g \in \mathbb{Z}[x]$ are associates in $\mathbb{Q}[x]$, then they are associates in $\mathbb{Z}[x]$.

3.9.1. Let A be a unique factorization domain with quotient field F.
(1) If $f \in A[x]$ and $f = ap$, where $a \in A$ and $p \in A[x]$ is a primitive polynomial, prove that a is a greatest common divisor of the coefficients of f.
(2) If $f \in F[x]$, then $f = \frac{a}{b}p$, where $a, b \in A$ and p is a primitive polynomial in $A[x]$.

3.9.2. Prove that the polynomials
(1) $14x^3 + 280x^2 - 420x + 15$
(2) $\frac{2}{3}x^5 + 4x^4 - 12x^3 + 6x^2 + 2x + 14$
are irreducible over $\mathbb{Q}$.

3.9.3. Let A be a unique factorization domain, and let $f = \alpha_1 \ldots \alpha_s p_1 \ldots p_h = \beta_1 \ldots \beta_t q_1 \ldots q_k$ where $\alpha_1, \ldots, \alpha_s; \beta_1, \ldots, \beta_t$ are irreducible elements of A, and $p_1, \ldots, p_h; q_1, \ldots, q_k$ are irreducible polynomials of positive degree in $A[x]$. If $\alpha_1 \ldots \alpha_s \sim \beta_1 \ldots \beta_t$ in A and $p_1 \ldots p_h \sim q_1 \ldots q_k$ in $A[x]$, prove that the two products representing f are equivalent (see the proof of Theorem 3.9.1).

3.9.4. Let A be a unique factorization domain, and let F be its field of quotients. Prove the following Corollary of Theorem 3.9.2 (Eisenstein's Criterion):
$$\text{If} \quad f = \sum_{i=0}^{n} a_i x^i,\ a_n \neq 0,\ a_i \in A \quad (i = 0, \ldots, n),$$
with f not necessarily primitive, and if p is an irreducible element of A such that
(1) $p \mid a_i$ for each $i = 0, \ldots, n-1$
(2) $p \nmid a_n$
(3) $p^2 \nmid a_0$
then f is not expressible as the product of two polynomials, each of degree $< n$, in $F[x]$.

3.9.5. Prove that the polynomials $3x^3 + x^2 + 1$, $2x^5 + 3x^4 - 2x^3 + 5x^2 + 1$ and $2x^7 + 1$ are irreducible over $\mathbb{Q}$.

3.9.6. Let A be a unique factorization domain and let $f(x)$ be a primitive polynomial in $A[x]$. Prove: if $f(x + k)$ is irreducible in $A[x]$ for some $k \in A$, then $f(x)$ is irreducible in $A[x]$, hence in $F[x]$, where F is the quotient field of A.

3.9.7. Use the result in Exercise 3.9.6 to prove that the polynomial $\Phi_5(x) = x^4 + x^3 + x^2 + x + 1$ is irreducible over $\mathbb{Q}$ by finding an integer k such that $\Phi_5(x + k)$ is an Eisenstein polynomial.

$\left(\text{Note: } \Phi_5(x) = \dfrac{x^5 - 1}{x - 1}.\right)$

3.9.8. Generalize the result of Exercise 3.9.7 to

$$\Phi_p(x) = \frac{x^p - 1}{x - 1} = x^{p-1} + x^{p-2} + \ldots + x + 1,$$

where p is any prime in $\mathbb{Z}$. (Φ_p is the "pth cyclotomic polynomial." Its zeros are the $p - 1$ complex numbers, different from 1, which are pth roots of 1:

$$e^{2\pi i/p}, e^{4\pi i/p}, \ldots, e^{(p-1)2\pi i/p}.$$

(See Section 4.11.)

3.9.9. Let F be a field. Prove that the polynomial domain $F[x_1, x_2]$ in two indeterminates x_1 and x_2 is not a principal ideal domain. Generalize your result to an arbitrary number of indeterminates. (Recall that these domains *do* have unique factorization.)

3.9.10. Prove that, for each positive integer n, there are irreducible polynomials of degree n in $\mathbb{Q}[x]$.

3.9.11. Let $\mathbb{Q}_{(2)} = \left\{ \dfrac{a}{s} \,\middle|\, 2 \nmid s,\ a, s \in \mathbb{Z} \right\}$. ($\mathbb{Q}_{(2)}$ is the local ring of $\mathbb{Z}$ at 2—see Exercise 3.4.7).

(1) Find all irreducible elements in $\mathbb{Q}_{(2)}$.

(2) Prove that $\mathbb{Q}_{(2)}$ is a unique factorization domain.

3.9.12. If $f \in \mathbb{Q}_{(2)}[x]$ is a primitive polynomial, prove that f is irreducible in $\mathbb{Q}_{(2)}[x]$ if and only if f is irreducible in $\mathbb{Q}[x]$.

3.9.13. *(1)* Prove that $1 + 2i$ is an irreducible element of $\mathbb{Z}[i]$.

(2) Use the result in (1) to prove that the polynomial

$$(1 - i)x^3 + (3 + 6i)x^2 + (2 - i)x - 1 + 3i$$

is irreducible over $\mathbb{Q}[i]$.

3.9.14. Prove that the polynomial

$$\frac{57}{69}x^{10} - \frac{18}{17}x^5 + \frac{40}{19}x + \frac{98}{101} \in \mathbb{Q}_{(2)}[x]$$

is irreducible over $\mathbb{Q}$.

3.9.15. Let $f \in \mathbb{Q}_{(p)}[x]$ (p prime in $\mathbb{Z}$) be an Eisenstein polynomial. Prove that there is a unit u in $\mathbb{Q}_{(p)}$ such that uf is an Eisenstein polynomial in $\mathbb{Z}[x]$.

3.9.16. Prove that every Eisenstein polynomial f in $\mathbb{Z}[x]$ is an Eisenstein polynomial in $\mathbb{Q}_{(p)}[x]$ for some prime $p \in \mathbb{Z}$.

3.9.17. Let $I_{-7} = \{a + b\sqrt{-7} | a, b \in \mathbb{Z}\}$.

(1) Prove that $2 + \sqrt{-7}$ is an irreducible element of I_{-7}.

(2) Prove that the polynomial

$$x^4 - 11x^3 + (-3 + 4\sqrt{-7})x + (4 + 2\sqrt{-7})$$

is irreducible over the field

$$\mathbb{Q}(\sqrt{-7}) = \{p + q\sqrt{-7} | p, q \in \mathbb{Q}\}.$$

3.10

Groups with Operators; Modules

If M, X are sets and $\mu : M \times X \to X$ is a binary operation, then μ is an *external composition on* X. (In this case, for $m \in M, x \in X$, $\mu(s, x)$ will usually be denoted by "mx" or by "xm.")

Definition 3.10.1: Let $\langle G, \circ \rangle$ be a group, M any set, $\mu : M \times G \to G$ an external composition on G, with $\mu(m, x)$ denoted by mx for each $m \in M, x \in G$. Then the ordered quadruple $\langle G, \circ, M, \mu \rangle$ is a *group with operators* if $m(x \circ y) = mx \circ my$ for all $m \in M$, $x, y \in G$. (More simply, we may say: "G forms a group with operator set M" if there is no doubt as to the operations.)

Example 1:

Let $\langle G, \circ \rangle$ be a group and let $\mathscr{E}$ be the set of all endomorphisms of G. Let $\mu : \mathscr{E} \times G \to G$ be defined by

$$\mu(\sigma, x) = \sigma(x)$$

for each $\sigma \in \mathscr{E}$, $x \in G$. Then $\langle G, \circ, \mathscr{E}, \mu \rangle$ is a group with operator set $\mathscr{E}$ since $\sigma(x \circ y) = \sigma(x) \circ \sigma(y)$ for all $\sigma \in \mathscr{E}$, $x, y \in G$.

Example 2:

Let $\langle A, +, \cdot \rangle$ be a ring. Define $\mu : A \times A \to A$ by $\mu(a, x) = a \cdot x$ for each $a, x \in A$. Since $a \cdot (x + y) = a \cdot x + a \cdot y$ for each $a, x, y \in A$, $\langle A, +, A, \mu \rangle$ is a group with operators.

(An analogous example may be obtained by defining $\mu(a, x) = x \cdot a$ for each $x, a \in A$.)

Definition 3.10.2: Let $\langle G, \circ, M, \mu \rangle$ be a group with operators, and let $H < G$. Then H is an *M-subgroup* of G if $\mu(m, h) \in H$ for each $m \in M$, $h \in H$.

Example 1:

Let $\langle G, \circ\rangle$ be a group, $\mathscr{E}$ the set of all endomorphisms, $\mathscr{A}$ the set of all automorphisms, and $\mathscr{I}$ the set of all inner automorphisms of G. (Thus, $\mathscr{I} \subset \mathscr{A} \subset \mathscr{E}$.) Define $\mu(\sigma, x) = \sigma(x)$ for each $\sigma \in \mathscr{E}$. Then a subgroup H of G is an $\mathscr{I}$-subgroup if it is normal, an $\mathscr{A}$-subgroup if it is characteristic, and an $\mathscr{E}$-subgroup if it is fully unvariant. (See Section 2.11.)

Example 2:

Let $\langle A, +, \cdot\rangle$ be a ring. If $\mu : A \times A \to A$ is defined by

$$\mu(a, x) = a \cdot x \quad (a, x \in A),$$

then the A-subgroups of $\langle A, +\rangle$ are the left ideals of the ring A. On the other hand, if $\mu : A \times A \to A$ is defined by

$$\mu(a, x) = x \cdot a \quad (a, x \in A),$$

then the A-subgroups of $\langle A, +\rangle$ are the right ideals of A.

These examples illustrate that the concept "group with operators" may serve to clarify and unify our thinking about algebraic structures.

The concept of homomorphism extends readily to groups with operators.

Definition 3.10.3: Let $\langle G, \cdot, M, \mu\rangle$ and $\langle G', \cdot', M', \mu'\rangle$ be groups with the same operator set M. Write $\mu(s, x) = sx$ and $\mu'(m, x') = mx'$ $(m \in S, x \in G, x' \in G')$. Let $\phi : G \to G'$ be a group homomorphism. Then ϕ is an *M-homomorphism* if

$$\phi(mx) = m\phi(x)$$

for all $m \in M, x \in G$.

We sketch briefly the generalization to groups with operators of our results on group homomorphisms.

If G, G' are M-groups and $\phi : G \to G'$ is an M-homomorphism, then $\operatorname{Im} \phi$ is an M-subgroup of G' and $\operatorname{Ker} \phi = \phi^{-1}\{0'\}$ is a normal M-subgroup of G'. Conversely, given a normal M-subgroup H of G, the factor group G/H can be made into an M-group by defining $m(Hx) = H(mx)$ for each $x \in G$, $m \in M$. The canonical epimorphism $\nu : G \to G/H$ is an M-homomorphism with kernel H. Thus, a subgroup H of an M-group G is the kernel of an M-homomorphism of G if and only if it is a normal M-subgroup. The *Fundamental Theorem on M-Homomorphism* states: if $\phi : G \to G'$ is an M-homomorphism, then $\operatorname{Im} \phi$ is M-isomorphic to $G/\operatorname{Ker} \phi$.

We now turn our attention to a special class of groups with operators: modules (including vector spaces). Briefly: a module is an abelian group with operators, where the operator set is a ring, and the various operations are well-behaved relative to each other.

Definition 3.10.4: Let A be a ring, and let $\langle X, +\rangle$ be an abelian group. Let $\mu : A \times X \to X$ be an external composition, with $\mu(a, x)$ denoted by "ax" for each $a \in A$, $x \in X$. Then X forms a *left A-module* if

(1) $a(x + y) = ax + ay$ for all $a \in A$, $x, y \in X$;

(2) $(a + b)x = ax + bx$ for all $a \in A$, $x \in X$;

(3_l) $(ab)x = a(bx)$ for all $a, b \in A$, $x \in X$

If A has an identity 1, and the additional condition

(4) $1x = x$ for all $x \in X$

is fulfilled, then X forms a *unitary* (or *unital*) left A-moduie.

Remark: With no necessary change of notation, a *right A-module* may be defined similarly, merely by replacing condition 3_l by condition

$$(3_r)\ (ab)x = b(ax) \quad \text{for all} \quad a, b \in A, x \in X.$$

Since condition 3_r appears more natural if $\mu(a, x)$ is denoted by xa for $a \in A$, $x \in X$, this notation would normally be used for right A-modules. Condition 3_r then reads

$$(3_r)\ x(ab) = (xa)b \quad \text{for all} \quad a, b \in A, x \in X.$$

Of course, if A is a commutative ring, then $ab = ba$ for all $a, b \in A$; hence $(ab)x = (ba)x$. This implies that, if A is commutative, 3_l and 3_r are equivalent: for $a, b \in A$, $x \in X$, if 3_l holds, then $a(bx) = (ab)x = (ba)x = b(ax)$, hence 3_r holds; and conversely. The distinction between right and left A-modules is thus purely notational if A is commutative.

We shall confine ourselves to left modules in our discussions. Unless otherwise specified, the term *A-module* shall mean "left A-module."

Example 1:

Every abelian group is a $\mathbb{Z}$-module. Let $\langle G, +\rangle$ be an abelian group. For $x \in G$, $k \in \mathbb{Z}$, define kx as in Chapter 2, p. 40 ("coefficientation" is the additive analogue of exponentiation). Since $+$ is commutative, $k(x + y) = kx + ky$ for each $k \in \mathbb{Z}$, $x, y \in G$. Thus (1) of Definition 3.10.4 holds. Conditions (2), (3_l) and (4) hold, independent of the commutativity of $+$.

Example 2:

Every ring A is a left (or right) A-module. (See Example 2 following Definition 3.10.1.) We make $\langle A, +\rangle$ into a group with operator set A by defining $ax = a \cdot x$ for each $a, x \in A$, and observe that (1), (2), and 3_l of Definition 3.10.4 hold. If A has identity 1, then A is a *unitary* left A-module. (Defining $xa = x \cdot a$ will make A into a right A-module.)

Example 3:

Let $\langle G, +\rangle$ and $\langle G', +'\rangle$ be abelian groups and let $X = \mathrm{Hom}_{\mathbb{Z}}(G, G')$ be the set of all group homomorphisms from G to G'. Then X is an abelian group, hence a $\mathbb{Z}$-module (see Exercise 3.10.7).

Definition 3.10.5: *(1)* If A is a ring, X an A-module and Y a subset of X, then Y is a *submodule* of X if Y is an A-subgroup of $\langle X, +\rangle$.

(2) If X and X' are (left) A-modules over the same ring A, then a group homomorphism ϕ from $\langle X, +\rangle$ to $\langle X', +'\rangle$ is a *module homomorphism* if it is an A-homomorphism in the sense of Definition 3.10.3, i.e., if

$$\phi(ax) = a\phi(x)$$

for each $a \in A$, $x \in X$.

From the remarks following Definition 3.10.3, we infer that the kernel of a module homomorphism $\phi : X \to X'$ is a submodule of X and that *every* submodule of an A-module X is the kernel of some module homomorphism of X. (For: *all* A-subgroups of $\langle X, +\rangle$ are normal.)

Example 3 may now be generalized.

Example:

Let A be a ring and let X, X' be left A-modules. Then the set

$$Y = \mathrm{Hom}_A(X, X')$$

of all A-module homomorphisms from X to X' is itself an A-module (see Exercise 3.10.7).

Exercises 3.10

True or False

1. Every group can be regarded as a group with operators.
2. Every group can be regarded as a $\mathbb{Z}$-module.
3. Every ring A can be regarded as a right A-module.
4. Every submodule of an A-module X is the kernel of a module homomorphism of X.
5. Every module is a group with operators.

3.10.1. Prove that every subgroup of an abelian group G is a submodule of G, considered as a $\mathbb{Z}$-module.

3.10.2. Let G be a group with operator set M, and let $\phi : G \to G'$ be an M-homomorphism. Prove that $\mathrm{Ker}\,\phi$ is a (normal) M-subgroup of G, and that $\mathrm{Im}\,\phi$ is an M-subgroup of G'.

3.10.3. If H is a normal M-subgroup of an M-group G, show how the factor group G/H can be made into an M-group such that the canonical epimorphism $\nu : G \to G/H$ is an M-epimorphism.

3.10.4. Let A be a ring and let X be a left A-module. Denote by "0" the 0-element of A, and by "$\bar{0}$" the 0-element of X. Prove that
(1) $0x = \bar{0}$ for each $x \in X$
(2) $(-1)x = -x$ for each $x \in X$.
(Note the same symbol "0" usually denotes the additive identities of $\langle A, +\rangle$ and of $\langle X, +\rangle$. We use "$\bar{0}$" here only to emphasize the distinction between the two zero elements.)

3.10.5. Let A be a ring, K a right ideal of A, and let X be a left A-module. Prove that the set $Y_X = \{x \in X \mid kx = 0 \text{ for all } k \in K\}$ is a submodule of A.

3.10.6. Interpret a ring as a group with operator set M in such a way that the M-subgroups are the 2-sided ideals.

3.10.7. Let A be a ring and let X, X' be left A-modules. Let $Y = \mathrm{Hom}_A(X, X')$ be the set of all module homomorphisms from X to X'. On Y, define $+$ by $(\sigma + \tau)(x) = \sigma(x) + \tau(x)$ $(\sigma, \tau \in Y; x \in X)$, and define $a\sigma$ by $(a\sigma)(x) = a\sigma(x)$ $(\sigma \in Y, a \in A, x \in X)$. Show that Y is a left A-module.

3.11

Vector Spaces, Subspaces, and Linear Independence

The structure of a ring A imposes powerful constraints on the structure of A-modules. For example, by the Fundamental Theorem of Abelian Groups, every finitely generated $\mathbb{Z}$-module (i.e., abelian group) is a direct product of finitely many $\mathbb{Z}$-modules, each isomorphic either to the $\mathbb{Z}$-module $\mathbb{Z}$, or to one of the homomorphic images of $\mathbb{Z}$. (The special case of *finite* $\mathbb{Z}$-modules was treated in Section 2.13.)

We now turn our attention to a particularly interesting class of modules, known as vector spaces: they are modules of division rings and, in particular, of fields.

Definition 3.11.1: Let F be a division ring. Then a unitary left F-module V is called a *left vector space over* V. Thus, explicitly, V is a left vector space over F if $\langle V, +\rangle$ is an abelian group, $\mu : F \times V \to V$ is an external composition (with $\mu(a, x)$ denoted by ax for each $a \in F, x \in V$) satisfying

(1) $a(x + y) = ax + ay$ for each $a \in F, x, y \in V$;

(2) $(a + b)x = ax + bx$ for each $a, b \in F, x \in V$;

(3) $(ab)x = a(bx)$ for each $a, b \in F, x \in V$;

(4) $1x = x$ for each $x \in V$.

(The elements of V are referred to as *vectors*, the operation $+$ on V as *vector addition*; the elements of F are called *scalars* and the external composition $\mu : (a, x) \mapsto ax$ is called *scalar multiplication*.)

In the following, we omit the adjective *left*; it will be understood that *vector space* means "left vector space."

Example 1:

Let P be a Cartesian plane. Two real numbers $\rho > 0$ and θ determine a *geometric vector* v in the plane, of magnitude ρ and inclination θ relative to the positive x-axis. The vector v may be represented by *any* arrow in the plane, of length ρ and inclination θ. Let V be the set of all geometric vectors in plane P. On V, define vector addition by the parallelogram law, or (equivalently) by the triangle law:

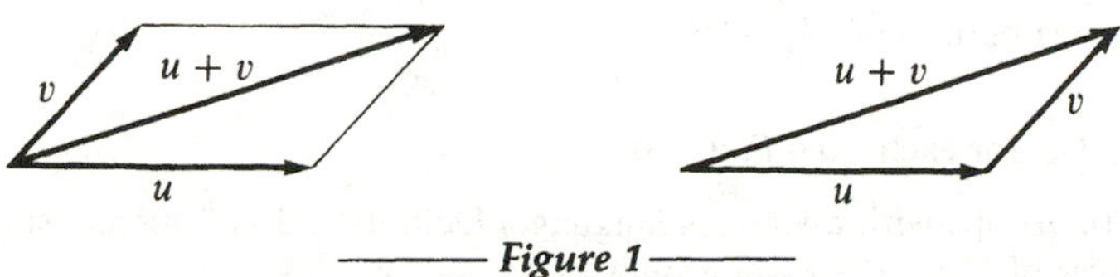

Figure 1

Define scalar multiplication as follows. Let $v \in V$ be a vector of magnitude ρ and inclination θ, and let $\alpha \in \mathbb{R}$. Then αv is a vector of magnitude $|\alpha|\rho$. If $\alpha \geq 0$, the inclination of αv is θ. If $\alpha < 0$, the inclination of v is $\theta + \pi$.

Then v forms a vector space over $\mathbb{R}$ (see Exercise 3.11.5).

Example 2:

Let F be a division ring and let $F^n = \left\{ \begin{pmatrix} a_1 \\ \vdots \\ a_n \end{pmatrix} \middle| a_i \in F \right\}$.

(Note: the elements of F^n are just n-tuples over F, written vertically, as *column vectors*.) On F^n, define vector addition and scalar multiplication componentwise, i.e.,

$$\begin{pmatrix} a_1 \\ \vdots \\ a_n \end{pmatrix} + \begin{pmatrix} b_1 \\ \vdots \\ b_n \end{pmatrix} = \begin{pmatrix} a_1 + b_1 \\ \vdots \\ a_n + b_n \end{pmatrix}, \quad c\begin{pmatrix} a_1 \\ \vdots \\ a_n \end{pmatrix} = \begin{pmatrix} ca_1 \\ \vdots \\ ca_n \end{pmatrix}, \quad (c, a_i, b_i \in F, \quad 1 \leq i \leq n).$$

Then F^n forms a vector space over F.

Example 3:

Let E be a division ring, F a division ring that is a subring of E. Then E forms a vector space over F, with vector addition in E defined as addition in the ring E, and scalar multiplication defined by

$$xy = x \cdot y$$

for $x \in F$, $y \in E$.

(This fact is very important in field theory—see Section 4.7.)

We have assumed throughout this text that the reader has some prior acquaintance with linear algebra and thus with vector spaces, at least over the real field. In the following, our purpose will be to review some of the basic properties of vector spaces, now couched within an abstract algebraic setting. We continue to assume that the scalars form a division ring, not necessarily commutative. In most applications, however, the scalars do form a (commutative) field.

Definition 3.11.2: Let V be a (left) vector space over a division ring F, and let W be a subset of V. Then W is a *subspace* of V if

(1) W is a subgroup of $\langle V, +\rangle$

and

(2) $aw \in W$ for each $a \in F, w \in W$.

(Note: In groups-with-operators language, Definition 3.11.2 merely states that the subspaces of V are the F-subgroups of $\langle V, +\rangle$.)

A handy subspace test is provided by the following corollary.

Corollary: Let V be a vector space over F, and let W be a subset of V. Then W is a subspace of V if and only if

(1) $W \neq \phi$

(2) $w_1, w_2 \in W \Rightarrow w_1 + w_2 \in W$

(3) $a \in F, w \in W \Rightarrow aw \in W$.

Proof: Clearly, if W is a subspace of V, then (1), (2), and (3) must hold.

Conversely, suppose (1), (2), and (3) hold. By (3), if $w \in W$, then $(-1)w \in W$. By Exercise 3.10.4, $(-1)w = -w \in W$. But then, by Theorem 2.3.1, W is a subgroup of $\langle W, +\rangle$. Together with (3), this implies that W is a subspace of V. ■

Obviously, not every subset of a vector space *is* a subspace. But every subset *generates* a subspace, in the usual sense:

Definition 3.11.3: If V is a vector space over a division ring F and S is a subset of V, then the *subspace* of V *generated* (or *spanned*) by S is

$$[S] = \bigcap_{\substack{W \text{ subspace} \\ \text{of } V}} W.$$

Briefly, $[S]$ is often called the *span* of S.

Remark 1: The intersection of any collection of subspaces is, clearly, a subspace, by the Corollary of Definition 3.11.2.

Remark 2: If $S = \varnothing$, then $[S] = \{0\}$—the subspace consisting of the 0-vector. Clearly, any subspace of V containing a non-empty subset S of V must contain all scalar multiples, hence all sums of scalar multiples, of elements of S.

Definition 3.11.4: If $v_1, \ldots, v_t$ are elements of a vector space V over F, then $\sum_{i=1}^{t} a_i v_i$ $(a_i \in F, i = 1, \ldots, t)$ is a *linear combination*, over F, of $a_1, \ldots, a_t$.

We can characterize the subspace spanned by a nonempty subset S of a vector space V in terms of linear combinations.

Theorem 3.11.1: If V is a vector space over a division ring F and S is a nonempty subset of V, then the subspace of V spanned by S is

$$[S] = \{c_1v_1 + \cdots + c_tv_t | c_i \in F, v_i \in S, t \geq 1\}.$$

Proof: Let $L_S = \{c_1v_1 + \cdots + c_tv_t | c_i \in F, v_i \in S, t \geq 1\}$. By the Corollary of Definition 3.11.2, L_S is clearly a subspace of V. For each $v \in S$, $v = 1v \in L_S$. Hence $S \subset L_S$. Since L_S is a subspace of V, containing S, L_S contains the intersection of *all* such subspaces, i.e., $L_S \supset [S]$.

On the other hand, since $[S]$ is itself a subspace of V, $[S]$ contains all linear combinations of elements of S, i.e., $[S] \supset L_S$. It follows that $[S] = L_S$. ∎

Of course, for any vector V, V itself is a spanning set, i.e., $[V] = V$—but this is a matter of little interest. More interesting is the question: how *small* a spanning set can a vector space have? For example, the sets

$$Y_1 = \left\{u_1 = \begin{pmatrix}1\\2\end{pmatrix}, u_2 = \begin{pmatrix}3\\1\end{pmatrix}, u_3 = \begin{pmatrix}4\\3\end{pmatrix}, u_4 = \begin{pmatrix}2\\-1\end{pmatrix}\right\}$$

$$Y_2 = \left\{v_1 = \begin{pmatrix}3\\1\end{pmatrix}, v_2 = \begin{pmatrix}0\\3\end{pmatrix}, v_3 = \begin{pmatrix}3\\7\end{pmatrix}\right\}$$

$$Y_3 = \left\{w_1 = \begin{pmatrix}1\\-1\end{pmatrix}, w_2 = \begin{pmatrix}2\\0\end{pmatrix}\right\}$$

are all spanning sets (see Exercise 3.11.8) for the vector space $\mathbb{R}^2$. Obviously, *no fewer* than two vectors can span $\mathbb{R}^2$ since there is no single vector in $\mathbb{R}^2$ of which *all* the vectors in $\mathbb{R}^2$ are scalar multiples. Thus, Y_3 is a minimal spanning set for $\mathbb{R}^2$. An interesting feature of minimal spanning sets is that they have no redundant vectors, i.e., no vectors that are contained in the span of the remaining vectors. This leads to the notions of linear dependence and independence. In the following, we shall call a linear combination $\sum_{i=1}^{t} c_iv_i$ *non-trivial* if at least one of the scalars c_i is non-zero, and *trivial* if all of the c_i are equal to zero.

Definition 3.11.5: Let V be a vector space over a division ring F. If S is a *finite* subset of S, then S is *linearly dependent over F* if some non-trivial linear combination of the elements of S is equal to the zero vector; S is *linearly independent over F* if S is not linearly dependent, i.e., if no non-trivial linear combination of the elements of S is equal to the zero vector. (Thus, in particular, if $S = \emptyset$, then S is linearly independent over F.)

If S is an *infinite* subset of V, then S is *linearly dependent over F* if some finite subset of S is linearly dependent over F; S is *linearly independent over F* if S is not linearly dependent over F, i.e., if all finite subsets of S are linearly independent over F.

Remark: Suppose the vectors $v_1, \ldots, v_n \in V$ are known to be linearly independent over F. If for some $c_1, \ldots, c_n \in F$, we have $c_1v_1 + \cdots + c_nv_n = 0$, we must conclude that the linear combination $c_1v_1 + \cdots + c_nv_n$ is trivial, i.e., $c_1 = c_2 = \cdots = c_n = 0$.

Theorem 3.11.2: Let V be a vector space over a division ring F, and let S be a linearly independent subset of V. Then every subset of S is linearly independent.

Proof: If $S = \varnothing$, the only subset, $\varnothing$, of S is linearly independent. Suppose $S \neq \varnothing$. Then certainly the subsets S and $\varnothing$ are linearly independent. Let S' be a subset of S such that $S' \neq \varnothing$, $S' \neq S$. If S is infinite, then S has no linearly independent finite subsets; hence neither does S'. It follows that S' is linearly independent, by Definition 3.11.5. If S is finite, we can label its elements so that $S = \{v_1, \ldots, v_t\}$ and $S' = \{v_1, \ldots, v_m\}$, $1 \leq m < t$. Suppose S' is linearly dependent. Then there are scalars $c_1, \ldots, c_m$ not all zero such that

$$c_1v_1 + \cdots + c_mv_m = 0.$$

But then $c_1v_1 + \cdots + c_mv_m + 0v_{m+1} + \cdots + 0v_t = 0$, and so (since the c_i are not all zero) $S = \{v_1, \ldots, v_t\}$ is linearly dependent—contradiction! It follows that S' is linearly independent. ∎

Corollary 1: If S is a non-empty subset of a vector space V and $v \in [S]$, then $S \cup \{v\}$ is linearly dependent.

Proof: If $S = \varnothing$, then $[S] = \{0\}$. Thus, if $v \in [S]$, then $v = 0$, hence $S \cup \{v\} = \varnothing \cup \{0\} = \{0\}$, a linearly dependent set. Suppose $S \neq \varnothing$. Since $v \in [S]$, there are vectors $v_1, \ldots, v_t \in S$ such that $v = c_1v_1 + \cdots + c_tv_t$ for some scalars $c_1, \ldots, c_t$. But then $1v + (-c_1)v_1 + \cdots + (-c_t)v_t = 0$, and so $\{v, v_1, \ldots, v_t\}$ is a linearly dependent subset of $S \cup \{v\}$. It follows that $S \cup \{v\}$ is linearly dependent. ∎

Corollary 2: Let S be a non-empty subset of a vector V over a division ring F. Then S is linearly dependent over F if and only if there is some element $x \in S$ such that $x \in [S \backslash \{x\}]$.

Proof: If there is some $x \in S$ such that $x \in [S\backslash\{x\}]$, then $S = (S\backslash\{x\}) \cup \{x\}$ is linearly dependent, by Corollary 1. Conversely, suppose S is linearly dependent. By Definition 3.11.5, S has a finite linearly dependent subset S' (if S is finite, we may, of course, let $S' = S$). Suppose $S' = \{u_1, \ldots, u_r\}$. Then there are scalars $c_1, \ldots, c_r$ not all zero such that $c_1u_1 + \cdots + c_ru_r = 0$. Without loss of generality, $c_1 \neq 0$. Then $u_1 = -(c_1)^{-1}[c_2u_2 + \cdots + c_ru_r] \in [u_2, \ldots, u_r]$. Thus $u_1 \in [S'\backslash\{u_1\}] \subset [S\backslash\{u_1\}]$. ∎

Returning to the example preceding Definition 3.11.5, we can now observe that sets Y_1 and Y_2 are linearly dependent spanning sets for $\mathbb{R}^2$, while Y_3 is a linearly independent spanning set for $\mathbb{R}^2$. In Y_1, we have $u_3 = u_1 + u_2 \in [Y_1\backslash\{u_3\}]$, and $u_4 = u_2 - u_1 \in [Y_1\backslash\{u_4\}]$. In Y_2, $v_3 = v_1 + 2v_2 \in [Y_2\backslash\{v_3\}]$. Y_3, on the other hand, is linearly independent for, if $c_1w_1 + c_2w_2 = 0 \quad (c_1, c_2 \in \mathbb{R})$, then $\left.\begin{matrix} c_1 + 2c_2 = 0 \\ -c_1 = 0 \end{matrix}\right\}$, a linear system having only the trivial solution $c_1 = c_2 = 0$.

A linearly independent spanning set for a vector space V is called a *basis* for V. We proceed to discuss bases in the following section.

Exercises 3.11

True or False

1. The scalars of a vector space must form a (commutative) field.
2. Every subset of a vector space is a subspace.
3. Every subset of a vector space spans a subspace.
4. If $c_1v_1 + \cdots + c_kv_k = 0$ for some set of scalars $c_1, \ldots, c_k$, then the vectors $v_1, \ldots, v_k$ are linearly dependent.
5. If $v_1, \ldots, v_k$ are linearly independent vectors, then no linear combination of $v_1, \ldots, v_k$ is equal to 0.

3.11.1. Let V be a vector space over a division ring F. Prove: if $av = 0$ ($a \in F$, $v \in V$), then either $a = 0$ or $v = 0$. Give an example to show that the corresponding result does not hold for modules generally.

3.11.2. Let V be a vector space over a division ring F, and let $v \in V$. Prove: if $v \neq 0$, then $\{v\}$ is linearly independent over F.

3.11.3. Let V be a vector space over a division ring F. Prove that any subset S of V which includes the 0-vector is linearly dependent.

3.11.4. Let v, w be non-zero vectors in a vector space V over a division ring F. Prove: v and w are linearly dependent over F if and only if each is a scalar multiple of the other.

3.11.5. Verify that the geometric vectors in the plane form a vector space, V, over $\mathbb{R}$ under the operations introduced in Example 1 on p. 195.

3.11.6. With V as defined in the preceding exercise, prove:
(1) Any two non-collinear vectors are linearly independent.
(2) Any two non-collinear vectors span V.

3.11.7. Let W_1, W_2 be subspaces of a vector space V over a division ring F. Prove:
(1) $W_1 \cap W_2$ is a subspace of V.
(2) $W_1 + W_2 = \{w_1 + w_2 | w_1 \in W_1, w_2 \in W_2\}$ is a subspace of V.
(3) $W_1 \cup W_2$ is *not* a subspace of V unless either $W_1 \subset W_2$ or $W_2 \subset W_1$.

3.11.8. Prove that the sets Y_1, Y_2, and Y_3 on p. 197 are all spanning sets for $\mathbb{R}^2$.

3.12

Basis and Dimension

Definition 3.12.1: Let V be a vector space over a division ring F. A linearly independent spanning set for V is called a *basis* for V.

Note: If $V = \{0\}$, then $\varnothing$ is a basis for V.

Like all important concepts in mathematics, a vector space basis can be characterized in a variety of ways: for example, as a minimal spanning set for V, or as a maximal linearly independent subset of V.

Definition 3.12.2: Let V be a vector space over a division ring F, and let $S \subset V$. Then S is a *minimal spanning set* for V if

(a) S spans V; and

(b) if $S' \subsetneqq S$, then S' fails to span V.

S is a *maximal linearly independent subset* of V if

(a) S is linearly independent;

and

(b) if $S \subsetneqq S' \subset V$, then S' is linearly dependent.

Theorem 3.12.1: Let V be a vector space over a division ring F, and let S be a subset of V. Then the following statements are equivalent:

(1) S is a minimal spanning set for V;

(2) S is a basis for V;

(3) S is a maximal linearly independent subset of V.

Proof: If $V = \{0\}$, then $\varnothing$ is the only subset of V satisfying (1), (2), and (3); hence in this case the three assertions of the theorem are trivially equivalent. In the remainder of the proof, we therefore assume $V \neq \{0\}$.

$1 \Rightarrow 2$. Suppose S is a minimal spanning set for V. Since $V \neq \{0\}$, we have $S \neq \varnothing$. Suppose S is linearly dependent. By Corollary 2 of Theorem 3.11.2, there is some $v \in S$ such that $v \in [S \backslash \{v\}]$. This implies that $S \backslash \{v\}$ is a spanning set for V, contrary to the hypothesis that S is a *minimal* spanning set for V. Thus, S is linearly independent, and so S is a basis for V.

$2 \Rightarrow 3$. If S is a basis for V, then S is linearly independent. Suppose $S \subsetneqq S' \subset V$. If $w \in S' - S$, then $w \in [S]$ since S spans V. From $S \subset S' \backslash \{w\}$, it follows that $w \in [S' \backslash \{w\}]$. But then, by Corollary 2 of Theorem 3.11.2, S' is linearly dependent. Hence S is a maximal linearly independent subset of V.

$3 \Rightarrow 1$. Suppose S is a maximal linearly independent subset of V. Then S spans V. For, suppose $v \in V$. If $v \notin [S]$, then $v \notin S$, and $S \cup \{v\}$ is linearly dependent, by the maximality of S. This implies that some finite subset T of $S \cup \{v\}$ is linearly dependent. Since S is linearly independent, $T = S' \cup \{v\}$ for some $S' \subset S$. Hence there are vectors $v_1, \ldots, v_t$ in S' and scalars $c_1, \ldots, c_t$, not all zero, such that

$$c_1 v_1 + \cdots + c_t v_t + cv = 0.$$

Since S is linearly independent, so is S'; hence $c \neq 0$. But then $v = -c^{-1} \sum_{i=1}^{t} c_i v_i \in [S]$. Contradiction. It follows that $v \in [S]$, and so S is a spanning set (hence a basis) for V. Now suppose $S' \subsetneqq S$ and S' is also a spanning set for V. If

$w \in S \backslash S'$, then $w \in [S']$; hence $S' \cup \{w\}$ is a linearly dependent subset of S, by Corollary 1 of Theorem 3.11.2. This is impossible since S is linearly independent. It follows that S is a minimal spanning set for V. ■

From Theorem 3.12.1, it follows immediately that every vector space V *has* a basis. For, using Zorn's Lemma of set theory, it may be shown that the set of all linearly independent subsets of V includes a maximal linearly independent subset, i.e., a basis for V. For finitely generated vector spaces, i.e., for vector spaces having a finite spanning set, the existence of a basis may be established without the use of Zorn's Lemma.

Theorem 3.12.2: Let V be a vector space over a division ring F. If V has a finite spanning set, then V has a basis.

Proof: Let S be a finite spanning set for V, consisting of k elements. Then all subsets of S that span V have $\leq k$ elements. By the well-ordering property of $\mathbb{Z}^+ \cup \{0\}$, there is at least $t \geq 0$ such that S has a subset T that spans V and has t elements. Then T is clearly a minimal spanning set, hence a basis, for V. ■

If F is an infinite division ring, then any vector space over F will have infinitely many bases (see Exercise 3.12.8). However, all bases of a given vector space have the same number of elements. This number is called the dimension of the space (see Definition 3.12.3). We have more work to do before we can formally define dimension.

Lemma: Let V be a vector space over a division ring F, and let S be a finite spanning set for V, consisting of n elements. Then every linearly independent subset T of V is finite and has $\leq n$ elements.

Proof: If $V = \{0\}$, the assertion of the theorem is obvious since $\varnothing$ is the only linearly independent subset of V.

Suppose $V \neq \{0\}$. Then $S \neq \varnothing$, and we may write $S = \{v_1, \ldots, v_n\}$ $(n \geq 1)$. We first prove the following *Exchange Property*: If T is a non-empty finite linearly independent subset of V consisting of m vectors $w_1, \ldots, w_m$, then the m vectors w_i can be exchanged for m vectors in S to produce a new spanning set $S^{(m)}$ for V, consisting of n vectors. We proceed by induction on m.

For $m = 1$, we have $T = \{w_1\}$. Since $[S] = V$, $w_1 \in [S]$, and so $w_1 = c_1 v_1 + \cdots + c_n v_n$ $(c_i \in F)$. Since T is linearly independent, $w_1 \neq 0$; hence $c_i \neq 0$ for some $i (1 \leq i \leq n)$. Without loss of generality, $c_1 \neq 0$. But then

$$v_1 = -c_1^{-1} \left[w_1 + \sum_{i=2}^{m} c_i v_i \right].$$

From $V = [v_1, \ldots, v_n] \subset [w_1, v_2, \ldots, v_n] \subset V$, we infer that $V = \{w_1, v_2, \ldots, v_n\}$, and so $S' = [w_1, v_2, \ldots, v_n]$ is a spanning set for V, consisting of n vectors.

Proceeding with the induction, suppose that, for some $m \in \mathbb{Z}^+$, the Exchange Property holds. Let $T = \{w_1, \ldots, w_{m+1}\}$ be a linearly independent subset of V. By the induction hypothesis, there is a spanning set $S^{(m)}$ for V consisting of $w_1, \ldots, w_m$

and $n - m$ of the vectors of S. Without loss of generality, we may label the vectors of S so that

$$S^{(m)} = \{w_1, \ldots, w_m, v_{m+1}, \ldots, v_n\}.$$

Then $w_{m+1} \in [S^{(m)}]$; hence $w_{m+1} = d_1 w_1 + \cdots + d_m w_m + c_{m+1} v_{m+1} + \cdots + c_n v_n$ $(c_i, d_j \in F)$. Since $\{w_1, \ldots, w_{m+1}\}$ is a linearly independent set, the c_j cannot all be zero. Hence there is some j $(m + 1 \leq j \leq n)$ such that $c_j \neq 0$. Relabeling, if necessary, we may assume that $j = m + 1$. Then

$$v_{m+1} = -c_{m+1}^{-1}\left[w_{n+1} - \sum_{i=1}^{m} d_i w_i - \sum_{j=m+2}^{n} c_j v_j\right].$$

But then

$$V = [w_1, \ldots, w_m, v_{m+1}, \ldots, v_n] \subset [w_1, \ldots, w_m, w_{m+1}, v_{m+2}, \ldots, v_n] \subset V,$$

and so $S^{(m+1)} = \{w_1, \ldots, w_{m+1}, v_{m+2}, \ldots, v_n\}$ is a new spanning set for V, consisting of n vectors.

By the Principle of Induction, we conclude that the Exchange Property holds for each positive integer m. But then every non-empty finite linearly independent subset of V has $\leq n$ elements. This implies that, in fact, every linearly independent subset of V *is finite and* has $\leq n$ elements. (For: an infinite linearly independent subset of V would have finite linearly independent subsets of *every* possible finite cardinality!) ■

Theorem 3.12.3: Let V be a vector space over a division ring F, and let S_1 and S_2 be bases for V. Then S_1 and S_2 have the same cardinality. (In the infinite case, as in the finite case, this means that there is a bijection from S_1 to S_2.)

Proof: Suppose one of the bases, say S_1, is finite, of cardinality n. By the Lemma, since S_1 is a finite spanning set for V, and S_2 is linearly independent, S_2 is finite and has cardinality $m \leq n$. But then, again by the Lemma, since S_2 is a finite spanning set for V, and S_1 is linearly independent, we have $n \leq$ m. It follows that $n = m$.

If S_1 and S_2 are both infinite, it can be shown fairly easily that S_1 and S_2 have the same cardinality. However, the proof requires more background in set theory than we have assumed in this book. (Once the theory of transfinite cardinals is known, one can argue for example as follows: if $\alpha = $ Card F, $\beta_1 = $ Card S_1 and $\beta_2 = $ Card S_2, then $\alpha^{\beta_1} = $ Card $V = \alpha^{\beta_2}$. For β_1, β_2 infinite, this implies that $\beta_1 = \beta_2$.) ■

We are now ready to define dimension.

Definition 3.12.3: Let V be a vector space over a division ring F. The *dimension of* V is the cardinality of any basis for V.

Notation: We denote the dimension of V by "dim V."

Remarks: ***(1)*** If $V = \{0\}$, then dim $V = 0$ (the cardinality of $\varnothing$).

(2) If V has a finite spanning set of k elements, then V is finite-dimensional, with dim $V \leq k$. (For, by the Lemma preceding Theorem 3.12.2, any linearly independent subset of V must have $\leq k$ elements.)

Example 1:

If F is any division ring and F^n is the set of all column vectors $\begin{pmatrix} a_1 \\ \vdots \\ a_n \end{pmatrix}$, $a_i \in F$, then F^n has dimension n. For example, the *unit vectors* $e_1 = \begin{pmatrix} 1 \\ 0 \\ \vdots \\ 0 \end{pmatrix}$, $e_2 = \begin{pmatrix} 0 \\ 1 \\ 0 \\ \vdots \\ 0 \end{pmatrix}, \ldots, e_n = \begin{pmatrix} 0 \\ 0 \\ \vdots \\ 1 \end{pmatrix}$ form a basis for F^n. (Clearly $\begin{pmatrix} a_1 \\ \vdots \\ a_n \end{pmatrix} = a_1e_1 + \cdots + a_ne_n$; and if $c_1e_1 + \cdots + c_ne_n = 0$, then $\begin{pmatrix} c_1 \\ \vdots \\ c_n \end{pmatrix} = \begin{pmatrix} 0 \\ \vdots \\ 0 \end{pmatrix}$, hence $c_i = 0$ for each $i = 1, \ldots, n$.)

Example 2:

For n a positive integer, F a field, let

$$P_n = \{f \in F[x] \mid f = 0 \quad \text{or} \quad \deg f < n\}.$$

Then P_n forms a vector space over F, with addition of polynomials as vector addition and multiplication of polynomials by field elements (i.e., by 0-degree polynomials, or by 0) as scalar multiplication. The polynomials $1, x, \ldots, x^{n-1}$ form a basis for P_n. (They form a spanning set since each polynomial $f \in P_n$ is equal to $a_01 + a_1x + a_2x^2 + \cdots + a_{n-1}x^{n-1}$, $a_i \in F$. They are linearly independent since, for $c_0, \ldots, c_{n-1} \in F$, $c_01 + c_1x + \cdots + c_{n-1}x^{n-1} = 0 = 0 \cdot 1 + 0x + \cdots + 0x^{n-1}$ implies that $c_i = 0$ for each $i = 0, \ldots, n-1$, by our definition of polynomials as sequences.) Thus $\dim P_n = n$.

Example 3:

For F a field, the polynomial domain $F[x]$ itself is a vector space over F, with addition and scalar multiplication defined as in Example 2. The polynomials

$$1, x, x^2, \ldots, x^{n-1}$$

form a linearly independent subset of n elements for each positive integer n. This implies that $F[x]$ is infinite dimensional. (For, if $F[x]$ has finite dimension k, then $n \leq k$ for *each* positive integer n, by the Lemma preceding Theorem 3.12.3. This is impossible!)

In the next theorem, we collect several useful assertions regarding finite-dimensional vector spaces.

Theorem 3.12.4: Let V be a vector space over a division ring F. Suppose V has finite dimension n.

(1) Every linearly independent subset of V has $\leq n$ elements.

(2) Every spanning set for V has $\geq n$ elements.

(3) Every linearly independent subset of V is contained in a basis for V.

(4) Every spanning set for V contains a basis for V.

(5) Every linearly independent subset of V consisting of n vectors is a basis for V.

(6) Every spanning set for V consisting of n vectors is a basis for V.

(7) Every subspace $W \subsetneq V$ has dimension $< n$.

Proof: *(1)* Since dim $V = n$, V has a basis, hence a spanning set, consisting of n elements. By the Lemma preceding Theorem 3.12.3, it follows that every linearly independent subset of V has $\leq n$ elements.

(2) Again, since dim $V = n$, V has a basis, hence a linearly independent subset, of n elements. By the Lemma preceding Theorem 3.12.3, it follows that every spanning set for V has $\geq n$ elements.

(3) Let L be a linearly independent subset of V. By (1), L has $\leq n$ elements. Hence there is a largest positive integer $m \leq n$ such that some linearly independent subset $M \supset L$ has m elements. But then adjoining another vector to M will yield a linearly dependent set, i.e., M is a maximal linearly independent subset of V. By Theorem 3.12.1, M is a basis (and thus, by Theorem 3.12.2, $m = n$).

(4) Let S be a spanning set for V. If $V = \{0\}$, then either $S = \{0\}$ or $S = \varnothing$. In either case, $S \supset \varnothing$, a basis for V. Suppose $V \neq \{0\}$. Then S contains some non-zero vector, hence S has a linearly independent subset T. By (1), all linearly independent subsets of S which contain T have $\leq n$ elements. Hence there is a largest positive integer $q \leq n$ such that some linearly independent subset Q, $T \subset Q \subset S$, has q elements. If $Q = S$, then Q is a linearly independent spanning set, hence a basis, for V. If $Q \subsetneq S$, then for each $s \in S \backslash Q$, $Q \cup \{s\}$ is linearly dependent, hence $s \in [Q]$. But then $S \subset [Q]$, hence $V = [S] \subset [[Q]] = [Q] \subset V$, and so $V = [Q]$. Thus, Q is a basis for V.

(5) Let L be a linearly independent subset of V consisting of n vectors. By (3), L is contained in a basis, B, for V. But B also consists of n vectors (n finite), hence $L \subset B$ implies $L = B$. Thus, L is a basis for V.

(6) Let S be a spanning set for V, consisting of n vectors. By (4), S contains a basis, B, for V. But B also consists of n vectors (n finite), hence $S \supset B$ implies $S = B$. Thus S is a basis for V.

(7) Since $W \subset V$, any linearly independent subset of W has $\leq n$ elements. In particular, any maximal linearly independent subset of W, i.e., any basis of W, has $\leq n$ elements. Thus, dim $W \leq$ dim V. If $W \subsetneq V$, then dim $W <$ dim V. For: if dim $W =$ dim $V = n$, then W has a basis, B, consisting of n elements. But then, by (5), B is a basis for V, and so $W = [B] = V$, contrary to hypothesis. ∎

If U and W are subspaces of a vector space V, then both $U \cap W$ and $U + W = \{u + w \mid u \in U, w \in W\}$ are easily seen to be subspaces of V, by the Corollary of Definition 3.11.2. For finite-dimensional subspaces U and W, we obtain a useful formula relating the dimensions of U, W, $U \cap W$ and $U + W$.

Theorem 3.12.5: Let U, W be finite-dimensional subspaces of a vector space V over a division ring F. Then

$$\dim(U + W) = \dim U + \dim W - \dim(U \cap W).$$

Proof: Let $\{t_1, \ldots, t_m\}$, $m \geq 0$, be a basis for $U \cap W$. Since $U \cap W$ is a subspace of U and of W, and $\{t_1, \ldots, t_m\}$ is linearly independent, property (3) of Theorem 3.12.4 guarantees that there is a basis $\{u_1, \ldots, u_h, t_1, \ldots, t_m\}$ for U, and there is a basis $\{t_1, \ldots, t_m, v_1, \ldots, v_k\}$ for V. We claim that

$$S = \{u_1, \ldots, u_h, t_1, \ldots, t_m, v_1, \ldots, v_k\}$$

is a basis for $U + W$. For, if $u \in U$, $w \in W$, then

$$u = c_1u_1 + \cdots + c_hu_h + d_1t_1 + \cdots + d_mt_m$$

$$w = e_1t_1 + \cdots + e_mt_m + f_1v_1 + \cdots + f_kv_k$$

$(c_1, \ldots, c_h; d_1, \ldots, d_n; e_1, \ldots, e_m; f_1, \ldots, f_k$ all elements of $F)$. Hence

$$u + w = c_1u_1 + \cdots + c_hu_h + (d_1 + e_1)t_1 + \cdots + (d_m + e_m)t_m + f_1v_1 + \cdots + f_kv_k,$$

and so $u + w \in [S]$. Thus $U + W \subset [S]$.

On the other hand, since $S \subset U \cup W \subset U + W$ (a subspace of V), $[S] \subset U + W$, and so $[S] = U + W$. Thus, S spans $U + W$. Suppose S is linearly dependent. Then there are scalars $\alpha_1, \ldots, \alpha_h, \beta_1, \ldots, \beta_m, \gamma_1, \ldots, \gamma_k$, not all zero, such that

$$\alpha_1u_1 + \cdots + \alpha_hu_h + \beta_1t_1 + \cdots + \beta_mt_m + \gamma_1v_1 + \cdots + \gamma_kv_k = 0.$$

Since $\{t_1, \ldots, t_m, v_1, \ldots, v_k\}$ is a linearly independent set, the α_i cannot all be zero. From $\alpha_1u_1 + \cdots + \alpha_hu_h = -(\beta_1t_1 + \cdots + \beta_mt_m + \gamma_1v_1 + \cdots + \gamma_kv_k)$, we conclude that $\alpha_1u_1 + \cdots + \alpha_hu_h$ is in W, hence in $U \cap W$. Thus

$$\alpha_1u_1 + \cdots + \alpha_hu_h = \delta_1t_1 + \cdots + \delta_mt_m$$

for some scalars $\delta_1, \ldots, \delta_m$, and

$$\alpha_1u_1 + \cdots + \alpha_hu_h + (-\delta_1)t_1 + \cdots + (-\delta_m)t_m = 0.$$

Since the α_i are not all zero, this implies that the set $\{u_1, \ldots, u_h, t_1, \ldots, t_m\}$ is linearly dependent, contradicting its choice as a basis for U. It follows that S is linearly independent, hence forms a basis for $U + V$. Counting, we obtain $\dim(U + V) = h + m + k = (h + m) + (m + k) - m = \dim U + \dim W - \dim(U \cap W)$. ■

Exercises 3.12

True or False

1. Every vector space has a unique basis.
2. Every linearly independent subset S of a vector space V forms a basis for $[S]$.
3. Every non-zero vector in a vector space V forms part of a basis for V.
4. Every spanning set for a finite-dimensional vector space V forms a basis for V.
5. If S_1, S_2 are subsets of a vector space V, then $S_1 \subsetneq S_2$ implies $[S_1] \subsetneq [S_2]$.

3.12.1. Find the dimension of each of the following:

(1) The subspace of $\mathbb{R}^3$ spanned by the vectors

$$v_1 = \begin{pmatrix} 1 \\ 2 \\ 3 \end{pmatrix}, v_2 = \begin{pmatrix} 2 \\ 0 \\ 1 \end{pmatrix}, v_3 = \begin{pmatrix} 4 \\ 4 \\ 7 \end{pmatrix}.$$

(2) The subspace of a vector space V spanned by a single non-zero vector.

(3) The subspace of $\mathbb{R}[x]$ spanned by $f_1 = x^2 + x + 3$, $f_2 = 3x + 1$, $f_3 = 5x + 2$.

(4) The subspace of $\mathbb{R}[x]$ spanned by the set $\{x^i | i \geq 20\}$.

(5) $\mathbb{C}$, considered as a vector space over $\mathbb{R}$ (with scalar multiplication defined as multiplication: $rz = r \cdot z$ for $r \in \mathbb{R}, z \in \mathbb{C}$).

3.12.2. In the vector space $P_4 = \{f \in \mathbb{R}[x] | f = 0 \text{ or } \deg f < 4\}$, let $U = \{f \in P_4 | f(1) = 0\}$ and let $V = \{f \in P_4 | f(-1) = 0\}$. Show that U and V are subspaces of P_4. Find dim U, dim V, dim$(U \cap V)$ and dim$(U + V)$.

3.12.3. Find a basis for $\mathbb{R}^4$ which includes the vectors

$$v_1 = \begin{pmatrix} 1 \\ 2 \\ 0 \\ 1 \end{pmatrix}, v_2 = \begin{pmatrix} 3 \\ 1 \\ 4 \\ 2 \end{pmatrix}, v_3 = \begin{pmatrix} 4 \\ 3 \\ 4 \\ 4 \end{pmatrix}.$$

3.12.4. Let $\mathbb{Q}(\sqrt{2}) = \{a + b\sqrt{2} | a, b \in \mathbb{Q}\}$. Interpret $\mathbb{Q}(\sqrt{2})$ as a vector space over $\mathbb{Q}$ and find its dimension. Find a basis for $\mathbb{Q}(\sqrt{2})$ over $\mathbb{Q}$.

3.12.5. Let $\mathbb{Q}(\sqrt[3]{2}) = \{a + b\sqrt[3]{2} + c\sqrt[3]{4} | a, b, c \in \mathbb{Q}\}$. Interpret $\mathbb{Q}(\sqrt[3]{2})$ as a vector space over $\mathbb{Q}$ and find its dimension. Find a basis for $\mathbb{Q}(\sqrt[3]{2})$ over $\mathbb{Q}$.

3.12.6. In the vector space $\mathbb{R}[x]$ over $\mathbb{R}$, let $S_1 = \{x^i | i \geq 0 \text{ in } \mathbb{Z}\}$ and let $S_2 = \{x^{2i} | i \geq 0 \text{ in } \mathbb{Z}\}$. Prove

(1) S_1 and S_2 are linearly independent sets;

(2) S_1 is a basis for $\mathbb{R}[x]$ over $\mathbb{R}$;

(3) S_2 is not a basis for $\mathbb{R}[x]$ over $\mathbb{R}$;

(4) S_1 and S_2 have the same cardinality, i.e., there is a bijection of S_1 onto S_2. (Compare this result with (5) of Theorem 3.12.4.)

3.12.7. Let $\mathbb{H}$ be the ring of all real quaternions. Interpret $\mathbb{H}$ as a vector space over $\mathbb{R}$, and find its dimension. Find a basis for $\mathbb{H}$ over $\mathbb{R}$.

3.12.8. Let F be a division ring consisting of infinitely many elements, and let V be a vector space over F.

(1) Prove that either $V = \{0\}$, or V is an infinite set.

(2) Prove that, if $V \neq \{0\}$, then V has infinitely many bases.

3.12.9. Give an example of an infinite-dimensional vector space over a finite field F.

3.12.10. If the definitions of *linear independence* and *basis* (given here for vector spaces) are applied to an A-module, where A is not necessarily a division ring, which implications in Theorem 3.12.1 remain valid?

Linear Transformations

We now examine homomorphisms of vector spaces.

Definition 3.13.1: Let V, W be vector spaces over a division ring F. Then $\phi : V \to W$ is a *linear transformation* if it is a module homomorphism, i.e., if

(1) $\phi(x + y) = \phi(x) + \phi(y)$ for each $x, y \in V$;
(2) $\phi(cx) = c\phi(x)$ for each $c \in F$, $x \in V$.

(Linear transformations are also called *linear operators.* In particular, the endomorphisms $\phi : V \to V$ are called *linear operators on* V.)

If $\phi : V \to W$ is a linear transformation, then Ker ϕ is called the *nullspace* of ϕ and Im ϕ is called the *range* of ϕ. The dimension of the nullspace of ϕ is the *nullity* of ϕ; the dimension of the range of ϕ is the *rank* of ϕ. ϕ is *singular* if its nullity is non-zero, *non-singular* if its nullity is zero.

Notation: We use "$\nu(\phi)$" to denote the nullity, "$\rho(\phi)$" to denote the rank, of ϕ.

For linear transformations of finite-dimensional domain, a handy formula relates rank and nullity.

Theorem 3.13.1: Let V, W be vector spaces over a division ring F, with V of finite dimension. Then

$$\rho(\phi) + \nu(\phi) = n.$$

Proof: Since $\dim V$ is finite, so is $\dim \operatorname{Ker} \phi = \nu(\phi)$. Write $s = \nu(\phi)$. Let $\{v_1, \ldots, v_s\}$ be a basis for Ker ϕ. By Theorem 3.12.4(3), $\{v_1, \ldots, v_s\}$ is part of a basis $\{v_1, \ldots, v_s, v_{s+1}, \ldots, v_n\}$ for V. Consider the set $\{\phi v_1, \ldots, \phi v_s, \phi v_{s+1}, \ldots, \phi v_n\} = \{\phi v_{s+1}, \ldots, \phi v_n\}$. If $w \in \operatorname{Im} \phi$, then $w = \phi(v)$ for some $v \in V$. There are scalars $c_1, \ldots, c_s, c_{s+1}, \ldots, c_n$ such that $v = \sum_{i=1}^{n} c_i v_i$; hence

$$w = \phi(v) = \phi\left(\sum_{i=1}^{n} c_i v_i\right) = \sum_{i=1}^{n} c_i \phi(v_i) = \sum_{i=s+1}^{n} c_i \phi(v_i).$$

Thus $\{\phi v_{s+1}, \ldots, \phi v_n\}$ is a spanning set for Im ϕ. We show that the set $\{\phi v_{s+1}, \ldots, \phi v_n\}$ is linearly independent. Suppose, for $d_{s+1}, \ldots, d_n \in F$, we have $d_{s+1}\phi(v_{s+1}) + \cdots + d_n\phi(v_n) = 0$. Then $\phi(d_{s+1}v_{s+1} + \cdots + d_n v_n) = 0$; hence $d_{s+1}v_{s+1} + \cdots + d_n v_n \in \operatorname{Ker} \phi$. But then there are scalars $e_1, \ldots, e_s$ in F such that $d_{s+1}v_{s+1} + \cdots + d_n v_n = e_1 v_1 + \cdots + e_s v_s$; hence $(-e_1)v_1 + \cdots + (-e_s)v_s + d_{s+1}v_{s+1} + \cdots + d_n v_n = 0$. Since $\{v_1, \ldots, v_n\}$ is a basis for V, it follows that all the coefficients are zero, in particular, $d_{s+1} = \cdots = d_n = 0$, and so $\phi v_{s+1}, \ldots, \phi v_n$ are linearly independent. But then $\{\phi v_{s+1}, \ldots, \phi v_n\}$ is a basis for Im ϕ. Hence $\dim \operatorname{Im} \phi = n - s$, and so

$$n = (\dim \operatorname{Im} \phi) + s = \dim \operatorname{Im} \phi + \dim \operatorname{Ker} \phi = \rho(\phi) + \nu(\phi). \qquad \blacksquare$$

Trivially, the identity map ι_V on any vector space V is a linear operator on V.

Definition 3.13.2: Let V be a vector space over a division ring F, and let $\phi : V \to V$ be a linear operator. Then ϕ is *invertible* if there is a linear operator $\psi : V \to V$ such that $\psi\phi = \iota_V = \phi\psi$.

As an immediate consequence of Theorem 3.13.1, we can characterize non-singular linear operators on a finite-dimensional vector space in a variety of ways.

Theorem 3.13.2: Let V be a finite-dimensional vector space over a division ring F, and let ϕ be a linear operator on V. Then the following statements are equivalent:

(1) ϕ is non-singular.

(2) ϕ is 1-1.

(3) ϕ is bijective.

(4) ϕ is onto.

(5) there is a linear operator $\psi : V \to V$ such that $\psi\phi = \iota_V$.

(6) ϕ is invertible.

(7) there is a linear operator $\psi : V \to V$ such that $\phi\psi = \iota_V$.

Proof: $1 \Rightarrow 2$. If ϕ is non-singular, then, by Definition 3.13.1, $\dim \operatorname{Ker} \phi = \nu(\phi) = 0$; hence $\operatorname{Ker} \phi = \{0\}$. This implies that ϕ is 1-1.

$2 \Rightarrow 3$. Suppose ϕ is 1-1. Then $\operatorname{Ker} \phi = \{0\}$; hence, $\nu(\phi) = 0$. By Theorem 3.13.1, it follows that $\rho(\phi) = \dim \operatorname{Im} \phi = \dim V$. Since V is finite-dimensional, this implies that $\operatorname{Im} \phi = V$ (Theorem 3.12.4(7)). But then ϕ is onto, hence bijective.

$3 \Rightarrow 4$. If ϕ is bijective, then, of course, ϕ is onto.

$4 \Rightarrow 5$. Suppose ϕ is onto. By Theorem 1.2.3, there is a *mapping* $\psi : V \to V$ such that $\psi\phi = \iota_V$. We show that ψ is linear. Let $u, v \in V$. Then there are $x, y \in V$ such that $u = \phi(x)$, $v = \phi(y)$. Hence $\psi(u) = \psi(\phi(x)) = (\psi\phi)(x) = \iota_V(x) = x$ and, similarly, $\psi(v) = y$. Hence

$$\begin{aligned}\psi(u + v) = \psi(\phi(x) + \phi(y)) = \psi(\phi(x + y)) = (\psi\phi)(x + y) = \iota_V(x + y) = x + y \\ = \psi(u) + \psi(v)\end{aligned}$$

and, for $c \in F$,

$$\psi(cu) = \psi(c\phi(x)) = \psi(\phi(cx)) = (\psi\phi)(cx) = \iota_V(cx) = cx = c\psi(u).$$

Thus, ψ is a linear operator on V, with $\psi\phi = \iota_V$.

$5 \Rightarrow 6$. Suppose there is a linear operator $\psi : V \to V$ such that $\psi\phi = \iota_V$. By Theorem 1.2.3, ϕ is 1-1; hence, by Theorem 3.13.1, ϕ is onto. But then, by Theorem 1.2.3, there is a mapping $\psi' : V \to V$ such that $\phi\psi' = \iota_V$. From $(\psi\phi)\psi' = \psi(\phi\psi')$, we have $\iota_V\psi' = \psi\iota_V$; hence $\psi' = \psi$. But then

$$\psi\phi = \iota_V = \phi\psi,$$

and so ϕ is invertible.

$6 \Rightarrow 7$. If ϕ is invertible, then there is a linear operator $\psi : V \to V$ such that $\psi\phi = \iota_V = \phi\psi$, hence, certainly, (7) holds.

$7 \Rightarrow 1$. Suppose there is a linear operator $\psi : V \to V$ such that $\phi\psi = \iota_V$. By Theorem 1.2.3, ϕ is onto. Hence by Theorem 3.13.1, ϕ is 1-1. But then Ker $\phi = \{0\}$, and $\nu(\phi) = \dim \text{Ker } \phi = 0$, and so ϕ is non-singular. ■

Remark: The equivalence of the assertions in Theorem 3.13.2 does not hold for infinite-dimensional vector spaces. For instance, for F a field, let $V = F[x]$ (the vector space described in Example 3 of Section 3.12). Define $\phi : V \to V$ by $\phi\left(\sum_{i=0}^{n} a_i x^i\right) = \sum_{i=0}^{n} a_i x^{i+1}$ $(a_i \in F, i = 0, \dots, n, n \geq 0)$. Then ϕ is a linear operator, with Ker $\phi = \{0\}$, i.e., ϕ is 1-1. But ϕ is not onto since, e.g., the constant polynomial 1 is not in Im ϕ.

Dimensionality is a powerful property of vector spaces. In fact, if two vector spaces have the same dimension, then they are isomorphic. We give a proof of this fact for the finite-dimensional case only (since the infinite-dimensional case would once more take us somewhat beyond our set-theoretic boundaries).

Theorem 3.13.3: Let V, W be vector spaces of the same finite dimension n over a division ring F. Then V is isomorphic to W, i.e., there is a bijective linear transformation from V to W.

Proof: Isomorphism is an equivalence relation on any set of vector spaces (a property inherited from groups via groups with operators). Thus, to prove V and W isomorphic to each other, it suffices to find a vector space to which both V and W are isomorphic. We propose the space F^n consisting of all ordered n-tuples $\begin{pmatrix} x_1 \\ \vdots \\ x_n \end{pmatrix}$, $x_i \in F$. Let $\{v_1, \dots, v_n\}$ be a basis for V. Define $\psi : F^n \to V$ by $\psi\begin{pmatrix} x_1 \\ \vdots \\ x_n \end{pmatrix} = x_1 v_1 + \cdots + x_n v_n$. ψ is clearly well-defined. For, if $\begin{pmatrix} x_1 \\ \vdots \\ x_n \end{pmatrix} = \begin{pmatrix} y_1 \\ \vdots \\ y_n \end{pmatrix}$, then $x_i = y_i$ for each $i = 1, \dots, n$, by a basic property of n-tuples. Thus $\psi\begin{pmatrix} x_1 \\ \vdots \\ x_n \end{pmatrix} = \sum_{i=1}^{n} x_i v_i = \psi\begin{pmatrix} y_1 \\ \vdots \\ y_n \end{pmatrix}$. ψ is onto since $\{v_1, \dots, v_n\}$ spans V; hence if $v \in V$, then $v = \sum_{i=1}^{n} x_i v_i = \psi\begin{pmatrix} x_1 \\ \vdots \\ x_n \end{pmatrix}$ for some $x_i \in F$, $i = 1, \dots, n$. ψ is 1-1; for, suppose $\psi\begin{pmatrix} x_1 \\ \vdots \\ x_n \end{pmatrix} = x_1 v_1 + \cdots + x_n v_n = 0$. Then $x_1 = \cdots = x_n = 0$ since the set $\{v_1, \dots, v_n\}$ is linearly independent; hence $\begin{pmatrix} x_1 \\ \vdots \\ x_n \end{pmatrix} = \begin{pmatrix} 0 \\ \vdots \\ 0 \end{pmatrix}$, the zero vector in F^n. Finally, ψ is linear; for, if $x_i, y_i \in F$,

$i = 1, \ldots, n$, and $c \in F$, then

$$\psi\left(\begin{pmatrix} x_1 \\ \vdots \\ x_n \end{pmatrix} + \begin{pmatrix} y_1 \\ \vdots \\ y_n \end{pmatrix}\right) = \psi\begin{pmatrix} x_1 & + & y_1 \\ & \vdots & \\ x_n & + & y_n \end{pmatrix}$$

$$= \sum_{i=1}^{n} (x_i + y_i)v_i = \sum_{i=1}^{n} x_i v_i + \sum_{i=1}^{n} y_i v_i = \psi\begin{pmatrix} x_1 \\ \vdots \\ x_n \end{pmatrix} + \psi\begin{pmatrix} y_1 \\ \vdots \\ y_n \end{pmatrix},$$

and

$$\psi\left(c\begin{pmatrix} x_1 \\ \vdots \\ x_n \end{pmatrix}\right) = \psi\begin{pmatrix} cx_1 \\ \vdots \\ cx_n \end{pmatrix} = \sum_{i=1}^{n} (cx_i)v_i = \sum_{i=1}^{n} c(x_i v_i) = c\sum_{i=1}^{n} x_i v_i = c\psi\begin{pmatrix} x_1 \\ \vdots \\ x_n \end{pmatrix}.$$

We have shown that ψ is a bijective linear transformation, i.e., an isomorphism, from F^n to V.

Similarly, there is an isomorphism ψ' from F^n to W, whence $\psi'\psi^{-1}$ is an isomorphism from V to W. ∎

Exercises 3.13

True or False

1. Every linear transformation preserves linear combinations.
2. Every linear transformation preserves linear independence.
3. If a linear transformation is 1-1, then it is onto.
4. If ϕ is a linear transformation of V onto W, then ϕ maps any basis for V to a basis for W.
5. If a linear operator on a finite-dimensional vector space is onto, then it is 1-1.

3.13.1. Let $C[a, b]$ be the set of all continuous real valued functions defined on a non-zero interval $[a, b]$ on the real line.

(1) Interpret $C[a, b]$ as a vector space over $\mathbb{R}$.

(2) Define $\phi : C[a, b] \to \mathbb{R}$ by $\phi(f) = \int_a^b f(x)\,dx$. Prove that ϕ is a linear transformation. Find its nullspace and its range.

3.13.2. Let F be a field. Consider $F[x]$ as a vector space over F. Define $\phi : F[x] \to F[x]$ by

$$\phi(a_0 + a_1 x + \cdots + a_n x^n) = a_1 + 2a_2 x + 3a_3 x^2 + \cdots + na_n x^{n-1},$$

i.e., ϕ maps each polynomial in $F[x]$ to its *formal derivative*.

(1) Prove that ϕ is a (well-defined) mapping.

(2) Prove that ϕ is a linear transformation.

(3) Find the nullspace and the range of ϕ.

(4) Note that ϕ is onto, but not 1-1. Does this contradict Theorem 3.13.2?

3.13.3. Let V, W be vector spaces over a division ring F, and let $S = \{v_1, \ldots, v_n\}$ be a basis for V. If $w_1, \ldots, w_n$ are any n vectors in W, prove that there is a unique linear transformation $\phi : V \to W$ such that $\phi(v_i) = w_i$ $(i = 1, \ldots, n)$.

3.13.4. In view of Exercise 3.13.3, note that each of the following assignments of images to basis vectors defines a linear operator, ϕ, on V. Find the rank and the nullity of ϕ.

(1) $\phi : \mathbb{R}^3 \to \mathbb{R}^3$ such that

$$\phi\begin{pmatrix}1\\0\\0\end{pmatrix} = \begin{pmatrix}1\\2\\2\end{pmatrix}, \quad \phi\begin{pmatrix}0\\1\\0\end{pmatrix} = \begin{pmatrix}2\\4\\6\end{pmatrix}, \quad \phi\begin{pmatrix}0\\0\\1\end{pmatrix} = \begin{pmatrix}1\\0\\0\end{pmatrix}.$$

(2) $\phi : P_3 \to P_3$, where $P_3 = \{f \in \mathbb{R}[x] \mid f = 0 \text{ or } \deg f < 3\}$, and $\phi(6) = x$, $\phi(x) = 5$, $\phi(5 + x^2) = 1$.

3.13.5. Let V, W be vector spaces over a division ring F, and let $\phi : V \to W$ be a non-singular linear operator. If S is a linearly independent subset of V, prove that $\phi S = \{\phi(s) \mid s \in S\}$ is a linearly independent subset of W.

3.13.6. Let V be a finite-dimensional vector space and let $\phi : V \to V$ be a non-singular linear operator. Prove that for any basis $S = \{v_1, \ldots, v_n\}$ of V, $\{\phi v_1, \ldots, \phi v_n\}$ is another basis for V.

3.13.7. Prove that $\mathbb{Q}(\sqrt{2}) = \{a + b\sqrt{2} \mid a, b \in \mathbb{Q}\}$ and $\mathbb{Q}(i) = \{a + bi \mid a, b \in \mathbb{Q}\}$ are isomorphic as vector spaces over $\mathbb{Q}$.

3.13.8. Let V be a vector space over a division ring F, and let $\mathrm{End}_F V$ be the set of all linear operators on V. On $\mathrm{End}_F V$, define $+$ by $(\phi + \psi)(v) = \phi(v) + \psi(v)$ $(\phi, \psi \in \mathrm{End}_F V, v \in V)$. Prove that $\mathrm{End}_F V$ forms a ring with respect to $+$ and composition.

3.13.9. Let U, V, W be finite-dimensional vector spaces over a division ring F, and let $\phi : U \to V$ and $\psi : V \to W$ be linear transformations. Prove that $\rho(\psi\phi) \le \rho(\phi)$ and $\nu(\psi\phi) \ge \nu(\phi)$.

3.13.10. Let V be a finite-dimensional vector space over a division ring F, and let $A = \mathrm{End}_F V$. Prove that, for $\phi, \psi \in A$, $\phi\psi = \iota$ implies $\psi\phi = \iota$. Conclude that the ring A contains no one-sided units.

3.13.11. Let V be a vector space over a division ring F and let $\phi : V \to V$ be a linear operator. Prove that the following assertions are equivalent:

(a) if $\phi^2(x) = 0$, then $\phi(x) = 0$;

(b) $\mathrm{Im}\,\phi \cap \mathrm{Ker}\,\phi = \{0\}$.

3.13.12. Let ϕ be an idempotent linear operator, i.e., a linear operator such that $\phi^2 = \phi$. Prove that $\mathrm{Im}\,\phi \cap \mathrm{Ker}\,\phi = \{0\}$. (Hint: See the preceding exercise.)

3.14

Coordinate Vectors, Matrices, and Determinants

Until now, we have assumed that the scalars of the vector spaces under discussion formed a division ring F, not necessarily commutative. There is no *mathematical* reason for us to begin at this point to assume commutativity of F. With notations suitably chosen, all the results of this section prior to those involving determinants are valid for non-commutative division rings. With the scalars written (comfortably) on the left, however, non-commutativity would force us to use row vectors instead of column vectors. In view of current fashion, this might cause undue confusion for readers interested primarily in the commutative case. Thus, we henceforth assume that F is a commutative division ring, i.e., a field.

Given a vector space basis S consisting of n vectors ($n \in \mathbb{Z}^+$), we can permute the vectors in S in $n!$ ways to obtain n-tuples of basis vectors. We shall refer to an n-tuple of basis vectors as an *ordered basis.* Script capital letters will denote ordered bases.

Theorem 3.14.1: Let V be an n-dimensional vector space over a field F ($n \in \mathbb{Z}^+$), and let $\mathscr{B} = (v_1, \ldots, v_n)$ be an ordered basis for V. Then for each $v \in V$, there is a unique n-tuple $\begin{pmatrix} x_1 \\ \vdots \\ x_n \end{pmatrix} \in F^n$ such that $v = \sum_{i=1}^{n} x_i v_i$.

Proof: Since $\{v_1, \ldots, v_n\}$ spans V, there are scalars $x_i \in F$, $i = 1, \ldots, n$, such that $v = \sum_{i=1}^{n} x_i v_i$. If, also, $v = \sum_{i=1}^{n} y_i v_i$, then from $\sum_{i=1}^{n} x_i v_i = \sum_{i=1}^{n} y_i v_i$, we infer that

$$\sum_{i=1}^{n} (x_i - y_i)v_i = 0.$$

Since $\{v_1, \ldots, v_n\}$ is a linearly independent set, it follows that $x_i - y_i = 0$; hence, $x_i = y_i$, for each $i = 1, \ldots, n$. But then $\begin{pmatrix} x_1 \\ \vdots \\ x_n \end{pmatrix} = \begin{pmatrix} y_1 \\ \vdots \\ y_n \end{pmatrix}$. ■

Definition 3.14.1: Let $\mathscr{B} = (v_1, \ldots, v_n)$ be an ordered basis for an n-dimensional vector space V over a field F ($n \in \mathbb{Z}^+$). If $v = \sum_{i=1}^{n} x_i v_i \in V$, then the n-tuple

$X = \begin{pmatrix} x_1 \\ \vdots \\ x_n \end{pmatrix}$ is the *coordinate vector* of v relative to $\mathscr{B}$.

Notation: We write "$v_{\mathscr{B}}$" for the coordinate vector of v relative to $\mathscr{B}$.

Example:

If $\mathscr{B} = (v_1, v_2, v_3)$ is an ordered basis for a 3-dimensional vector space over $\mathbb{R}$, and if $v = 3v_1 - 5v_2 + v_3$, then $v_{\mathscr{B}} = \begin{pmatrix} 3 \\ -5 \\ 1 \end{pmatrix}$. (On the other hand, if $\mathscr{B}' = (v_2, v_1, v_3)$, then $v_{\mathscr{B}'} = \begin{pmatrix} -5 \\ 3 \\ 1 \end{pmatrix}$.)

Theorem 3.14.2: Let V be an n-dimensional vector space ($n \in \mathbb{Z}^+$) over a field F, and let $\mathscr{B} = (v_1, \ldots, v_n)$ be an ordered basis for V. Then $\phi : V \to F^n$ defined by $\phi(v) = v_{\mathscr{B}}$ $(v \in V)$ is an isomorphism of the vector space V onto the vector space F^n.

Proof: ϕ is simply the inverse of the isomorphism $\psi : F^n \to V$ in the proof of Theorem 3.13.3.

Or, explicitly, ϕ is well-defined, by Theorem 3.14.1. If $\phi(v) = \begin{pmatrix} 0 \\ \vdots \\ 0 \end{pmatrix}$ for some $v \in V$, then $v = 0v_1 + \cdots + 0v_n = 0$. Thus ϕ is 1-1. Since $\nu(\phi) = 0$, $\rho(\phi) = n$ (by Theorem 3.13.1), hence ϕ is a bijection. Let $u, v \in V$, $c \in F$. Then $u = \sum_{i=1}^{n} c_i v_i$, $v = \sum_{i=1}^{n} d_i v_i$ $(c_i, d_i \in F, i = 1, \ldots, n)$. Hence

$$\phi(u + v) = \phi\left(\sum_{i=1}^{n} c_i v_i + \sum_{i=1}^{n} d_i v_i\right) = \phi\left(\sum_{i=1}^{n} (c_i + d_i) v_i\right)$$

$$= \begin{pmatrix} c_1 + d_1 \\ \vdots \\ c_n + d_n \end{pmatrix} = \begin{pmatrix} c_1 \\ \vdots \\ c_n \end{pmatrix} + \begin{pmatrix} d_1 \\ \vdots \\ d_n \end{pmatrix} = \phi(u) + \phi(v);$$

$$\phi(cu) = \phi\left(c \sum_{i=1}^{n} c_i v_i\right) = \phi\left(\sum_{i=1}^{n} c(c_i v_i)\right) = \phi\left(\sum_{i=1}^{n} (cc_i) v_i\right)$$

$$= \begin{pmatrix} cc_1 \\ \vdots \\ cc_n \end{pmatrix} = c \begin{pmatrix} c_1 \\ \vdots \\ c_n \end{pmatrix} = c\phi(u).$$

Thus, ϕ is a bijective linear transformation, i.e., an isomorphism, from V to F^n. ■

Every ordered basis thus induces an isomorphism of an n-dimensional vector space V over F onto F^n. (A single basis *set* alone gives rise to $n!$ ordered bases, hence to $n!$ isomorphisms of V onto F^n.)

The importance of Theorem 3.14.2 lies in the fact that it is usually easier to compute in F^n than in a given n-dimensional F-space V. In view of the fact that the correspondence $v \mapsto v_{\mathscr{B}}$ ($\mathscr{B}$ any ordered basis) is an isomorphism, one can replace the vectors of V by their coordinate vectors for computational purposes.

In order to describe the effect of linear transformations on coordinate vectors, we introduce *matrices.*

Definition 3.14.2: Let F be a field. An $m \times n$ array

$$(a_{ij})_{m \times n} = \begin{pmatrix} a_{11} & \cdots & a_{1n} \\ \vdots & & \vdots \\ a_{m1} & \cdots & a_{mn} \end{pmatrix} \quad (a_{ij} \in F, 1 \le i \le m, 1 \le j \le n)$$

is called an *$m \times n$ matrix with entries in F*. Two matrices $(a_{ij})_{m\times n}$ and $(b_{ij})_{m\times n}$ with entries in F are *equal* if $a_{ij} = b_{ij}$ for each $i = 1, \dots, m$; $j = 1, \dots, n$. (Thus, $m \times n$ matrices are merely mn-tuples, written in rectangular form.)

Notation: We denote by $M_{m\times n}(F)$ the set of all $m \times n$ matrices with entries in F. For $M_{m\times n}(F)$, we usually write more simply $M_n(F)$.

The operations we are about to introduce are tailor-made for the use of matrices in describing linear transformations.

Definition 3.14.3: Let F be a field and let m and n be positive integers.

(1) Multiplication of a matrix by a scalar. If $A = (a_{ij})_{m\times n} \in M_{m\times n}(F)$ and $c \in F$, then $cA = (ca_{ij})_{m\times n} \in M_{m\times n}(F)$.

(2) Matrix Addition. If $A = (a_{ij})_{m\times n}$, $B = (b_{ij})_{m\times n}$ in $M_{m\times n}(F)$, then $A + B = (a_{ij} + b_{ij})_{m\times n} \in M_{m\times n}(F)$.

(3) Matrix Multiplication. If $A = (a_{ij})_{m\times n} \in M_{m\times n}(F)$, and $B = (b_{ij})_{n\times r} \in M_{n\times r}(F)$, then $AB = C$, where

$$C = (c_{ij})_{m\times r} \in M_{m\times r}(F), \quad \text{with} \quad c_{ij} = \sum_{h=1}^{n} a_{ih}b_{hj}$$

for each $i = 1, \dots, m$; $j = 1, \dots, r$.

(4) Zero matrices, identity matrices, and inverse matrices. We denote by $0_{m\times n}$ the matrix in $M_{m\times n}(F)$ all of whose entries are zero. If $A = (a_{ij})_{m\times n}$, we write $-A = (-a_{ij})_{m\times n}$. The matrix $I_n \in M_n(F)$ given by $I_n = (\delta_{ij})_{n\times n}$, where $\delta_{ij} = 1$ for $i = j$, $\delta_{ij} = 0$ for $i \neq j$, is the $n \times n$ *identity* matrix in $M_n(F)$.

If, for $A \in M_n(F)$, there is a matrix $B \in M_n(F)$ such that $AB = I_n = BA$, then A is an *invertible matrix*. In this case, we write $B = A^{-1}$. [The associativity asserted in Theorem 3.14.3 will ensure the uniqueness of inverses. See Exercise 3.14.4(1).]

Theorem 3.14.3: Let A, B, C be matrices with entries in a field F, and let $c \in F$. In each of the following statements, we assume that the matrices referred to are of the proper size in order for the operations involved to be meaningful.

(1) $A + B = B + A$

(2) $A + 0_{m\times n} = A = 0_{m\times n} + A \quad (A \in M_{m\times n}(F))$

(3) $A + (-A) = 0_{m\times n} = -A + A \quad (A \in M_{m\times n}(F))$

(4) $A(BC) = (AB)C$

(5) $A(B + C) = AB + AC$, $(B + C)A = BA + CA$

(6) $c(AB) = (cA)B = A(cB)$

(7) $I_m A = A$, $AI_n = A \quad (A \in F_{m\times n})$.

(In general, $AB \neq BA$.)

We leave the proof as an exercise (Exercise 3.14.3).

Corollary: Let F be a field. Then: (1) $\langle M_n(F), +, \cdot\rangle$ is a ring with identity I_n; (2) $M_n(F)$ is non-commutative except for $n = 1$.

Proof: The first assertion follows immediately from Theorem 3.14.3. To prove the second assertion, we observe that, for $n \geq 2$, the $n \times n$ matrices

$$A = \begin{pmatrix} 0 & 1 & 0 & \cdots & 0 \\ 0 & 0 & 0 & \cdots & 0 \\ \vdots & \vdots & \vdots & & \\ 0 & 0 & 0 & \cdots & 0 \end{pmatrix} \quad \text{and} \quad B = \begin{pmatrix} 1 & 0 & 0 & \cdots & 0 \\ 0 & 0 & 0 & \cdots & 0 \\ \vdots & \vdots & \vdots & & \\ 0 & 0 & 0 & \cdots & 0 \end{pmatrix}$$

do not commute. On the other hand, for $n = 1$, the 1×1 matrices $(a)_{1 \times 1}$, $a \in F$, obviously form a field, isomorphic to F. ■

Associated with every $n \times n$ matrix A over a field F is a field element known as the determinant of A. After giving one of the less sophisticated definitions of *determinant*, we collect in a single theorem some of the most useful properties of determinants, which we state without proof.

Definition 3.14.4: Let F be a field, $n \in \mathbb{Z}^+$, and let $A \in M_n(F)$. If $A = (a_{ij})_{m \times n}$, then the *determinant* of A is given by

$$\det A = \sum_{\sigma \in S_n} (-1)^{s_\sigma} a_{1\sigma 1} a_{2\sigma 2} \cdots a_{n\sigma n},$$

where $s_\sigma = 0$ if σ is an even permutation, and $s_\sigma = 1$ if σ is an odd permutation.

For each i, j $(1 \leq i, j \leq n)$, the determinant M_{ij} of the $(n-1) \times (n-1)$ submatrix of A obtained by deleting the i-th row and the j-th column of A is the *minor* of a_{ij}, and the field element $A_{ij} = (-1)^{i+j} M_{ij}$ is the *cofactor* of a_{ij}.

Example:

If $A = (a_{ij})_{3 \times 3}$, then

$$\det A = a_{11}a_{22}a_{33} + a_{12}a_{23}a_{31} + a_{13}a_{21}a_{32} - a_{13}a_{12}a_{31} - a_{11}a_{23}a_{32} - a_{12}a_{21}a_{33}.$$

Remark: The expansion of $\det(a_{ij})_{n \times n}$ has $n!$ terms.

Theorem 3.14.4: Let F be a field, $n \in \mathbb{Z}^+$.

(1) *Cofactor Expansion.* If $A = (a_{ij})_{n \times n} \in M_n(F)$, then $\det A = \sum_{i=1}^{n} a_{ij} A_{ij}$ for each $j = 1, \ldots, n$, and $\det A = \sum_{j=1}^{n} a_{ij} A_{ij}$ for each $i = 1, \ldots, n$.

(2) *Multiplicativity.* If $A = (a_{ij})_{n \times n}$, $B = (b_{ij})_{n \times n} \in M_n(F)$, then $\det AB = \det A \det B$.

(3) *Elementary Operations.* Let $A, B \in M_n(F)$.

(a) If B is obtained from A by interchanging two rows, or two columns, of A, then $\det B = -\det A$.

(b) If B is obtained from A by multiplying a single row, or a single column, by $k \neq 0$ in F, then $\det B = k \det A$.

(c) If B is obtained from A by adding a scalar multiple of row i to row j, or by adding a scalar multiple of column i to column j, where $i \neq j$, then $\det B = \det A$.

Exercises 3.14

True or False

1. A determinant is a matrix.
2. For each positive integer $n > 1$, the expansion (by Definition 3.14.4.) of $\det(a_{ij})_{n \times n}$ has $2n$ terms.
3. If $A, B \in M_n(F)$, then $\det(A + B) = \det A + \det B$.
4. If $A = (a_{ij})_{n \times n} \in M_n(F)$ and $k \in F$, then $\det(kA) = k^n \det A$.
5. The mapping $\phi : M_n(F) \to F$ defined by $\phi(A) = \det A \quad (A \in M_n(F))$ is a ring homomorphism.

3.14.1. If $A = (a_{ij})_{m \times n} \in M_{m \times n}(F)$ (F a field), define $A^T = (a_{ji})_{n \times m} \in M_{n \times m}(F)$ to be the *transpose* of A.

(1) Let A, B be matrices for which the product AB is defined. Prove that $(AB)^T = B^T A^T$.

(2) Prove that, for any matrix A, $(A^T)^T = A$.

(3) Prove that, for any matrix $A \in M_n(F)$, $\det A^T = \det A$.

3.14.2. If A is an $n \times n$ matrix with two equal rows, or with two equal columns, prove that $\det A = 0$.

3.14.3. Prove Theorem 3.14.3.

3.14.4. *(1)* If $A \in M_n(F)$ is invertible, prove that A has exactly one inverse.

(2) If $A, B \in M_n(F)$ are invertible matrices, prove that $(AB)^{-1} = B^{-1}A^{-1}$.

3.14.5. If $A \in M_n(F)$ is invertible, prove that $\det A \neq 0$. (The converse is also true. See Exercise 3.14.7.)

3.14.6. Let $A = (a_{ij})_{n \times n} \in M_n(F)$, F a field. Prove:

(1) for $i \neq h \quad (1 \leq i, h \leq n)$, $\displaystyle\sum_{j=1}^{n} a_{ij}A_{hj} = 0$;

(2) for $j \neq k \quad (1 \leq j, k \leq n)$, $\displaystyle\sum_{i=1}^{n} a_{ij}A_{ik} = 0$.

(Capital letters denote cofactors.)

3.14.7. For $A = (a_{ij})_{n \times n} \in M_n(F)$, F a field, define the *adjoint matrix of A* by:

$$\operatorname{Adj} A = (A_{ij})^T_{n \times n},$$

where A_{ij} is the cofactor of a_{ij} for each $i, j \quad (1 \leq i, j \leq n)$. Prove that

$$A \cdot \operatorname{Adj} A = \begin{pmatrix} \det A & 0 & \cdots & 0 \\ 0 & \det A & \cdots & 0 \\ \vdots & & \ddots & \\ 0 & 0 & & \det A \end{pmatrix} = \operatorname{Adj} A \cdot A.$$

Conclude that, if $\det A \neq 0$, then A is invertible, with $A^{-1} = \frac{1}{\det A} \cdot \operatorname{Adj} A$.

3.14.8. Use Exercise 3.14.7 to find a formula for the inverse of any invertible 2×2 matrix with entries in a field F.

3.14.9. Prove Cramer's Rule: let F be a field, and let

$$\begin{aligned} a_{11}x_1 + \cdots + a_{1n}x_n &= b_1 \\ &\vdots \\ a_{n1}x_1 + \cdots + a_{nn}x_n &= b_n \end{aligned} \tag{1}$$

be a system of n equations in n unknowns ($a_{ij} \in F$, $b_i \in F$, $1 \leq i, j \leq n$). Let $A = (a_{ij})_{n \times n}$. For each $j = 1, \ldots, n$, let A_j be the matrix obtained from A by replacing the j-th column by $\begin{pmatrix} b_1 \\ \vdots \\ b_n \end{pmatrix}$. Prove that the unique solution of (1) is given by

$$x_1 = \frac{\det A_1}{\det A}, x_2 = \frac{\det A_2}{\det A}, \cdots, x_n = \frac{\det A_n}{\det A}.$$

(Hint: Expand $\det A_j$ ($j = 1, \ldots, n$) by cofactors of the j-th column.)

3.14.10. Prove Theorem 3.14.4.

3.15

Representation of Linear Transformations by Matrices

Let F be a field. Then the elements of F^n and of $M_{n \times 1}(F)$ are the n-tuples (written as columns) over F. Thus, we may simply regard F^n and $M_{n \times 1}(F)$ as equal sets. This makes it possible to form the matrix product of $A \in M_{m \times n}(F)$ and $x \in F^n$, to obtain $Ax \in F^m$.

We shall refer to $(e_1, \ldots, e_n)$, where $e_i = \begin{pmatrix} 0 \\ \vdots \\ 0 \\ 1 \\ 0 \\ \vdots \\ 0 \end{pmatrix}$ (1 in i-th position) as the *standard ordered basis*, $\mathscr{B}_0$, for F^n.

Theorem 3.15.1: Let F be a field, m, n positive integers, and let $A \in M_{m \times n}(F)$. Define

$$\phi : F_n \to F_m$$

by

$$\phi(x) = Ax$$

for each $x \in F^n$. Then ϕ is a linear transformation. The columns of A are the images $\phi(e_i)$ $(i = 1, \ldots, n)$ of the standard basis vectors $e_1, \ldots, e_n$ of F^n.

Proof: First note that since $A \in F_{m \times n}$ and $x \in F^n = M_{n \times 1}(F)$, the product Ax is defined for each $x \in F^n$, and $Ax \in M_{m \times 1}(F) = F^m$. For $x, y \in F^n$, $c \in F$, we have

$$\phi(x + y) = A(x + y) = Ax + Ay = \phi(x) + \phi(y)$$

and

$$\phi(cx) = A(cx) = c(Ax) = c\phi(x),$$

by Theorem 3.14.3.

For each $i = 1, \ldots, n$, $\phi(e_i) = Ae_i$ is the i-th column of A:

$$\begin{pmatrix} a_{11} & \cdots & a_{1i} & \cdots & a_{1n} \\ a_{21} & \cdots & a_{2i} & \cdots & a_{2n} \\ \vdots & & \vdots & & \vdots \\ a_{m1} & \cdots & a_{mi} & \cdots & a_{mn} \end{pmatrix} \begin{pmatrix} 0 \\ \vdots \\ 0 \\ 1 \\ 0 \\ \vdots \\ 0 \end{pmatrix} (ith) = \begin{pmatrix} a_{1i} \\ a_{2i} \\ \vdots \\ a_{mi} \end{pmatrix}.$$

Note that any vector $x = \begin{pmatrix} x_1 \\ \vdots \\ x_n \end{pmatrix}$ in F^n is equal to $x_1 e_1 + \cdots + x_n e_n$, where the e_i are the standard basis vectors. Hence the coordinate vector $x_{\mathscr{B}_0}$ of x with respect to the standard basis $\mathscr{B}_0 = (e_1, \ldots, e_n)$ is simply x itself: $x_{\mathscr{B}_0} = x$. In view of this, the columns of matrix A in the preceding theorem can be regarded as $(\phi e_1)_{\mathscr{B}'_0}, \ldots, (\phi e_n)_{\mathscr{B}'_0}$, where $e_1, \ldots, e_n$ are the standard basis vectors of F^n, and $\mathscr{B}'_0 = (e'_1, \ldots, e'_m)$ is the standard basis for F^m. ∎

We bear this in mind in defining the matrix of any linear transformation with respect to a pair of ordered bases.

Definition 3.15.1: Let F be a field and let m, n be positive integers. Let V, V' be vector spaces, with $\dim V = n$, $\dim V' = m$. Let $\mathscr{B} = (v_1, \ldots, v_n)$ be an ordered basis for V, and let $\mathscr{B}' = (w_1, \ldots, w_m)$ be an ordered basis for V'. Let $\phi : V \to V'$ be a linear transformation. Then the matrix $A \in F_{m \times n}$ whose i-th column is the coordinate vector $(\phi v_i)_{\mathscr{B}'}$, for each $i = 1, \ldots, n$, is *the matrix of* ϕ *with respect to the pair of ordered bases* $(\mathscr{B}, \mathscr{B}')$.

Notations: We write $A = \phi_{\mathscr{B}, \mathscr{B}'}$. Matrix A may be written as a *partitioned matrix* (partitioned into its columns): $A = ((\phi v_1)_{\mathscr{B}'}|\cdots|(\phi v_n)_{\mathscr{B}'})$. Thus

$$\phi_{\mathscr{B}, \mathscr{B}'} = ((\phi v_1)_{\mathscr{B}'}|\cdots|(\phi v_n)_{\mathscr{B}'}).$$

Using the terminology and notation just introduced, we observe that the matrix A in Theorem 3.15.1 is $\phi_{\mathscr{B}_0, \mathscr{B}'_0} = ((\phi e_1)_{\mathscr{B}'_0}|\cdots|(\phi e_n)_{\mathscr{B}'_0})$, where ϕ is the linear transformation from F^n to F^m defined by $\phi(x) = Ax \quad (x \in F^n)$. Since $x = x_{\mathscr{B}_0}$ in F^n and $Ax = (Ax)_{\mathscr{B}_0'}$ in F^m, we can restate this relation as

$$(Ax)_{\mathscr{B}_0} = (\phi x)_{\mathscr{B}_0'} \quad (x \in F^n).$$

This formula is generalized in the following theorem.

Theorem 3.15.2: Let F be a field, m, n positive integers. Let V, V' be vector spaces, with $\dim V = n$, $\dim V' = m$. Let $\mathscr{B}$ be an ordered basis for V, $\mathscr{B}'$ an ordered basis for V'. Let $\phi : V \to V'$ be a linear transformation, and let A be the matrix of ϕ with respect to the pair of bases $(\mathscr{B}, \mathscr{B}')$. Then, for each $x \in V$,

$$Ax_{\mathscr{B}} = (\phi x)_{\mathscr{B}'}.$$

A is the *only* matrix satisfying this condition.

Proof: Let $\mathscr{B} = (v_1, \ldots, v_n)$, $\mathscr{B}' = (w_1, \ldots, w_n)$. Then $A = \phi_{\mathscr{B}, \mathscr{B}'} = ((\phi v_1)_{\mathscr{B}'}|\cdots|(\phi v_n)_{\mathscr{B}'})$. If $x \in V$, then $x = \sum_{i=1}^{n} x_i v_i \quad (x_i \in F)$, hence $x_{\mathscr{B}} = \begin{pmatrix} x_1 \\ \vdots \\ x_n \end{pmatrix}$, and

$$Ax_{\mathscr{B}} = ((\phi v_1)_{\mathscr{B}'}|\cdots|(\phi v_n)_{\mathscr{B}'})\begin{pmatrix} x_1 \\ \vdots \\ x_n \end{pmatrix} = x_1(\phi v_1)_{\mathscr{B}'} + \cdots + x_n(\phi v_n)_{\mathscr{B}'}.$$

By Theorem 3.14.2, the right-hand side is equal to $(x_1\phi v_1 + \cdots + x_n\phi v_n)_{\mathscr{B}'}$. Since ϕ is linear, we therefore have

$$Ax_{\mathscr{B}} = (\phi(x_1v_1 + \cdots + x_nv_n))_{\mathscr{B}'} = (\phi x)_{\mathscr{B}'}.$$

To prove the uniqueness of A, suppose M is another matrix such that

$$Mx_{\mathscr{B}} = (\phi x)_{\mathscr{B}'}$$

for each $x \in V$. Then $M \in F_{m \times n}$, and, in particular, $M(v_i)_{\mathscr{B}} = (\phi v_i)_{\mathscr{B}'}$ for each $i = 1, \ldots, n$. Since $v_i = 0v_1 + \cdots + 0v_{i-1} + 1v_i + 0v_{i+1} + \cdots + 0v_n$ for each $i = 1, \ldots, n$, we have $(v_i)_{\mathscr{B}} = e_i \quad (i = 1, \ldots, n)$. But then $M(v_i)_{\mathscr{B}} = Me_i$ is the i-th column of M. Thus, for each $i = 1, \ldots, n$, $(\phi v_i)_{\mathscr{B}'}$ is the i-th column of M. But then $M = A$. ■

Example 1:

Let $P_3 = \{f \in \mathbb{R}[x] | f = 0 \text{ or } \deg f < 3\}$. Define $\phi : P_3 \to P_3$ by $\phi(f) = f'$ (the derivative of f). Let $\mathscr{B} = \mathscr{B}' = \{1, x, x^2\}$. Then $A = \phi_{\mathscr{B}, \mathscr{B}'} = \phi_{\mathscr{B}, \mathscr{B}} = ((\phi(1))_{\mathscr{B}}|(\phi(x))_{\mathscr{B}}|(\phi(x^2))_{\mathscr{B}})$. Since

$$\phi(1) = 0 = 0.1 + 0x + 0x^2$$
$$\phi(x) = 1 = 1.1 + 0x + 0x^2$$
$$\phi(x^2) = 2x = 0.1 + 2x + 0x^2,$$

we have

$$A = \begin{pmatrix} 0 & 1 & 0 \\ 0 & 0 & 2 \\ 0 & 0 & 0 \end{pmatrix}.$$

Now let $g = 2 + 3x + 5x^2$. Then $g_{\mathscr{B}} = \begin{pmatrix} 2 \\ 3 \\ 5 \end{pmatrix}$, and $Ag_{\mathscr{B}} = \begin{pmatrix} 0 & 1 & 0 \\ 0 & 0 & 2 \\ 0 & 0 & 0 \end{pmatrix}\begin{pmatrix} 2 \\ 3 \\ 5 \end{pmatrix} = \begin{pmatrix} 3 \\ 10 \\ 0 \end{pmatrix} = (\phi g)_{\mathscr{B}'}$. Thus, $\phi(g) = 3.1 + 10x$, which we recognize as g'.

Example 2:

If we modify Example 1 by a change of codomain so that $\phi : P_3 \to P_2$ is defined by $\phi(f) = f' \quad (f \in P_3)$, with $\mathscr{B} = (1, x, x^2)$ and $\mathscr{B}' = (1, x)$, then $\phi_{\mathscr{B}, \mathscr{B}'} = \begin{pmatrix} 0 & 1 & 0 \\ 0 & 0 & 2 \end{pmatrix}$, and $(\phi g)_{\mathscr{B}'} = \begin{pmatrix} 0 & 1 & 0 \\ 0 & 0 & 2 \end{pmatrix}\begin{pmatrix} 2 \\ 3 \\ 5 \end{pmatrix} = \begin{pmatrix} 3 \\ 10 \end{pmatrix} = (\phi g)_{\mathscr{B}'}$.

Example 3:

Let $\mathscr{B}_0$ be the standard ordered basis for F^n and let $\mathscr{B}_0'$ be the standard ordered basis for $F^m \quad (m, n \in \mathbb{Z}^+)$. Let $\phi : F^n \to F^m$ be such that

$$\phi\begin{pmatrix} x_1 \\ \vdots \\ x_n \end{pmatrix} = \begin{pmatrix} a_{11}x_1 + \cdots + a_{1n}x_n \\ a_{21}x_1 + \cdots + a_{2n}x_n \\ \vdots \qquad\qquad \vdots \\ a_{m1}x_1 + \cdots + a_{mn}x_n \end{pmatrix} \qquad (a_{ij} \in F).$$

If $e_j \quad (1 \leq j \leq m)$ is the j-th vector in $\mathscr{B}_0$, then $\phi(e_j) = \begin{pmatrix} a_{1j} \\ \vdots \\ a_{mj} \end{pmatrix} = (\phi e_j)_{\mathscr{B}_0'}$; hence $\phi_{\mathscr{B}_0, \mathscr{B}_0'} = (a_{ij})_{m \times n}$.

In the following corollaries, we explore further the properties of the correspondence of a linear transformation and its matrix with respect to a given pair of ordered bases.

Corollary 1: Let V, V', V'' be finite-dimensional non-zero vector spaces over a field F, with ordered bases $\mathscr{B}, \mathscr{B}', \mathscr{B}''$, respectively. Let $\phi : V \to V'$ and $\psi : V' \to V''$ be linear transformations. Then

$$(\psi\phi)_{\mathscr{B}, \mathscr{B}''} = \psi_{\mathscr{B}', \mathscr{B}''} \cdot \phi_{\mathscr{B}, \mathscr{B}'}.$$

Proof: For $v \in V$, we have

$$(\psi\phi)_{\mathscr{B},\mathscr{B}''}v_{\mathscr{B}} = ((\psi\phi)v)_{\mathscr{B}''}.$$

Also,

$$(\psi_{\mathscr{B}',\mathscr{B}''} \cdot \phi_{\mathscr{B},\mathscr{B}'})v_{\mathscr{B}} = \psi_{\mathscr{B}',\mathscr{B}''}(\phi_{\mathscr{B},\mathscr{B}'}v_{\mathscr{B}}) = \psi_{\mathscr{B}',\mathscr{B}''}(\phi v)_{\mathscr{B}'} = (\psi(\phi v))_{\mathscr{B}''} = (\psi\phi)v_{\mathscr{B}''}.$$

By the uniqueness in Theorem 3.15.2, it follows that

$$(\psi\phi)_{\mathscr{B},\mathscr{B}''} = \psi_{\mathscr{B}',\mathscr{B}''} \cdot \phi_{\mathscr{B},\mathscr{B}'}. \qquad \blacksquare$$

Corollary 2: Let V be a vector space of finite dimension $n > 0$ over a field F, and let $\mathscr{B}$ be an ordered basis for V.

(1) If $\iota : V \to V$ is the identity operator on V, then $\iota_{\mathscr{B},\mathscr{B}} = I_n$.

(2) If $\phi : V \to V$ is an invertible linear operator, then $\phi_{\mathscr{B},\mathscr{B}}$ is an invertible matrix, and $(\phi^{-1})_{\mathscr{B},\mathscr{B}} = (\phi_{\mathscr{B}\mathscr{B}})^{-1}$.

Proof: *(1)* For $v \in V$,

$$\iota_{\mathscr{B},\mathscr{B}}v_{\mathscr{B}} = (\iota v)_{\mathscr{B}} = v_{\mathscr{B}} = I_n v_{\mathscr{B}}.$$

By the uniqueness in Theorem 3.15.2, it follows that $\iota_{\mathscr{B},\mathscr{B}} = I_n$.

(2) Since ϕ is invertible, there is a linear operator $\psi = \phi^{-1}$ such that $\phi\psi = \iota = \psi\phi$; hence $(\phi\psi)_{\mathscr{B},\mathscr{B}} = \iota_{\mathscr{B},\mathscr{B}} = (\psi\phi)_{\mathscr{B},\mathscr{B}}$. But then, by Corollary 1, and (1),

$$\phi_{\mathscr{B},\mathscr{B}}\psi_{\mathscr{B},\mathscr{B}} = I_n = \psi_{\mathscr{B},\mathscr{B}}\phi_{\mathscr{B},\mathscr{B}}.$$

It follows that $\phi_{\mathscr{B},\mathscr{B}}$ is an invertible matrix with inverse $\psi_{\mathscr{B},\mathscr{B}}$. Thus, $(\phi^{-1})_{\mathscr{B},\mathscr{B}} = (\phi_{\mathscr{B},\mathscr{B}})^{-1}$. ■

Corollary 3: Let V, V' be finite-dimensional non-zero vector spaces over a field F, and let $\phi : V \to V'$ and $\psi : V \to V'$ be linear transformations. Let $\mathscr{B}$ be an ordered basis for V, and let $\mathscr{B}'$ be an ordered basis for V'. Then

$$\phi_{\mathscr{B},\mathscr{B}'} + \psi_{\mathscr{B},\mathscr{B}'} = (\phi + \psi)_{\mathscr{B},\mathscr{B}'}$$

and

$$(c\phi)_{\mathscr{B},\mathscr{B}'} = c\phi_{\mathscr{B},\mathscr{B}'}.$$

($\phi + \psi : V \to V'$ is defined by $(\phi + \psi)(v) = \phi(v) + \psi(v)$ for each $v \in V$. It is easy to show that $\phi + \psi$ is linear—see Exercise 3.13.8.)

We leave the proof as an exercise (Exercise 3.15.3).

Corollary 4: Let V be a vector space of finite dimension $n > 0$ over a field F. Each ordered basis $\mathscr{B}$ of V determines an isomorphism $\phi \mapsto \phi_{\mathscr{B},\mathscr{B}}$ of the endomorphism ring $\text{End}_F V$ onto the matrix ring $M_n(F)$.

Proof: Given an ordered basis $\mathscr{B} = (v_1, \ldots, v_n)$ for V, define $\mu : \text{End}_F V \to M_n(F)$ by $\mu(\phi) = \phi_{\mathscr{B},\mathscr{B}}$ for each $\phi \in \text{End}_F V$. Then μ is well-defined since the columns

$(\phi v_i)_{\mathscr{B}}$ of $\phi_{\mathscr{B},\mathscr{B}}$ are uniquely determined for each $\phi \in \text{End}_F V$. μ is 1-1, for, if $\mu(\phi) = \phi_{\mathscr{B},\mathscr{B}}$ is the $n \times n$ zero matrix, then

$$(\phi v_i)_{\mathscr{B}} = \begin{pmatrix} 0 \\ \vdots \\ 0 \end{pmatrix} \text{ for each } i = 1, \ldots, n;$$

hence $\phi v = 0$ for each $v \in V$, and so ϕ is the 0-map, hence the 0-element of the ring $\text{End}_F V$. By Corollaries (1) and (3), μ preserves addition and multiplication. Finally, μ maps $\text{End}_F V$ *onto* $M_n(F)$. For, given matrix $A \in M_n(F)$, there is a unique linear operator ϕ such that $(\phi v_i)_{\mathscr{B}}$ is the i-th column of A, for each $i = 1, \ldots, n$ (see Exercise 3.13.3). But then $A = \phi_{\mathscr{B},\mathscr{B}} = \mu(\phi)$. Thus, μ is an isomorphism. ∎

Exercises 3.15

True or False

1. Every linear transformation can be represented by a unique matrix.
2. If $A \in M_{m \times n}(F)$, $\mathscr{B}$ is a basis for F^n, and $\mathscr{B}'$ is a basis for F^m, then there is a linear transformation $\phi : F^n \to F^m$ such that $\phi_{\mathscr{B},\mathscr{B}'} = A$.
3. If $\phi : V \to V$ is linear and $\phi_{\mathscr{B},\mathscr{B}} = \begin{pmatrix} 0 & 1 \\ 1 & 0 \end{pmatrix}$, for some basis $\mathscr{B}$ of V, then $\phi^2 = \iota_V$.
4. If $\phi : V \to V$ is linear and $\phi_{\mathscr{B},\mathscr{B}'} = \begin{pmatrix} 1 & 1 & 0 \\ 2 & 0 & 0 \\ 0 & 0 & 1 \end{pmatrix}$ with respect to some basis $B = (v_1, v_2, v_3)$ of V, then $\phi(v_1) = v_1 + 2v_2$, $\phi v_2 = v_1$ and $\phi v_3 = v_3$.
5. If $\mathscr{B}$ is an ordered basis for an n-dimensional vector space V over a field F, then every permutation on the vectors in $\mathscr{B}$ induces a linear operator on V.

3.15.1. Let $\phi : \mathbb{R}^3 \to \mathbb{R}^3$ be the linear operator such that $\phi(e_1) = 3e_1 + 2e_2 + e_3$, $\phi(e_2) = e_1 + e_2 + e_3$, $\phi(e_3) = 3e_1 - e_2 + 2e_3$, where e_1, e_2, e_3 are the standard basis vectors for $\mathbb{R}^3$. Find $\phi_{\mathscr{B}_0,\mathscr{B}_0}$, where $\mathscr{B}_0$ is the standard basis for $\mathbb{R}^3$.

3.15.2. Let V be a vector space of finite dimension n over a field F, and let $\phi : V \to V$ be a nonsingular linear operator. Prove that there is a pair of ordered bases $\mathscr{B}$ and $\mathscr{B}'$ for V such that $\phi_{\mathscr{B},\mathscr{B}'} = I_n$.

3.15.3. Prove Corollary 3 of Theorem 3.15.2.

3.15.4. Use Corollary 1 of Theorem 3.15.2 to prove that matrix multiplication is associative.

3.15.5. For F a field, n a positive integer, and V an n-dimensional vector space over F, prove that the group of units of the ring $\text{End}_V F$ is isomorphic to $GL_n(F)$ (the group of all invertible $n \times n$ matrices with entries in F).

3.15.6. Let $\mathscr{B} = (v_1, v_2, v_3, v_4, v_5)$ be an ordered basis for a vector space V. Suppose that a linear operator $\phi : V \to V$ permutes the vectors v_i according to the permutation $(v_1 v_4 v_3)(v_2 v_5)$ (in cycle notation). Find the matrix $\phi_{\mathscr{B},\mathscr{B}}$.

3.15.7. Let V be an n-dimensional vector space over a field F $(n > 0)$ and let ϕ be an idempotent linear operator on V, of rank r. (Recall that ϕ is idempotent if $\phi^2 = \phi$.) Prove that there is an ordered basis $\mathscr{B}$ for V such that $\phi_{\mathscr{B},\mathscr{B}} = (a_{ij})_{n \times n}$, where $a_{ii} = 1$ for $i = 1, \ldots, r$, and all other entries of $\phi_{\mathscr{B},\mathscr{B}}$ are equal to 0. (Hint: See Exercise 3.13.12.)

3.15.8. Let V be an n-dimensional vector space over a field F $(n > 0)$ and let ϕ be a linear operator on V such that $\phi^n = 0$ while $\phi^{n-1} \neq 0$. Prove that there is an ordered basis $\mathscr{B}$ for V such that $\phi_{\mathscr{B},\mathscr{B}} = (a_{ij})_{n \times n}$, where $a_{i\,i-1} = 1$ for $i = 2, \ldots, n$, and all other entries of $\phi_{\mathscr{B},\mathscr{B}}$ are equal to 0.

3.16

Non-Singular Matrices, Change of Basis, and Similarity

If A is an $m \times n$ matrix over a field F, then $\phi : F^n \to F^m$ defined by $\phi(x) = Ax$ $(x \in F^n)$ is a linear operator with nullspace

$$\operatorname{Ker} \phi = \{x \in F^n | Ax = 0\}$$

and range

$$\operatorname{Im} \phi = \{Ax | x \in F^n\}.$$

If A has columns $\gamma_1, \ldots, \gamma_n \in F^m$, and $x = \begin{pmatrix} x_1 \\ \vdots \\ x_n \end{pmatrix} \in F^n$, then

$$Ax = (\gamma_1 | \ldots | \gamma_n) \begin{pmatrix} x_1 \\ \vdots \\ x_n \end{pmatrix} = x_1\gamma_1 + \cdots + x_n\gamma_n.$$

Thus, $\operatorname{Im} \phi$ is the subspace of F^m spanned by the columns of A.

Definition 3.16.1: Let F be a field, $A \in F_{m \times n}$ $(m, n \in \mathbb{Z}^+)$. Then the *nullspace of A* is the subspace

$$N(A) = \{x \in F^n | Ax = 0\}$$

of F^n, and the *column space of A* is the subspace

$$R(A) = \{Ax | x \in F^n\}$$

of F_m.

The dimension of $N(A)$ is the *nullity*, $\nu(A)$, of A and the dimension of $R(A)$ is the *rank*, $\rho(A)$, *of A*.

Corollary: If A is an $m \times n$ matrix $(m, n \in \mathbb{Z}^+)$ with entries in a field F, then

$$n = \rho(A) + \nu(A).$$

Proof: If $\phi : F^n \to F^m$ is the linear operator defined by $\phi(x) = Ax \quad (x \in F^n)$, then $\rho(A) = \rho(\phi)$ and $v(A) = v(\phi)$. By Theorem 3.13.1, it follows immediately that

$$n = v(A) + \rho(A). \qquad \blacksquare$$

Definition 3.16.2: Let A be an $n \times n$ matrix over a field $F \quad (n \in \mathbb{Z}^+)$. Then A is *singular* if $v(A) \neq 0$, and A is *non-singular* if $vA = 0$.

Theorem 3.16.1: Let A be an $n \times n$ matrix over a field $F \quad (n \in \mathbb{Z}^+)$. Then the following statements are equivalent:

(1) A is non-singular.
(2) There is an $n \times n$ matrix B such that $BA = I_n$.
(3) A is invertible.
(4) There is an $n \times n$ matrix B such that $AB = I_n$.
(5) The columns of A are linearly independent.

Proof: $1 \Rightarrow 2$. Suppose A is non-singular. Let $\phi : F^n \to F^n$ be the linear operator such that $\phi(x) = Ax \quad (x \in F^n)$. Then $v(A) = v(\phi) = 0$. By Theorem 3.13.2, there is a linear operator $\psi : F^n \to F^n$ such that $\psi\phi = \iota_{F^n}$. If $\mathscr{B}_0$ is the standard basis for F^n, then $A = \phi_{\mathscr{B}_0, \mathscr{B}_0}$. Let $B = \psi_{\mathscr{B}_0, \mathscr{B}_0}$. By Corollary 4 of Theorem 3.15.2., it follows that $BA = I_n$.

$2 \Rightarrow 3$. Suppose B is an $n \times n$ matrix such that $BA = I_n$. If $\phi : F^n \to F^n$ is defined by $\phi(x) = Ax \quad (x \in F^n)$, and $\psi : F^n \to F^n$ is defined by $\psi x = Bx \quad (x \in F^n)$, then $\phi_{\mathscr{B}_0, \mathscr{B}_0} = A$ and $\psi_{\mathscr{B}_0, \mathscr{B}_0} = B$, where $\mathscr{B}_0$ is the standard basis for F^n. By Corollary 4 of Theorem 3.15.2, we conclude that $\psi\phi = \iota_V$. But then, by Theorem 3.13.2, ϕ is invertible. Hence, by Corollary 4 of Theorem 3.15.2, A is invertible.

$3 \Rightarrow 4$, of course.

$4 \Rightarrow 5$. Suppose there is an $n \times n$ matrix B such that $AB = I_n$. *If* ϕ and ψ are the linear operators defined by $\phi(x) = Ax$, $\psi(x) = Bx \quad (x \in F^n)$, we have $A = \phi_{\mathscr{B}_0, \mathscr{B}_0}$, $B = \psi_{\mathscr{B}_0, \mathscr{B}_0}$, where $\mathscr{B}_0$ is the standard basis for F^n. By Corollary 4 of Theorem 3.15.2, we conclude that $\phi\psi = \iota_V$. But then, by Theorem 3.13.2, ϕ is non-singular. By Definition 3.16.1, $v(A) = v(\phi) = 0$. Thus A is non-singular. Let $\gamma_1, \ldots, \gamma_n \in F^n$ be the columns of A. Suppose $x_1\gamma_1 + \cdots + x_n\gamma_n = 0$ $(x_i \in F, i = 1, \ldots, n)$. Write $x = \begin{pmatrix} x_1 \\ \vdots \\ x_n \end{pmatrix}$. Then $Ax = (\gamma_1 | \ldots | \gamma_n) \begin{pmatrix} x_1 \\ \vdots \\ x_n \end{pmatrix} = \begin{pmatrix} 0 \\ \vdots \\ 0 \end{pmatrix}$. Since $v(A) = 0$, this implies that $x = 0$ in F^n, hence $x_1 = x_2 = \cdots = x_n = 0$. But then $\gamma_1, \ldots, \gamma_n$ are linearly independent.

$5 \Rightarrow 1$. Suppose that the columns $\gamma_1, \ldots, \gamma_n$ of A are linearly independent. If $x = \begin{pmatrix} x_1 \\ \vdots \\ x_n \end{pmatrix} \in N(A)$, then $Ax = 0$; hence

$$x_1\gamma_1 + \cdots + x_n\gamma_n = (\gamma_1 | \cdots | \gamma_n)\begin{pmatrix} x_1 \\ \vdots \\ x_n \end{pmatrix} = Ax = 0.$$

Since $\gamma_1, \ldots, \gamma_n$ are linearly independent, it follows that $x_1 = x_2 = \cdots = x_n = 0$; hence $x = 0$ in F^n. But then $N(A) = \{0\}$, and so $\nu(A) = 0$. By Definition 3.16.2, A is non-singular. ■

Now let $\mathscr{B}_1, \mathscr{B}_2$ be ordered bases for a finite-dimensional vector space V over a field F, and let $\iota : V \to V$ be the identity operator. Then the matrix $\iota_{\mathscr{B}_1, \mathscr{B}_2}$ serves a very useful purpose: it allows us to find the coordinates of a given vector with respect to $\mathscr{B}_2$ from its coordinates with respect to $\mathscr{B}_1$.

For, if $v \in V$, then

$$\iota_{\mathscr{B}_1, \mathscr{B}_2} v_{\mathscr{B}_1} = (\iota v)_{\mathscr{B}_2} = v_{\mathscr{B}_2},$$

by Theorem 3.15.2.

Definition 3.16.3: Let V be a vector space of finite dimension $n > 0$ over a field F and let $\mathscr{B}_1 = (v_1, \ldots, v_n)$ and $\mathscr{B}_2 = (w_1, \ldots, w_n)$ be ordered bases for V. Then the matrix

$$\iota_{\mathscr{B}_1, \mathscr{B}_2} = ((v_1)_{\mathscr{B}_2} | \cdots | (v_n)_{\mathscr{B}_2})$$

is the *transition matrix from* $\mathscr{B}_1$ *to* $\mathscr{B}_2$.

Note: By Theorem 3.15.2, if M is the transition matrix from $\mathscr{B}_1$ to $\mathscr{B}_2$, then M is the *only* matrix with the property

$$Mv_{\mathscr{B}_1} = v_{\mathscr{B}_2}$$

for each $v \in V$.

Corollary: Let V be a vector space of dimension $n > 0$ over a field F, and let $\mathscr{B}_1, \mathscr{B}_2$ be ordered bases for V. If M is the transition matrix from $\mathscr{B}_1$ to $\mathscr{B}_2$, then M is invertible, and the transition matrix from $\mathscr{B}_2$ to $\mathscr{B}_1$ is M^{-1}.

Proof: By Corollary 1 of Theorem 3.15.2 we have

$$\iota_{\mathscr{B}_1, \mathscr{B}_2} \cdot \iota_{\mathscr{B}_2, \mathscr{B}_1} = \iota_{\mathscr{B}_1, \mathscr{B}_1} = I_n, \quad \text{and} \quad \iota_{\mathscr{B}_2, \mathscr{B}_1} \cdot \iota_{\mathscr{B}_1, \mathscr{B}_2} = \iota_{\mathscr{B}_2, \mathscr{B}_2} = I_n.$$

Thus, $\iota_{\mathscr{B}_2, \mathscr{B}_1} = (\iota_{\mathscr{B}_1, \mathscr{B}_2})^{-1} = M^{-1}$. ■

Now suppose ϕ is a linear transformation from a vector space V to a vector space V'. Relative to a given pair of ordered bases $\mathscr{B}_1, \mathscr{B}'_1$ for V and V', respectively, ϕ has matrix $\phi_{\mathscr{B}_1, \mathscr{B}'_1}$. If $\mathscr{B}_2, \mathscr{B}'_2$ is another pair of bases for V and V', respectively, how is $\phi_{\mathscr{B}_2, \mathscr{B}'_2}$ related to $\phi_{\mathscr{B}_1, \mathscr{B}'_1}$?

Theorem 3.16.2: Let V be a vector space of finite dimension $n > 0$. Let $\mathscr{B}_1, \mathscr{B}_2$ be ordered bases for V, and let $\mathscr{B}'_1, \mathscr{B}'_2$ be ordered bases for V'. Let P be the transition matrix from $\mathscr{B}_2$ to $\mathscr{B}_1$ and let Q be the transition matrix from $\mathscr{B}'_2$ to $\mathscr{B}'_1$. Let $\phi : V \to V'$ be a linear transformation, with $A = \phi_{\mathscr{B}_1, \mathscr{B}'_1}$ and $B = \phi_{\mathscr{B}_2, \mathscr{B}'_2}$. Then

$$B = Q^{-1}AP.$$

Proof: $(Q^{-1}AP)v_{\mathscr{B}_2} = (Q^{-1}A)Pv_{\mathscr{B}_2} = (Q^{-1}A)v_{\mathscr{B}_1} = Q^{-1}(Av_{\mathscr{B}_1}) = Q^{-1}(\phi v)_{\mathscr{B}'_1} = (\phi v)_{\mathscr{B}'_2}$. By Theorem 3.15.2 (uniqueness), it follows that $Q^{-1}AP = \phi_{\mathscr{B}_2, \mathscr{B}'_2} = B$. ■

In the important special case where $V = V'$, $\mathscr{B}_1 = \mathscr{B}'_1$ and $\mathscr{B}_2 = \mathscr{B}'_2$, we have the following important result.

Corollary 1: Let V be a vector space of finite dimension $n > 0$ over a field F. Let $\mathscr{B}_1, \mathscr{B}_2$ be ordered bases for V, and let P be the transition matrix from $\mathscr{B}_2$ to $\mathscr{B}_1$. Let ϕ be a linear operator on V, with $A = \phi_{\mathscr{B}_1, \mathscr{B}_1}$ and $B = \phi_{\mathscr{B}_2, \mathscr{B}_2}$. Then

$$B = P^{-1}AP.$$

Proof: The matrix $Q = \iota_{\mathscr{B}'_2, \mathscr{B}'_1}$ of Theorem 3.16.2 here is equal to $P = \iota_{\mathscr{B}_2, \mathscr{B}_1}$. ■

Definition 3.16.4: Let A, B be $n \times n$ matrices over a field F $(n \in \mathbb{Z}^+)$. Then B is *similar to A over F* if there is an $n \times n$ invertible matrix $P \in M_n(F)$ such that $B = P^{-1}AP$.

It is easy to show that similarity of matrices is an equivalence relation (see Exercise 3.16.8).

Corollary 2: Let V be a vector space of finite dimension $n > 0$ over a field F and let A, B be $n \times n$ matrices with entries in F. If A represents some linear operator ϕ relative to an ordered basis $\mathscr{B}_1$ of V, then B represents the same linear operator ϕ relative to an ordered basis $\mathscr{B}_2$ of V if and only if B is similar to A over F.

Proof: By Corollary 1, if $A = \phi_{\mathscr{B}_1, \mathscr{B}_1}$, $B = \phi_{\mathscr{B}_2, \mathscr{B}_2}$, and $P = \iota_{\mathscr{B}_2, \mathscr{B}_1}$, then $B = P^{-1}AP$; hence B is similar to A over F.

Conversely, suppose $A = \phi_{\mathscr{B}_1, \mathscr{B}_1}$ and $B = P^{-1}AP$ for some non-singular matrix P in $M_n(F)$. If $\mathscr{B}_1 = (v_1, \ldots, v_n)$, define $w_1, \ldots, w_n$ by $w_j = p_{1j}v_1 + \cdots + p_{nj}v_n$ for each $j = 1, \ldots, n$, where $(p_{ij})_{n \times n} = P$, hence $\begin{pmatrix} p_{1j} \\ \vdots \\ p_{nj} \end{pmatrix}$ is the j-th column of P for each $j = 1, \ldots, n$. But then $P = ((w_1)_{\mathscr{B}_1} | \cdots | (w_n)_{\mathscr{B}_1})$. Since P is non-singular, the columns of P are linearly independent over F (by Theorem 3.16.1); hence the vectors $w_1, \ldots, w_n$ are linearly independent over F (by Theorem 3.14.2). But then, by Theorem 3.12.4(5), $\{w_1, \ldots, w_n\}$ is a basis for V. Let $\mathscr{B}_2 = (w_1, \ldots, w_n)$. Then $P = ((w_1)_{\mathscr{B}_1} | \cdots | (w_n)_{\mathscr{B}_1}) = \iota_{\mathscr{B}_2, \mathscr{B}_1}$; hence

$$B = P^{-1}AP = \phi_{\mathscr{B}_2, \mathscr{B}_2}. \qquad ■$$

Example:

In $\mathbb{R}^2$, let $\mathscr{B}_0 = (e_1, e_2)$ (the standard ordered basis), and let $\mathscr{B} = \left(\begin{pmatrix} 1 \\ 2 \end{pmatrix}, \begin{pmatrix} 3 \\ 1 \end{pmatrix}\right)$. (The two vectors in $\mathscr{B}$ are clearly linearly independent, hence form a basis for $\mathbb{R}^2$.) Let $\phi : \mathbb{R}^2 \to \mathbb{R}^2$ be the linear operator given by $\phi(e_1) = 3e_1 + 2e_2$,

$\phi(e_2) = e_1 - 2e_2$. Then $A = \phi_{\mathscr{B}_0, \mathscr{B}_0} = \begin{pmatrix} 3 & 1 \\ 2 & -2 \end{pmatrix}$. The transition matrix $P = \iota_{\mathscr{B}, \mathscr{B}_0} = \begin{pmatrix} 1 & 3 \\ 2 & 1 \end{pmatrix}$, with inverse $\begin{pmatrix} -1/5 & 3/5 \\ 2/5 & -1/5 \end{pmatrix} = P^{-1}$. Hence $B = \phi_{\mathscr{B}, \mathscr{B}} =$ $P^{-1}AP = \begin{pmatrix} -1/5 & 3/5 \\ 2/5 & -1/5 \end{pmatrix}\begin{pmatrix} 3 & 1 \\ 2 & -2 \end{pmatrix}\begin{pmatrix} 1 & 3 \\ 2 & 1 \end{pmatrix} = \begin{pmatrix} -1/5 & 3/5 \\ 2/5 & -1/5 \end{pmatrix}\begin{pmatrix} 5 & 10 \\ -2 & 4 \end{pmatrix} =$ $\begin{pmatrix} -11/5 & 2/5 \\ 12/5 & 16/5 \end{pmatrix}$.

Check: Write $v_1 = \begin{pmatrix} 1 \\ 2 \end{pmatrix}, v_2 = \begin{pmatrix} 3 \\ 1 \end{pmatrix}$. Then $v_1 = e_1 + 2e_2, v_2 = 3e_1 + e_2$; hence

$$e_1 = \frac{\begin{vmatrix} v_1 & 2 \\ v_2 & 1 \end{vmatrix}}{\begin{vmatrix} 1 & 2 \\ 3 & 1 \end{vmatrix}} = \frac{v_1 - 2v_2}{-5}, \quad e_2 = \frac{\begin{vmatrix} 1 & v_1 \\ 3 & v_2 \end{vmatrix}}{\begin{vmatrix} 1 & 2 \\ 3 & 1 \end{vmatrix}} = \frac{-3v_1 + v_2}{-5}.$$

Thus,

$$\begin{aligned} \phi v_1 &= \phi e_1 + 2\phi e_2 = 3e_1 + 2e_2 + 2e_1 - 4e_2 = 5e_1 - 2e_2 \\ &= -v_1 + 2v_2 + \frac{2}{5}(-3v_1 + v_2) \\ &= -\frac{11}{5}v_1 + \frac{12}{5}v_2 \\ \phi v_2 &= 3\phi e_1 + \phi e_2 = 9e_1 + 6e_2 + e_1 - 2e_2 = 10e_1 + 4e_2 \\ &= -2v_1 + 4v_2 + \frac{12}{5}v_1 - \frac{4}{5}v_2 \\ &= \frac{2}{5}v_1 + \frac{16}{5}v_2. \end{aligned}$$

Hence $B = \phi_{\mathscr{B}, \mathscr{B}} = \begin{pmatrix} -\frac{11}{5} & \frac{2}{5} \\ \frac{12}{5} & \frac{16}{5} \end{pmatrix}$.

Exercises 3.16

True or False

1. The transition matrix from one basis $\mathscr{B}$ of a vector space to another basis $\mathscr{B}'$ is non-singular.
2. Every non-singular matrix in $M_n(F)$ (F a field, $n \in \mathbb{Z}^+$) is the transition matrix from a given basis $\mathscr{B}$ of F^n to some other basis $\mathscr{B}'$ of F^n.

3. If two matrices in $M_3(\mathbb{R})$ are similar over $\mathbb{C}$, then they are similar over $\mathbb{R}$.
4. The matrices $\begin{pmatrix} 1 & 1 \\ 0 & 0 \end{pmatrix}$ and $\begin{pmatrix} 1 & 0 \\ 0 & 0 \end{pmatrix}$ are similar over $\mathbb{R}$.
5. If V is a finite-dimensional vector space over a field F, $\mathscr{B}$ a basis for V, and ϕ a linear operator on V, then $\phi_{\mathscr{B},\mathscr{B}'}$ is similar to $\phi_{\mathscr{B},\mathscr{B}}$ for each basis $\mathscr{B}'$ of V.

3.16.1. Let $P_3 = \{f \in \mathbb{R}[x] \mid f = 0 \text{ or } \deg f < 3\}$. Let $\mathscr{B}_1 = \{x, 2x^2 + 1, 3\}$ and let $\mathscr{B}_2 = \{1, x, x^2\}$. Let $\phi : P_3 \to P_3$ be a linear operator such that $\phi_{\mathscr{B}_1,\mathscr{B}_1} = \begin{pmatrix} 1 & 0 & 0 \\ 2 & 1 & 0 \\ 1 & 2 & 3 \end{pmatrix}$. Find $\phi_{\mathscr{B}_2,\mathscr{B}_2}$.

3.16.2. Let F be a field, and let A be a non-singular matrix in $M_n(F)$. Prove that there are non-singular matrices K, L in $M_n(F)$ such that $KAL = I_n$.

3.16.3. Let F be a field and let m, n be positive integers. Let $\phi : F^n \to F^m$ be a linear operator. Prove that there is a basis $\mathscr{B}$ for F^n and a basis $\mathscr{B}'$ for F^m such that $\phi_{\mathscr{B},\mathscr{B}'} = (e_1 | \cdots | e_r | 0 | \cdots | 0)$ where $r = \rho(\phi)$, and the e_i are the standard basis vectors in F^m. (Hint: start with a basis for the nullspace of ϕ.)

3.16.4. From Exercise 3.16.3, conclude that, for each $m \times n$ matrix A of rank r over a field F, there are non-singular matrices $K \in M_m(F)$ and L in $M_n(F)$ such that $KAL = (e_1 | \cdots | e_r | 0 | \cdots | 0)$.

3.16.5. Use Exercise 3.16.4 to prove that, for each $A \in M_n(F)$ (F a field, $n \in \mathbb{Z}^+$), there are non-singular matrices $K, L \in M_n(F)$ such that $KAL = \operatorname{diag}(1, \ldots, 1, 0, \ldots, 0)$, where $r = \rho(A)$. Note that $(KAL)^2 = KAL$, i.e., KAL is idempotent.

3.16.6. Let F be a field, n a positive integer. Use Exercise 3.16.5 to prove that, for each $A \in M_n(F)$, there is a unit X in $M_n(F)$ such that $AXA = A$. (This makes $M_n(F)$ a special kind of *von Neumann regular ring*, sometimes referred to as *unit-regular*.)

3.16.7. Let F be a field, $n \in \mathbb{Z}^+$, and let $A, B \in M_n(F)$. Define B to be *equivalent* to A over F if there are non-singular matrices $K, L \in M_n(F)$ such that $B = KAL$.

(1) Prove that equivalence of matrices is an equivalence relation.

(2) Use Exercise 3.16.5 to prove: two matrices in $M_n(F)$ are equivalent if and only if they have the same rank.

3.16.8. Prove that similarity of matrices (see Definition 3.16.4) is an equivalence relation. Give an example of two equivalent matrices (see Exercise 3.16.7) that are not similar. (Note that similar matrices are equivalent.)

3.16.9. Let $A = (a_{ij})$ be an $m \times n$ matrix over a field F. For $x = \begin{pmatrix} x_1 \\ \vdots \\ x_n \end{pmatrix} \in F^n$, $b = \begin{pmatrix} b_1 \\ \vdots \\ b_m \end{pmatrix} \in F^m$, the matrix equation

$$Ax = b \tag{1}$$

represents the linear system

$$\begin{aligned} a_{11}x_1 + \cdots + a_{1n}x_n &= b_1 \\ &\vdots \\ a_{m1}x_1 + \cdots + a_{mn}x_n &= b_m. \end{aligned} \tag{2}$$

Prove:

(a) For $b = 0$, the solution set of the (homogeneous) linear system $Ax = 0$ is the nullspace, $N(A)$, of A. (Hence the solutions of (2) depend on $\nu(A) = n - \rho(A)$ parameters.)

(b) The linear system (2), represented by $Ax = b$ (b not necessarily equal to 0 in F^m), is consistent (i.e., *has* solutions) if and only if $b \in R(A)$ (the column space of A).

(c) If the linear system (2) is consistent, and $x_0 \in F^n$ is one solution of (2), then the set of all solutions of (2) is the additive coset $N(A) + x_0 \subset F^n$. (Hence the solutions of (2) depend on $\nu(A) = n - \rho(A)$ parameters.)

3.16.10. Given an $m \times n$ linear system

$$Ax = b,$$

and a non-singular matrix $K \in M_m(F)$, prove that the linear system

$$KAx = Kb$$

has the same solution set as the linear system $Ax = b$.

3.16.11. Let F be a field, $n \in \mathbb{Z}^+$. Prove that an $n \times n$ matrix A over F is either a unit or a zero divisor in the ring $M_n(F)$. Conclude: if V is an n-dimensional vector space over F, then a linear operator $\phi : V \to V$ is either a unit or a zero divisor in the ring $\mathrm{End}_F V$.

3.16.12. Let F be a field and let $B \in M_n(F)$. Prove that B is singular if and only if $\det B = 0$. (Equivalently, B is non-singular if and only if $\det B \neq 0$.)

3.16.13. Let F be a field and let $E = E^2$ in $M_n(F)$ be an idempotent matrix of rank r. Prove that E is similar to the matrix $(a_{ij})_{n \times n}$ with $a_{ii} = 1$ for $i = 1, \ldots, r$, and all other entries equal to 0. Conclude that all idempotent matrices in $M_n(F)$ of the same rank are similar. (See Exercise 3.15.7.)

3.16.14. Let F be a field and let $T \in M_n(F)$ be a matrix such that $T^n = 0$ while $T^{n-1} \neq 0$. Prove that T is similar to the matrix $A = (a_{ij})_{n \times n}$ where $a_{i\,i-1} = 1$ for $i = 2, \ldots, n$, while all other entries of A are equal to 0. (See Exercise 3.15.8.)

3.17

Eigenvalues and Diagonalization

As the last example in the preceding section illustrates, the choice of a basis is crucial in finding a *simple* matrix for a linear operator. We now investigate, in particular, under what conditions a given linear operator can be represented by a diagonal matrix, i.e., by a matrix all of whose off-diagonal entries are equal to 0. In view of Theorem 3.16.2, Corollary 2, this is equivalent to determining under what conditions a given $n \times n$ matrix is similar to a diagonal matrix.

Notation: We write $\operatorname{diag}(\lambda_1, \ldots, \lambda_n) = \begin{pmatrix} \lambda_1 & 0 & \ldots & 0 \\ 0 & \lambda_2 & \ldots & 0 \\ 0 & 0 & \ldots & \lambda_n \end{pmatrix}$.

Lemma: Let V be a vector space of finite dimension $n > 0$ over a field F, and let $\phi : V \to V$ be a linear operator. Let $\mathscr{B} = (v_1, \ldots, v_n)$ be an ordered basis for V. Then $\phi_{\mathscr{B},\mathscr{B}}$ is diagonal if and only if there are scalars $\lambda_1, \ldots, \lambda_n \in F$ (not necessarily distinct) such that $\phi v_i = \lambda_i v_i$ for each $i = 1, \ldots, n$.

Proof: For each $i = 1, \ldots, n$, we have

$$v_i = 0v_1 + \cdots + 0v_{i-1} + 1v_i + 0v_{i+1} + \cdots + 0v_n,$$

hence $(v_i)_{\mathscr{B}} = e_i$. But then $((\phi v_1)_{\mathscr{B}}|\ldots|(\phi v_n)_{\mathscr{B}}) = \operatorname{diag}(\lambda_1, \ldots, \lambda_n)(\lambda_i \in F)$ $\Leftrightarrow ((\phi v_1)_{\mathscr{B}}|\ldots|(\phi v_n)_{\mathscr{B}}) = (\lambda_1(v_1)_{\mathscr{B}}|\ldots|\lambda_n(v_n)_{\mathscr{B}}) = ((\lambda_1 v_1)_{\mathscr{B}}|\ldots|(\lambda_n v_n)_{\mathscr{B}})(\lambda_i \in F) \Leftrightarrow$ $(\phi v_i)_{\mathscr{B}} = (\lambda_i v_i)_{\mathscr{B}}$ $(i = 1, \ldots, n;\ \lambda_i \in F) \Leftrightarrow \phi v_i = \lambda_i v_i \quad (i = 1, \ldots, n;\ \lambda_i \in F)$. ■

Definition 3.17.1: Let V be a vector space of finite dimension $n > 0$ over a field F, and let $\phi : V \to V$ be a linear operator. Then $\lambda \in F$ is an *eigenvalue* of ϕ if $\phi(v) = \lambda v$ for some $v \neq 0$ in V. In this case, v is an *eigenvector* of ϕ, associated with the eigenvalue λ.

(The German word-fragment *eigen* relates to ownership. The eigenvalues and eigenvectors of a linear operator ϕ *belong* to the operator. At times, eigenvalues are referred to as *characteristic values*, or *characteristic roots*, and eigenvectors are referred to as *characteristic vectors*.)

We can now reformulate the Lemma:

Theorem 3.17.1: Let V be a vector space of finite dimension $n > 0$ over a field F, and let $\phi : V \to V$ be a linear operator. Then ϕ can be represented by a diagonal matrix in $M_n(F)$ relative to some basis $\mathscr{B}$ if and only if there exist n linearly independent eigenvectors for ϕ, associated with eigenvalues in F.

Proof: If ϕ can be represented by a diagonal matrix in $M_n(F)$, then there is a basis $\mathscr{B}$ such that $\phi_{\mathscr{B},\mathscr{B}}$ is diagonal. By the Lemma and Definition 3.17.1, $\mathscr{B}$ consists of n eigenvectors of ϕ, certainly linearly independent, associated with eigenvalues in F.

Conversely, suppose there exist n linearly independent eigenvectors $v_1, \ldots, v_n$ of ϕ, associated with eigenvalues in F. By Theorem 3.12.4(5), $\{v_1, \ldots, v_n\}$ is a basis for V. Let $\mathscr{B} = (v_1, \ldots, v_n)$. Then $\phi v_i = \lambda_i v_i \quad (i = 1, \ldots, n;\ \lambda_i \in F)$; hence $\phi_{\mathscr{B},\mathscr{B}} = \operatorname{diag}(\lambda_1, \ldots, \lambda_n) \in M_n(F)$. ■

If $\phi : V \to V$ is a linear operator and $A = \phi_{\mathscr{B},\mathscr{B}}$ with respect to some ordered basis $\mathscr{B}$ of V, and if λ is an eigenvalue of ϕ, then $\phi v = \lambda v$ for some $v \neq 0$ in V. But then $(\phi v)_{\mathscr{B}} = (\lambda v)_{\mathscr{B}}$; hence $\phi_{\mathscr{B},\mathscr{B}} v_{\mathscr{B}} = \lambda v_{\mathscr{B}}$. This suggests defining eigenvalues and eigenvectors directly for matrices.

Definition 3.17.2: Let A be an $n \times n$ matrix over a field F $(n \in \mathbb{Z}^+)$. Then $\lambda \in F$ is an *eigenvalue of* A if there is some $x \neq 0$ in F^n such that $Ax = \lambda x$. In this case, x is an *eigenvector of* A associated with the eigenvalue λ.

A is *diagonalizable over* F if there is a non-singular matrix $P \in M_n(F)$ such that $P^{-1}AP$ is a diagonal matrix.

Corollary 1: Let V be a vector space of finite dimension $n > 0$ over a field F, and let $\phi : V \to V$ be a linear operator. Let $\lambda \in F$. Then the following statements are equivalent.

(1) λ is an eigenvalue of ϕ;

(2) λ is an eigenvalue of $\phi_{\mathscr{B},\mathscr{B}}$ for *every* ordered basis $\mathscr{B}$ of V;

(3) λ is an eigenvalue of $\phi_{\mathscr{B},\mathscr{B}}$ for *some* ordered basis $\mathscr{B}$ of V.

Proof: $1 \Rightarrow 2$. Suppose λ is an eigenvalue of ϕ. Let v be an eigenvector of ϕ associated with λ, and let $\mathscr{B}$ be an ordered basis for V. Then $\phi_{\mathscr{B},\mathscr{B}} v_{\mathscr{B}} = (\phi v)_{\mathscr{B}} = (\lambda v)_{\mathscr{B}} = \lambda v_{\mathscr{B}}$. Since $v \neq 0$ in V, we have $v_{\mathscr{B}} \neq 0$ in F^n; hence λ is an eigenvalue of $\phi_{\mathscr{B},\mathscr{B}}$.

$2 \Rightarrow 3$, of course.

$3 \Rightarrow 1$. Suppose λ is an eigenvalue of $\phi_{\mathscr{B},\mathscr{B}}$ for some ordered basis $\mathscr{B} = (v_1, \ldots, v_n)$ of V. Then there is some $x = \begin{pmatrix} x_1 \\ \vdots \\ x_n \end{pmatrix} \neq 0$ in F^n such that $\phi_{\mathscr{B},\mathscr{B}} x = \lambda x$. Let $v = \sum_{i=1}^{n} x_i v_i$. Since $x \neq 0$ and the v_i are linearly independent, we have $v \neq 0$. Since $x = v_{\mathscr{B}}$, $\phi_{\mathscr{B},\mathscr{B}} v_{\mathscr{B}} = \lambda v_{\mathscr{B}}$; hence $(\phi v)_{\mathscr{B}} = (\lambda v)_{\mathscr{B}}$. But then $\phi v = \lambda v$, and so λ is an eigenvalue of ϕ. ∎

Corollary 2: Let A, B be $n \times n$ matrices over a field F $(n \in \mathbb{Z}^+)$. If A and B are similar over F, then A and B have the same set of eigenvalues in F.

Proof: Let $\mathscr{B}_0$ be the standard basis for F^n and let $\phi : F^n \to F^n$ be the linear operator given by

$$\phi(x) = Ax$$

for each $x \in F^n$. Then $A = \phi_{\mathscr{B}_0,\mathscr{B}_0}$. If $B \in M_n(F)$ is similar to A over F, then $B = \phi_{\mathscr{B},\mathscr{B}}$ for some ordered basis $\mathscr{B}$ of F_n (by Corollary 2 of Theorem 3.16.2). But then, by Corollary 1, A and B have the same set of eigenvalues in F. ∎

By Corollary 1, to find the eigenvalues of a linear operator $\phi : V \to V$, it suffices to find the eigenvalues of any matrix representing ϕ relative to some ordered basis for V.

Theorem 3.17.2: Let A be an $n \times n$ matrix over a field F $(n \in \mathbb{Z}^+)$, and let $\lambda \in F$. Then the following statements are equivalent:

(1) λ is an eigenvalue of A;
(2) the matrix $A - \lambda I_n$ is singular;
(3) $\det(A - \lambda I_n) = 0$.

Proof: $1 \Rightarrow 2$. If λ is an eigenvalue of A, then $Ax = \lambda x$ for some $x \neq 0$ in F^n. Hence $0 = (A - \lambda I_n)x = Ax - (\lambda I_n)x = Ax - \lambda(I_n x) = Ax - \lambda x$. But then the nullity of $A - \lambda I_n$ is non-zero, and so $A - \lambda I_n$ is singular, by Definition 3.16.1.

$2 \Rightarrow 3$. If $A - \lambda I_n$ is singular, then $\det(A - \lambda I_n) = 0$, by Exercise 3.16.12.

$3 \Rightarrow 1$. Suppose $\det(A - \lambda I_n) = 0$. By Exercise 3.16.12, $A - \lambda I_n$ is singular, hence $(A - \lambda I_n)x = 0$ for some $x \neq 0$ in F^n. But then $Ax = \lambda x$ $(x \neq 0$ in $F^n)$, and so λ is an eigenvalue of A. ∎

Remark: If $A \in M_n(F)$, and t is an indeterminate over F, then $\det(A - tI_n)$ is a polynomial of degree n in the polynomial domain $F[t]$.

Definition 3.17.3: If A is an $n \times n$ matrix over a field F $(n \in \mathbb{Z}^+)$ then the (monic) polynomial $p(t) = (-1)^n \det(A - tI_n) \in F[t]$ is the *characteristic polynomial* of A.

If λ is an eigenvalue of A, then the nullspace of the matrix $A - \lambda I$ is the *eigenspace* of A associated with λ. (Thus, the eigenspace associated with λ is the set of all eigenvectors associated with λ, together with the 0-vector.)

By Theorem 3.17.2, $\lambda \in F$ is an eigenvalue of $A \in M_n(F)$ if and only if λ is a zero of the characteristic polynomial, $p(t)$, of A. In general, not all of the zeros of $p(t)$ will be in F. (For example, if $F = \mathbb{R}$, $p(t)$ may have some imaginary zeros.) However, every field, F, is contained in an *algebraically closed field*, E, i.e., in a field E such that every polynomial in $E[x]$ splits into linear factors over E. We may then regard $A \in M_n(F)$ as a matrix in $M_n(E)$, whose eigenvalues are precisely the zeros, in E, of $p(t) = (-1)^n \det(A - tI_n)$.

The following theorem is a useful tool for building a basis consisting of eigenvectors of a given matrix.

Theorem 3.17.3: Let A be an $n \times n$ matrix over a field F, and let $\lambda_1, \ldots, \lambda_k$ be the distinct eigenvalues of A in F. For each $i = 1, \ldots, k$, let v_i be an eigenvector associated with λ_i. Then the vectors $v_1, \ldots, v_k$ are linearly independent.

Proof: Suppose $v_1, \ldots, v_k$ are linearly dependent. Then $k \geq 1$ and there is a least positive integer h such that

$$c_1 v_{i_1} + \cdots + c_h v_{i_h} = 0 \tag{1}$$

for some subset $\{v_{i_1}, \ldots, v_{i_h}\}$ of $\{v_1, \ldots, v_k\}$ and some scalars $c_1, \ldots, c_h$, *all non*-zero. Then

$$A(c_1 v_{i_1} + \cdots + c_h v_{i_h}) = c_1 A v_{i_1} + \cdots + c_h A v_{i_h} = A0 = 0; \tag{2}$$

hence

$$c_1\lambda_{i_1}v_{i_1} + \cdots + c_h\lambda_{i_h}v_{i_h} = 0. \tag{3}$$

Multiplying (1) by λ_{i_h}, we have

$$c_1\lambda_{i_h}v_{i_1} + \cdots + c_h\lambda_{i_h}v_{i_h} = 0. \tag{4}$$

Subtracting (3) from (4) yields

$$c_1(\lambda_{i_h} - \lambda_{i_1})v_{i_1} + \cdots + c_{h-1}(\lambda_{i_h} - \lambda_{i_{h-1}})v_{i_{h-1}} = 0. \tag{5}$$

Since the λ_i are all distinct ($i = 1, \ldots, k$), and the field F contains no zero divisors, the coefficients $c_h(\lambda_{i_h} - \lambda_{i_j})$ $(j = 1, \ldots, h-1)$ are all non-zero. This contradicts the choice of h as minimal.

It follows that $v_1, \ldots, v_k$ are linearly independent. ■

Corollary 1: Let F be a field, A an $n \times n$ matrix over F, with distinct eigenvalues $\lambda_1, \ldots, \lambda_m \in F$. For each $i = 1, \ldots, m$, let S_i be a basis for the eigenspace $W_i = N(A - \lambda_i I)$. Then $S = \bigcup_{i=1}^{k} S_i$ is a linearly independent set.

Proof: Suppose S is linearly dependent. If $S_i = \{w_{i1}, \ldots, w_{ik_i}\}$, $(i = 1, \ldots, m)$, then there are scalars c_{ij}, not all zero, such that $(c_{11}w_{11} + \cdots + c_{1k_1}w_{1k_1}) + (c_{21}w_{21} + \cdots + c_{2k_2}w_{2k_2}) + \cdots + (c_{m1}w_{m1} + \cdots + c_{mk_m}w_{mk_m}) = 0$. Write $u_i = c_{i1}w_{i1} + \cdots + c_{ik_i}w_{ik_i}$, for each $i = 1, \ldots, m$. Each u_i is in W_i; hence either $u_i = 0$, or u_i is an eigenvector of A, associated with λ_i. If $u_i = 0$, then (since $w_{i1}, \ldots, w_{ik_i}$ are linearly independent) $c_{i1}, \ldots, c_{ik_i}$ are all equal to 0. But $c_{ij} \neq 0$ for at least one i, j $(1 \leq i \leq m, 1 \leq j \leq k_i)$; hence at least one of the u_i is non-zero. Since $u_1 + \cdots + u_m = 0$, at least two of the u_i must be non-zero. But then the non-zero vectors among the u_i are linearly dependent eigenvectors belonging to distinct eigenvalues of A. By Theorem 3.17.3, this is impossible. It follows that $S = \bigcup_{i=1}^{m} S_i$ is a linearly independent set. ■

Corollary 2: Let A be an $n \times n$ matrix over a field F. Then A is diagonalizable over F if and only if the sum of the dimensions of the eigenspaces of A belonging to eigenvalues in F is equal to n.

Proof: Let $\lambda_1, \ldots, \lambda_m$ be the distinct eigenvalues of A in F, and let $W_1, \ldots, W_m$ be the corresponding eigenspaces. For each $i = 1, \ldots, m$, let S_i be a basis for W_i, and let $m_i = \dim W_i$. By Corollary 1, $S = \bigcup_{i=1}^{m} S_i$ is a linearly independent set; hence the S_i are pairwise disjoint. But then the number of vectors in S is equal to $\sum_{i=1}^{m} n_i$. Thus, if $\sum_{i=1}^{m} \dim W_i = n$, then S is a linearly independent set consisting of n eigenvectors of A. By Theorem 3.17.1, it follows that A is diagonalizable over F.

Conversely, suppose A is diagonalizable over F. By Theorem 3.17.1, there are n linearly independent eigenvectors for A, belonging to eigenvalues of A in F, hence contained in $\bigcup_{i=1}^{m} W_i \subset \sum_{i=1}^{m} W_i$. Thus $n \le \dim \sum_{i=1}^{m} W_i \le n$, and so $\dim \sum_{i=1}^{m} W_i = n$, and $\sum_{i=1}^{m} W_i = V$. From Theorem 3.12.5, it follows that $\dim \sum_{i=1}^{m} W_i \le \sum_{i=1}^{m} \dim W_i$; hence $n \le \sum_{i=1}^{m} \dim W_i \le n$. But then $\sum_{i=1}^{m} \dim W_i = n$. ■

Example 1:

Let $A = \begin{pmatrix} 0 & 0 & 0 \\ 1 & 0 & -1 \\ 0 & 1 & 2 \end{pmatrix} \in M_3(\mathbb{R})$. Then $A - \lambda I = \begin{pmatrix} -\lambda & 0 & 0 \\ 1 & -\lambda & -1 \\ 0 & 1 & 2-\lambda \end{pmatrix}$; hence $p(\lambda) = (-1)^3 \det \begin{pmatrix} -\lambda & 0 & 0 \\ 1 & -\lambda & -1 \\ 0 & 1 & 2-\lambda \end{pmatrix} = \lambda^3 - 2\lambda^2 + \lambda = \lambda(\lambda - 1)^2$. Thus, the eigenvalues of A are 0 and 1 (a double zero of $p(\lambda)$).

We find the corresponding eigenspaces: $W_1 = N(A - 0I_3) = \{x \in \mathbb{R}^3 \mid Ax = 0\}$. Setting $\begin{pmatrix} 0 & 0 & 0 \\ 1 & 0 & -1 \\ 0 & 1 & 2 \end{pmatrix}\begin{pmatrix} x_1 \\ x_2 \\ x_3 \end{pmatrix} = \begin{pmatrix} 0 \\ 0 \\ 0 \end{pmatrix}$, we have

$$x_1 - x_3 = 0$$

$$x_2 + 2x_3 = 0;$$

hence $x_1 = x_3$, $x_2 = -2x_3$. Thus $W_1 = \left\{ \begin{pmatrix} x_3 \\ -2x_3 \\ x_3 \end{pmatrix} \middle| x_3 \in \mathbb{R} \right\}$, the subspace of $\mathbb{R}^3$ spanned by the vector $\begin{pmatrix} 1 \\ -2 \\ 1 \end{pmatrix}$. $W_2 = (NA - 1I_3) = \{x \in \mathbb{R}^3 \mid (A - 1I_3)x = 0\}$.

Setting $\begin{pmatrix} -1 & 0 & 0 \\ 1 & -1 & -1 \\ 0 & 1 & 1 \end{pmatrix}\begin{pmatrix} x_1 \\ x_2 \\ x_3 \end{pmatrix} = \begin{pmatrix} 0 \\ 0 \\ 0 \end{pmatrix}$, we have

$$-x_1 \qquad = 0$$

$$x_1 - x_2 - x_3 = 0$$

$$x_2 + x_3 = 0;$$

hence $x_1 = 0$, $x_2 = -x_3$. Thus $W_2 = \left\{ \begin{pmatrix} 0 \\ -x_3 \\ x_3 \end{pmatrix} \middle| x_3 \in \mathbb{R} \right\}$, the subspace of $\mathbb{R}^3$ spanned by the vector $\begin{pmatrix} 0 \\ -1 \\ 1 \end{pmatrix}$. But then $\dim W_1 = \dim W_2 = 1$; hence $\dim W_1 + \dim W_2 = 2 < 3$, and so, by Corollary 2, A is not diagonalizable.

Example 2:

Let $A = \begin{pmatrix} 1 & -2 & 0 \\ 0 & 0 & 0 \\ 0 & 0 & 1 \end{pmatrix} \in M_3(\mathbb{R})$. The characteristic polynomial of A is again $p(t) = \lambda(\lambda - 1)^2 = \lambda^3 - 2\lambda^2 + \lambda$; hence the eigenvalues are $\lambda = 0$ and $\lambda = 1$ (a double zero). Let $W_1 = N(A - 0I)$. Setting

$$\begin{pmatrix} 1 & -2 & 0 \\ 0 & 0 & 0 \\ 0 & 0 & 1 \end{pmatrix}\begin{pmatrix} x_1 \\ x_2 \\ x_3 \end{pmatrix} = \begin{pmatrix} 0 \\ 0 \\ 0 \end{pmatrix},$$

we have

$$\begin{aligned} x_1 - 2x_2 &= 0 \\ x_3 &= 0; \end{aligned}$$

hence

$$\begin{aligned} x_1 &= 2x_2 \\ x_3 &= 0. \end{aligned}$$

Thus, $W_1 = \left\{ \begin{pmatrix} 2x_2 \\ x_2 \\ 0 \end{pmatrix} \middle| x_2 \in \mathbb{R} \right\}$, the span of the vector $\begin{pmatrix} 2 \\ 1 \\ 0 \end{pmatrix}$, and so $\dim W_1 = 1$.

Let $W_2 = N(A - 1I)$. Setting

$$\begin{pmatrix} 0 & -2 & 0 \\ 0 & -1 & 0 \\ 0 & 0 & 0 \end{pmatrix}\begin{pmatrix} x_1 \\ x_2 \\ x_3 \end{pmatrix} = \begin{pmatrix} 0 \\ 0 \\ 0 \end{pmatrix},$$

we have

$$\begin{aligned} -2x_2 &= 0 \\ -x_2 &= 0; \end{aligned}$$

hence $x_2 = 0$ and $W_2 = \left\{ \begin{pmatrix} x_1 \\ 0 \\ x_3 \end{pmatrix} \middle| x_1, x_3 \in \mathbb{R} \right\} = \{x_1 e_1 + x_3 e_3 \mid x_1, x_3 \in \mathbb{R}\}$. Thus, $\dim W_2 = 2$.

Since $\dim W_1 + \dim W_2 = 1 + 2 = 3$, we conclude that A is diagonalizable.

By Corollary 1, since $S_1 = \left\{ \begin{pmatrix} 2 \\ 1 \\ 0 \end{pmatrix} \right\}$ is a basis for W_1 and $S_2 = \left\{ \begin{pmatrix} 1 \\ 0 \\ 0 \end{pmatrix}, \begin{pmatrix} 0 \\ 0 \\ 1 \end{pmatrix} \right\}$ is a basis for W_2, the set $S = S_1 \cup S_2 = \left\{ \begin{pmatrix} 2 \\ 1 \\ 0 \end{pmatrix}, \begin{pmatrix} 1 \\ 0 \\ 0 \end{pmatrix}, \begin{pmatrix} 0 \\ 0 \\ 1 \end{pmatrix} \right\}$ is a basis for $\mathbb{R}^3$, consisting of eigenvectors of A. Let $\mathscr{B} = \{v_1, v_2, v_3\}$, where $v_1 = \begin{pmatrix} 2 \\ 1 \\ 0 \end{pmatrix}$, $v_2 = \begin{pmatrix} 1 \\ 0 \\ 0 \end{pmatrix}$, $v_3 = \begin{pmatrix} 0 \\ 0 \\ 1 \end{pmatrix}$. If $\phi : V \to V$ is the linear operator represented by A relative to the standard basis $\mathscr{B}_0$, then the transition matrix from $\mathscr{B}$ to $\mathscr{B}_0$ is $P = \begin{pmatrix} 2 & 1 & 0 \\ 1 & 0 & 0 \\ 0 & 0 & 1 \end{pmatrix}$; hence, by Theorem 3.16.2, Corollary 1, $\phi_{\mathscr{B}, \mathscr{B}} = P^{-1}AP$. On the other hand, $\phi_{\mathscr{B}, \mathscr{B}} = ((\phi v_1)_{\mathscr{B}} | (\phi v_2)_{\mathscr{B}} | (\phi v_3)_{\mathscr{B}}) = ((0v_1)_{\mathscr{B}} | (1v_2)_{\mathscr{B}} | 1v_3)_{\mathscr{B}}) = \begin{pmatrix} 0 & 0 & 0 \\ 0 & 1 & 0 \\ 0 & 0 & 1 \end{pmatrix}$. It follows that $P^{-1}AP = \text{diag}(0, 1, 1)$.

Exercises 3.17

True or False

1. Every square matrix over a field F is diagonalizable over F.
2. If $A \in M_n(F)$ is diagonalizable over F, then A has n distinct eigenvalues in F.
3. The eigenspace of a matrix A, associated with an eigenvalue λ, consists of all eigenvectors of A associated with λ.
4. If two $n \times n$ matrices over a field F have the same characteristic polynomial, then they are similar.
5. The characteristic polynomial of a matrix $A \in M_2(\mathbb{R})$ may have no zeros in $\mathbb{R}$.

3.17.1. Give an example of each of the following:

(1) a matrix $A \in M_2(\mathbb{R})$ that is diagonalizable over $\mathbb{R}$ and has two distinct eigenvalues;

(2) a matrix $A \in M_2(\mathbb{R})$ that is diagonalizable over $\mathbb{R}$ and has only one eigenvalue in $\mathbb{R}$;

(3) a matrix $A \in M_2(\mathbb{R})$ that is *not* diagonalizable over $\mathbb{R}$;

(4) a matrix $A \in M_2(\mathbb{R})$ that is diagonalizable over $\mathbb{C}$ but not over $\mathbb{R}$.

3.17.2. Let A be an $n \times n$ matrix over a field F. Prove: if A has n distinct eigenvalues in F, then A is diagonalizable over F. (Is the converse true?)

3.17.3. Let $A = \begin{pmatrix} 0 & 0 & 3 \\ 1 & 0 & 1 \\ 0 & 1 & -3 \end{pmatrix} \in M_3(\mathbb{R})$. Find a matrix $P \in M_3(\mathbb{R})$ such that $P^{-1}AP$ is diagonal.

3.17.4. Prove that similar matrices have the same rank. Give an example of two 2×2 matrices that have the same rank but are not similar.

3.17.5. Let A be an $n \times n$ matrix over a field F. Prove that A^T and A have the same eigenvalues. For each eigenvalue λ of A (and of A^T), is the eigenspace $N(A - \lambda I)$ equal to the eigenspace $N(A^T - \lambda I)$?

3.17.6. *(1)* Let $p = a_0 + a_1x + \cdots + a_{n-1}x^{n-1} + x^n \in F[x]$ (F a field), and let $A \in M_n(F)$ be defined by

$$A = (e_2|e_3|\ldots|e_n|a),$$

where e_i ($i = 2, \ldots, n$) are unit vectors in F^n and $a = \begin{pmatrix} -a_0 \\ -a_1 \\ \vdots \\ -a_{n-1} \end{pmatrix} \in F^n$.

Prove that p is the characteristic polynomial of A. (A is called the *companion matrix* of the polynomial p.)

(2) Find the companion matrix, A, in $M_3(\mathbb{R})$, of the polynomial $p = 2 + x + 2x^2 + x^3$. Find the eigenvalues of A and determine whether A is diagonalizable

(a) over $\mathbb{R}$;

(b) over $\mathbb{C}$.

3.17.7. Let $A = \begin{pmatrix} 1 & 2 \\ 3 & 4 \end{pmatrix}$. Find the characteristic polynomial p of A and verify that $p(A) = 0$. (This illustrates the *Cayley-Hamilton Theorem*, according to which every square matrix satisfies its characteristic equation.)

3.17.8. Prove that a matrix $A \in M_n(F)$ ($n \in \mathbb{Z}^+$, F a field) is singular if and only if one of its eigenvalues is equal to 0.

3.17.9. Let λ be an eigenvalue of $A \in M_n(F)$. Prove that, for each positive integer k, λ^k is an eigenvalue of A^k.

3.17.10. Let $E = E^2$ be a non-zero idempotent matrix in $M_n(F)$.

(1) Prove that the only eigenvalues of E are 0 and 1.

(2) Without using the result of Exercise 3.16.3, prove that every idempotent matrix in $M_n(F)$ is diagonalizable.

3.17.11. A matrix in $M_n(F)$ is *nilpotent* if $A^k = 0$ for some positive integer k.

(1) Prove that the only eigenvalue of a nilpotent matrix is 0.

(2) Prove that a non-zero nilpotent matrix is not diagonalizable.

3.17.12. Let $A \in M_n(\mathbb{R})$ be a matrix whose only eigenvalues are 1 and -1. Prove that A is diagonalizable if and only if A is *involutory*, i.e., if and only if $A^2 = I$.

3.17.13. For A, B in $M_n(F)$, prove that the matrices AB and BA have the same eigenvalues.

3.17.14. Let F be a field and let k be a fixed element of F. Let $A \in M_n(F)$ be a matrix such that the sum of the entries in each row of A is equal to k. Prove that k is an eigenvalue of A.

3.17.15. Let $A \in M_n(\mathbb{C})$ be a diagonalizable matrix. Prove that, for each positive integer k, there is a matrix $B \in M_n(\mathbb{C})$ such that $B^k = A$.

4

Fields

4.1

Subfields, Extensions, Prime Fields, and Characteristic

If a subring, E, of field F is itself a field, then E is a *subfield* of F, and F is an *extension field* (or simply an extension) of E. More generally, if $\varepsilon : E \to K$ is a ring monomorphism of a field E into a field K, we call the ordered pair $\langle K, \varepsilon \rangle$ an *extension* of E. Such a monomorphism $\varepsilon : E \to K$ can be extended in a natural way to a monomorphism $\hat{\varepsilon} : E[x] \to K[x]$ which sends each

$$f = \sum_{i=0}^{n} a_i x^i \in E[x] \quad \text{to} \quad f^{\varepsilon} = \sum_{i=0}^{n} (\varepsilon a_i) x^i \in K[x].$$

(See Exercise 4.1.8.)

If F is a field with identity 1_F and A is a non-zero subring of F with identity 1_A, then A is an integral domain, and $1_A = 1_F$. For, A is a commutative ring without zero divisors, and from $1_A 1_A = 1_A 1_F$, by cancellation of 1_A, we have $1_A = 1_F$ (hence $1_A \neq 0_A = 0_F$). In particular, the identity of any subfield E of F is equal to 1_F.

Every subset S of a field F generates a subring, and a subfield, of F, in the sense of the following definition:

Definition 4.1.1: Let F be a field and let S be a subset of F. Then the *subring of F generated by S* is the intersection of all sub*rings* of F which contain S; and the *subfield of F generated by S* is the intersection of all sub*fields* of F which contain S.

Example:

The subring of $\mathbb{R}$ generated by $S = \{1\}$ is $\mathbb{Z}$. The subfield of $\mathbb{R}$ generated by $S = \{1\}$ is $\mathbb{Q}$.

Lemma: Let F be a field, S a subset of F, with $1_F \in S$. If A is the sub*ring* of F generated by S, and K is the sub*field* of F generated by S, then A is an integral domain and K is the field of quotients of A in F.

Proof: Since fields are rings, the intersection of all sub*rings* of F, containing S, is a subset of the intersection of all sub*fields* of F, containing S. Thus, $A \subset K$. Since $1_F \in S$, we have $1_F \in A$; hence A is an integral domain. Let

$$\bar{K} = \{ab^{-1} | a, b \neq 0 \quad \text{in} \quad A\},$$

the field of quotients of A in F. Since $A \subset K$, clearly $\bar{K} \subset K$. But $\bar{K}$ is a subfield of F, containing S; hence $\bar{K}$ contains the intersection of *all* such subfields of F, i.e., $\bar{K} \supset K$. It follows that $K = \bar{K}$, and so K is the field of quotients of A in F. ∎

For an arbitrary field F, we now take a closer look at the subring, and the subfield, generated by 1_F (i.e., generated by the set $S = \{1_F\}$).

Theorem 4.1.1: Let F be a field.

(1) The subring of F generated by 1_F is $\mathbb{Z}1_F = \{k1_F | k \in \mathbb{Z}\}$. The subfield of F generated by 1_F is the quotient field, in F, of $\mathbb{Z}1_F$.

(2) Let $\phi : \mathbb{Z} \to F$ be defined by: $\phi(k) = k1_F$ for each $k \in \mathbb{Z}$. Then ϕ is a ring homomorphism, with $\operatorname{Im} \phi = \mathbb{Z}1_F$ and $\operatorname{Ker} \phi$ either equal to $\{0\}$, or equal to $p\mathbb{Z}$ for some prime $p \in \mathbb{Z}$.

(3) If $\operatorname{Ker} \phi = \{0\}$, then the subring, $\mathbb{Z}1_F$, of F generated by 1_F is isomorphic to $\mathbb{Z}$, and the subfield of F generated by 1_F is isomorphic to $\mathbb{Q}$.

(4) If $\operatorname{Ker} \phi = p\mathbb{Z}$ for some prime $p \in \mathbb{Z}$, then the subring, $\mathbb{Z}1_F$, of F generated by 1_F is a field, isomorphic to $\mathbb{Z}/p\mathbb{Z}$, hence equal to the subfield of F generated by 1_F.

Proof: *(1)* Every subring A of F containing 1_F must contain $k1_F$ for each $k \in \mathbb{Z}$. Hence the intersection of all subrings containing 1_F must contain $\mathbb{Z}1_F$. But $\mathbb{Z}1_F$ is itself a subring containing 1_F, hence it contains the intersection of *all* such subrings. It follows, by Definition 4.1.1, that $\mathbb{Z}1_F$ is the subring of F generated by 1_F. Its quotient field in F is the subfield generated by 1_F.

(2) We leave as an exercise the verification that ϕ is a homomorphism. (See Exercise 4.1.3.) Clearly, by the definition of ϕ, $\operatorname{Im} \phi = \mathbb{Z}1_F$, an integral domain. By Theorem 3.4.2, it follows that $\operatorname{Ker} \phi$ is a prime ideal in $\mathbb{Z}$. By Corollary 2 of Theorem 3.4.2, either $\operatorname{Ker} \phi = \{0\}$, or $\operatorname{Ker} \phi = p\mathbb{Z}$ for some prime $p \in \mathbb{Z}$.

(3) If $\operatorname{Ker} \phi = \{0\}$, then ϕ is a monomorphism, hence $\mathbb{Z}1_F = \operatorname{Im} \phi$ is isomorphic to $\mathbb{Z}$ and its quotient field is isomorphic to $\mathbb{Q}$. Thus, the subring of F generated by 1_F is isomorphic to $\mathbb{Z}$, and the subfield of F generated by 1_F is isomorphic to $\mathbb{Q}$.

(4) If $\operatorname{Ker} \phi = p\mathbb{Z}$ for some prime $p \in \mathbb{Z}$, then $\mathbb{Z}1_F = \operatorname{Im} \phi \cong \mathbb{Z}/p\mathbb{Z}$ is a field, hence equal to its own quotient field: the subfield of F generated by 1_F. ∎

Theorem 4.1.1 points up an important distinction between two kinds of fields. Note that the subfield of a field F generated by 1_F is, in fact, the intersection of *all* subfields of F.

Definition 4.1.2: The intersection of all subfields of a field F is called the *prime field* of F.

Using this terminology, the results of Theorem 4.1.1 can be reworded as follows:

Corollary: Let F be a field and let $\phi : \mathbb{Z} \to F$ be defined by $\phi(k) = k1_F$ for each $k \in \mathbb{Z}$. Then either $\text{Ker } \phi = \{0\}$, or $\text{Ker } \phi = p\mathbb{Z}$ for some prime $p \in \mathbb{Z}$. If $\text{Ker } \phi = \{0\}$, then the prime field of F is isomorphic to $\mathbb{Q}$. If $\text{Ker } \phi = p\mathbb{Z}$ (p prime in $\mathbb{Z}$), then the prime field of F is isomorphic to $\mathbb{Z}/p\mathbb{Z}$.

Definition 4.1.3: A field F has *characteristic 0* if its prime field is isomorphic to $\mathbb{Q}$.

A field F has *characteristic p* (p prime in $\mathbb{Z}$) if its prime field is isomorphic to $\mathbb{Z}/p\mathbb{Z}$.

Notation: We write "Char F" to denote the characteristic of a field F.

Corollary: Let F be a field.

I. The following statements are equivalent:

(1) Char $F = 0$;

(2) $k1_F = 0_F \Leftrightarrow k = 0 \quad (k \in \mathbb{Z})$.

II. The following statements are equivalent:

(1) Char $F = p \quad$ (p prime in $\mathbb{Z}$);

(2) $k1_F = 0_F \Leftrightarrow p|k \quad (k \in \mathbb{Z})$.

Example 1:

$\mathbb{Q}$, $\mathbb{R}$, $\mathbb{C}$ all have characteristic 0. So do the quotient fields $\mathbb{Q}(x)$, $\mathbb{R}(x)$, $\mathbb{C}(x)$ of the polynomial domains $\mathbb{Q}[x]$, $\mathbb{R}[x]$, $\mathbb{C}[x]$.

Example 2:

Any finite field has characteristic $p > 0$ for some prime $p \in \mathbb{Z}$. (For, any field of characteristic 0 has a subfield isomorphic to $\mathbb{Q}$, and is therefore infinite!)

Example 3:

The fields $\mathbb{Z}/p\mathbb{Z}$ (p prime), of course, have characteristic p. So do the quotient fields $(\mathbb{Z}/p\mathbb{Z})(x)$ of the polynomial domains $(\mathbb{Z}/p\mathbb{Z})[x]$. Thus, *please note that there exist infinite fields* of characteristic $p > 0$.

Example 4:

As an example of a finite field not having a prime number of elements, consider the set $F = \{0, 1, \alpha, 1 + \alpha\}$, with addition and multiplication defined by the following tables.

$+$	0	1	α	$1+\alpha$
0	0	1	α	$1+\alpha$
1	1	0	$1+\alpha$	α
α	α	$1+\alpha$	0	1
$1+\alpha$	$1+\alpha$	α	1	0

and

$\cdot$	0	1	α	$1+\alpha$
0	0	0	0	0
1	0	1	α	$1+\alpha$
α	0	α	$1+\alpha$	1
$1+\alpha$	0	$1+\alpha$	1	α

One can verify directly that $\langle F, +, \cdot\rangle$ is a field, consisting of 4 elements. The tedium of verification (of associative and distributive laws) is unnecessary if one recognizes that F is the field obtained by "adjoining" to $\mathbb{Z}/2\mathbb{Z}$ a zero, α, of the irreducible polynomial $1 + x + x^2$ in $(\mathbb{Z}/2\mathbb{Z})[x]$. We shall address ourselves to this problem in Section 4.3. As we shall see (Theorem 4.4.5), the number of elements in a finite field is always a power of a prime.

Remark: If Char $F = p > 0$, then $px = 0$ for all $x \in F$. For: $px = p(1_F x) = (p1_F)x = 0x = 0 \quad (x \in F)$.

Exercises 4.1

True or False

1. There is a field of characteristic 10.
2. If F is a field of characteristic $p > 0$, then F is a finite field.
3. Every finite field has p elements for some prime p.
4. All subfields of a field F have the same characteristic.
5. Every field of characteristic 0 is infinite.

4.1.1. *(1)* Find the subring of $\mathbb{R}$ generated by $\varnothing$.
(2) Find the subfield of $\mathbb{R}$ generated by $\varnothing$.
(3) Find the subring of $\mathbb{R}$ generated by $\{0\}$.
(4) Find the subfield of $\mathbb{R}$ generated by $\{0\}$.

4.1.2. Find the subring and the subfield of $\mathbb{R}$ generated by S, where
(1) $S = \{2\}$;
(2) $S = \{2, 3\}$;
(3) $S = \{1, 2, 3\}$.

4.1.3. Let F be a field, and let $\phi : \mathbb{Z} \to F$ be defined by

$$\phi(k) = k1_F$$

for each $k \in \mathbb{Z}$ (see Theorem 4.1.1). Prove that ϕ is a ring homomorphism.

4.1.4. Prove the "Freshman Binomial Theorem" for a field of characteristic p, i.e., prove: if Char $F = p$ (p prime), then $(x + y)^p = x^p + y^p$ for each $x, y \in F$. Conclude that $(x + y)^{p^\alpha} = x^{p^\alpha} + y^{p^\alpha}$ for each $\alpha \geq 1$ in $\mathbb{Z}$ and each $x, y \in F$.

4.1.5. If E is a subfield of a field F, prove that Char E = Char F.

4.1.6. Let F be field. Prove that the groups $\langle F, +\rangle$ and $\langle F\backslash\{0\}, \cdot\rangle$ are not isomorphic. (The case where Char $F = 2$ requires special handling!)

4.1.7. Prove that every integral domain contains an isomorphic copy either of $\mathbb{Z}$, or of $\mathbb{Z}/p\mathbb{Z}$ for some prime $p \in \mathbb{Z}$.

4.1.8. Let E, K be fields, and let $\varepsilon : E \to K$ be a monomorphism. Prove that there is a monomorphism $\hat{\varepsilon} : E[x] \to K[x]$ and such that $\hat{\varepsilon}|_E = \varepsilon$.

4.1.9. Prove that $\mathbb{Q}$ has no proper subfield (i.e., no subfield different from $\mathbb{Q}$ itself).

4.1.10. Prove that, for p prime in $\mathbb{Z}$, $\mathbb{Z}/p\mathbb{Z}$ has no proper subfield.

4.11.1. Let F be a field having no proper subfield. Prove that either $F \cong \mathbb{Q}$, or $F \cong \mathbb{Z}/p\mathbb{Z}$ for some prime $p \in \mathbb{Z}$.

4.1.12. If F is a field of characteristic $p > 0$ and k is a positive integer, prove that an element $\alpha \in F$ has at most one p^k-th root in any extension field of F.

4.1.13. Let E, K be fields and let $\phi : E \to K$ be a homomorphism. Prove that either $\phi(x) = 0$ for all $x \in E$, or ϕ is a monomorphism.

4.2

Adjunctions; Algebraic and Transcendental Elements

Consider now a field K, a subfield F, and a subset Y of K. By the Lemma preceding Theorem 4.1.1, the subfield of K generated by $F \cup Y$ is the field of quotients, in K, of the subring of K generated by $F \cup Y$.

> ***Notation:*** We denote by "$F[Y]$" the sub*ring* of K generated by $F \cup Y$, and by "$F(Y)$" the sub*field* of K generated by $F \cup Y$. In the special case where $Y = \{\alpha\}$ ($\alpha \in K$), we write "$F[\alpha]$," and "$F(\alpha)$," respectively, for the subring and the subfield of K generated by $F \cup \{\alpha\}$.
>
> $F[\alpha]$ and $F(\alpha)$ are called the *subring* and the *subfield* of K, *obtained by adjoining* α *to* F. $F(\alpha)$ is called a *simple* extension of F.

We now examine more closely the relationship of an element, α, of a field K to a subfield F of K. In particular, we raise the question: is α a zero of some non-zero polynomial with coefficients in F? It will turn out that the answer is "yes" if and only if $F[\alpha]$ is a field, i.e., if and only if $F[\alpha] = F(\alpha)$.

Our main tool will be the *substitution homomorphism* $\phi_{\alpha/F} : F[x] \to K$ defined by: $\phi_{\alpha/F}(f) = f(\alpha)$ $(f \in F[x])$. (See Exercise 3.5.6.)

Definition 4.2.1: Let F be a subfield of a field K, and let $\alpha \in K$. Then α is *algebraic over* F if there is some $f \in F[x]$, $f \neq 0$, such that $f(\alpha) = 0$; α is *transcendental over* F if there is no non-zero polynomial $f \in F[x]$ such that $f(\alpha) = 0$.

Example:

$\alpha = \sqrt[3]{5}$ is algebraic over $\mathbb{Q}$ since it is a zero of $f = x^3 - 5 \in \mathbb{Q}[x]$; $\alpha = \dfrac{3 - \sqrt[6]{2}}{9}$ is algebraic over $\mathbb{Q}$ since it is a zero of $f = (9x - 3)^6 - 2 \in \mathbb{Q}[x]$; π and e are transcendental over $\mathbb{Q}$. (The transcendence of e was proved by Hermite in 1873; the transcendence of π was proved by Lindemann in 1882.) $\sqrt{\pi}$ is algebraic over $\mathbb{Q}(\pi)$ since it is a zero of $x^2 - \pi \in \mathbb{Q}(\pi)[x]$.

The following corollary is an immediate consequence of Definition 4.2.1.

Corollary: Let F be a subfield of K, $\alpha \in K$. Then α is algebraic over F if and only if Ker $\phi_{\alpha/F} \neq \{0\}$; α is transcendental over F if and only if Ker $\phi_{\alpha/F} = \{0\}$.

Proof: It suffices to note that Ker $\phi_{\alpha/F} = \{f \in F[x] \mid f(\alpha) = 0\}$, and to apply Definition 4.2.1. ∎

Theorem 4.2.1: Let F be a subfield of a field K, and let $\alpha \in K$.

I. The following statements are equivalent:

(1) α is algebraic over F.

(2) Ker $\phi_{\alpha/F} = p\,F[x]$ for some irreducible polynomial $p \in F[x]$.

(3) Im $\phi_{\alpha/F} = F[\alpha]$ is a field, hence equal to its quotient field $F(\alpha)$.

II. The following statements are equivalent:

(1) α is transcendental over F.

(2) Ker $\phi_{\alpha/F} = \{0\}$,

(3) Im $\phi_{\alpha/F} = F[\alpha]$ is isomorphic to the polynomial domain $F[x]$.

Proof: First note that Im $\phi_{\alpha/F} = \{f(\alpha) \mid f \in F[x]\}$ is a subring of K, containing α, which is contained in every subring of K that contains α. Hence Im $\phi_{\alpha/F} = F[\alpha]$.

I. $1 \Rightarrow 2$. If α is algebraic over F, then Ker $\phi_{\alpha/F} \neq \{0\}$. Since Im $\phi_{\alpha/F} = F[\alpha]$ is a subring of K, hence an integral domain, Ker $\phi_{\alpha/F}$ is a prime ideal in $F[x]$, by Theorem 3.4.2. But then, by Theorem 3.6.2 and the Corollary on p. 166, Ker $\phi_{\alpha/F}$ is a maximal ideal in $F[x]$, and there is some irreducible polynomial $p \in F[x]$ such that Ker $\phi_{\alpha/F} = p\,F[x]$.

$2 \Rightarrow 3$. If Ker $\phi_{\alpha/F} = p\,F[x]$, p an irreducible polynomial in $F[x]$, then, since Ker $\phi_{\alpha/F}$ is a maximal ideal in $F[x]$, $F[\alpha] = $ Im $\phi_{\alpha/F}$ is a field, hence equal to its field of quotients, $F(\alpha)$. (See the Corollary, p. 153 and Theorem 3.4.1.)

$3 \Rightarrow 1$. If $F[\alpha] = F(\alpha)$, then, since Im $\phi_{\alpha/F}$ is a field, Ker $\phi_{\alpha/F}$ is a maximal ideal of $F[x]$. The zero ideal in $F[x]$ is *not* maximal, and so Ker $\phi_{\alpha/F} \neq \{0\}$. It follows that α is algebraic over F.

II. $1 \Rightarrow 2$. If α is transcendental over F, then Ker $\phi_{\alpha/F} = \{0\}$.

$2 \Rightarrow 3$. If Ker $\phi_{\alpha/F} = \{0\}$, then $\phi_{\alpha/F}$ is a monomorphism. Hence $F[\alpha] = \text{Im } \phi_{\alpha/F}$ is isomorphic to $F[x]$.

$3 \Rightarrow 1$. If $F[\alpha] \cong F[x]$, then $F[\alpha]$ is not a field; hence Ker $\phi_{\alpha/F} \neq p \quad F[x]$ for p irreducible in $F[x]$. It follows that the other alternative holds: Ker $\phi_{\alpha/F} = \{0\}$. Thus, α is transcendental over F. ■

A polynomial with leading coefficient 1 is called a *monic* polynomial.

Corollary 1: Let F be a subfield of a field K, and let $\alpha \in K$ be algebraic over F. Then there is a unique monic irreducible polynomial $q \in F[x]$ such that $q(\alpha) = 0$.

Proof: By Theorem 4.2.1. (I), there is an irreducible polynomial $p \in F[x]$ such that Ker $\phi_{\alpha/F} = p \quad F[x]$. Since $p \in \text{Ker } \phi_{\alpha/F}$, we have $p(\alpha) = 0$.

The polynomial p has a unique monic associate $q \in F[x]$, clearly also irreducible, with $q(\alpha) = 0$. By the Dictionary Lemma, p. 165, $q \quad F[x] = p\,F[x]$. Any monic irreducible polynomial $q' \in F[x]$ such that $q'(\alpha) = 0$ would also be a generator of the (maximal) ideal $p\,F[x] = \text{Ker } \phi_{\alpha/F}$, hence also an associate of p, hence equal to q (see Exercise 4.2.11). ■

Definition 4.2.2: If F is a subfield of K, $\alpha \in K$ algebraic over F, then the monic irreducible polynomial in $F[x]$ having α as a zero is called the *minimal polynomial*, $m_{\alpha/F}$, of α over F.

Note that $m_{\alpha/F}$ is the unique monic generator of the ideal Ker $\phi_{\alpha/F}$.

Corollary 2: Let F be a subfield of K, and let $\alpha \in K$ be algebraic over F. Then the minimal polynomial, $m_{\alpha/F}$, has the following properties:

(1) $m_{\alpha/F}$ is the unique monic polynomial of least degree in $F[x]$ having α as a zero.

(2) For $f \in F[x]$, $f(\alpha) = 0$ if and only if $m_{\alpha/F} | f$.

(3) $F(\alpha) \cong F[x]/m_{\alpha/F}\,F[x]$.

(4) $F(\alpha) = \{r(\alpha) | r \in F[x], r = 0 \quad \text{or} \quad \deg r < \deg m_{\alpha/F}\}$.

Proof: *(1)* We have Ker $\phi_{\alpha/F} = \{f \in F[x] | f(\alpha) = 0\}$. Since Ker $\phi_{\alpha/F} = m_{\alpha/F}\,F[x]$, deg $m_{\alpha/F} \leq \deg f$, for each $f \in \text{Ker } \phi_{\alpha/F}$. Uniqueness follows from the fact that $m_{\alpha/F}$ is the unique monic generator of Ker $\phi_{\alpha/F}$.

(2) For $f \in F[x]$, we have $f(\alpha) = 0 \Leftrightarrow f \in \text{Ker } \phi_{\alpha/F} \Leftrightarrow f \in m_{\alpha/F}\,F[x] \Leftrightarrow m_{\alpha/F} | f$.

(3) Since α is algebraic over F, by Theorem 4.2.1 (I), $F(\alpha) = \text{Im } \phi_{\alpha/F} \cong F[x]/\text{Ker } \phi_{\alpha/F} = F[x]/m_{\alpha/F}\,F[x]$.

(4) Let $f \in F[x]$. Then, by Theorem 3.5.3, $f = m_{\alpha/F}q + r$, where $q, r \in F[x]$, $r = 0$ or $\deg r < \deg m_{\alpha/F}$. If $\deg m_{\alpha/F} = n$, then $n \geq 1$, and $r = a_0 + a_1x + \cdots + a_{n-1}x^{n-1}$ for some $a_0, \ldots, a_{n-1} \in F$. But then $f(\alpha) = m_{\alpha/F}(\alpha)q(\alpha) + r(\alpha) = 0 + r(\alpha) = r(\alpha)$. Thus, $F(\alpha) = F[\alpha] \subset \{r(\alpha) | r \in F[x], r = 0 \text{ or } \deg r < \deg m_{\alpha/F}\}$. The opposite inclusion is obvious, and so we have equality. ■

Example 1:

In $\mathbb{R}$, let $\alpha = \sqrt{2}$. Then $m_{\sqrt{2}/\mathbb{Q}} = x^2 - 2$, and

$$\operatorname{Ker} \phi_{\sqrt{2}/\mathbb{Q}} = \{f \in \mathbb{Q}[x] \mid f(\sqrt{2}) = 0\} = (x^2 - 2)\,\mathbb{Q}[x].$$

$$\operatorname{Im} \phi_{\sqrt{2}/\mathbb{Q}} = \mathbb{Q}[\sqrt{2}] = \mathbb{Q}(\sqrt{2}) = \{f(\sqrt{2}) \mid f \in \mathbb{Q}[x]\} = \{a + b\sqrt{2} \mid a, b \in \mathbb{Q}\}.$$

Example 2:

In $\mathbb{C}$, let $\alpha = i = \sqrt{-1}$. Then $m_{i/\mathbb{R}} = x^2 + 1$, and

$$\operatorname{Ker} \phi_{i/\mathbb{R}} = \{f \in \mathbb{R}[x] \mid f(i) = 0\} = (x^2 + 1)\,\mathbb{R}[x].$$

$$\operatorname{Im} \phi_{i/\mathbb{R}} = \mathbb{R}[i] = \mathbb{R}(i) = \mathbb{C} = \{f(i) \mid f \in \mathbb{R}[x]\} = \{a + bi \mid a, b \in \mathbb{R}\}.$$

Example 3:

In $\mathbb{R}$, let $\alpha = \dfrac{3 - \sqrt[6]{3}}{5}$. Then $(5\alpha - 3)^6 - 3 = 0$. Since $(5x - 3)^6 - 3$ is an Eisenstein polynomial, $m_{\alpha/\mathbb{Q}}$ is an associate of $(5x - 3)^6 - 3$; hence we have

$$\operatorname{Ker} \phi_{\alpha/\mathbb{Q}} = \{f \in \mathbb{Q}[x] \mid f(\alpha) = 0\} = [(5x - 3)^6 - 3]\mathbb{Q}[x]$$

and

$$\begin{aligned}\operatorname{Im} \phi_{\alpha/\mathbb{Q}} &= \mathbb{Q}[\alpha] = \mathbb{Q}(\alpha) = \{f(\alpha) \mid f \in \mathbb{Q}[x]\} \\ &= \{a_0 + a_1\alpha + a_2\alpha^2 + a_3\alpha^3 + a_4\alpha^4 + a_5\alpha^5 \mid \alpha_i \in \mathbb{Q}, i = 0, \ldots, 5\}.\end{aligned}$$

Exercises 4.2

True or False

1. $\mathbb{Q}(\pi) = \mathbb{Q}[\pi]$.
2. $\mathbb{Q}(1 + \sqrt[4]{3}) = \mathbb{Q}[1 + \sqrt[4]{3}]$.
3. $\mathbb{R}(i) = \mathbb{C}$.
4. $\mathbb{C}$ is a homomorphic image of $\mathbb{R}[x]$.
5. Every complex number is algebraic over $\mathbb{R}$.

4.2.1. Find the minimal polynomial of

(1) $\dfrac{1 + \sqrt{2}}{3}$ over $\mathbb{Q}$;

(2) $\dfrac{1 + \sqrt{2}}{3}$ over $\mathbb{Q}(\sqrt{2})$;

(3) $e^{2\pi i/3}$ over $\mathbb{R}$;

(4) $e^{2\pi i/3}$ over $\mathbb{C}$.

4.2.2. Prove that $\mathbb{Q}(i\sqrt{3}) = \mathbb{Q}(e^{2\pi i/3})$.

4.2.3. Let F be a subfield of a field K and let $\gamma \in K$ be transcendental over F. If f is a polynomial of positive degree in $F[x]$, prove that $f(\gamma)$ is transcendental over F.

4.2.4. Assuming as known that π is transcendental over $\mathbb{Q}$, prove that, for each positive integer n, π^n is transcendental over $\mathbb{Q}$.

4.2.5. Let F be a subfield of a field K, and let $\alpha_1, \ldots, \alpha_n \in K$. If $S = \{\alpha_1, \ldots, \alpha_n\}$, prove that
(1) $F[S] = F[\alpha_1][\alpha_2] \ldots [\alpha_n]$ and
(2) $F(S) = F(\alpha_1)(\alpha_2) \ldots (\alpha_n)$.

4.2.6. Let F be a subfield of a field K, and let S, T be subsets of K. Prove that $F(S \cup T) = F(S)(T) = F(T)(S)$.

4.2.7. Find the minimal polynomial, $m_{\alpha/\mathbb{Q}}$, of $\alpha = \sqrt{2} + \sqrt{3}$ over $\mathbb{Q}$. Find all zeros of $m_{\alpha/\mathbb{Q}}$ in $\mathbb{C}$.

4.2.8. Prove that, if $\zeta = e^{2\pi i/p}$, p prime, then $m_{\zeta/\mathbb{Q}} = \Phi_p = x^{p-1} + x^{p-2} + \cdots + x + 1$. (See Exercise 3.9.8.)

4.2.9. Prove that $\mathbb{Q}(e^3)$ is isomorphic to $\mathbb{Q}(1 + \pi^2)$.

4.2.10. Prove that $\mathbb{Q}(\sqrt{2})$ is *not* isomorphic to $\mathbb{Q}(\sqrt{3})$.

4.2.11. If F is a field, prove that every non-zero polynomial in $F[x]$ has a unique monic associate in $F[x]$.

4.3

Finding an Extension in Which a Given Polynomial Has a Zero

So far, we have examined zeros lying in a known extension field of a polynomial's coefficient field. For example, $\sqrt{2}$ is already known to us as a real number, and i is known to us as a complex number. But what if we start with a polynomial such as $f = x^2 + x + 1$ in $(\mathbb{Z}/2\mathbb{Z})[x]$ (which clearly has no zero in $\mathbb{Z}/2\mathbb{Z}$)? Can we be certain that there *is* an extension of $\mathbb{Z}/2\mathbb{Z}$ in which f has a zero?

More generally, consider a field F and let $p \in F[x]$ be an irreducible polynomial. If we already *had* an extension field K of F such that p has a zero, α, in K, then $F(\alpha) = F[\alpha]$ would be isomorphic to $F[x]/\text{Ker } \phi_{\alpha/F} = F[x]/p\,F[x]$, by Theorem 4.2.1. In the absence of a known extension field, we are at liberty to create one *by forming* $E = F[x]/p\,F[x]$. This field will contain an isomorphic copy εF of F and a zero of the polynomial, p^ε, corresponding to p in $(\varepsilon F)[x]$. We can then, without loss, identify εF with F, hence p^ε with p, and regard E as the desired extension field of F, containing a zero of p.

Theorem 4.3.1: Let F be a field and let $p = \sum_{i=0}^{n} a_i x^i$ $(a_i \in F, a_n \neq 0)$ be an irreducible polynomial in $F[x]$. Then there is a field E and a monomorphism $\varepsilon : F \to E$ such that the polynomial $p^\varepsilon = \sum_{i=0}^{n} (\varepsilon a_i) x^i \in (\varepsilon F)[x]$ has a zero in E.

Proof: Let $E = F[x]/p\,F[x]$. Since p is an irreducible polynomial in $F[x]$, $p\,F[x]$ is a maximal ideal in $F[x]$, hence $E = F[x]/p\,F[x]$ is a field. Let $\nu : F[x] \to$

$F[x]/p\,F[x]$ be the natural homomorphism modulo the ideal $p\,F[x]$. Then $\nu f = p\,F[x] + f$ for each $f \in F[x]$. Consider the restriction, ε, of ν to F. We claim that ε is a monomorphism. For, if $\varepsilon a = 0_E \quad (a \in F)$, then $p\,F[x] + a = p\,F[x]$; hence $a \in p\,F[x]$. The only constant polynomial that is a multiple of p is the zero polynomial. Hence $a = 0$. Im $\varepsilon = \{\varepsilon a \mid a \in F\}$ is therefore a field, εF, isomorphic to F.

Let $p^\varepsilon = \sum_{i=0}^{n} (\varepsilon\, a_i)x^i$. Since $p \in \text{Ker } \nu$, $\nu p = 0_E$; hence $0_E = \nu\left(\sum_{i=0}^{n} a_i x^i\right) = \sum_{i=0}^{n} (\nu a_i)(\nu x)^i = \sum_{i=0}^{n} (\varepsilon a_i)(\nu x)^i = p^\varepsilon(\nu x)$. Thus, νx is a zero, in E, of p^ε. ■

Remark: In fact, with $E = F[x]/p\,F[x]$, and $\alpha = \nu x$ (see the preceding proof), we have $E \doteq (\varepsilon F)(\alpha)$, for: $E = \nu F[x] = \{(\nu f)(\nu x) \mid f \in F[x]\} = \{f^\varepsilon(\alpha) \mid f \in F[x]\} = \{g(\alpha) \mid g \in (\varepsilon F)[x]\} = (\varepsilon F)(\alpha)$.

Figure 1 may serve to clarify the situation:

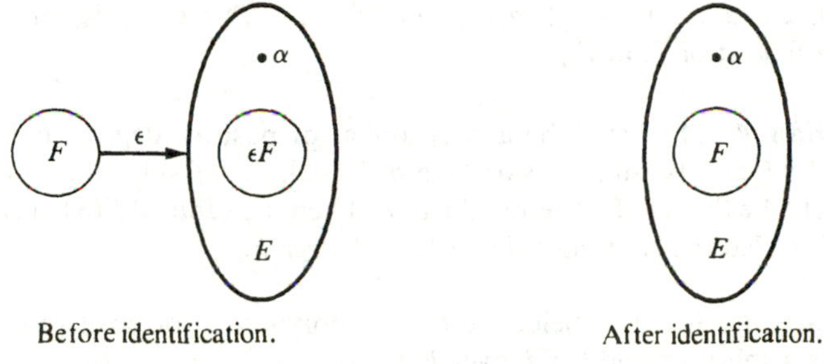

Figure 1

Remark: The reason we can "identify" F with εF is the following: given a monomorphism $\varepsilon : F \to E$ of a field F into a field E, it is possible to "wrap around" F a field $\bar{E}$ isomorphic to E (under an isomorphism whose restriction to F is ε). (Try proving it!)

Example:

With $F = \mathbb{Z}/2\mathbb{Z}$, let $p = x^2 + x + 1$, clearly irreducible in $F[x]$ since it has no zeros in F, hence no linear factors in $F[x]$. Forming $E = F[x]/p\,F[x]$, we have (in the notation of Theorem 4.3.1)

$$\nu p = (\nu 1)(\nu x)^2 + (\nu 1)\nu x + \nu 1 = (\varepsilon 1)(\nu x)^2 + (\varepsilon 1)\nu x + \varepsilon 1$$
$$= p^\varepsilon(\nu x) = 0_E.$$

With $\alpha = \nu x$, we have $E = (\varepsilon F)(\alpha) = \{f^\varepsilon(\alpha) \mid f \in F[x]\} = \{\varepsilon a_0 + \varepsilon a_1 \alpha \mid a_0, a_1 \in F\}$—the last equality by Corollary 2, part (4), p. 244. If we identify each $a \in F$ with its image εa, we have simply $E = F(\alpha) = \{a_0 + a_1\alpha \mid a_0, a_1 \in F\}$. (Using $1 + \alpha + \alpha^2 = 0$, we can compute the addition and multiplication tables previously presented in Example 4 of Section 4.1.)

As a consequence of Theorem 4.3.1, every polynomial of positive degree in $F[x]$ can be expressed as a product of linear factors in some extension field of F.

Theorem 4.3.2: Let F be a field, and let $f \in F[x]$ be a polynomial of degree $n > 0$. Then there is a field E containing F as a subfield such that $f = a_n(x - \alpha_1) \dots (x - \alpha_n)$, $a_n \in F$, $\alpha_i \in E$ (not necessarily distinct), $i = 1, \dots, n$.

Proof: We proceed by induction on n. If $n = 1$, then f *is* linear and we obtain the required representation $f = a_1(x - \alpha_1)$ where $\alpha_1 = a_0 a_1^{-1} \in F$.

For $n > 1$, suppose the theorem holds for polynomials of degree $n - 1$. By Theorem 3.6.5, f has an irreducible divisor $p \in F[x]$. Write $f = pg$, $g \in F[x]$. By Theorem 4.3.1, there is a field E_p, containing F as a subfield, such that p has a zero, α_1, in E_p. By the Factor Theorem (recall Exercise 3.6.17), $(x - \alpha_1)|p$ in $E_p[x]$. Hence $f = pg = (x - \alpha_1)hg$ for some $h \in E_p[x]$. By the induction hypothesis, since $\deg hg = n - 1$, E_p has an extension field E such that $hg = a_n(x - \alpha_2) \dots (x - \alpha_n)$, $\alpha_i \in E$, $i = 2, \dots, n$. But then E is an extension field of F such that $f = a_n(x - \alpha_1) \dots (x - \alpha_n)$, $\alpha_i \in E$, $i = 0, \dots, n$. (Note: a_n is, of course, the leading coefficient of f, in F.) ■

Definition 4.3.1: Let f be a polynomial of positive degree in $F[x]$ (F a field), and let $E \supset F$ be an extension field of F such that f splits into linear factors in $E[x]$. Let S be the set of all zeros of f in E. Then the subfield $F(S)$ of E generated by $F \cup S$ is called the *splitting field, in E, of f over F*.

Corollary: **(1)** If F is a field, $f \in F[x]$ a polynomial of positive degree, then there exists a splitting field for f over F.

(2) If $K \supset F$ is a splitting field for f over F and $E \supset F$ is a subfield of K such that f splits over E, then $E = K$.

Proof: **(1)** follows immediately from Theorem 4.3.2 and Definition 4.3.1.

(2) If f splits over E, then $S \subset E$, where S is the set of all zeros of f in K; hence $K = F(S) \subset E \subset K$, and so $E = K$. ■

Example 1:

The splitting field, in $\mathbb{C}$, of $f = x^2 + x + 1$ over $\mathbb{Q}$ is $\mathbb{Q}(\sqrt{-3})$.

Example 2:

The splitting field, in $\mathbb{R}$, of $f = x^2 - 5$ over $\mathbb{Q}$ is $\mathbb{Q}(\sqrt{5})$. The splitting field, in $\mathbb{R}$, of $x^2 - 5$ over $\mathbb{R}$ is $\mathbb{R}$.

Example 3:

The splitting field, in $\mathbb{C}$, of $f = x^2 + 1$ over $\mathbb{Q}$ is $\mathbb{Q}(i)$; its splitting field, in $\mathbb{C}$, over $\mathbb{R}$, is $\mathbb{C}$.

Using Theorem 4.3.1, combined with the power of set theory, it is possible to prove that every field F has an algebraic extension $\bar{F} \supset F$ such that *every* polynomial in $F[x]$ (and, indeed, in $\bar{F}[x]$) splits into linear factors in $\bar{F}[x]$. $\bar{F}$ is called an *algebraic closure* of F. (cf. [3], [8].)

Exercises 4.3

True or False

1. $\mathbb{Z}[x]/(x^2 - 2)\mathbb{Z}[x]$ is a field.
2. $\mathbb{Q}[x]/(x^2 - 4)\mathbb{Q}[x]$ is a field.
3. If F is a field and $f \in F[x]$ is irreducible over F, then some homomorphic image of $F[x]$ contains a zero of f.
4. There is *no* field F such that the polynomial $f = \sqrt{\pi}x^5 + ix^4 - (1 + \sqrt[3]{2})x^3 + e^3x^2 + \sqrt[5]{7}x + \sqrt{17}$ splits over F.
5. If $F = \mathbb{Q}[x]/(x^2 + x + 1)\mathbb{Q}[x]$, then F contains an element α such that $\alpha^3 = 1_F$.

4.3.1. Prove that the polynomial $f = x^4 - 2$ is irreducible over $\mathbb{Z}/5\mathbb{Z}$. Find an extension field E of $\mathbb{Z}/5\mathbb{Z}$ such that f has a zero in E, and show that, in fact, f splits over E. How many elements are in a splitting field for f over $\mathbb{Z}/5\mathbb{Z}$?

4.3.2. In contrast to the preceding exercise, find an extension field E of $\mathbb{Q}$ such that the polynomial $x^4 - 2$ has a zero in E, but does not split over E.

4.3.3. Let $p \in \mathbb{Z}$ be a prime, and let E_p be a splitting field, in $\mathbb{C}$, of $x^p - 1$ over $\mathbb{Q}$. Prove that E_p is isomorphic to $\mathbb{Q}[x]/\Phi_p\mathbb{Q}[x]$, where $\Phi_p = x^{p-1} + x^{p-2} + \cdots + x + 1$. (See Exercise 4.2.8.). Conclude that $(E_p : \mathbb{Q}] = p - 1$.

4.3.4. Construct a splitting field for $x^2 + x + 1$ over $(\mathbb{Z}/5\mathbb{Z})[x]$.

4.3.5. Find the splitting field, in $\mathbb{C}$, of $x^3 - 5$ over $\mathbb{Q}$.

4.3.6. Let F be a field, with Char $F \neq 2$, and let $f = ax^2 + bx + c$ be polynomial in $F[x]$ $(a \neq 0, b, c \in F)$. If $E \supset F$ is a splitting field for f over F, prove:

(1) $E = F(\alpha)$, where $\alpha^2 = b^2 - 4ac$;

(2) the zeros of f in E are $(-b + \alpha)(2a)^{-1}$ and $(-b - \alpha)(2a)^{-1}$.

Conclude: if $K \supset F$ is a field containing a zero of f, then f splits over K.

4.3.7. Let F be a field of characteristic 2. Prove that the final conclusion of Exercise 4.3.6 still holds: if f is a quadratic polynomial in $F[x]$ such that f has a zero in $K \supset F$, then f splits over K.

4.3.8. Let p be a prime, k a positive integer, and let F be a field of characteristic p. Let K be an extension field of F and let $\alpha \in K$. If $\alpha^{p^k} = a \in F$, prove that $x^{p^k} - a$ has splitting field $F(\alpha)$ in K. Factor $x^{p^k} - a$ over $F(\alpha)$. (Recall Exercise 4.1.12.)

4.3.9. Let F be a field of characteristic $p \neq 0$, and let $F(u)$ be an extension field of F, with u transcendental over F.

(1) If α is a zero of f in an extension field K of $F(u)$, prove that $F(u)(\alpha)$ is a splitting field for f over $F(u)$. Factor f over $F(u)$.

(2) Prove that $f = x^p - u \in F(u)[x]$ is irreducible over $F(u)$.

4.3.10. Let $f = x^5 + 3x^4 + 6x^3 + 9x^2 - 12x + 15$.

(1) Construct an extension field $\mathbb{Q}(\alpha)$ of $\mathbb{Q}$, where α is a zero of f.

(2) Express $\alpha^{-1} \in \mathbb{Q}(\alpha)$ as a polynomial in α.

4.3.11. Let F be a field, and let p be an irreducible polynomial in $F[x]$. Let α_1, α_2 be zeros of p in an extension field E of F. Prove that there is an isomorphism $\phi : F(\alpha_1) \to F(\alpha_2)$ such that $\phi(\alpha_1) = \alpha_2$ and $\phi|_F$ is the identity map on F. (Such an isomorphism ϕ is called an *F-isomorphism.*)

4.3.12. Using Exercise 4.3.11, conclude that the fields $\mathbb{Q}(\sqrt[3]{2})$, $\mathbb{Q}(\sqrt[3]{2}\,\omega)$, and $\mathbb{Q}(\sqrt[3]{2}\,\omega^2)$ are $\mathbb{Q}$-isomorphic $\left(\text{where } \omega = e^{2\pi i/3} = \dfrac{-1 + i\sqrt{3}}{2}\right)$.

4.4

Algebraic and Transcendental Extensions; Degree of an Extension; Finite Fields

Definition 4.4.1: Let F be a subfield of a field K. Then K is an *algebraic extension* of F if every element of K is algebraic over F. Otherwise, K is a *transcendental extension* of F.

It is easy enough to give examples of transcendental extensions. Granting the existence of transcendental irrational numbers, we know that $\mathbb{R}$ is transcendental over $\mathbb{Q}$. For any field, F, the quotient field $F(x)$ of the polynomial domain $F[x]$ is transcendental over F (what non-zero polynomial in $F[x]$ has x as a zero?).

But even the simplest questions regarding algebraicity at first seem difficult to answer. For example, if $F \subset E$ and $\alpha \in E$ is algebraic over F, is $F(\alpha)$ an algebraic extension of F? Or: if $F \subset E \subset K$, with E algebraic over F and K algebraic over E, is K algebraic over F? Is the sum, or the product, of two algebraic elements algebraic?

The key to coming to grips with these and other questions is the concept of the degree of an extension.

Let F be a subfield of a field K. Then $\langle K, + \rangle$ can be regarded as a vector space over the scalar field F, with field multiplication as scalar multiplication, i.e., for $x \in K$ and $a \in F$, the scalar multiple ax of x is the field product $a \cdot x$ in K. (See Exercise 4.4.1.) Every vector space, V, has dimension, either finite or infinite, equal to the cardinality of any basis of V.

Definition 4.4.2: Let F be a subfield of a field K. Then the dimension of K, considered as a vector space over F, is the *degree*, $[K : F]$, *of K over F*. K is a *finite extension* of F if $[K : F]$ is finite; K is an *infinite extension* of F if $[K : F]$ is infinite.

Notation: Various abbreviations are used to state quickly, for example, that a field K is algebraic (or transcendental, or finite) over a subfield F. We shall write "K/F is algebraic," etc., when brevity seems desirable. (Warning: in this connection, do not confuse "K/F" with a quotient structure!)

Theorem 4.4.1: Every finite extension is algebraic.

Proof: Let F be a subfield of K, with $[K : F] = n$, a positive integer, and let $\alpha \in K$. Then the $n + 1$ elements $1, \alpha, \alpha^2, \ldots, \alpha^n$ are linearly dependent over F. Hence there are elements $a_0, \ldots, a_n \in F$, not all zero, such that $\sum_{i=0}^{n} a_i\alpha^i = 0$. But then $f = \sum_{i=0}^{n} a_i x^i$ is a non-zero polynomial (of degree $\leq n$) in $F[x]$ such that $f(\alpha) = 0$, and so α is algebraic over F. It follows that K is an algebraic extension of F. ∎

Corollary: If F is a subfield of K, and $\alpha \in K$ is algebraic over F, with minimal polynomial $m_{\alpha/F}$ of degree n, then $[F(\alpha) : F] = n$; and, thus, $F(\alpha)$ is algebraic over F.

Proof: By Corollary 2 of Theorem 4.2.1, $F(\alpha) = \{r(\alpha) | r \in F[x],\ r = 0$ or $\deg r < n\}$; hence the elements $1, \alpha, \ldots, \alpha^{n-1}$ span the subspace $F(\alpha)$ of the vector space K over F. Moreover, $1, \alpha, \ldots, \alpha^{n-1}$ are linearly independent over F. For, if $\sum_{i=0}^{n-1} c_i\alpha^i = 0$, then the polynomial $g = \sum_{i=0}^{n-1} c_i x^i$ has α as a zero. If $g \neq 0$, then $n = \deg m_{\alpha/F} \leq \deg g \leq n - 1$. Contradiction! Hence $g = 0$, and so $c_i = 0$ for each $i = 0, \ldots, n - 1$.

We conclude that $1, \alpha, \ldots, \alpha^{n-1}$ form a basis for $F(\alpha)$ over F, and so $[F(\alpha) : F] = n$. From Theorem 2.4.1, it now follows that $F(\alpha)$ is an algebraic extension of F. ∎

> ***Warning:*** Theorem 4.4.1 asserts that every finite extension is algebraic. *The converse is false!* An algebraic extension need not be finite. For example, the set of all elements of $\mathbb{R}$ that are algebraic over $\mathbb{Q}$ forms an infinite algebraic extension of $\mathbb{Q}$ — see Theorem 4.4.3 and Exercise 4.4.10.

Definition 4.4.3: If F is a subfield of K and $\alpha \in K$ is algebraic over F, then $F(\alpha)$ is called a *simple algebraic extension* of F. The degree, $[F(\alpha) : F]$, of $F(\alpha)$ over F is called the *degree*, $[\alpha : F]$, *of* α *over* F.

Example:

$[\sqrt[3]{5} : \mathbb{Q}] = 3$, and $[\sqrt[3]{5} : \mathbb{R}] = 1$.

Theorem 4.4.2: Let $F \subset E \subset K$ be a *field tower*, i.e., let F be a subfield of E and let E be a subfield of K. Suppose $[E : F]$ and $[K : E]$ are finite. Then K is a finite extension of F, with

$$[K : F] = [E : F] \cdot [K : E].$$

Proof: Let $\{x_i\}_{i=1}^{s}$ be a basis for E over F, and let $\{y_j\}_{j=1}^{t}$ be a basis for K over E. We show that $\mathscr{B} = \{x_i y_j\}_{i=1,\ldots,s;\, j=1,\ldots,t}$ is a basis for K over F.

Let $x \in K$. Then there are elements $e_1, \ldots, e_t \in E$ such that

$$x = \sum_{j=1}^{t} e_j y_j.$$

For each $j = 1, \ldots, t$, there are elements $f_{1j}, \ldots, f_{sj} \in F$ such that

$$e_j = \sum_{i=1}^{s} f_{ij} x_i.$$

But then

$$x = \sum_{j=1}^{t} e_j y_j = \sum_{j=1}^{t} \left(\sum_{i=1}^{s} f_{ij} x_i \right) y_j = \sum_{i=1}^{s} \sum_{j=1}^{t} f_{ij} x_i y_j.$$

Thus, $\mathscr{B} = \{x_i y_j\}_{i=1,\ldots,s;\, j=1,\ldots,t}$ spans the vector space K over F.

$\mathscr{B}$ is linearly independent over F. For: suppose, for elements $a_{ij} \in F$,

$$\sum_{i=1}^{s} \sum_{j=1}^{t} a_{ij} x_i y_j = 0.$$

Then $\sum_{j=1}^{t} \left(\sum_{i=1}^{s} a_{ij} x_i \right) y_j = 0$. Since the y_j are linearly independent over E, $\sum_{i=1}^{s} a_{ij} x_i = 0$ for each $j = 1, \ldots, t$. But then, since the x_i are linearly independent over F, $a_{ij} = 0$ for each $i = 1, \ldots, s; j = 1, \ldots, t$.

It follows that $\mathscr{B} = \{x_i y_j\}_{i=1,\ldots,s;\, j=1,\ldots,t}$ forms a basis for K over F. But then, since the st elements $x_i y_j \in \mathscr{B}$ are all distinct (being linearly independent over F), we have

$$[K : F] = st = [E : F] \cdot [K : E]. \qquad \blacksquare$$

An easy induction extends the result of Theorem 4.4.2 to arbitrary finite field towers.

Corollary 1: If $F = F_0 \subset F_1 \subset F_2 \subset \cdots \subset F_n = E$ is a field tower such that $[F_i : F_{i-1}]$ is finite for each $i = 1, \ldots, n$, then $[E : F]$ is finite, with $[E : F] = \prod_{i=1}^{n} [F_i : F_{i-1}]$.

The following corollary characterizes finite extensions as being "finitely generated" by algebraic elements.

Corollary 2: Let F, E be fields, $F \subset E$. The following statements are equivalent:

(1) E/F is finite.

(2) There are finitely many elements $\alpha_1, \ldots, \alpha_n$ of E, each algebraic over F, such that $E = F(\alpha_1, \ldots, \alpha_n)$.

(3) There is a field tower

$$F = F_0 \subset F_1 \subset \cdots \subset F_n = E$$

such that $F_i = F_{i-1}(\alpha_i)$, with $\alpha_{i/F_{i-1}}$ algebraic for each $i = 1, \ldots, n$.

Proof: $1 \Rightarrow 2$: If E/F is finite, then there is a vector space basis $\alpha_1, \ldots, \alpha_n$ for E over F, and we have $E = F(\alpha_1, \ldots, \alpha_n)$. By Theorem 4.4.1, each α_i is algebraic over F.

$2 \Rightarrow 3$: If $E = F(\alpha_1, \ldots, \alpha_n)$, with each α_i algebraic over F, then

$$F \subset F(\alpha_1) \subset F(\alpha_1, \alpha_2) \subset \cdots \subset F(\alpha_1, \ldots, \alpha_n) = E. \tag{1}$$

Writing $F_0 = F$, $F_i = F(\alpha_1, \ldots, \alpha_i)$ for $i = 1, \ldots, n$, we have $F_i = F_{i-1}(\alpha_i)$, with α_i algebraic over F_{i-1}, for each $i = 1, \ldots, n$. Thus, (1) is a tower with the required properties.

$3 \Rightarrow 1$: If $F = F_0 \subset F_1 \subset \cdots \subset F_n = E$ is a tower such that $F_i = F_{i-1}(\alpha_i)$, with α_i algebraic over F_{i-1} $(i = 1, \ldots, n)$, then $[F_i : F_{i-1}]$ is finite for each $i = 1, \ldots, n$. By Corollary 1, it follows that $[E : F]$ is finite. ■

Remark: Theorem 4.4.2 holds more generally. The requirement that $[E : F]$ and $[K : E]$ be finite is not needed to produce the conclusion $[K : F] = [K : E]\,[E : F]$, provided that the notions of basis, dimension, and the product of two cardinal numbers are suitably extended (cf. [43], [45]).

Theorem 4.4.3: Let F be a subfield of a field K.

(1) If $\alpha, \beta \in K$ are algebraic over F, then $\alpha + \beta$, $\alpha - \beta$, and $\alpha\beta$ are algebraic over F; $\alpha\beta^{-1}$ is algebraic over F provided $\beta \neq 0$.

(2) If $K_A = \{\alpha \in K \mid \alpha \text{ is algebraic over } F\}$, then K_A is a subfield of K, containing F.

Proof: *(1)* If $\alpha, \beta \in K$ are algebraic over F, then β is algebraic over $F(\alpha)$ (see Exercise 4.4.3); hence (by the Corollary of Theorem 4.4.1) we have $F \subset F(\alpha) \subset F(\alpha)(\beta)$, with $[F(\alpha) : F]$ and $[F(\alpha)(\beta) : F(\alpha)]$ both finite. But then, by Theorem 4.4.2, $[F(\alpha)(\beta) : F]$ is finite; hence $F(\alpha)(\beta)/F$ is algebraic, by Theorem 4.4.1. Since $\alpha, \beta \in F(\alpha)(\beta)$, we conclude that $\alpha \pm \beta$, $\alpha\beta$ and $\alpha\beta^{-1}$ (in case $\beta \neq 0$) are all algebraic over F.

(2) follows immediately from (1). ■

Note that, in contrast, the set of all transcendental elements in an extension field is *not* closed under field operations: we have $(1 + \pi) + (1 - \pi) = 2 \in \mathbb{Q}$, and $\pi \cdot \dfrac{1}{\pi} = 1 \in \mathbb{Q}$.

Theorem 4.4.4: If $F \subset E \subset K$ is a field tower, with E/F and K/E algebraic, then K/F is algebraic.

Proof: Let $\alpha \in K$. Then $m_{\alpha/E} = a_0 + a_1 x + \cdots + a_{n-1}x^{n-1} + x^n$ for some $a_0, \ldots, a_{n-1} \in E$. Let $S = \{a_0, a_1, \ldots, a_{n-1}\}$. Then $F(S) = F(a_0)(a_1)\ldots(a_{n-1})$(see Exercise 4.2.5), and $[F(S) : F]$ is finite (see Exercise 4.4.9). Since α is a zero of $m_{\alpha/E} \in F(S)[x]$, α is algebraic over $F(S)$; hence $[F(S)(\alpha) : F(S)]$ is finite by the Corollary of Theorem 4.4.1. From the field tower

$$F \subset F(S) \subset F(S)(\alpha),$$

it follows, by Theorem 4.4.2, that $[F(S)(\alpha) : F]$ is finite, hence algebraic (by Theorem 4.4.1). Thus, α is algebraic over F. We conclude that K is algebraic over F.

■

Theorem 4.4.2 has, thus, enabled us to answer several questions regarding algebraic extensions. It also enables us to characterize completely the structure of all finite fields.

Theorem 4.4.5: Let F be a finite field. Then there is a prime p and a positive integer n such that the number of elements in F is p^n. *If* E is a subfield of F, then E has p^m elements for some positive divisor, m, of n. For each positive integer m such that $m|n$, F has a unique subfield, E, having p^m elements.

Proof: Since F is finite, it has characteristic p for some prime $p \in \mathbb{Z}$. Let P be the prime field of F. Then P is isomorphic to $\mathbb{Z}/p\mathbb{Z}$, hence has p elements. Being finite, F is finite-dimensional as a vector space over P. Let $n = [F : P]$. If $\{x_1, \ldots, x_n\}$ is a basis for F over P, then $F = \left\{\sum_{i=1}^{n} a_i x_i | a_i \in P\right\}$, and each $x \in F$ is *uniquely* expressible as a linear combination of the x_i. Since, for each $i = 1, \ldots, n$, there are exactly p choices of coefficients for x_i, F has p^n elements.

Now suppose E is a subfield of F. Since $1 \in E$, $P = \mathbb{Z}1 \subset E$, and E, as a vector space over P, forms a subspace of F, of degree $[E : P] = m \leq n = [F : P]$. But then, as above, we may conclude that E has p^m elements. By Theorem 4.4.2, $m|n$.

Conversely, suppose m is a positive divisor of $n = [F : P]$. Then $n = ms$ for some positive integer s. Since $\langle F - \{0\}, \cdot \rangle$ is a group of order $p^n - 1$, $a^{p^n - 1} = 1$, for each $a \in F^*$. Hence the polynomial $f = x^{p^n - 1} - 1$ in $P[x]$ has the $p^n - 1$ non-zero elements of F as its (distinct) zeros in F. Now, $p^n - 1 = (p^m)^s - 1 = (p^m - 1) \cdot k$, where $k = (p^m)^{s-1} + \cdots + p^m + 1 \in \mathbb{Z}$. But then $x^{p^n - 1} - 1 = x^{(p^m - 1)k} - 1 = f = (x^{p^m - 1} - 1)g$, where $g = (x^{p^m - 1})^{k-1} + \cdots + x^{p^m - 1} + 1$ in $P[x]$. By Theorem 3.6.5, since f splits into distinct linear factors in $F[x]$, so does each factor of f. Hence $x^{p^m - 1} - 1$ has $p^m - 1$ distinct zeros in F. But then the polynomial $h = x^{p^m} - x = x(x^{p^m - 1} - 1)$ has p^m distinct zeros in F. Let E be the set of all zeros in F of $x^{p^m} - x$. Let α, $\beta \in E$. Then $(\alpha + \beta)^{p^m} = \alpha^{p^m} + \beta^{p^m} = \alpha + \beta$ (see Exercise 4.1.4), whence $\alpha + \beta \in E$. Likewise, $(\alpha - \beta)^{p^m} = \alpha^{p^m} - \beta^{p^m} = \alpha - \beta$ (for p odd as well as for $p = 2$), whence $\alpha - \beta \in E$; $(\alpha\beta)^{p^m} = \alpha^{p^m}\beta^{p^m} = \alpha\beta$, whence $\alpha\beta \in E$. Finally, for $\beta \neq 0$, $(\alpha\beta^{-1})^{p^m} = \alpha^{p^m}\beta^{-p^m} = \alpha^{p^m}(\beta^{p^m})^{-1} = \alpha\beta^{-1}$, whence $\alpha\beta^{-1} \in E$. It follows that E is a subfield of F, having p^m elements.

If E' is another subfield of F, having p^m elements, then since $E' - \{0\}$ has $p^m - 1$ elements, $\alpha^{p^m - 1} = 1$ for each $\alpha \neq 0$ in E'. But then, $\alpha^{p^m} = \alpha$ for each $\alpha \in E'$, and so the elements of E' are all zeros, in F, of the polynomial $x^{p^m} - x$. This implies that $E' \subset E$. Since E' and E have the same (finite) number of elements, it follows that $E' = E$; this establishes the uniqueness claimed in the theorem. ■

Because it is relevant to the structure of finite fields and will also be useful on several occasions later on, we include here the following:

Theorem 4.4.6: Every finite subgroup of the multiplicative group of any field is cyclic.

Proof: Let F be a field, and let G be a finite subgroup of F^*, with $|G| = n$. If $n = 1$, then of course $G = \{1\}$ is cyclic. Suppose $n > 1$. Since $|G|$ is finite, there is an element, a, of maximal order, m, in G. If $m = n$, then $G = [a]$, hence cyclic. Suppose $m < n$. Then there is some $y \in G\backslash[a]$. For each $x \in [a]$, by Corollary 2 of Theorem 2.5.2, $x^m - 1 = 0$ in F; hence the elements of $[a]$ are m distinct zeros of the polynomial $x^m - 1$ in F. Since $x^m - 1$ cannot have more than m zeros in F, y is *not* a zero of $x^m - 1$, hence $o(y) \nmid m$. Write $t = o(y)$. Then there are positive integers t' and m' such that $t = dt'$ and $m = dm'$, where $d = \gcd(t, m)$, and thus $\gcd(t', m') = 1$. Since $t \nmid m$, we have $t' > 1$. The elements y^d and a^d have relatively prime orders t' and m', respectively, hence (recall Exercise 2.3.16) $o(y^d a^d) = t'm'$. But then $o(ya) = dt'm' > dm' = m$, contrary to the hypothesis that m is maximal among the orders of elements of G. Contradiction! We conclude that $m = n$, and so $G = [a]$ is cyclic. ■

Corollary: If F is a finite field, then F^* is cyclic.

Exercises 4.4

True or False

1. Every algebraic extension is a finite extension.
2. If $F \subset E \subset K$, with E/F algebraic and K/E transcendental, then K/F is transcendental.
3. There is a finite field consisting of 12 elements.
4. If Char $F = p > 0$, then $(\alpha + \beta)^p = \alpha^p + \beta^p$ for all $\alpha, \beta \in F$.
5. If E/F is a transcendental extension, then every element $\alpha \in E\backslash F$ is transcendental over F.

4.4.1. Let K be a field, and let F be a subfield of K. Prove that K forms a vector space over F, with vector addition defined as addition in K, and scalar multiplication defined by:

$$ax = a \cdot x$$

for each $a \in F$, $x \in K$.

4.4.2. Let $F \subset E \subset K$ be a field tower. Prove: if K is algebraic over F, then K is algebraic over E.

4.4.3. Let $F \subset E \subset K$ be a field tower. Prove: if E is transcendental over F, then K is transcendental over F.

4.4.4. Find:

(1) $[\mathbb{Q}(\sqrt[4]{2}) : \mathbb{Q}]$
(2) $[\mathbb{Q}(\sqrt[4]{2}) : \mathbb{Q}(\sqrt{2})]$
(3) $[\mathbb{Q}(\sqrt[4]{2}, \sqrt{3}) : \mathbb{Q}(\sqrt{2}, \sqrt{3})]$
(4) $[\mathbb{Q}(\sqrt{2}, \sqrt{3}) : \mathbb{Q}]$.

4.4.5. Find a basis for

(1) $\mathbb{Q}(\sqrt{2})$ over $\mathbb{Q}$;
(2) $\mathbb{Q}(\sqrt{2}, \sqrt{3})$ over $\mathbb{Q}$;
(3) $\mathbb{Q}(\sqrt{2}, \sqrt{3}, \sqrt{5})$ over $\mathbb{Q}$.

4.4.6. *(1)* Find the degree of $\sqrt{2} + \sqrt{3}$ over $\mathbb{Q}$.
(2) Prove that $\mathbb{Q}(\sqrt{2} + \sqrt{3}) = \mathbb{Q}(\sqrt{2}, \sqrt{3})$. (See Exercise 4.2.7.)

4.4.7. Determine under what conditions the subfields of a finite field can be represented by a single field tower.

4.4.8. Draw the subfield lattice for
(1) a field of 64 elements;
(2) a field of 59049 elements;
(3) a field of 125 elements.
Indicate the degree of each subfield over its immediate predecessors in the lattice.

4.4.9. Let E/F be a finite extension of odd degree, with $E = F(\alpha)$. Prove that $F(\alpha^2) = F(\alpha)$.

4.4.10. Let $p_1, p_2, \ldots$ be the primes in $\mathbb{Z}$, in increasing order (i.e., $p_1 = 2, p_2 = 3$, etc.). Prove that $\mathbb{Q}(\sqrt{p_1}, \sqrt{p_2}, \ldots)$ is an infinite algebraic extension of $\mathbb{Q}$. (Hint: Proceeding by induction on n, prove that $\mathbb{Q}(\sqrt{p_1}, \ldots, \sqrt{p_n})$ is of degree 2^n over $\mathbb{Q}$, for each positive integer n.)

4.4.11. For each $k \in \mathbb{Q}$, $k \neq 0$, let $\alpha = \sqrt{2} + k\sqrt{3}$.
(1) Find $m_{\alpha/\mathbb{Q}}$.
(2) Find all zeros of $m_{\alpha/\mathbb{Q}}$ in $\mathbb{C}$.
(3) Prove that $\mathbb{Q}(\sqrt{2} + k\sqrt{3}) = \mathbb{Q}(\sqrt{2}, \sqrt{3})$.

4.4.12. Prove that $\mathbb{Q}(\sqrt{2} + \sqrt[4]{2}) = \mathbb{Q}(\sqrt{2}, \sqrt[4]{2})$.

4.4.13. Let F be a field, with Char $F \neq 2$. Let E be an extension field of F such that $[E : F] = 2$. Prove:
(1) For each $\beta \in E$, $\beta \notin F$, $[F(\beta) : F] = 2$, and $E = F(\beta)$.
(2) There is some $\alpha \in E$ such that $\alpha^2 \in F$ and $E = F(\alpha)$. (See Exercise 4.3.6.)

4.4.14. Define $\tau : \mathbb{Q}(\sqrt{2}) \to \mathbb{Q}(\sqrt{2})$ by: $\tau(a + b\sqrt{2}) = a - b\sqrt{2}$ for each $a, b \in \mathbb{Q}$.
(1) Prove that τ is an automorphism of the field $\mathbb{Q}(\sqrt{2})$ such that $\tau|_{\mathbb{Q}} = \iota_{\mathbb{Q}}$.
(2) Prove that τ is an automorphism (i.e., a non-singular linear operator) of the vector space $\mathbb{Q}(\sqrt{2})$ over $\mathbb{Q}$.
(3) Find the matrix of τ relative to the ordered basis $[1, \sqrt{2}]$ for $\mathbb{Q}(\sqrt{2})$ over $\mathbb{Q}$.

4.4.15. Let F be a field with prime field P such that $n = [F : P]$ is finite. Prove:
(1) If Char $F = p > 0$, then F is finite, and its additive group is the internal direct sum of n subgroups, each isomorphic to $\langle \mathbb{Z}_p, \oplus \rangle$;
(2) if Char $F = 0$, then F is infinite, and its additive group is the internal direct sum of n subgroups, each isomorphic to $\langle \mathbb{Q}, + \rangle$.

4.4.16. Prove that the additive group of $\mathbb{C}$ is the internal direct sum of two subgroups, each isomorphic to $\langle \mathbb{R}, + \rangle$.

4.4.17. Assuming that π is transcendental over $\mathbb{Q}$, prove that $\sqrt[n]{\pi}$ is transcendental over $\mathbb{Q}$ for each positive integer n.

4.4.18. Modify the proof of Theorem 4.4.6 to obtain the following purely group-theoretic result:
Let G be a finite abelian group with the property: for every positive integer m, there are at most m elements $a \in G$ such that $a^m = e$. Then G is cyclic.

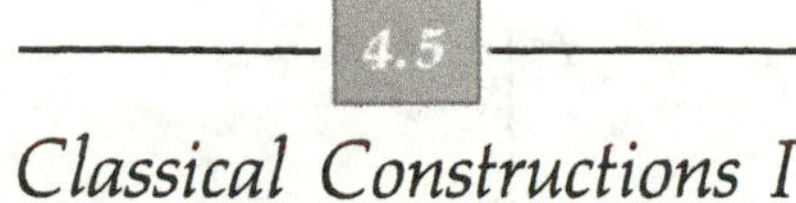

Classical Constructions I

In this section, we demonstrate further the power of the few results in field theory which we have thus far developed.

The ancient Greeks, who were masters of synthetic geometry, attempted many constructions using only two simple tools: an unmarked straightedge and a pair of compasses. With these tools, they were able to perform many interesting constructions — but they were unable, in particular, to accomplish the following three tasks:

(1) The trisection of an arbitrary angle.

(2) The "duplication of the cube," i.e., the construction of the edge of a cube with volume twice the volume of any given cube.

(3) The "squaring of the circle," i.e., the construction of the edge of a square with area equal to the area of any given circle.

Lacking sufficient algebraic tools, they were also unable to prove the impossibility of these constructions. As a result, much energy has been expended ever since, particularly by amateurs, in attempting to accomplish these tasks. Only part of the small amount of field theory which we developed in the preceding sections is required to prove that each of the three proposed tasks is impossible. (Despite the fact that the impossibility of these constructions has been known for over a century, occasional purported solutions by amateurs are still being submitted to mathematics departments.)

The two basic constructions that can be performed using the classical tools are:

(1) drawing a straight line through two given points and

(2) drawing a circle with given center and radius.

To set the stage for our algebraic discussion, we begin by constructing a coordinate system.

Mark two points O, P arbitrarily in the plane; draw a line through O and P using the straightedge (see Figure 2).

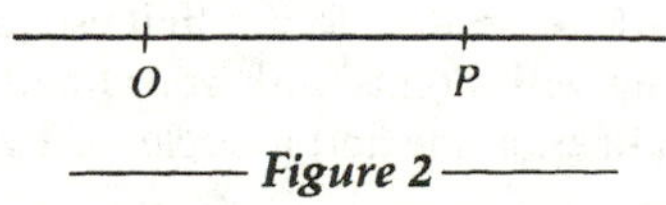

Figure 2

Declare the line to be the x-axis with O as origin, and P as unit point (i.e., the point with x-coordinate 1). Using straightedge and compasses, construct the perpendicular to the x-axis at O, declare it to be the y-axis, and, on it, mark the unit point, P'. You now have a Cartesian coordinate system (see Figure 3). Any point with integer coordinates, on as well as off the axes, can now be constructed in the obvious way, using compasses.

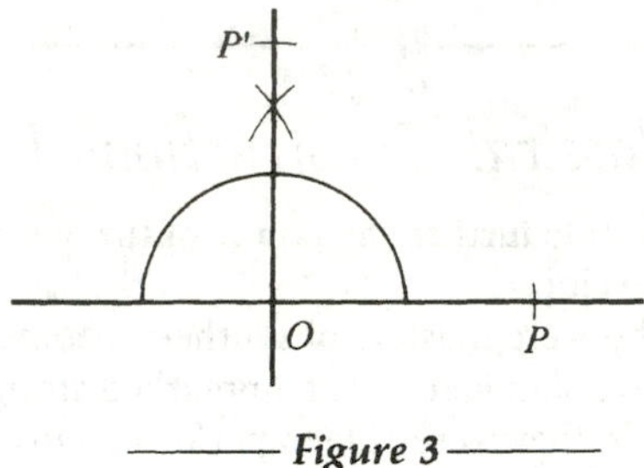

Figure 3

Using the construction illustrated in Figure 4 for dividing a line segment into n equal parts (for n any positive integer), any rational point on the coordinate axes, with x or y coordinate $\frac{m}{n}$, can now be constructed. From this, one can proceed to construct any point in the plane if both of its coordinates are rational.

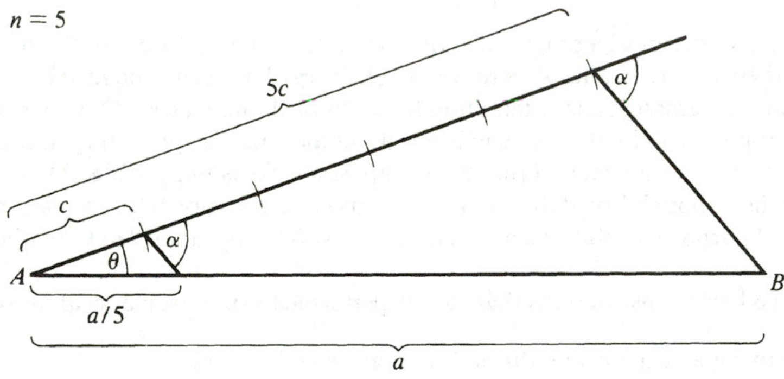

c: an arbitrary length
θ: an arbitrary acute angle

Figure 4

Definition 4.5.1: We shall say that a *point* P with real coordinates is *constructible* over a subfield K, $\mathbb{Q} \subset K \subset \mathbb{R}$, if a finite sequence of basic constructions (either A or B), starting with points with coordinates in K, leads to P as a point of intersection of two lines, a line and a circle, or two circles.

A *real number* ρ is *constructible* over K if the point $(\rho, 0)$ is constructible over K.

Lemma: Let $P_1, \ldots, P_n$ be points with real coordinates. For each $i = 1, \ldots, n$, let E_i be the subfield of $\mathbb{R}$ obtained by adjoining to $\mathbb{Q} = E_o$ the coordinates of $P_1, \ldots, P_i$. Suppose that, for each $i = 1, \ldots, n$, the point P_i is obtainable by a single basic construction from points with coordinates in E_{i-1}. Then the field tower

$$\mathbb{Q} = E_o \subset E_1 \subset \cdots \subset E_n$$

has $[E_i : E_{i-1}]$ equal to 1 or 2 for each $i = 1, \ldots, n$, whence $[E_n : \mathbb{Q}]$ is a power of 2.

Proof: For each $i = 1, \ldots, n$, let (x_i, y_i) be the coordinates of P_i.

Case 1: P_i is the point of intersection of two straight lines L_1 and L_2, each passing through two points with coordinates in E_{i-1}. Then the coefficients of the equations for L_1 and L_2 are also elements of E_{i-1}, and therefore the simultaneous solution (x_i, y_i) has both x_i and y_i in E_{i-1}. Thus, in this case, $E_i = E_{i-1}$, and we have $[E_i : E_{i-1}] = 1$.

Case 2: P_i is a point of intersection of a line L through two points with coordinates in E_{i-1} and a circle C whose center has coordinates in E_{i-1}, and whose radius is in E_{i-1}. Again, the equations representing L and C have their coefficients in E_{i-1}. Solving them simultaneously, one obtains a quadratic equation with coefficients in E_{i-1}. Hence the simultaneous solution (x_i, y_i) has both coordinates either in E_{i-1} or in $E_{i-1}(\alpha)$ where α^2 is an element of E_{i-1}. Thus, $[E_i : E_{i-1}]$ is either 1 or 2.

Case 3: P_i is a point of intersection of two circles C_1 and C_2, each with center coordinates and radius in E_{i-1}. Again, the equations representing C_1 and C_2 have all their coefficients in E_{i-1}. Solving simultaneously the equations

$$x^2 + y^2 + a_1x + b_1y + c_1 = 0 \quad \text{and} \quad x^2 + y^2 + a_2x + b_2y + c_2 = 0$$

is equivalent to solving simultaneously either one of these equations and the equation

$$(a_1 - a_2)x + (b_1 - b_2)y + c_1 - c_2 = 0$$

representing the line through the points of intersection of C_1 and C_2 (again with coefficients in E_{i-1}). Case 3 thus reduces to the previous case, and the outcome is the same.

It follows immediately that $[E_n : \mathbb{Q}]$ is a power of 2. ∎

Theorem 4.5.1 (Fundamental Theorem on Constructibility): A real number ρ is constructible over $\mathbb{Q}$ if and only if there is a field tower

$$\mathbb{Q} = E_0 \subset E_1 \subset \cdots \subset E_t \subset \mathbb{R} \qquad (t \geq 0)$$

such that $\rho \in E_t$ and $[E_i : E_{i-1}] = 2$ for each $i = 1, \ldots, t$.

Proof: By Definition 4.5.1, if $\rho \in \mathbb{R}$ is constructible over $\mathbb{Q}$, then the point $P : (\rho, 0)$ is constructible over $\mathbb{Q}$. Hence there is a finite sequence of basic constructions, A or B, leading from rational points to P. The (finitely many) points of intersection obtained at the various stages in the construction of P form a sequence $P_1, \ldots, P_n = P$ of points satisfying the conditions of the Lemma. Deleting repetitions from the field tower guaranteed by the Lemma, we obtain a tower

$$\mathbb{Q} = E_0 \subset E_1 \subset \cdots \subset E_t \subset \mathbb{R} \qquad (t \geq 0),$$

with $\rho \in E_t$, $[E_i : E_{i-1}] = 2 \quad (i = 1, \ldots, t)$.

Conversely, suppose there is a field tower

$$\mathbb{Q} = E_0 \subset E_1 \subset E_2 \subset \cdots \subset E_t \subset \mathbb{R}, \quad [E_i : E_{i-1}] = 2$$

for each $i = 1, \ldots, t$. Let $\rho \in E_t$. We prove, by induction on t, that ρ is constructible over $\mathbb{Q}$. For $t = 0$, we have $\rho \in \mathbb{Q}$, hence ρ is constructible over $\mathbb{Q}$. For $t > 0$, suppose our assertion holds for $t - 1$. Since $\rho \in E_t$ and $[E_t : E_{t-1}] = 2$, ρ has degree 1 or 2 over E_{t-1}. If ρ has degree 1 over E_{t-1}, then $\rho \in E_{t-1}$; hence ρ is constructible over $\mathbb{Q}$, by the induction hypothesis. If ρ has degree 2 over E_{t-1}, then $m_{\rho/E_{t-1}} = x^2 + px + q$ for some $p, q \in E_{t-1}$, and $\rho = 2^{-1}(-p + \alpha)$ where $0 < \alpha^2 = p^2 - 4q \in E_{t-1}$. By the induction hypothesis, α^2 is constructible over $\mathbb{Q}$. Since real square roots of constructible numbers are constructible, it follows that α is constructible over $\mathbb{Q}$. But then, since $2 \in \mathbb{Q}$ and $p \in E_{t-1}$, $\rho = 2^{-1}(-p + \alpha)$ is constructible over $\mathbb{Q}$. ■

Corollary: If $\rho \in \mathbb{R}$ is constructible over $\mathbb{Q}$, then ρ is algebraic, of degree a power of 2 over $\mathbb{Q}$.

Proof: If $\rho \in \mathbb{R}$ is constructible over $\mathbb{Q}$, then there is a field tower

$$\mathbb{Q} = E_0 \subset E_1 \subset \cdots \subset E_t \qquad (t \geq 0)$$

with $\rho \in E_t$ and $[E_i : E_{i-1}] = 2$ for each $i = 1, \ldots, t$. Hence $[E_t : \mathbb{Q}]$ is a power of 2. Since $\mathbb{Q} \subset \mathbb{Q}(\rho) \subset E_t$, we conclude that ρ is algebraic, of degree a power of 2, over $\mathbb{Q}$. ■

This corollary is the only tool required to settle the Grecian problems.

Theorem 4.5.2: Using only straightedge and compasses, it is impossible

(1) to duplicate an arbitrary cube;

(2) to square an arbitrary circle;

(3) to trisect an arbitrary angle.

(Put another way: there exist cubes that cannot be duplicated, circles that cannot be squared, and angles that cannot be trisected.)

Proof: *(1)* To duplicate a cube of side length 1, we would need to construct the side of a cube whose volume is $2 \cdot 1^3 = 2$. This is impossible, by Theorem 4.5.1, since $\sqrt[3]{2}$ is of degree 3, not a power of 2, over $\mathbb{Q}$.

(2) To square a circle of radius 1, we would need to construct the edge of a square whose area is $\pi \cdot 1^2 = \pi$. This is impossible, by Theorem 4.5.1, since $\sqrt{\pi}$ is transcendental over $\mathbb{Q}$.

(3) Let us define an angle θ, $0 \leq \theta < \pi$, to be constructible if its cosine is a constructible real number. (This is a sensible definition. For, given $\cos\theta$ and the initial side of θ, we can construct a point on the terminal side of θ—see Figure 5.)

Now let $\phi = 60°$. We prove that $\theta = 20°$ is not constructible. Since $\cos 60° = \frac{1}{2}$, and $\cos 3\theta = 4\cos^3\theta - 3\cos\theta$ for each $\theta \in \mathbb{R}$, we have

$$\tfrac{1}{2} = 4\cos^3 20° - 3\cos 20°.$$

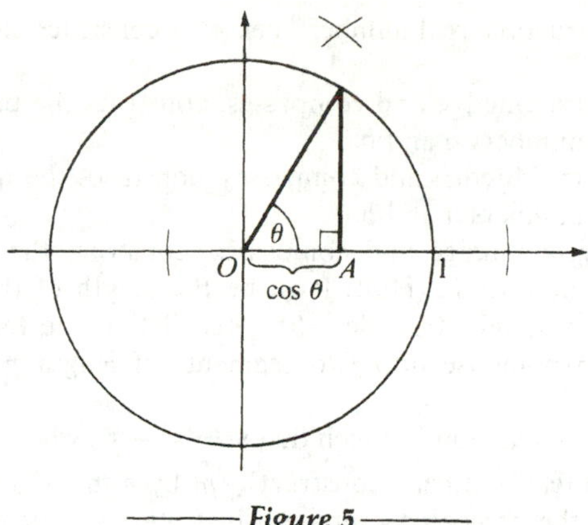

Figure 5

Letting $\alpha = 2\cos 20°$, we obtain

$$\alpha^3 - 3\alpha - 1 = 0.$$

Thus, α is a zero of $f = x^3 - 3x - 1 \in \mathbb{Q}[x]$.

If f were reducible over $\mathbb{Q}$, one of its factors would be linear, and so f would have a rational zero. Since the only possible candidates, ± 1, for rational zeros of f do not work, f is irreducible over $\mathbb{Q}$. But then f is the minimal polynomial of α over $\mathbb{Q}$, and so $\alpha = 2\cos 20°$ has degree 3 over $\mathbb{Q}$. By Theorem 4.5.1, it follows that $\alpha = 2\cos 20°$ is not constructible — hence neither is $\frac{\alpha}{2} = \cos 20°$. But then 60° is not trisectable. ■

Remark: (1) Clearly, some angles *are* trisectable, e.g., 180°, 90°, 135°. (2) Also, some cubes *can* be duplicated, and some circles *can* be squared: you should have no trouble duplicating a cube of side length $\sqrt[3]{4}$, or squaring a circle of radius $\frac{1}{\sqrt{\pi}}$.

Exercises 4.5

True or False

1. Someday, someone will figure out how to trisect an arbitrary angle using only a straightedge and compasses.
2. Some angles can be trisected, some cubes can be duplicated, and some circles can be squared.
3. $\sqrt[6]{5}$ is constructible using only straightedge and compasses.
4. An angle of 40° is constructible using only a straightedge and compasses.

5. If ρ is a constructible real number, then ρ^n is constructible for each positive integer n.

4.5.1. Using only straightedge and compasses, construct the product, ab, of two positive real numbers a and b.

4.5.2. Using only straightedge and compasses, construct the quotient, $\frac{a}{b}$, of two positive real numbers a and b.

4.5.3. Using only straightedge and compasses, construct the square root of a positive real number h. (Hint: let h be the length of the altitude on the hypotenuse of a right triangle ABC. Recall that the foot of the altitude divides the hypotenuse into two segments of length p and q such that $pq = h^2$.)

4.5.4. If m is a positive real number such that $m = s^2 + t^2$, where s and t are known constructible real numbers, construct $\sqrt{m}$ by a method simpler than that suggested in the preceding exercise. Illustrate your method by constructing $\sqrt{5}$.

4.5.5. Let $\zeta = e^{2\pi i/5} = \cos 72° + i \sin 72°$. By Exercise 4.2.8, $m_{\zeta/\mathbb{Q}} = \Phi_5 = x^4 + x^3 + x^2 + x + 1$. Prove:

(1) $\zeta + \zeta^{-1} = 2\cos 72°$.

(2) $(\zeta + \zeta^{-1})^2 + (\zeta + \zeta^{-1}) - 1 = 0$.

(3) $\cos 72° = \dfrac{-1 + \sqrt{5}}{4}$.

4.5.6. Use the result of Exercise 4.5.4 to construct a regular pentagon inscribed in a circle of radius 1.

4.5.7. The proportion most favored in classical antiquity, known as the Golden Section, was $2 : (1 + \sqrt{5})$. Construct a rectangle whose sides are proportioned according to the Golden Section. (Then look at a picture of the Acropolis in Athens.)

4.5.8. (1) Prove that the angles 2°, 4°, 5°, and 10° are not constructible. (Hint: use Theorem 4.5.2.2.)

(2) Conclude that the angles 6°, 12°, 15°, and 30° are not trisectable.

(3) Is a regular 72-gon constructible?

4.5.9. (1) Prove directly that cos 40° is not constructible. (Hint: following the proof of Theorem 4.5.2 (2), prove that 120° is not trisectable.)

(2) Alternatively, deduce the non-constructibility of cos 40° from the result of Theorem 4.5.2(2).

(3) Is a regular 9-gon constructible?

Remark: For the convenience of those not proceeding to Section 4.17, we state here the theorem of Gauss regarding constructibility of regular n-gons:
Let n be a positive integer. Then a regular n-gon is constructible if and only if $n = 2^s p_1 \dots p_t$ where $s \geq 0$ and $p_1, \dots, p_t$ are distinct Fermat primes. (A prime, p, of form $p = 2^{2^m} + 1 \quad (m \geq 0)$ is called a Fermat prime. It is not known whether the number of Fermat primes is infinite. In fact, only five Fermat primes are known to date.)

4.6

Extension of Isomorphisms

We have already observed (p. 238 and Exercise 4.1.8) that any field monomorphism $\varepsilon : F \to K$ extends to a ring monomorphism $\hat{\varepsilon} : F[x] \to K[x]$. In particular, a field isomorphism $\varepsilon : F \to K$ extends to a ring isomorphism $\hat{\varepsilon} : F[x] \to K[x]$ such that, if $f = \sum_{i=0}^{n} a_i x^i \in F[x]$, then $\hat{\varepsilon}(f) = f^{\varepsilon} = \sum_{i=0}^{n} (\varepsilon a_i)x^i \in K[x]$. If $p \in F[x]$ is irreducible over F, then $p^{\varepsilon} \in K[x]$ is irreducible over K.

We now consider certain extensions of isomorphisms associated with a single polynomial.

Theorem 4.6.1: Let $\varepsilon : F \to K$ be a field isomorphism and let $p \in F[x]$ be irreducible over F. Let $E \supset F$ be an extension field of F containing a zero, α, of p, and let $L \supset K$ be an extension field of K containing a zero, β, of p^{ε}. Then ε extends to an isomorphism $\varepsilon_1 : F(\alpha) \to K(\beta)$ such that $\varepsilon_1(\alpha) = \beta$.

Proof: Note that $p \sim m_{\alpha/F}$ and $p^{\varepsilon} \sim m_{\beta/K}$ (where "$\sim$" denotes "is an associate of"). Since α is algebraic over F and β is algebraic over K, we have $F(\alpha) = F[\alpha]$ and $K(\beta) = K[\beta]$ (by Theorem 4.2.1).

Define $\varepsilon_1 : F(\alpha) \to K(\beta)$ by: $\varepsilon_1(f(\alpha)) = f^{\varepsilon}(\beta)$. Then ε_1 is well-defined and 1-1. For: if $f, g \in F[x]$, then $f(\alpha) = g(\alpha) \Leftrightarrow (f - g)(\alpha) = 0 \Leftrightarrow p \mid f - g \Leftrightarrow p^{\varepsilon} \mid \hat{\varepsilon}(f - g) \Leftrightarrow p^{\varepsilon} \mid f^{\varepsilon} - g^{\varepsilon} \Leftrightarrow (f^{\varepsilon} - g^{\varepsilon})(\beta) = 0 \Leftrightarrow f^{\varepsilon}(\beta) = g^{\varepsilon}(\beta)$.

For $f, g \in F[x]$, $\varepsilon_1(f(\alpha) + g(\alpha)) = \varepsilon_1((f + g)(\alpha)) = (f + g)^{\varepsilon}(\beta) = f^{\varepsilon}(\beta) + g^{\varepsilon}(\beta) = \varepsilon_1(f(\alpha)) + \varepsilon_1(g(\alpha))$, and $\varepsilon_1(f(\alpha)g(\alpha)) = \varepsilon_1((fg)(\alpha)) = (fg)^{\varepsilon}(\beta) = f^{\varepsilon}(\beta)g^{\varepsilon}(\beta) = \varepsilon_1(f(\alpha))\varepsilon_1(g(\alpha))$. Thus, ε_1 is a homomorphism. For each $\bar{f} \in K[x]$, there is some $f \in F[x]$ such that $\bar{f} = f^{\varepsilon}$; hence $\bar{f}(\beta) = f^{\varepsilon}(\beta) = \varepsilon_1(f(\alpha))$. Thus, ε_1 is an isomorphism of $F(\alpha)$ onto $K(\beta)$. Clearly, $\varepsilon_1(\alpha) = \beta$. ■

Corollary: Let F be a field, $p \in F[x]$ an irreducible polynomial, and let $E \supset F$ be an extension field of F. If $\alpha, \beta \in E$ are both zeros of p, then there is an isomorphism of $F(\alpha)$ onto $F(\beta)$ which sends α to β, and leaves each element of F fixed.

Proof: This follows immediately from the preceding theorem, with $\varepsilon = \iota_F$. ■

The substance of this corollary lies close to the heart of Galois theory: structurally, $F(\alpha)$ and $F(\beta)$ are *indistinguishable* if α and β are zeros of the same irreducible polynomial over F. We shall restate the corollary after introducing some new language.

Definition 4.6.1: Let E be an extension field of F, with $E \supset F$. If $K \supset F$ is another extension field of F, and $\phi : E \to K$ is a homomorphism such that $\phi(a) = a$

for each $a \in F$, then ϕ is an *F-homomorphism* from E to K. (The terms *F-monomorphism*, *F-isomorphism*, and *F-automorphism* have the obvious meanings.)

Remark: Under the hypotheses of Definition 4.6.1, a field homomorphism $\phi : E \to K$ is an F-homomorphism if and only if ϕ is a linear transformation of the vector space E over F into the vector space K over F (see Exercise 4.6.5).

Our corollary can now be worded as follows.

Corollary (restated): Let F be a field, $p \in F[x]$ an irreducible polynomial, and let $E \supset F$ be an extension field of F. If $\alpha, \beta \in E$ are zeros of p, then there is an F-isomorphism of $F(\alpha)$ onto $F(\beta)$ which sends α to β.

Example 1:

There is a $\mathbb{Q}$-automorphism of $\mathbb{Q}(\sqrt{2}) = \mathbb{Q}(-\sqrt{2})$, sending $\sqrt{2}$ to $-\sqrt{2}$.

Example 2:

There is an $\mathbb{R}$-automorphism of $\mathbb{R}(i) = \mathbb{R}(-i)$, sending i to $-i$ (generally known as complex conjugation).

Example 3:

The fields $\mathbb{Q}(\sqrt[3]{2})$, $\mathbb{Q}(\sqrt[3]{2}\omega)$ and $\mathbb{Q}(\sqrt[3]{2}\omega^2)$ are $\mathbb{Q}$-isomorphic $\left(\text{where } \omega = \dfrac{-1+\sqrt{-3}}{2},\ \omega^2 = \dfrac{-1-\sqrt{-3}}{2}\right)$. No two of these fields are equal. (See Exercise 4.6.1.)

Another natural consequence of Theorem 4.6.1 is the following theorem.

Theorem 4.6.2: Let F be a field, f a polynomial of positive degree in $F[x]$, and $\varepsilon : F \to K$ a field isomorphism. If $E \supset F$ is a splitting field of f over F and $L \supset K$ is a splitting field of f^{ε} over K, then ε extends to an isomorphism $\bar{\varepsilon} : E \to L$.

Proof: We proceed by induction on $\deg f$. For $\deg f = 1$, we have $E = F$, $L = K$, and $\bar{\varepsilon} = \varepsilon$.

For $n > 1$, suppose the theorem holds for polynomials of degree $n - 1$. Let p be an irreducible divisor of f in $F[x]$ and let α be a zero of p in E. (Note that p splits over E.) Then p^{ε} is an irreducible divisor of f^{ε} in $K[x]$. If β is a zero of p^{ε} in K, then, by Theorem 4.6.1, ε extends to an isomorphism $\varepsilon_1 : F(\alpha) \to K(\beta)$ sending α to β. By the Factor Theorem (see Exercise 3.6.17), $f = (x - \alpha)f_1$; hence $f^{\varepsilon} = (x - \beta)f_1^{\varepsilon_1}$, where $f_1 \in F(\alpha)[x]$ and $f_1^{\varepsilon_1} \in K(\beta)[x]$. Then f_1 splits over E (why?) and $f_1^{\varepsilon_1}$ splits over L, whence E contains a splitting field E_1 for f_1 over $F(\alpha)$ and L contains a splitting field L_1 for $f_1^{\varepsilon_1}$ over $K(\beta)$. But $\deg f_1 = n - 1$; hence, by the induction hypothesis, ε_1 extends to an isomorphism $\bar{\varepsilon} : E_1 \to L_1$. Clearly, f splits over E_1 and

f^{ε} splits over L_1, whence $E_1 = E$, $L_1 = L$ (by the Corollary on p. 248), and $\bar{\varepsilon}$ is an isomorphism of E onto L, with $\bar{\varepsilon}|_F = \varepsilon$, as required. ∎

Corollary: Let F be a field, $f \in F[x]$. If $E \supset F$ and $K \supset F$ are splitting fields of f over F, then there is an F-isomorphism of E onto K.

Proof: Let $\varepsilon = \iota_F$, and apply Theorem 4.6.2. ∎

Exercises 4.6

True or False

1. If α and β have the same degree over F, then $F(\alpha)$ and $F(\beta)$ are isomorphic, as fields.
2. If α and β have the same degree over F, then $F(\alpha)$ and $F(\beta)$ are isomorphic, as vector spaces over F.
3. $\mathbb{Q}(\sqrt[3]{2})$ and $\mathbb{Q}\left(\frac{\sqrt[3]{2}}{2}(-1 + \sqrt{-3})\right)$ are isomorphic fields.
4. If K is a prime field of characteristic 0 and L is a splitting field of $1_K x^2 - 2_K$ over K, then L is isomorphic to $\mathbb{Q}(\sqrt{2})$.
5. If α and β are zeros of $f \in F[x]$, then $F(\alpha)$ is isomorphic to $F(\beta)$.

4.6.1. Verify the details of Examples 1, 2, and 3 following Definition 4.6.1.

4.6.2. Prove that the fields $\mathbb{Q}(\sqrt[4]{2})$ and $\mathbb{Q}(\sqrt[4]{2}i)$ are $\mathbb{Q}$-isomorphic. Are they equal?

4.6.3. Prove that the fields $\mathbb{Q}(\sqrt[6]{5})$ and $\mathbb{Q}(\sqrt[6]{5}(1 + i\sqrt{3}))$ are $\mathbb{Q}$-isomorphic. Are they equal?

4.6.4. Let α be a zero of $f = x^3 + x + 1 \in \mathbb{Z}_5[x]$. Prove that the fields $\mathbb{Z}_5(\alpha)$ and $\mathbb{Z}_5(-\alpha + \sqrt{1 + 2\alpha^2})$ are $\mathbb{Z}_5$-isomorphic (where the radical represents either of the square roots of $1 + 2\alpha^2$). Are they equal?

4.6.5. Let F be a field and let $E \supset F$ and $K \supset F$ be extension fields of F. Let $\phi : E \to K$ be a field homomorphism. Prove that ϕ is an F-homomorphism if and only if ϕ is a linear transformation of the vector space E over F into the vector space K over F.

4.6.6. Prove that $\mathbb{Q}(\pi)[x]$ is isomorphic to $\mathbb{Q}(e)[x]$. (Assume as known that π and e are transcendental over $\mathbb{Q}$.)

4.6.7. In the matrix ring $M_2(\mathbb{R})$, let

$$F = \left\{\begin{pmatrix} a & 0 \\ 0 & a \end{pmatrix} \middle| \, a \in \mathbb{R}\right\}$$

and let

$$E = \left\{\begin{pmatrix} a & b \\ -b & a \end{pmatrix} \middle| \, a, b \in \mathbb{R}\right\}.$$

Prove that F, E are fields, with F a subfield of E, isomorphic to $\mathbb{R}$. Prove that E is a splitting field, over F, of the polynomial $x^2 + I \in F[x]$ and conclude: E is isomorphic to $\mathbb{C}$. (Note: here, I denotes the identity matrix in $M_2(\mathbb{R})$.)

4.7

Normal Extensions

In general, if $p \in F[x]$ is an irreducible polynomial, and $E \supset F$ is an extension field of F which contains a zero of p, there is no reason to expect p to split over E. However, there is an important class of extensions where this is precisely what does occur.

Definition 4.7.1: Let F, E be fields, with $E \supset F$. Then E is a *normal extension* of F (or: E/F is normal) if

(1) E/F is algebraic

and

(2) if $p \in F[x]$ is an irreducible polynomial which has a zero in E, then p splits over E.

It is easy to give examples of *non*-normal extensions: e.g., $\mathbb{Q}(\sqrt[3]{2})/\mathbb{Q}$ is not normal since $x^3 - 2$ is an irreducible polynomial in $\mathbb{Q}[x]$ which has a zero in $\mathbb{Q}(\sqrt[3]{2})$, but fails to split over $\mathbb{Q}(\sqrt[3]{2})$ since its other two zeros, $\sqrt[3]{2}\omega$ and $\sqrt[3]{2}\omega^2$ $\left(\omega = \dfrac{-1 + i\sqrt{3}}{2}\right)$ are imaginary, hence *not* in $\mathbb{Q}(\sqrt[3]{2})$.

What about $\mathbb{Q}(\sqrt{2})/\mathbb{Q}$? This is one of a class of extensions, easily proved to be normal.

Theorem 4.7.1: Every extension of degree 2 is normal.

Proof: Let $E \supset F$, with $[E : F] = 2$, and let $p \in F[x]$ be irreducible. Clearly, E/F is algebraic. Suppose E contains a zero, α, of p. (Note that p is an associate of $m_{\alpha/F}$ in $F[x]$.) Then $F \subset F(\alpha) \subset E$; hence either $[F(\alpha) : F] = 1$ or $[F(\alpha) : F] = 2$ (by Theorem 4.4.2). In the first case, $F(\alpha) = F$ and p has degree 1, hence splits over F (and certainly over E). In the second case, $F(\alpha) = E$ and p has degree 2. By the Factor Theorem, $p = (x - \alpha)q$, where $q \in E[x]$. But q is linear, hence equal to $a(x - \beta)$, where a is the leading coefficient of p, and $a\beta \in E$. But then $\beta \in E$, and so p splits over E. It follows that E is a normal extension of F. ■

You may be wondering why the term *normal* is used both in group theory and in field theory; indeed, you may have noticed a vague analogy between the preceding theorem and Corollary 2, on p. 83, which states that every subgroup of index 2 is normal. Is the analogy accidental? The answer will be a resounding "no" as soon as, in Section 4.10, we introduce the Galois correspondence, relating extension fields to groups.

The following terminology will reinforce the analogy with normality in group theory.

Definition 4.7.2: Let $E \supset F$ be an algebraic extension, and let $\alpha, \beta \in E$. Then β is an *F-conjugate* of α if $m_{\alpha/F} = m_{\beta/F}$.

F-conjugacy is, clearly, an equivalence relation on E if E/F is an algebraic extension.

Corollary: Let $E \supset F$ be an algebraic extension, and let $\alpha, \beta \in E$. Then β is an F-conjugate of α if and only if there is an F-isomorphism of $F(\alpha)$ onto $F(\beta)$, sending α to β.

Proof: If β is an F-conjugate of α, then $m_{\alpha/F} = m_{\beta/F}$. By the Corollary of Theorem 4.6.1 there is an F-isomorphism of $F(\alpha)$ onto $F(\beta)$, sending α to β.

Conversely, suppose $\phi : F(\alpha) \to F(\beta)$ is an F-isomorphism sending α to β. From $m_{\alpha/F}(\alpha) = 0$, we have $m_{\alpha/F}(\beta) = m^{\phi}_{\alpha/F}(\phi\alpha) = \phi(m_{\alpha/F})(\alpha)) = 0$. Hence $m_{\beta/F} = m_{\alpha/F}$, and so β is an F-conjugate of α. ∎

Using this terminology, we can characterize normality as follows;

Theorem 4.7.2: An extension E/F is normal if and only if

(1) E/F is algebraic
and
(2) if $\alpha \in E$, and β is an F-conjugate of α, contained in some extension $K \supset E$, then $\beta \in E$.

Proof: Suppose E/F is normal. By Definition 4.7.1, E/F is algebraic and, if $\alpha \in E$, then $m_{\alpha/F}$ splits over E. Let $\alpha \in E$ and let $\beta \in K$ be an F-conjugate of α. Then $m_{\beta/F} = m_{\alpha/F}$; hence $m_{\beta/F}$ splits over E, and so $\beta \in E$. Thus (1) and (2) hold.

Conversely, suppose (1) and (2) hold. Let $\alpha \in E$, and let K be a splitting field of $m_{\alpha/F}$ over E. Then K is an extension field of E over which $m_{\alpha/F}$ splits. By (2), since the zeros of $m_{\alpha/F}$ in K are F-conjugates of α, they are contained in E, and so $m_{\alpha/F}$ splits over E. But then E/F is normal. ∎

We now turn to finite normal extensions.

Theorem 4.7.3: An extension $E \supset F$ is finite normal if and only if E is a splitting field, over F, of some polynomial $f \in F[x]$.

Proof: Suppose E/F is finite normal. Since E/F is finite, there are elements $\alpha_1, \ldots, \alpha_k \in E$ such that $E = F(\alpha_1, \ldots, \alpha_k)$. Since E/F is normal, $m_{\alpha_i/F}$ splits over E, for each $i = 1, \ldots, k$. Let $f = \prod_{i=1}^{k} m_{\alpha_i/F}$. Clearly, f splits over E and E is generated by the zeros of f over F. Thus, E is a splitting field for f over F.

Conversely, suppose E is a splitting field for some polynomial f over F. Then E is generated over F by the (finitely many) zeros of f, each algebraic over F, hence $[E : F]$ is finite. Let $p \in F[x]$ be irreducible, and let $\alpha \in E$ be a zero of p. Let L be a splitting field for p over E, and let β be a zero of p in L. Then $m_{\alpha/F} = a^{-1}p = m_{\beta/F}$, where $a \in F$ is the leading coefficient of p. By the Corollary of Theorem 4.6.1, there is an F-isomorphism $\varepsilon_1 : F(\alpha) \to F(\beta)$, sending α to β. Since E is a splitting field for f

over F, $E(\alpha)$ is a splitting field for f over $F(\alpha)$, and $E(\beta)$ is a splitting field for f over $F(\beta)$. (See Exercise 4.7.11) Hence, by Theorem 4.6.2, ε_1 extends to an F-isomorphism $\bar{\varepsilon} : E(\alpha) \to E(\beta)$. But then $E(\alpha)$ and $E(\beta)$ are isomorphic as vector spaces over F, and so $[E(\alpha) : F] = [E(\beta) : F]$. Thus, $[E(\alpha) : E] \cdot [E : F] = [E(\beta) : E] \cdot [E : F]$, and so $[E(\alpha) : E] = [E(\beta) : E]$. Since $\alpha \in E$, $[E(\alpha) : E] = 1$; hence $[E(\beta) : E] = 1$, and $\beta \in E$. It follows that p splits over E. Hence E/F is normal. ■

Most important from the point of view of Galois theory is the following characterization of normality of finite extensions.

Theorem 4.7.4: Let $E \supset F$ be a finite extension of a field F. Then E/F is normal if and only if every F-monomorphism, ϕ, of E into an extension field $K \supset E$ has $\operatorname{Im} \phi = E$ (and thus defines an F-automorphism of E).

Proof: Suppose E/F is normal. Let $\phi : E \to K$ be an F-monomorphism, where $K \supset E$ is some extension field of E. Then, for each $\alpha \in E$, $\phi(\alpha)$ is an F-conjugate of α, hence contained in E, by Theorem 4.7.2. Thus, $\operatorname{Im} \phi \subset E$. Since $\operatorname{Im} \phi$ and E are isomorphic as vector spaces over F, we have $[\operatorname{Im} \phi : F] = [E : F]$. Since $[E : F]$ is finite, E is a finite-dimensional vector space over F, and so $\operatorname{Im} \phi \subset E$ implies $\operatorname{Im} \phi = E$.

Conversely, suppose that every F-monomorphism $\phi : E \to K$, where $K \supset E$ is some extension field of E, has $\operatorname{Im} \phi = E$. Since $[E : F]$ is finite, there are elements $\alpha_1, \ldots, \alpha_k \in E$ such that $E = F(\alpha_1, \ldots, \alpha_k)$. Let $f = m_{\alpha_1/F} \ldots m_{\alpha_k/F}$ and let L be a splitting field for f over E. Since E is generated over F by a subset, $\{\alpha_1, \ldots, \alpha_k\}$, of the set, S, of all zeros of f in L, we have $L = E(S) = F(\alpha_1, \ldots, \alpha_k)(S) = F(S)$, and so L is a splitting field for f over F. Let $\beta \in S$, i.e., let β be a zero of f in L. Then $m_{\alpha_i/F}(\beta) = 0$ for some i, $1 \le i \le k$, and, by Theorem 4.6.1, there is an F-isomorphism, ε_i, of $F(\alpha_i)$ onto $F(\beta)$, sending α_i to β. But L is a splitting field for f over $F(\alpha_i)$ and also over $F(\beta)$. Thus, by Theorem 4.6.2, ε_i extends to an F-isomorphism $\bar{\varepsilon} : L \to L$, i.e., to an F-automorphism, $\bar{\varepsilon}$, of L, sending α_i to β. The restriction, $\bar{\varepsilon}|_E$, of $\bar{\varepsilon}$ to E is an F-monomorphism of E into L, which, by our hypothesis, has image E. But then $\beta \in E$. Thus, $S \subset E$, and so $L = F(S) = E$, i.e., E is a splitting field of f over F. By Theorem 4.7.3, it follows that E/F is normal. ■

In the proof of the preceding theorem, we constructed a "smallest" normal extension of a given finite extension.

Definition 4.7.3: Let E/F be an extension and let $F \subset E \subset L$. Then L is a *normal closure* of E/F if

(1) L/F is normal

and

(2) if $F \subset E \subset K \subset L$, with K/F normal, then $K = L$.

Theorem 4.7.5: Every finite extension E/F has a normal closure, L, with $[L : F]$ finite. If L_1 and L_2 are both normal closures of E/F, then L_1 is F-isomorphic to L_2.

Proof: As in the proof of Theorem 4.7.4, we have $E = F(\alpha_1, \ldots, \alpha_k)$, $\alpha_i \in E \quad (1 \leq i \leq k)$. If $f = \prod_{i=1}^{k} m_{\alpha_i/F}$ and L is a splitting field of f over E, then L is a splitting field of f over F, and so L/F is normal, with $F \subset E \subset L$.

Suppose $F \subset E \subset K \subset L$, with K/F normal. Each $m_{\alpha_i/F}$ has a zero, α_i, in K, hence splits over K. But then f splits over K, and so $K = L$ (see the Corollary on p. 248). Thus, L is a normal closure of E/F. Since L is generated, over F, by the zeros of the polynomial f, $[L : F]$ is finite.

Next, suppose $L' \supset E$ is another normal closure of E/F. For each $i = 1, \ldots, k$, $\alpha_i \in L'$; hence $m_{\alpha_i/F}$ splits over L'. But then f splits over L', and so L' contains a splitting field, M, of f. Since $\alpha_1, \ldots, \alpha_k$ are zeros of f, in L', we have $E \subset M$. By Theorem 4.7.3, M/F is normal; hence, by Definition 4.7.3, we conclude that $M = L'$. By Theorem 4.6.2, the splitting fields L and L' are F-isomorphic.

We have shown that every normal closure of E/F is F-isomorphic to L. It follows that every normal closure of E/F is finite, and any two normal closures L_1 and L_2 of E/F are F-isomorphic. ∎

Theorem 4.7.6: If $F \subset E \subset L$ with L/F finite normal, then every F-monomorphism of E into L extends to an F-automorphism of L.

Proof: Since L/F is finite normal, so is L/E (see Exercise 4.7.3). By Theorem 4.7.3, L is a splitting field over E of some polynomial f in $E[x]$. Let $\sigma : E \to L$ be an F-monomorphism. Then σ defines an F-isomorphism, also denoted by σ, of E onto σE. If L' is a splitting field for F over σE, then (by Theorem 4.6.2) the F-isomorphism σ extends to an F-isomorphism $\bar{\sigma} : L \to L'$ which defines an F-monomorphism, $\bar{\sigma}$, of L into $M = L(L')$. By Theorem 4.7.4, since L/F is normal, $Im\, \bar{\sigma} = L$, and so $\bar{\sigma}$ defines an F-automorphism, $\bar{\sigma}$, of L. Since $\bar{\sigma}x = \sigma x$ for each $x \in E$, the F-automorphism $\bar{\sigma}$ is an extension of the F-monomorphism $\sigma : E \to L$. (See Figure 6.) ∎

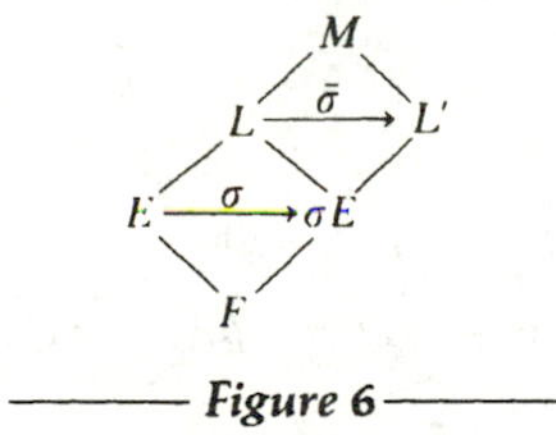

Figure 6

We conclude this section with a handy criterion for finding F-conjugates of sums. First, a lemma.

Lemma: Let E be a finite normal extension of F, and let $\alpha, \beta \in E$. Let $\alpha = \alpha_1, \ldots, \alpha_s$ be the F-conjugates of α in E, and let $\beta = \beta_1, \ldots, \beta_t$ be the F-conjugates of β in E. Suppose that $m_{\beta/F(\alpha)} = m_{\beta/F}$. Then, for each $i = 1, \ldots, s$; $j = 1, \ldots, t$,

$$m_{\beta_j/F(\alpha_i)} = m_{\beta_j/F}$$

and

$$m_{\alpha_i/F(\beta_j)} = m_{\alpha_i/F}.$$

Proof: Since $m_{\beta/F(\alpha)} = m_{\beta/F}$, we have $[F(\alpha, \beta) : F(\alpha)] = [F(\beta) : F]$; hence $[F(\alpha, \beta) : F(\beta)] = [F(\alpha) : F]$. Since $m_{\alpha/F(\beta)}$ divides $m_{\alpha/F}$, this implies: $m_{\alpha/F(\beta)} = m_{\alpha/F}$. For each $i = 1, \ldots, s$, since $m_{\alpha/F(\beta)} = m_{\alpha/F}$, α_i is not only an F-conjugate, but also an $F(\beta)$-conjugate, of α. Hence $m_{\alpha_i/F(\beta)} = m_{\alpha/F(\beta)} = m_{\alpha/F} = m_{\alpha_i/F}$. From this, again arguing on degrees, we obtain $m_{\beta/F(\alpha_i)} = m_{\beta/F}$. But then, for each $j = 1, \ldots, t$, β_j is not only an F-conjugate, but also an $F(\alpha_i)$-conjugate, of β, and we have $m_{\beta_j/F(\alpha_i)} = m_{\beta/F(\alpha_i)} = m_{\beta/F} = m_{\beta_j/F}$.

A final argument on degrees gives $m_{\alpha_i/F(\beta_j)} = m_{\alpha_i/F}$ (see Figure 7). ■

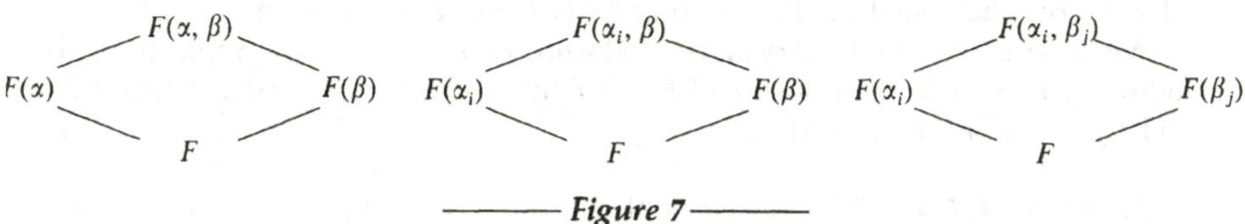

Figure 7

Theorem 4.7.7: Let E be a finite normal extension of F, with $\alpha, \beta \in E$. Let $\alpha_1 = \alpha, \alpha_2, \ldots, \alpha_s$ be the F-conjugates of α in E, and let $\beta_1 = \beta, \beta_2, \ldots, \beta_t$ be the F-conjugates of β in E. If $m_{\beta/F(\alpha)} = m_{\beta/F}$, then the F-conjugates of $\alpha + \beta$ are the elements $\alpha_i + \beta_j \quad (i = 1, \ldots, s; j = 1, \ldots, t)$.

Proof: We first observe that every F-conjugate of $\alpha + \beta$ is of the required form. For: if $\gamma \in E$ is an F-conjugate of $\alpha + \beta$, then there is an F-isomorphism ϕ of $F(\alpha + \beta)$ onto $F(\gamma)$ sending $\alpha + \beta$ to γ. By Theorem 4.7.6, ϕ extends to an F-automorphism, ϕ, of E. The restriction of ϕ to $F(\alpha)$ is an F-monomorphism of $F(\alpha)$ into E, whence $\phi(\alpha)$ is an F-conjugate of α. Similarly, $\phi(\beta)$ is an F-conjugate of β. But then $\gamma = \phi(\alpha + \beta) = \phi(\alpha) + \phi(\beta) = \alpha_i + \beta_j \quad (1 \le i \le s; 1 \le j \le t)$. (See Figure 8.)

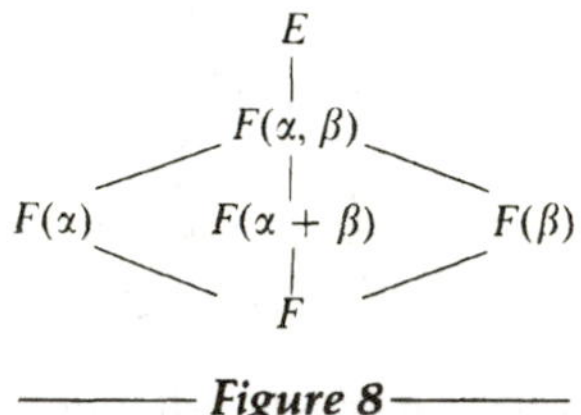

Figure 8

Conversely, for $1 \le i \le s$; $1 \le j \le t$, let $\delta = \alpha_i + \beta_j$. We use the Lemma to prove that δ is an F-conjugate of $\alpha + \beta$.

Since $m_{\alpha/F(\beta)} = m_{\alpha_i/F(\beta)}$, the identity map on $F(\beta)$ extends to an $F(\beta)$-isomorphism, hence an F-isomorphism, σ_i, of $F(\beta)(\alpha)$ onto $F(\beta)(\alpha_i)$, sending α to α_i. By the Lemma, we have

$$m_{\beta/F(\alpha_i)} = m_{\beta/F} = m_{\beta_j/F} = m_{\beta_j/F(\alpha_i)}.$$

The identity map on $F(\alpha_i)$ extends to an $F(\alpha_i)$-isomorphism, hence an F-isomorphism, τ_j, of $F(\alpha_i)(\beta)$ onto $F(\alpha_i)(\beta_j)$, sending β to β_j. Let $\phi_{ij} = \tau_j \sigma_i$. Then ϕ_{ij} is an F-isomorphism of $F(\alpha, \beta)$ onto $F(\alpha_i, \beta_j)$ sending α to α_i and β to β_j. But then $\phi_{ij}(\alpha + \beta) = \phi_{ij}(\alpha) + \phi_{ij}(\beta) = \alpha_i + \beta_j$. This implies that $\alpha_i + \beta_j$ is an F-conjugate of $\alpha + \beta$. (See Figure 9.) ■

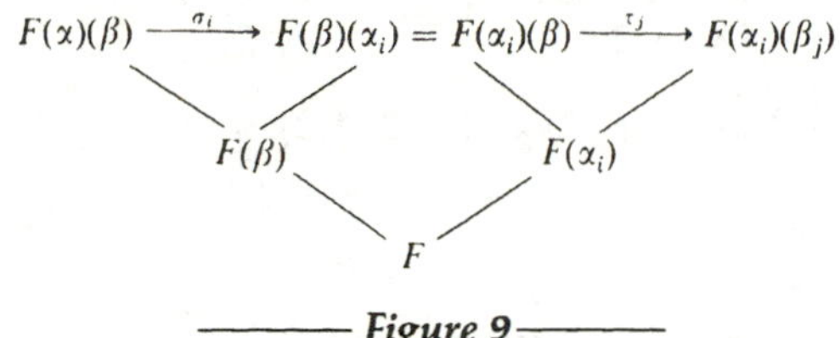

Figure 9

Example 1:

To illustrate Theorem 4.7.7, we first find the $\mathbb{Q}$-conjugates of $\sqrt{2} + \sqrt[3]{5}$. From the field tower $\mathbb{Q} \subset \mathbb{Q}(\sqrt[3]{5}) \subset \mathbb{Q}(\sqrt[3]{5})(\sqrt{2})$, it is clear that $\sqrt{2}$ must have either degree 2 or degree 1 over $\mathbb{Q}(\sqrt[3]{5})$. But if $\sqrt{2} \in \mathbb{Q}(\sqrt[3]{5})$, then $[\mathbb{Q}(\sqrt{2}) : \mathbb{Q}]$ divides $[\mathbb{Q}(\sqrt[3]{5}) : \mathbb{Q}]$, i.e., $2|3$. We conclude that $m_{\sqrt{2}/\mathbb{Q}}$ has degree 2 and is therefore equal to $x^2 - 2$. Thus the condition $m_{\sqrt{2}/\mathbb{Q}(\sqrt[3]{5})} = m_{\sqrt{2}/\mathbb{Q}}$ of Theorem 4.7.7 is met. Since the $\mathbb{Q}$-conjugates of $\sqrt{2}$ are $\sqrt{2}$ and $-\sqrt{2}$, and the $\mathbb{Q}$-conjugates of $\sqrt[3]{5}$ are $\sqrt[3]{5}$, $\sqrt[3]{5}\omega$ and $\sqrt[3]{5}\omega^2$, we conclude that the $\mathbb{Q}$-conjugates of $\sqrt{2} + \sqrt[3]{5}$ are the six (distinct) numbers $\pm\sqrt{2} + \sqrt[3]{5}$, $\pm\sqrt{2} + \sqrt[3]{5}\omega$, $\pm\sqrt{2} + \sqrt[3]{5}\omega^2$. From the field tower $\mathbb{Q} \subset \mathbb{Q}(\sqrt{2} + \sqrt[3]{5}) \subset \mathbb{Q}(\sqrt{2}, \sqrt[3]{5})$, we conclude that $m_{(\sqrt{2}+\sqrt[3]{5})/\mathbb{Q}}$ can have degree no greater than 6. Thus, $m_{(\sqrt{2}+\sqrt[3]{5})/\mathbb{Q}}$ is the polynomial given by $\prod_{i=1}^{6} (x - \alpha_i)$, where the α_i are the six $\mathbb{Q}$-conjugates we have found. Explicitly,

$$m_{(\sqrt{2}+\sqrt[3]{5})/\mathbb{Q}} = x^6 - 6x^4 - 10x^3 + 12x^2 - 60x + 17.$$

Since the degree of $\sqrt{2} + \sqrt[3]{5}$ over $\mathbb{Q}$ is equal to 6, we conclude that $\mathbb{Q}(\sqrt{2} + \sqrt[3]{5}) = \mathbb{Q}\sqrt{2}, \sqrt[3]{5})$.

Example 2:

We next find the $\mathbb{Q}$-conjugates of $\sqrt{2} + \sqrt[4]{2}$. Since $\sqrt{2} + \sqrt[4]{2}$ is an element of $\mathbb{Q}(\sqrt[4]{2})$, it has degree at most 4 over $\mathbb{Q}$, hence has at most 4 $\mathbb{Q}$-conjugates. The conditions of Theorem 4.7.7 are *not* met in this case, for: $m_{\sqrt[4]{2}/\mathbb{Q}(\sqrt{2})} = x^2 - \sqrt{2}$, while $m_{\sqrt[4]{2}/\mathbb{Q}} = x^4 - 2$. The $\mathbb{Q}$-conjugates of $\sqrt{2}$ are the two numbers $\pm\sqrt{2}$, the $\mathbb{Q}$-conjugates of $\sqrt[4]{2}$ are the four numbers $\pm\sqrt[4]{2}$, $\pm i\sqrt[4]{2}$, but only four of the eight numbers $\pm\sqrt{2} \pm \sqrt[4]{2}$, $\pm\sqrt{2} \pm i\sqrt[4]{2}$ are $\mathbb{Q}$-conjugates of $\sqrt{2} + \sqrt[4]{2}$. Any $\mathbb{Q}$-isomorphism of $\mathbb{Q}(\sqrt{2} + \sqrt[4]{2})$ into a normal closure, L, of $\mathbb{Q}(\sqrt[4]{2})/\mathbb{Q}$ extends to a $\mathbb{Q}$-automorphism ϕ of L which must assign consistent images to $\sqrt[4]{2}$ and $\sqrt{2}$. Thus, if $\phi(\sqrt[4]{2}) = \pm\sqrt[4]{2}$, then $\phi(\sqrt{2}) = \sqrt{2}$, and if $\phi(\sqrt[4]{2}) = \pm i\sqrt[4]{2}$, then $\phi(\sqrt{2}) = -\sqrt{2}$. This leads to the four $\mathbb{Q}$-conjugates $\sqrt{2} \pm \sqrt[4]{2}$ and $-\sqrt{2} \pm \sqrt[4]{2}\, i$. The minimal polynomial $m_{(\sqrt{2}+\sqrt[4]{2})/\mathbb{Q}}$ is equal to

$$(x - (\sqrt{2} + \sqrt[4]{2}))(x - (\sqrt{2} - \sqrt[4]{2}))(x - (-\sqrt{2} + \sqrt[4]{2}\, i))(x - (-\sqrt{2} - \sqrt[4]{2}\, i))$$
$$= x^4 - 4x^2 - 8x + 2.$$

Clearly, $\mathbb{Q}(\sqrt{2} + \sqrt[4]{2}) = \mathbb{Q}(\sqrt[4]{2}) = \mathbb{Q}(\sqrt{2}, \sqrt[4]{2})$.

Exercises 4.7

True or False

1. Every extension of degree 3 is normal.
2. If $\alpha = 2^{1/n}$ $(n \in \mathbb{Z}^+)$, then $\mathbb{Q}(\alpha)/\mathbb{Q}$ is normal if and only if $n = 1$, or $n = 2$.
3. If $E = F(\alpha_1, \ldots, \alpha_n)$ and $\prod_{i=1}^{n} (x - \alpha_i) \in F[x]$, then E/F is normal.
4. $\mathbb{Q}(\sqrt{1 + \sqrt[3]{5}})/\mathbb{Q}$ is a normal extension.
5. If $\beta \in \mathbb{C}$, then $\mathbb{R}(\beta)/\mathbb{R}$ is normal.

4.7.1. If n is a positive integer such that every extension of degree n is normal, prove that $n = 1$, or $n = 2$.

4.7.2. Let $C = \left\{ \begin{pmatrix} a & -b \\ b & a \end{pmatrix} \middle| a, b \in \mathbb{R} \right\}$.

(1) Prove that C is a subfield of the ring $M_2(\mathbb{R})$.
(2) Exhibit an isomorphism of C onto the complex number field $\mathbb{C}$.
(3) Find the subfield R of C that corresponds to the real number field $\mathbb{R}$ under the isomorphism in (2).
(4) Conclude that C is a splitting field of $x^2 + I$ in $R[x]$.

4.7.3. If $F \subset K \subset E$, with E/F normal, prove that E/K is normal.

4.7.4. Find all the $\mathbb{Q}$-conjugates, in $\mathbb{C}$, of $\alpha = \sqrt{3} + \sqrt[4]{2}$. Hence find $[\mathbb{Q}(\alpha) : \mathbb{Q}]$ and $m_{\alpha/\mathbb{Q}}$.

4.7.5. For $a, b \in \mathbb{R}$, find all the $\mathbb{R}$-conjugates of $a + bi$.

4.7.6. Determine whether each of the following extensions is normal.

(1) $\mathbb{Q}(\sqrt{2} + \sqrt{3})/\mathbb{Q}$.
(2) $\mathbb{Q}(\sqrt[3]{2} + \sqrt{-3})/\mathbb{Q}$.
(3) $\mathbb{Z}_3(u)(\alpha)/\mathbb{Z}_3(u)$, where $\alpha^3 = u$ (u transcendental over $\mathbb{Z}_3$).
(4) $\mathbb{Z}_3(u)(\alpha)/\mathbb{Z}_3(u)$, where $\alpha^4 = u$ (u transcendental over $\mathbb{Z}_3$).

4.7.7. Prove that normal inclusion of subfields is not transitive, i.e., if $F \subset K \subset E$, with K/F and E/K normal, prove that E/F need not be normal. (Hint: consider the field tower $\mathbb{Q} \subset \mathbb{Q}(\sqrt{a}) \subset \mathbb{Q}(\sqrt[4]{a})$, for $a > 0$ in $\mathbb{Q}$.) (Compare Exercise 2.8.10.)

4.7.8. If $F \subset K \subset E$, with E/F finite normal, prove that E contains a unique normal closure of K/F.

4.7.9. Find the normal closure, in $\mathbb{C}$, of each of the following extensions:

(1) $\mathbb{Q}(\sqrt[3]{5})/\mathbb{Q}$.

(2) $\mathbb{Q}(\sqrt{2} + \sqrt{3})/\mathbb{Q}$.

(3) $\mathbb{Q}(1 + \sqrt[4]{5})/\mathbb{Q}$.

4.7.10. Prove that $\mathbb{C}$ is a normal extension of $\mathbb{R}$.

4.7.11. Let $f \in F[x]$ be a polynomial with splitting field E over F, and let T be a subset of K, where K is some extension field of E. Prove that $E(T)$ is a splitting field for f over $F(T)$.

Separable Extensions

Let f be a polynomial in $F[x]$ (F a field) and let E be a splitting field for f over F. If $\alpha \in E$ is a zero of f, then α has *multiplicity* s if $x - \alpha$ occurs exactly s times as a factor of f in $E[x]$. If $s = 1$, then α is a *simple zero* of f; if $s \geq 2$, then α is a *multiple zero* of f.

For example, the polynomial $f = x^3 + x^2 - 5x + 3$ in $\mathbb{Q}[x]$ has $\alpha = 1$ as a zero of multiplicity $s = 2$. But note that f is reducible over $\mathbb{Q}$. Is there an irreducible polynomial in $\mathbb{Q}[x]$, with multiple zeros in its splitting field? Is there an irreducible polynomial over *any* field, with multiple zeros in its splitting field? We shall see that the answer to the first question is "no" — not only for $\mathbb{Q}$, but for *any* field of characteristic zero. For characteristic $p > 0$, there *are* such polynomials (see Exercise 4.3.9), but they are very special.

Definition 4.8.1: An irreducible polynomial $f \in F[x]$ is *separable* if it has no multiple zeros in its splitting field.

More generally, a *polynomial* $f \in F[x]$ *is separable* over F if each of its irreducible factors in $F[x]$ is separable over F.

If $E \supset F$, then an *element* $\alpha \in E$ *is separable* over F if $m_{\alpha/F}$ is separable over F.

An *algebraic extension* E/F *is separable* if each $\alpha \in E$ is separable over F. (For polynomials, field elements, or extension fields, we use the term *inseparable* to mean "not separable".)

To facilitate the study of separability, it is convenient to introduce formal derivatives of polynomials.

Definition 4.8.2: Let F be a field, and let $f = \sum_{i=0}^{n} a_i x^i$ in $F[x]$. Then the *formal derivative*, f', of f is the polynomial in $F[x]$ given by: $f' = \sum_{i=1}^{n} i a_i x^{i-1}$.

It is easy to check that formal differentiation, over any field F, obeys the familiar rules (known to us for derivatives over $\mathbb{R}$) regarding sums and products (see Exercise 4.8.2).

Theorem 4.8.1: Let F be a field, $f \in F[x]$, and let $K \supset F$ be a splitting field for f over F. Then $\alpha \in K$ is a multiple zero of f if and only if α is a common zero of f and f'.

Proof: Suppose $\alpha \in K$ is a multiple zero of f. Then $f = (x - \alpha)^s g$ for some integer $s \geq 2$ and some $g \in K[x]$. Differentiating, we have

$$f' = (x - \alpha)^s g' + s(x - \alpha)^{s-1} g.$$

Thus, $f'(\alpha) = f(\alpha) = 0$.

Conversely, suppose $\alpha \in K$ is a common zero of f and f'. Then $f = (x - \alpha)h$ for some $h \in K[x]$ and $f' = (x - \alpha)h' + h$. Hence $0 = f'(\alpha) = h(\alpha)$. But then $h = (x - \alpha)l$, $l \in K[x]$, and so $f = (x - \alpha)^2 l$. It follows that α is a multiple zero of f. ■

We are now ready to answer the questions with which we began this section.

Theorem 4.8.2:

(1) If F is a field of characteristic 0, then every polynomial in $F[x]$ is separable.

(2) If F is a field of characteristic $p > 0$, then an irreducible polynomial $f \in F[x]$ is *in*separable if and only if $f = \sum_{j=0}^{k} b_j x^{pj} = g(x^p)$ for some $k > 0$ in $\mathbb{Z}$, $b_j \in F$, $j = 0, \ldots, k$, $g \in F[x]$. In this case, if K is a splitting field for f over F, there are unique positive integers r and e such that

$$f = a[(x - \alpha_1) \ldots (x - \alpha_r)]^{p^e},$$

where the α_i are distinct elements of K, and $a \in F$ is the leading coefficient of f.

Proof: Let $f \in F[x]$, irreducible over F, and let $K \supset F$ be a splitting field for f over F. If $\alpha \in K$ is a multiple zero of f, then α is a common zero of f and f'. Since f is irreducible over F, $m_{\alpha/F}$ is an associate of f in $F[x]$. From $f'(\alpha) = 0$, we conclude that $m_{\alpha/F} \mid f'$ in $F[x]$. But then $f \mid f'$ in $F[x]$. Since $\deg f' < \deg f$, $f \mid f'$ holds if and only if f' is the 0-polynomial in $F[x]$. If $f = \sum_{i=0}^{n} a_i x^i$, then $f' = \sum_{i=1}^{n} i a_i x^{i-1}$ is the 0-polynomial if and only if $i a_i = 0$ for each $i = 1, \ldots, n$.

(1) If Char $F = 0$, $i a_i = 0$ for each $i = 1, \ldots, n$ implies $a_i = 0$ for each $i = 1, \ldots, n$ and so $f = a_0$, a constant polynomial, in contradiction to the hypothesis that f is irreducible (see Definition 3.6.2). It follows that any irreducible polynomial over a field of characteristic zero is separable.

(2) If Char $F = p > 0$, then $i a_i = 0$ for each $i = 1, \ldots, n$ if and only if, for each $i = 1, \ldots, n$, either $a_i = 0$, or $p \mid i$. This is true if and only if f is a polynomial of the form

$$f = \sum_{j=0}^{k} b_j x^{pj} \tag{1}$$

(where $n = pk$), i.e., f is a polynomial in x^p.

Thus, an irreducible polynomial $f \in F[x]$, where Char $F = p > 0$, is inseparable if and only if it is of form (1). Let p^e be the highest power of p that divides all the exponents of x in (1). Then f is a polynomial in x^{p^e}, hence $f(x) = g(x^{p^e})$, where $g \in F[x]$. Since f is irreducible over F, so is g. But then g can have no multiple zeros in K since, otherwise, $g(x) = h(x^p)$ for some polynomial $h \in F[x]$, whence $f(x) = g(x^{p^e}) = h(x^{p^{e+1}})$, contrary to the hypothesis that p^e is the highest power of p that divides all the exponents of x in f. It follows that there are distinct elements $\beta_1, \ldots, \beta_r$ (in an extension field of F) such that $g(x) = a(x - \beta_1)\ldots(x - \beta_r)$, $a \in F$, hence $f(x) = g(x^{p^e}) = a(x^{p^e} - \beta_1)\ldots(x^{p^e} - \beta_r)$. Since f splits over K, the β_i are, in fact, in K, and their (unique) p^e-th roots α_i (recall Exercise 4.1.12) are among the zeros of f in K. Thus $f(x) = a(x^{p^e} - \alpha_1{}^{p^e})\ldots(x^{p^e} - \alpha_r{}^{p^e}) = a\left[\prod_{i=1}^{r}(x - \alpha_i)\right]^{p^e}$. ■

Warning: Not every polynomial of form (1) is *irreducible*; for example, $f = x^{10} + x^5 + 2 \in \mathbb{Z}_5[x]$ has multiple zeros, but is reducible over $\mathbb{Z}_5$:

$$f = x^{10} + x^5 + 2 = (x^2 + x + 2)^5,$$

hence every zero of f in a splitting field K will be 5-fold. If $\alpha \in K$ is one zero, the other is $\alpha^2 + 1$, since $\alpha^2 + \alpha + 2 = 0$. Thus, over K, the factorization of f is: $f = (x - \alpha)^5(x - (\alpha^2 + 1))^5$.

On the other hand, the polynomial $f = x^5 - u \in \mathbb{Z}_5(u)[x]$, where u is transcendental over $\mathbb{Z}_5$, is irreducible and inseparable over $\mathbb{Z}_5(u)$ — its factorization in a splitting field K being $(x - \theta)^5$, where θ is a zero of f in K. (See Exercise 4.3.9.)

As an immediate consequence of Theorem 4.8.2, we have an important result.

Corollary: If F is a field of characteristic 0, then every algebraic extension of F is separable.

The following properties of separability are easy consequences of Definition 4.8.1:

Lemma: Let F, K, E be fields, with $F \subset K \subset E$.

(1) If E is separable over F, then K is separable over F.

(2) If E is separable over F, then E is separable over K.

Proof: *(1)* For each $\alpha \in K$, $m_{\alpha/F}$ is a separable polynomial.
(2) For each $\alpha \in E$, $m_{\alpha/K} | m_{\alpha/F}$, hence $m_{\alpha/K}$ is a separable polynomial. ■

For an algebraic element, α, of an extension E of a field F, the number of distinct F-conjugates of α in any normal extension of F containing α depends on whether or not $m_{\alpha/F}$ is a separable polynomial. Put another way, the number of F-monomorphisms of $F(\alpha)$ into any normal extension of F containing $F(\alpha)$ depends upon the separability, or inseparability, of $m_{\alpha/F}$. More generally, we now investigate the connection between the separability of a finite extension E/F and the number of F-monomorphisms of E into any normal extension of E. If M is a normal extension of

F containing E, and L is the normal closure, in M, of E/F, then all F-conjugates, in M, of elements of E lie in L. Thus, the number of F-monomorphisms of E into M is equal to the number of F-monomorphisms of E into L. It therefore suffices to investigate the number of F-monomorphisms of E into any normal closure, L, of E/F.

Definition 4.8.3: Let E/F be a finite extension, and let L be a normal closure of E/F. Let $\mathrm{Mon}_F(E, L)$ be the set of all F-monomorphisms of E into L, and let $[E:F]_s$ be the cardinality of the set $\mathrm{Mon}_F(E, L)$. Then $[E:F]_s$ is the *separability degree* of the extension E/F.

Note: Since all normal closures of E/F are F-isomorphic, the number $[E:F]_s$ is independent of the choice of normal closure. (Indeed, $[E:F]_s$ is equal to the number of F-monomorphisms of E into *any* normal extension N of F, with $E \subset N$.)

Separability degrees share with degrees of extensions the important property of multiplicativity.

Theorem 4.8.3: Let $F \subset E \subset K$, with $[E:F]_s$ and $[K:E]_s$ finite. Then

$$[K:E]_s \cdot [E:F]_s = [K:F]_s.$$

Proof: Let L be a normal closure of K/F, $\sigma_1, \ldots, \sigma_r$ the F-monomorphisms of E into L, and $\tau_1, \ldots, \tau_t$ the E-monomorphisms of K into L. By Theorem 4.7.6, each σ_i extends to an F-automorphism, $\bar{\sigma}_i$, of L, and each τ_j extends to an E-automorphism, hence an F-automorphism, $\bar{\tau}_j$, of L. For each (i, j), $1 \le i \le r$, $1 \le j \le t$, $\bar{\phi}_{ij} = \bar{\sigma}_i\bar{\tau}_j$ is an F-automorphism of L whose restriction, $\phi_{ij} = \bar{\phi}_{ij}|_K$ is an F-monomorphism of K into L. Every F-monomorphism, ϕ, of K into L is equal to one of the ϕ_{ij}, for: $\phi|_E$ is an F-monomorphism of E into L, hence equal to one of the σ_i. If $\bar{\phi}$ and $\bar{\sigma}_i$ are extensions of ϕ and σ_i, respectively, to F-automorphisms of L, then $\bar{\sigma}_i{}^{-1}\bar{\phi}$ is an F-automorphism of L whose restriction to K leaves every element of E fixed, hence is equal to one of the τ_j. Thus, $(\bar{\sigma}_i{}^{-1}\bar{\phi})|_K = \bar{\tau}_j|_K$ for some j, $1 \le j \le t$, whence $\bar{\sigma}_i{}^{-1}(\bar{\phi}|_K) = \bar{\tau}_j|_K$, and $\phi = \bar{\phi}|_K = \bar{\sigma}_i(\bar{\tau}_j|_K) = (\bar{\sigma}_i\bar{\tau}_j)|_K = \phi_{ij}$.

The ϕ_{ij} are all distinct, for: suppose $\phi_{ij} = \phi_{hk}$, $1 \le i, h \le r$, $1 \le j, k \le t$. Then

$$(\bar{\sigma}_i\bar{\tau}_j)|_K = (\bar{\sigma}_h\bar{\tau}_k)|_K$$

$$\bar{\sigma}_i(\bar{\tau}_j)|_K = \bar{\sigma}_h(\bar{\tau}_k)|_K$$

$$\bar{\sigma}_h{}^{-1}\bar{\sigma}_i(\bar{\tau}_j)|_K = (\bar{\tau}_k)|_K.$$

Since $\bar{\tau}_j$ and $\bar{\tau}_k$ fix all elements of E, so does $\bar{\sigma}_h{}^{-1}\bar{\sigma}_i$. But then $\sigma_h = \bar{\sigma}_h|_E = \bar{\sigma}_i|_E = \sigma_i$, whence $\tau_j = (\bar{\tau}_j)|_K = (\bar{\tau}_k)|_K = \tau_k$.

It follows that $[K:E]_s \cdot [E:F]_s = [K:F]_s$. ■

It is now easy to determine the connection between the separability of an extension and its separability degree.

Theorem 4.8.4: Let E/F be a finite extension. Then $[E:F]_s \le [E:F]$. The equality holds if and only if E/F is separable.

(Thus, E/F is separable if and only if $[E:F]_s = [E:F]$, and E/F is inseparable if and only if $[E:F]_s < [E:F]$.)

Proof: Since E/F is finite, there are elements $\alpha_1, \ldots, \alpha_k$ such that $E = F(\alpha_1, \ldots, \alpha_k)$. We first prove, by induction on k, that $[E:F]_s \leq [E:F]$, and that the equality holds if E/F is separable. (Assertion A).

For $k = 1$, we have $E = F(\alpha_1)$. In any normal closure, L, of E/F, let $\beta_1 = \alpha_1$, $\beta_2, \ldots, \beta_r$ be the (distinct) F-conjugates of α_1 in L. Then, for each $i = 1, \ldots, r$, there is an F-isomorphism of $F(\alpha_1)$ onto $F(\beta_i)$, hence an F-monomorphism, ϕ_i, of $E = F(\alpha_1)$ into L, sending α_1 to β_i. The ϕ_i are all distinct since they assign different images to α_1. If ϕ is any F-monomorphism of $E = F(\alpha_1)$ into L, then ϕ sends α_1 to one of its F-conjugates, β_i, hence ϕ coincides with one of the ϕ_i. Thus, the number, $[E:F]_s$, of F-monomorphisms of E into L is equal to r, the number of F-conjugates of α_1 in L. But $r \leq \deg m_{\alpha_i/F} = [E:F]$, and so we have $[E:F]_s \leq [E:F]$. If E/F is separable, then $m_{\alpha_1/F}$ is a separable polynomial, hence $r = \deg m_{\alpha_1/F}$, and so $[E:F]_s = [E:F]$.

For $k > 1$, suppose that Assertion A holds for $k - 1$. We have $E = F(\alpha_1, \ldots, \alpha_k) = F(\alpha_1)(\alpha_2, \ldots, \alpha_k)$. By the case $k = 1$, the induction hypothesis, and Theorem 4.8.3, it follows that $[E:F]_s = [E:F(\alpha_1)]_s \cdot [F(\alpha_1):F]_s \leq [E:F(\alpha_1)] \cdot [F(\alpha_1):F] = [E:F]$. If E/F is separable, then $E/F(\alpha_1)$ and $F(\alpha_1)/F$ are both separable; hence, again by the case $k = 1$, the induction hypothesis and Theorem 4.8.3, we have

$$[E:F]_s = [E:F(\alpha_1)]_s \cdot [F(\alpha_1):F]_s = [E:F(\alpha_1)] \cdot [F(\alpha_1):F] = [E:F].$$

Next suppose that E/F is inseparable. Then there is some $\alpha_1 \in E$ which is inseparable over F. The number, r, of F-conjugates of α_1 in L is equal to $[F(\alpha_1):F]_s$, but $r < \deg m_{\alpha_1/F}$. Thus, $[F(\alpha_1):F]_s < [F(\alpha_1):F]$. But then, since $[E:F(\alpha_1)]_s \leq [E:F(\alpha_1)]$, we have $[E:F]_s = [E:F(\alpha_1)]_s \cdot [F(\alpha_1):F]_s < [E:F(\alpha_1)] \cdot [F(\alpha_1):F] = [E:F]$. ■

A number of interesting consequences flow from Theorem 4.8.4.

Corollary 1: *(1)* If E/F is a finite extension, then $[E:F]_s$ is finite.
(2) If $F \subset E \subset K$, with E/F and K/E finite, then $[K:F]_s = [K:E]_s \cdot [E:F]_s$.

Proof: (1) follows immediately from Theorem 4.8.4; (2) follows from (1) and Theorem 4.8.3. ■

Corollary 2: Let K/F be a finite extension, $F \subset E \subset K$, with E/F and K/E separable. Then K/F is separable.

Proof: $[K:F]_s = [K:E]_s \cdot [E:F]_s = [K:E] \cdot [E:F] = [K:F]$, hence K/F is separable. ■

Corollary 3: If E/F is finite, with $E = F(\alpha)$, and if α is separable over F, then E/F is separable.

Proof: As in the proof of Theorem 4.8.4, in case $k = 1$, the separability of α over F implies that $[E:F]_s = [E:F]$, whence E/F is separable, by Theorem 4.8.4. ■

Corollary 4: If E/F is a finite extension, with $E = F(\alpha_1, \ldots, \alpha_k)$, where each α_i is separable over F, then E/F is separable.

Proof: Note that each α_i is separable over any field intermediate between F and E. (See Exercise 4.8.9.) Using Corollary 2, we have

$$\begin{aligned}[E:F]_s &= [F(\alpha_1):F]_s \cdot [F(\alpha_1, \alpha_2):F(\alpha_1)]_s \cdot \cdots \cdot [E:F(\alpha_1, \ldots, \alpha_{k-1})]_s \\ &= [F(\alpha_1):F] \cdot [F(\alpha_1, \alpha_2):F(\alpha_1)] \cdot \cdots \cdot [E:F(\alpha_1, \ldots, \alpha_{k-1})] \\ &= [E:F],\end{aligned}$$

hence E/F is separable. ■

Corollary 5: If E is a splitting field, over F, of a separable polynomial $f \in F[x]$, then E/F is separable.

Proof: If $\alpha_1, \ldots, \alpha_n$ are the zeros of f in E, then $E = F(\alpha_1, \ldots, \alpha_n)$, each α_i being separable over F. By Corollary 4, E/F is separable. ■

Corollary 6: Let E/F be a finite extension. Then the elements of E that are separable over F form a subfield, E_s, of E. E_s is separable over F and, if $F \subset K \subset E$, with K/F separable, then $K \subset E_s$.

Proof: Let E_s be the set of all elements of E that are separable over F. Note that $F \subset E_s$. If $\alpha, \beta \in E_s$, then $F(\alpha, \beta)/F$ is separable, by Corollary 4. Hence $\alpha \pm \beta$, $\alpha \cdot \beta$, and $\alpha\beta^{-1}$ (in case $\beta \neq 0$) are separable over F. It follows that E_s is a subfield of E. Obviously, E_s/F is separable and, if K is a subfield such that $F \subset K \subset E$ and K/F is separable, then $K \subset E_s$. ■

We conclude this section with a rather surprising property of finite separable extensions.

Theorem 4.8.5 (Primitive Element Theorem): Every finite, separable extension is simple.

Proof: Suppose E/F is finite separable. If F is a finite field, then so is E, and the unit group E^* of E is cyclic, generated by some element $\theta \in E$ (Theorem 4.4.6). But then $E = F(\theta)$, and so E/F is simple. Next, suppose F is an infinite field. Since $[E:F]$ is finite, there are elements $\alpha_1, \ldots, \alpha_n \in E$, each separable over F, such that $E = F(\alpha_1, \ldots, \alpha_n)$. We proceed by induction on n.

For $n = 1$, of course $F(\alpha_1)/F$ is simple. For $n > 1$, suppose the theorem holds for finite separable extensions generated over a given field by $n - 1$ elements. By Corollary 3 of Theorem 4.8.4, $F(\alpha_1, \ldots, \alpha_{n-1})$ is separable over F; hence, by the induction hypothesis, there is an element $\beta \in F(\alpha_1, \ldots, \alpha_{n-1})$ such that $F(\alpha_1, \ldots, \alpha_{n-1}) = F(\beta)$. Write $\alpha_n = \gamma$. Then $E = F(\beta)(\alpha_n) = F(\beta)(\gamma) = F(\beta, \gamma)$. We need a *primitive element*, θ, in E such that $E = F(\theta)$. Let us attempt to find such an element among the linear combinations $\beta + x\gamma$, $x \in F$.

Let L be a normal closure of E/F. Then the polynomials $f = m_{\beta/F}$ and $g = m_{\gamma/F}$ both split over L, each into distinct linear factors. Let $\beta = \beta_1, \dots, \beta_s$ be the F-conjugates of β in L, and let $\gamma = \gamma_1, \dots, \gamma_t$ be the F-conjugates of γ in L.

Now, for each $i = 1, \dots, s$ and each $j = 2, \dots, t$, there is at most one $x \in F$ such that

$$\beta_i + x\gamma_j = \beta + x\gamma. \tag{1}$$

Hence only finitely many $x \in F$ are solutions of (1) for some (i, j). Since F is infinite, there is some $c \in F$ such that

$$\beta + c\gamma \neq \beta_i + c\gamma_j \tag{2}$$

holds for all $i = 1, \dots, s$; $j = 2, \dots, t$.

Write $\theta = \beta + c\gamma$. Then $\beta = \theta - c\gamma$ and we have $f(\beta) = f(\theta - c\gamma) = 0$. Thus, the polynomial $h(x) = f(\theta - cx) \in F(\theta)[x]$ has γ as a zero. Since γ is also a zero of $g = m_{\gamma/F} \in F[x] \subset F(\theta)[x]$, $x - \gamma$ is a common divisor of g and h in $L[x]$. Let l be the monic greatest common divisor of g and h in $L[x]$. Then l splits into linear factors in L (since g does), and one of these linear factors is $x - \gamma$. We show: $l = x - \gamma$. If $\deg l > 1$, then there is another linear factor $x - \delta$ of l in $L[x]$. Thus, δ is one of the zeros of g, i.e., one of the F-conjugates of γ in L. Since γ is separable over F, $\delta \neq \gamma$. But if $\delta = \gamma_j$ for some $j = 2, \dots, t$, then since $0 = h(\delta) = f(\theta - c\delta) = f(\theta - c\gamma_j)$, $\theta - c\gamma_j$ is one of the β_i $(i = 1, \dots, s)$, and so $\theta = \beta_i + c\gamma_j$, contrary to the prohibition (2). It follows that $\deg l = 1$, and so $l = x - \gamma$. But $l \in F(\theta)[x]$, and so $\gamma \in F(\theta)$. Thus, we have $E = F(\beta, \gamma) = F(\theta - c\gamma, \gamma) \subset F(\theta) = F(\beta + c\gamma) \subset F(\beta, \gamma) = E$—i.e., $E = F(\theta)$.

Remark: If $E = F(\alpha)$, then α is sometimes referred to as a *primitive element* for E/F, hence the name of Theorem 4.8.5.

Corollary: Every finite extension of a field of characteristic 0 is simple.

Proof: This follows immediately from Theorem 4.8.5 and the corollary of Theorem 4.8.2. ■

Exercises 4.8

True or False

1. The polynomial $x^3 - 2x^2 + x \in \mathbb{Q}[x]$ is inseparable over $\mathbb{Q}$.
2. Every separable algebraic extension is simple.
3. If E/F is an inseparable extension, then the prime field of F is isomorphic to $\mathbb{Z}_p$ for some prime p.
4. If $E \supset F$ and $\alpha, \beta \in E$ are separable over F, then $\alpha + \beta$ is separable over F.
5. Every simple algebraic extension is separable.

4.8.1. Determine whether each of the following polynomials is separable over the given field:
(1) $x^3 + x^2 - x - 1$ over $\mathbb{Q}$;
(2) $x^4 + x^2 + 1$ over $\mathbb{Z}_2$;
(3) $x^{10} + x^5 + 3$ over $\mathbb{Z}_5$;
(4) $x^{10} + 4u$ over $\mathbb{Z}_5(u)$, u transcendental over $\mathbb{Z}_5$.

4.8.2. Prove that formal differentiation of polynomials over a field F obeys the usual rules regarding differentiation of sums and products.

4.8.3. Find a primitive element for the splitting field, in $\mathbb{C}$, of $f = (x^2 + 3)(x^3 - 2)$ over $\mathbb{Q}$.

4.8.4. Prove: a finite extension E/F is simple if and only if the number of intermediate fields K $(F \subset K \subset E)$ is finite.

4.8.5. Using Exercise 4.8.4, conclude: if E/F is finite separable, then there are only finitely many fields K such that $F \subset K \subset E$.

4.8.6. Let F be a field of characteristic 0, and let E/F be a finite extension. Prove that there are only finitely many intermediate fields K, $F \subset K \subset E$. (See Exercise 4.14.8 for an example of a finite extension with infinitely many intermediate fields.)

4.8.7. Let F be a field, n a positive integer. Prove that the polynomial $x^n + a$ $(a \in F, a \neq 0)$ is separable over F, provided Char $F = 0$, or Char $F = p, p \nmid n$.

4.8.8. If $f \in F[x]$ is a separable irreducible polynomial of degree n, and E is a splitting field of f over F, prove that the number of F-automorphisms of E is $\leq n!$

4.8.9. Let $F \subset K \subset E$ and let $\alpha \in E$. If $\alpha \in E$ is separable over F, prove that α is separable over K.

4.8.10. Let F be a field of characteristic zero, and let E/F be an extension of degree 2. Prove that $E = F(\alpha)$ for some $\alpha \in E$, with $\alpha^2 \in F$.

4.8.11. Let F be a field, and let α, β be elements of some extension field $E \supset F$. If α, β are separable over F, with $[\alpha : F] = p$, $[\beta : F] = q$, where p and q are distinct primes, and if $m_{\alpha/F(\beta)} = m_{\alpha/F}$, prove that $\alpha + \beta$ is a primitive element for $F(\alpha, \beta)$ over F. (Hint: recall Theorem 4.7.7.)

4.9

Galois Extensions and Galois Groups

If E is a splitting field over F of a polynomial $f \in F[x]$, then E/F is normal, and every F-automorphism of E sends each zero, α, of f to one of its F-conjugates in E, i.e., to another zero of the same irreducible factor, $m_{\alpha/F}$, of f. Thus, every F-automorphism of E induces a permutation on the zeros of f. The F-automorphisms of E form a group, G, isomorphic to the group of induced permutations on the zeros of f. It was Galois' brilliant discovery that a study of this group of permutations can reveal the essential properties of f. In particular, if f is a separable polynomial, then E/F turns out to be a separable extension and the order of the group, G, of all F-automorphisms of E is equal to $[E : F]$. Between the subgroups of G and the subfields intermediate between F and E, there is a 1-1 correspondence from which conclusions can be drawn regarding the nature of the zeros of f.

We thus turn our attention to the study of normal, separable finite extensions and their automorphisms.

Definition 4.9.1: An extension E/F which is both normal and separable is a *Galois extension.*

For any extension E/F, the F-automorphisms of E obviously form a subgroup of the group of *all* automorphisms of E.

Definition 4.9.2: If E/F is a Galois extension, then the group $G(E/F)$ of all F-automorphisms of E is the *Galois group of E over F.*

Theorem 4.9.1: If E/F is a finite Galois extension, then $|G(E/F)| = [E:F]$.

Proof: By Theorem 4.7.4, since E/F is normal, every F-monomorphism of E into a normal extension $L \supset E$ defines an F-automorphism of E. Thus, by Theorem 4.8.4, since E/F is separable, we conclude that $|G(E/F)| = [E:F]_s = [E:F]$. ∎

Our aim is to establish a 1-1 correspondence between the subgroups of the Galois group $G(E/F)$ of a finite Galois extension E/F and the subfields of E containing F.

Lemma 1: If E/F is a finite Galois extension and $F \subset K \subset E$, then E/K is a finite Galois extension.

Proof: E/K is finite (by Theorem 4.4.2), normal (by Exercise 4.7.3), and separable (by Lemma, p. 275). ∎

Thus, if E/F is finite Galois, and $F \subset K \subset E$, then $G(E/K)$ is a well-defined group, clearly a subgroup of $G(E/F)$, with $|G(E/K)| = [E:K]$. We want to establish that $\mu: K \mapsto G(E/K)$ defines a 1-1 correspondence between the set of all subfields of E containing F and the set of all subgroups of $G(E/F)$. The notion of "fixed field" will help us define an inverse map for μ.

It is easy to verify (see Exercise 4.9.1) that the fixed set of any set, A, of automorphisms of a field, E, forms a subfield of E.

Definition 4.9.3: If A is a set of automorphisms of a field E, then $\bar{A} = \{\alpha \in E \mid \sigma\alpha = \alpha \quad \text{for all} \quad \sigma \in A\}$ is the *fixed field* of A.

Theorem 4.9.2: If E/F is a finite Galois extension and $F \subset K \subset E$, then the fixed field of $G(E/K)$ is K.

Proof: E/K is, itself, finite Galois, hence $H = G(E/K)$ is defined, and the fixed field, $\bar{H}$, of H contains K. Suppose $\bar{H} \neq K$. Then there is some $\alpha \in E$, $\alpha \notin K$, such that α is fixed under every K-automorphism of E. Since $\alpha \notin K$, $m_{\alpha/K}$ has degree > 1.

Each K-conjugate, β, of α is in E and there is an isomorphism of $K(\alpha)$ onto $K(\beta)$, sending α to β, which extends to a K-automorphism of E. But α is fixed under each K-automorphism of E, and so α has no K-conjugate other than itself. It follows that α is the only zero, hence a multiple zero, of $m_{\alpha/K}$, and so α is inseparable over K. Contradiction! We conclude that $\bar{H} = K$. ■

Now note that any set, A, of automorphisms of a field E, with fixed field $K = \bar{A}$, is a set of endomorphisms of the additive group $\langle E, +\rangle$. Let $\mathscr{L}$ be the set of *all* endomorphisms of $\langle E, +\rangle$. In $\mathscr{L}$, we can define (vector) addition and (scalar) multiplication by elements of E as follows:

(1) for $\sigma, \tau \in \mathscr{L}$, $\sigma + \tau$ is defined by: $(\sigma + \tau)(\alpha) = \sigma(\alpha) + \tau(\alpha)$, for each $\alpha \in E$;

(2) for $\sigma \in \mathscr{L}$ and $a \in E$, $a\sigma$ is defined by: $(a\sigma)(\alpha) = a \cdot \sigma(\alpha)$ for each $\alpha \in E$.

It is easy to check that, for $\sigma, \tau \in \mathscr{L}$ and $a \in E$, $\sigma + \tau$ and $a\sigma$ are endomorphisms of $\langle E, +\rangle$ and that $\mathscr{L}$ forms a vector space over E with respect to these operations of addition and scalar multiplication. (See Exercise 4.9.5.) Linear combinations, linear dependence, and linear independence have the usual meanings in the vector space $\mathscr{L}$ over E. In particular, any linear combination of automorphisms in A is an endomorphism (though rarely an *auto*morphism) of $\langle E, +\rangle$.

It is in this setting that the following important lemma should be viewed.

Lemma 2 (Artin-Dedekind): Let E be a field, A a set of automorphisms of E. Then A is linearly independent over E.

Proof: Suppose A is linearly dependent over E. Then, among the non-trivial finite E-linear combinations of elements of A which equal the 0-endomorphism, we can select one that has the smallest possible number of non-zero coefficients. We label the elements of A so that this linear combination reads

$$c_1\sigma_1 + \cdots + c_t\sigma_t = 0, \tag{1}$$

with $c_1, \ldots, c_t$ in E all different from 0, and $\sigma_1, \ldots, \sigma_t \in A$. We shall obtain a contradiction by producing an even shorter non-trivial linear combination which equals 0.

By (1), for each $\alpha \in E$, we have

$$c_1\sigma_1(\alpha) + \cdots + c_t\sigma_t(\alpha) = 0. \tag{2}$$

The σ_i being all distinct, there is some $\beta \in E$ such that $\sigma_1(\beta) \neq \sigma_t(\beta)$. Then $\beta \neq 0$, and, using $\beta\alpha$ in place of α in (2), we obtain, for each $\alpha \in E$:

$$c_1\sigma_1(\beta\alpha) + \cdots + c_t\sigma_t(\beta\alpha) = 0 \tag{3}$$

whence

$$c_1\sigma_1(\beta)\sigma_1(\alpha) + \cdots + c_t\sigma_t(\beta)\sigma_t(\alpha) = 0. \tag{4}$$

Multiplying (2) by $c_t(\beta)$, we have

$$c_1\sigma_t(\beta)\sigma_1(\alpha) + \cdots + c_t\sigma_t(\beta)\sigma_t(\alpha) = 0. \tag{5}$$

Subtracting (5) from (4) yields

$$c_1[\sigma_1(\beta) - \sigma_t(\beta)]\sigma_1(\alpha) + \cdots + c_{t-1}[\sigma_{t-1}(\beta) - \sigma_t(\beta)]\sigma_{t-1}(\alpha) = 0, \tag{6}$$

for each $\alpha \in E$. Since the c_i are all non-zero and $\sigma_1(\beta) - \sigma_t(\beta) \neq 0$, by our choice of β, the endomorphism

$$c_1[\sigma_1(\beta) - \sigma_t(\beta)]\sigma_1 + \cdots + c_{t-1}[\sigma_{t-1}(\beta) - \sigma_t(\beta)]\sigma_{t-1} = 0, \tag{7}$$

contrary to the assumption that (5) represents the shortest such linear combination of elements of A. It follows that the automorphisms in A are linearly independent over E. ■

This leads to several interesting results.

Theorem 4.9.3: Let E be a field, H a finite group of automorphisms of E, with fixed field K. Then $|H| = [E : K]$.

Proof: Suppose $H = \{\sigma_1, \ldots, \sigma_t\}$, and $[E : K] = n$. If $n < t$, and $\{\alpha_1, \ldots, \alpha_n\}$ is a basis for E over K, then the linear system (with unknowns $c_1, \ldots, c_t$):

$$\begin{aligned} c_1\sigma_1(\alpha_1) + \cdots + c_t\sigma_t(\alpha_1) &= 0 \\ &\vdots \\ c_1\sigma_1(\alpha_n) + \cdots + c_t\sigma_t(\alpha_n) &= 0 \end{aligned} \tag{1}$$

is homogeneous and has fewer equations than unknowns, hence has a non-trivial solution $(c_1, \ldots, c_t)$ (c_i not all 0 in E). Since $\{\alpha_1, \ldots, \alpha_n\}$ is a K-basis for E, it follows from (1) that

$$c_1\sigma_1(\alpha) + \cdots + c_t\sigma_t(\alpha) = 0$$

for each $\alpha \in E$. But then $\sigma_1, \ldots, \sigma_t$ are linearly dependent over E, in contradiction to the Artin-Dedekind Lemma (p. 282). We conclude that $t \leq n$.

Suppose next that $t < n$. Then there are elements $\alpha_1, \ldots, \alpha_{t+1} \in E$, linearly independent over K. The linear system

$$\begin{aligned} c_1\sigma_1(\alpha_1) + \cdots + c_{t+1}\sigma_1(\alpha_{t+1}) &= 0 \\ &\vdots \\ c_1\sigma_t(\alpha_1) + \cdots + c_{t+1}\sigma_t(\alpha_{t+1}) &= 0 \end{aligned} \tag{2}$$

consists of t homogeneous linear equations in $t + 1$ unknowns, hence has a non-trivial solution $c_1, \ldots, c_{t+1} \in E$, not all 0. Among all non-trivial solutions, we choose one with the smallest number of non-zero elements and relabel, if necessary, so that $c_1, \ldots, c_s$ are non-zero while $c_{s+1} = \cdots = c_{t+1} = 0$.

For each $j = 1, \ldots, t$, we have

$$c_1\sigma_j(\alpha_1) + \cdots + c_s\sigma_j(\alpha_s) = 0. \tag{3}$$

Let $\sigma \in H$. Then, for each $j = 1, \ldots, t$,

$$\begin{aligned} \sigma(c_1\sigma_j(\alpha_1) + \cdots + c_s\sigma_j(\alpha_s)) &= \\ \sigma(c_1)(\sigma\sigma_j)(\alpha_1) + \cdots + \sigma(c_s)(\sigma\sigma_j(\alpha_s)) &= 0. \end{aligned} \tag{4}$$

But, as j runs from 1 to t, $\sigma\sigma_j$ takes on all possible values σ_i in the group H (why?). Hence (4) is equivalent to the system

$$\sigma(c_1)\sigma_j(\alpha_1) + \cdots + \sigma(c_s)\sigma_j(\alpha_s) = 0, \qquad 1 \leq j \leq t. \tag{5}$$

Now, multiplying (5) by c_1 and (3) by $\sigma(c_1)$, we obtain

$$c_1\sigma(c_1)\sigma_j(\alpha_1) + \cdots + c_1\sigma(c_s)\sigma_j(\alpha_s) = 0$$

and (6)

$$\sigma(c_1)c_1\sigma_j(\alpha_1) + \cdots + \sigma(c_1)c_s\sigma_j(\alpha_s) = 0.$$

Subtraction eliminates the first term and yields

$$[c_1\sigma(c_2) - \sigma(c_1)c_2]\sigma_j(\alpha_2) + \cdots + [c_1\sigma(c_s) - \sigma(c_1)c_s]\sigma_j(\alpha_s) = 0. \tag{7}$$

To avoid a contradiction to our choice of $c_1, \ldots, c_s, c_{s+1}, \ldots, c_{t+1}$ as a solution of (2) with the smallest number of non-zero elements, we must conclude that

$$c_1\sigma(c_i) - \sigma(c_1)c_i = 0 \tag{8}$$

for each $i = 2, \ldots, s$. But then

$$\sigma(c_ic_1^{-1}) = c_ic_1^{-1} \tag{9}$$

for each $i = 2, \ldots, s$, σ an arbitrary element of H, whence $c_ic_1^{-1}$ is in the fixed field, K, of H. Write $c_ic_1^{-1} = k_i \in K$, $i = 2, \ldots, s$. Then, from (3), with j chosen so that σ_j is the identity automorphism in H, we have $c_1\alpha_1 + c_1k_2\alpha_2 + \cdots + c_1k_s\alpha_s = 0$. Since $c_1 \neq 0$, $\alpha_1 + k_2\alpha_2 + \cdots + k_s\alpha_s = 0$, hence $\alpha_1, \ldots, \alpha_s$ are linearly dependent over K, contrary to their choice as elements of the linearly independent set $\alpha_1, \ldots, \alpha_{t+1}$. We conclude that $t \geq n$. But then $t = n$, i.e., $|H| = [E : K]$. ■

Theorem 4.9.4: Let E/F be a finite Galois extension, with Galois group $G = G(E/F)$, and let H be a subgroup of G. Then $H = G(E/K)$, where K is the fixed field of H.

Proof: If K is the fixed field of H, then $F \subset K \subset E$, and E/K is finite Galois. Clearly, $H \subset G(E/K)$, since the elements of H are K-automorphisms of E. By Theorem 4.9.1, $|G(E/K)| = [E : K]$. On the other hand, since K is the fixed field of H, by Theorem 4.9.3, $|H| = [E : K]$. Thus, $|H| = |G(E/K)|$, and so $H = G(E/K)$. ■

Exercises 4.9

True or False

1. If $F \subset K \subset E$, with E/K and K/F Galois, then E/F is Galois.
2. Every normal extension of a field of characteristic 0 is a Galois extension.
3. If E/F is a finite Galois extension, then the fixed field of $G(E/F)$ is F.
4. If E/F is a Galois extension of degree n, then $G(E/F)$ has order n.
5. If $F \subset K \subset E$, with E/F Galois, then both E/K and K/F are Galois extensions.

4.9.1. Let E be a field, and let A be a set of automorphisms of E. Prove that the fixed set, $\bar{A}$, of A defined by

$$\bar{A} = \{\alpha \in E \mid \sigma(\alpha) = \alpha \quad \text{for all} \quad \sigma \in A\}$$

forms a subfield of E.

4.9.2. Let G be a group of automorphisms of a field E and let S be a subset of E. If $H = \{\sigma \in G \mid \sigma(s) = s \quad \forall\, s \in S\}$, prove that $H < G$.

4.9.3. Let E be the splitting field, in $\mathbb{C}$, of $f = x^4 - 2$ over $\mathbb{Q}$. Find several examples of permutations on the zeros of f that do not extend to $\mathbb{Q}$-automorphisms of E. How many such permutations are there?

4.9.4. Let E be a splitting field of $f = x^3 - 2$ over $\mathbb{Q}$. Prove that *every* permutation on the zeros of f extends to a $\mathbb{Q}$-automorphism of E.

4.9.5. Let E be a field. Prove that the endomorphisms of $\langle E, +\rangle$ form a vector space over E, under the operations of addition and scalar multiplication defined on p. 282.

4.9.6. Let E be a field, and let $\sigma_1, \ldots, \sigma_n$ be automorphisms of E. Prove that there is some $t \in E$ such that

$$\sigma_1(t) + \cdots + \sigma_n(t) \neq 0.$$

4.10

The Fundamental Theorem of Galois Theory

Theorem 4.10.1: Let E/F be a finite Galois extension, Φ the set of all subfields K, $F \subset K \subset E$, and Γ the set of all subgroups of $G = G(E/F)$. Let $\mu : \Phi \to \Gamma$ be defined by

$$\mu(K) = G(E/K)$$

for each $K \in \Phi$, and let $\nu : \Gamma \to \Phi$ be defined by

$$\nu(H) = \bar{H} \quad \text{(the fixed field of } H\text{)}.$$

Then:

(1) μ and ν are inverse bijections, each reversing set inclusion.

(2) If $H \lhd G(E/F)$, then the fixed field of H is a normal extension of F.

(3) If $K \in \Phi$ and K/F is normal, then $G(E/K) \lhd G(E/F)$, and $G(K/F) \cong G(E/F)/G(E/K)$. (See Figure 10.)

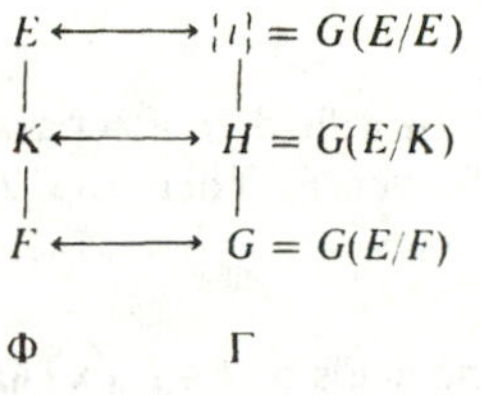

$$\begin{array}{ccl}
E & \longleftrightarrow & \{\iota\} = G(E/E) \\
| & & \quad | \\
K & \longleftrightarrow & H = G(E/K) \\
| & & \quad | \\
F & \longleftrightarrow & G = G(E/F) \\
 & & \\
\Phi & & \Gamma
\end{array}$$

Figure 10

Proof: (1) To prove that μ and ν are bijections, it suffices to show that they are one another's inverse maps. If $K \in \Phi$, then $\nu\mu K$ is the fixed field of $G(E/K)$, which is equal to K, by Theorem 4.9.2. Thus, $\nu\mu = \iota_\Phi$. If $H \in \Gamma$, then $\mu\nu H$ is $G(E/\nu H)$, which is equal to H, by Theorem 4.9.4. Thus, $\mu\nu = \iota_\Gamma$. We conclude that μ and ν are (inverse) bijections.

If $K_1, K_2 \in \Phi$, with $K_1 \subset K_2$, then $\mu K_1 = G(E/K_1) \supset G(E/K_2) = \mu K_2$ (since a larger subfield is fixed, elementwise, by fewer automorphisms). If $H_1, H_2 \in \Gamma$, with $H_1 \subset H_2$, then the fixed field, νH_1, of H_1 contains the fixed field νH_2 of H_2 (since a larger subgroup fixes fewer field elements). Thus, μ and ν both reverse set inclusion.

(2) Let $H \in \Gamma$. Suppose that $H \lhd G(E/F)$. Let $K = \nu H$, the fixed field of H. If $\tau \in G(E/F)$, then, for each $\sigma \in H$, $\tau\sigma\tau^{-1} \in H$. If $\alpha \in K$, then $(\tau\sigma\tau^{-1})(\tau\alpha) = \tau\sigma\alpha$. Since $\sigma \in H$ and $\alpha \in K = \nu H$, we have $\sigma\alpha = \alpha$. Thus, for each $\alpha \in K$ and each $\tau \in G(E/F)$, $\tau\alpha = (\tau\sigma\tau^{-1})(\tau\alpha)$. Since each element of H is $\tau\sigma\tau^{-1}$ for some $\sigma \in H$ (see Theorem 2.8.1), we conclude that $\tau\alpha$ is in the fixed field of H, i.e., $\tau\alpha \in K$. Thus, every F-conjugate of any element of K lies in K, and so K is a normal extension of F.

(3) Suppose $K \in \Phi$, and K/F is normal. For each $\tau \in G(E/F)$, $\tau|_K$ is an F-monomorphism of K into E. Since K/F is normal, $\tau|_K$ defines an F-automorphism $\rho \in G(K/F)$ such that $\rho\alpha = \tau\alpha$ for each $\alpha \in K$.

Conversely, if $\rho \in G(K/F)$, then (by Theorem 4.7.6) ρ extends to an F-automorphism of E, i.e., to an F-automorphism, τ, belonging to $G(E/F)$ such that $\tau\alpha = \rho\alpha$ for each $\alpha \in K$.

Define $\phi : G(E/F) \to G(K/F)$ by: $\phi(\tau) = \rho$ where ρ is the F-automorphism of K which agrees with τ on K $(\tau \in G(E/F))$. Then ϕ maps $G(E/F)$ *onto* $G(K/F)$ and is obviously a homomorphism. Since $\operatorname{Ker} \phi = \{\tau \in G(E/F) \mid \tau|_K = \iota_K\} = G(E/K)$, we conclude that $G(E/K) \lhd G(E/F)$ and that $G(K/F) = \operatorname{Im} \phi \cong G(E/F)/G(E/K)$. ■

Our immediate aim is to apply the Fundamental Theorem to the analysis of the zeros of specific polynomials.

Theorem 4.10.2: Let F be a field and let f be a separable polynomial in $F[x]$. If $E \supset F$ is a splitting field of f over F, then E/F is a finite Galois extension.

Proof: By Theorem 4.7.3, E/F is finite normal. By Corollary 4 of Theorem 4.8.4, E/F is separable. ■

Definition 4.10.1: Let F be a field, f a separable polynomial in $F[x]$, and $E \supset F$ a splitting field for f over F. Then $G(E/F)$ is the *Galois group of the polynomial* f over F.

Note: Since any two splitting fields of $f \in F[x]$ are F-isomorphic, the notion of "Galois group of a polynomial" is, thus, well-defined (see Exercise 4.10.1).

Example 1:

Let $f = x^3 - 5 \in \mathbb{Q}[x]$. The zeros of f in $\mathbb{C}$ are $\sqrt[3]{5}$, $\sqrt[3]{5}\,\omega$ and $\sqrt[3]{5}\,\omega^2$, where $\omega = \dfrac{-1 + i\sqrt{3}}{2}$, $\omega^2 = \bar{\omega} = \dfrac{-1 - i\sqrt{3}}{2}$. Thus, the splitting field, in $\mathbb{C}$, of f over $\mathbb{Q}$ is $E = \mathbb{Q}(\sqrt[3]{5}, \omega)$. At this stage, it is best to build field towers. We have

$$\mathbb{Q} \subset \mathbb{Q}(\sqrt[3]{5}) \subset \mathbb{Q}(\sqrt[3]{5}, \omega) = E.$$

Now, $m_{\sqrt[3]{5}/\mathbb{Q}} = x^3 - 5$, hence $[\mathbb{Q}(\sqrt[3]{5}) : \mathbb{Q}] = 3$. Since $m_{\omega/\mathbb{Q}} = x^2 + x + 1$ is irreducible over $\mathbb{Q}(\sqrt[3]{5})$, $m_{\omega/\mathbb{Q}(\sqrt[3]{5})} = x^2 + x + 1$, hence $[\mathbb{Q}(\sqrt[3]{5}, \omega) : \mathbb{Q}(\sqrt[3]{5})] = 2$. It follows that $[E : \mathbb{Q}] = 3 \cdot 2 = 6$, hence $G(E/\mathbb{Q})$ has order 6.

The elements of $G(E/\mathbb{Q})$ are, of course, the $\mathbb{Q}$-automorphisms of E. Now, every $\mathbb{Q}$-automorphism of E is completely determined by its effect on $\sqrt[3]{5}$ and on ω. If $\tau \in G(E/\mathbb{Q})$, then $\tau(\sqrt[3]{5})$ must be one of the $\mathbb{Q}$-conjugates, in E, of $\sqrt[3]{5}$, i.e., either $\sqrt[3]{5}$, $\sqrt[3]{5}\,\omega$, or $\sqrt[3]{5}\,\omega^2$. Similarly, $\tau\omega$ must be one of the $\mathbb{Q}$-conjugates of ω in E, i.e., either ω or ω^2. This analysis immediately yields the six automorphisms in $G(E/\mathbb{Q})$:

$$\tau_1 : \sqrt[3]{5} \mapsto \sqrt[3]{5}, \qquad \omega \mapsto \omega$$

$$\tau_2 : \sqrt[3]{5} \mapsto \sqrt[3]{5}, \qquad \omega \mapsto \omega^2$$

$$\tau_3 : \sqrt[3]{5} \mapsto \sqrt[3]{5}\,\omega, \qquad \omega \mapsto \omega$$

$$\tau_4 : \sqrt[3]{5} \mapsto \sqrt[3]{5}\,\omega, \qquad \omega \mapsto \omega^2$$

$$\tau_5 : \sqrt[3]{5} \mapsto \sqrt[3]{5}\,\omega^2, \qquad \omega \mapsto \omega$$

$$\tau_6 : \sqrt[3]{5} \mapsto \sqrt[3]{5}\,\omega^2, \qquad \omega \mapsto \omega^2.$$

It is easy to check (see Exercise 4.10.2) that the orders of the τ_i in the group $G(E/\mathbb{Q})$ are:

$$o(\tau_1) = 1$$

$$o(\tau_2) = 2$$

$$o(\tau_3) = 3$$

$$o(\tau_4) = 2$$

$$o(\tau_5) = 3$$

$$o(\tau_6) = 2$$

Thus, $G(E/\mathbb{Q})$ is not cyclic, and is therefore isomorphic to S_3 (see Exercise 2.6.16).

The subgroup lattice of $G(E/\mathbb{Q})$ is shown in Figure 11.

Since the Galois correspondence reverses inclusion, we draw an upside-down lattice diagram showing each subfield K of E, containing $\mathbb{Q}$, in a position corresponding to that of the Galois group $G(E/K)$, for which it serves as the fixed field. (See Figure 12.)

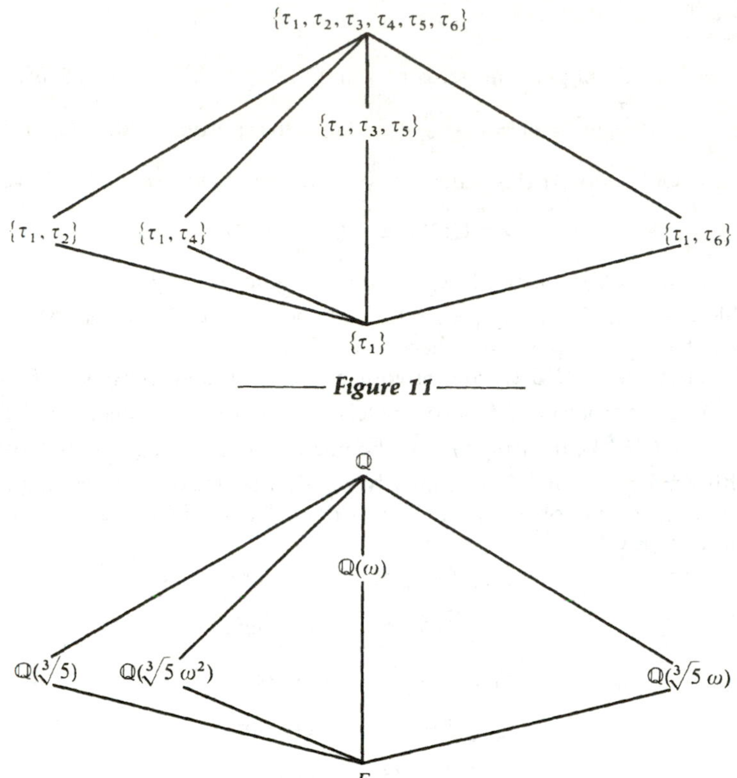

Figure 11

Figure 12

Note that

$$\tau_4(\sqrt[3]{5}\,\omega^2) = \tau_4(\sqrt[3]{5})\tau_4\omega^2 = \sqrt[3]{5}\,\omega(\tau_4\omega)^2 = \sqrt[3]{5}\,\omega\omega^4 = \sqrt[3]{5}\,\omega^2,$$

so that τ_4 fixes $\sqrt[3]{5}\,\omega^2$. Similarly, τ_6 fixes $\sqrt[3]{5}\,\omega$. For each subfield K, $\mathbb{Q} \subset K \subset E$, we have $|G(E/K)| = [E : K] = \frac{6}{[K : \mathbb{Q}]}$. Thus, to each subgroup of order 2 corresponds a subfield of degree 3 over $\mathbb{Q}$, and to each subgroup of order 3 corresponds a subfield of degree 2 over $\mathbb{Q}$.

The subgroup $\{\tau_1, \tau_3, \tau_5\}$ is the only normal subgroup of $G(E/\mathbb{Q})$ (recall that, for $n \geq 3$, S_n has no normal subgroup of order 2). Hence the only normal extension field of $\mathbb{Q}$ in E (other that $\mathbb{Q}$ and E themselves) is $\mathbb{Q}(\omega)$. This should be no surprise, for $[\mathbb{Q}(\omega) : \mathbb{Q}] = 2$, and every extension of degree 2 is normal!

More elegantly, one can describe the elements of $G(E/\mathbb{Q})$ in terms of "generators and relations." For example, we may let τ be the automorphism such that

$$\tau(\sqrt[3]{5}) = \sqrt[3]{5}\,\omega, \ \ \tau\omega = \omega \quad \text{(previously } \tau_3\text{)},$$

and σ the automorphism such that

$$\sigma(\sqrt[3]{5}) = \sqrt[3]{5}, \quad \sigma\omega = \omega^2 \quad \text{(previously } \tau_2\text{)}.$$

Then $\tau^3 = \iota, \sigma^2 = \iota$ (ι the identity automorphism) and the elements of $G(E/\mathbb{Q})$ are given by $\iota, \tau, \tau^2, \sigma, \tau\sigma, \tau^2\sigma$. (See Exercise 4.10.5.)

Example 2:

Let $\alpha = i + \sqrt[4]{2} \in \mathbb{C}$, and let E be the splitting field, in $\mathbb{C}$, of $m_{\alpha/\mathbb{Q}}$. We wish to find $G(E/\mathbb{Q})$ and exhibit the Galois correspondence.

Finding $m_{\alpha/\mathbb{Q}}$ is an easy, if tedious, task, which we ask you to perform in Exercise 4.10.7. For our purposes here, there is, however, no need to find $m_{\alpha/\mathbb{Q}}$. For, knowing one of the zeros of $m_{\alpha/\mathbb{Q}}$, we know *all* of its zeros in E: if $\tau \in G(E/\mathbb{Q})$, then $\tau\alpha = \tau i + \tau\sqrt[4]{2}$. The $\mathbb{Q}$-conjugates of i are $\pm i$; the $\mathbb{Q}$-conjugates of $\sqrt[4]{2}$ are $\pm\sqrt[4]{2}$ and $\pm\sqrt[4]{2}\,i$. Since $m_{i/\mathbb{Q}(\sqrt[4]{2})} = m_{i/\mathbb{Q}} = x^2 + 1$, Theorem 4.7.7 implies that the $\mathbb{Q}$-conjugates of α in E are the eight numbers

$$\pm i \pm \sqrt[4]{2} \text{ and } \pm i \pm i\sqrt[4]{2}. \tag{1}$$

Thus, $\deg m_{\alpha/\mathbb{Q}} = 8 = [\mathbb{Q}(\alpha) : \mathbb{Q}]$, and we have

$$\mathbb{Q} \subset \mathbb{Q}(\alpha) \subset E \subset \mathbb{Q}(\sqrt[4]{2}, i). \tag{2}$$

From the field tower

$$\mathbb{Q} \subset \mathbb{Q}(i) \subset \mathbb{Q}(i, \sqrt[4]{2}), \tag{3}$$

and the fact that $m_{i/\mathbb{Q}(\sqrt[4]{2})} = m_{i/\mathbb{Q}}$, we infer that $[\mathbb{Q}(\sqrt[4]{2}, i) : \mathbb{Q}] = 8$. But then, by (2),

$$8 = [\mathbb{Q}(\alpha) : \mathbb{Q}] \le [E : \mathbb{Q}] \le [\mathbb{Q}(\sqrt[4]{2}, i) : \mathbb{Q}] = 8, \tag{4}$$

and so $[E : \mathbb{Q}] = 8$, and $E = \mathbb{Q}(\sqrt[4]{2}, i) = \mathbb{Q}(\alpha)$.

Since $\alpha = i + \sqrt[4]{2}$ is a primitive element for the extension $E/\mathbb{Q}$, we could easily enumerate the elements of $G(E/\mathbb{Q})$ by characterizing each as sending α to a particular one of its $\mathbb{Q}$-conjugates.

Instead, we look for generators and relations. Let ρ be the $\mathbb{Q}$-automorphism of E such that $\rho(\sqrt[4]{2}) = \sqrt[4]{2}\,i$, $\rho(i) = i$. (Why does such a $\mathbb{Q}$-automorphism exist?) Then

$$\rho^2(\sqrt[4]{2}) = \rho(\sqrt[4]{2}\,i) = -\sqrt[4]{2}, \rho^3(\sqrt[4]{2}) = \rho(-\sqrt[4]{2}) = -\rho(\sqrt[4]{2}) = -\sqrt[4]{2}\,i,$$

and $\rho^4(\sqrt[4]{2}) = \rho(-\sqrt[4]{2}\,i) = (-\sqrt[4]{2}\,i)i = \sqrt[4]{2}$, while $\rho^4(i) = i$. Thus, ρ has order 4. Let σ be the $\mathbb{Q}$-automorphism of E such that $\sigma(i) = -i$, $\sigma(\sqrt[4]{2}) = \sqrt[4]{2}$. Then σ has order 2. The two automorphisms ρ and σ, along with the relations $\rho^4 = \iota$, $\sigma^2 = \iota$, and $\sigma\rho = \rho^{-1}\sigma$ suffice to determine the entire group. We obtain $G(E/F) = \{\iota, \rho, \rho^2, \rho^3, \sigma, \rho\sigma, \rho^2\sigma, \rho^3\sigma\}$, easily recognizable as an isomorphic copy of the dihedral group D_4, with the subgroup lattice shown in Figure 13.

The corresponding (upside down) lattice of intermediate fields is shown in Figure 14.

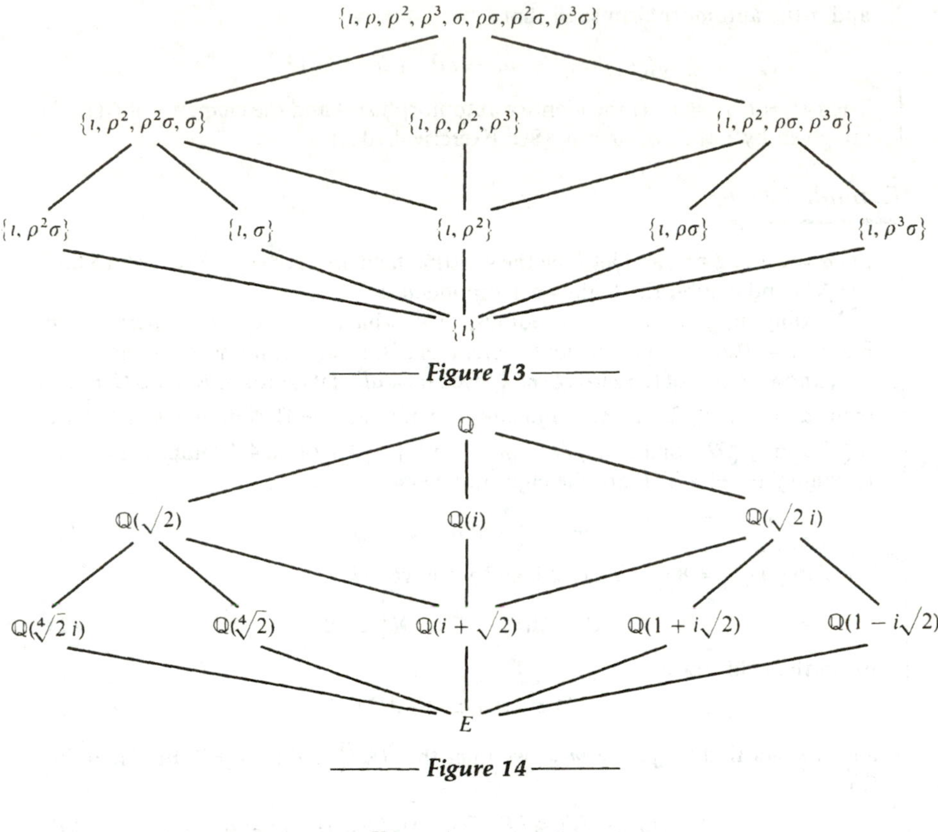

Figure 13

Figure 14

Exercises 4.10

True or False

Let E/F be a Galois extension, with $G = G(E/F)$. Then:

1. If $H < G$ and K is the fixed field of H, then $H = G(K/F)$.
2. If K/F is normal, then $G(K/F)$ is a homomorphic image of $G(E/F)$.
3. If $F \subset K \subset E$ and H is the group of all K-automorphisms of E, then K is the fixed field of H.
4. If $F \subset K \subset E$ and K is a normal extension of F, then $G(K/F) \lhd G$.
5. If $F \subset K_1 \subset K_2 \subset E$, then $G(E/K_1) \subset G(E/K_2)$.

4.10.1. If $E_1 \supset F$ and $E_2 \supset F$ are splitting fields for a separable polynomial $f \in F[x]$, prove that $G(E_1/F) \cong G(E_2/F)$.

4.10.2. Find the order in $G(E/\mathbb{Q})$ of each of the $\mathbb{Q}$-automorphisms τ_i in Example 1.

4.10.3. Let $F \subset K \subset E$, with E/F finite Galois and $[E : F] = n$. Prove that the order of $G(E/K)$ is equal to $\dfrac{n}{[K : F]}$.

4.10.4. Let E/F be finite Galois, with $G = G(E/F)$, and let $H < G$. Prove: if K is the fixed field of H, then $[G : H] = [K : F]$.

4.10.5. Verify that, in Example 1, $G(E/\mathbb{Q}) = \{\iota, \tau, \tau^2, \sigma, \tau\sigma, \tau^2\sigma\}$.

4.10.6. In Example 1, find a primitive element, α, for the extension $E/\mathbb{Q}$, and characterize the elements of $G(E/\mathbb{Q})$ in terms of their effect on α.

4.10.7. Find the minimal polynomial over $\mathbb{Q}$ of $\alpha = i + \sqrt[4]{2} \in \mathbb{C}$. (See Example 2, p. 289, and Theorem 4.7.7.)

4.10.8. Let $\alpha = \sqrt{2} + \sqrt[3]{2}$. Find the splitting field, in $\mathbb{C}$, of $m_{\alpha/\mathbb{Q}}$. Find the Galois group, G, of $m_{\alpha/\mathbb{Q}}$ over $\mathbb{Q}$. Draw the subgroup lattice of G, and the corresponding lattice of intermediate fields.

4.10.9. Let L and L' be lattices with partial order relations R and R', respectively. A bijection $\beta : L \to L'$ is a *lattice anti-isomorphism* if aRb implies $\beta(b)R'\,\beta(a)$ for each $a, b \in L$.

In Theorem 4.10.1, interpret Φ and Γ as lattices and observe that the mappings μ and ν are lattice anti-isomorphisms.

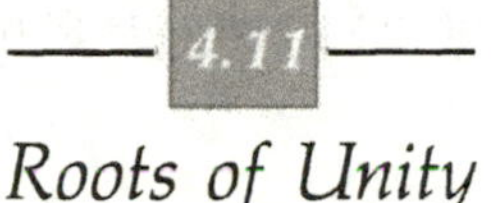

4.11 Roots of Unity

Our chief application of the Fundamental Theorem of Galois Theory will be to the classical problem of solvability of polynomials by radicals. Under what conditions can the zeros of a polynomial be expressed in terms of finitely many field operations and extractions of roots? We want to explore this question by determining what properties of the Galois group of a polynomial correspond to its solvability by radicals.

For n a positive integer, an "n-th root" is a zero, α, of a polynomial $x^n - a \in F[x]$, for some field F. If $x^n - a$ is irreducible over F and $\zeta^n = 1$ for some ζ in an extension field of F, then $(\alpha\zeta)^n = \alpha^n\zeta^n = a$; hence, $\alpha\zeta$ is a zero of $x^n - a$, and is thus an F-conjugate of α. Conversely, if β is an F-conjugate of α in some extension field of F, then $(\beta\alpha^{-1})^n = \beta^n(\alpha^n)^{-1} = aa^{-1} = 1$, whence $\zeta = \beta\alpha^{-1}$ is an n-th root of 1, and $\beta = \zeta\alpha = \alpha\zeta$. This suggests that the n-th roots of 1 will be important ingredients of any discussion relating the Galois group of a polynomial to the polynomial's solvability by radicals.

Definition 4.11.1: Let F be a field, and let $E \supset F$ be a splitting field of $x^n - 1$ over F. Then the zeros of $x^n - 1$ in E are the *n-th roots of unity* in E.

(Since any two splitting fields over F of $x^n - 1 \in F[x]$ are F-isomorphic, the relevant properties of n-th roots of unity will be independent of the choice of splitting field.)

Theorem 4.11.1: Let F be a field, and let E be a splitting field of $x^n - 1$ over F. If either Char $F = 0$, or Char $F = p > 0$, with $p \nmid n$, then the n-th roots of unity in E form a cyclic subgroup of E^*, of order n.

Proof: If Char $F = 0$, or if Char $F = p > 0$, with $p \nmid n$, then $x^n - 1$ is separable over F since $x^n - 1$ and its derivative nx^{n-1} clearly have no zeros in common (see Theorem 4.8.1.). Hence there are exactly n distinct n-th roots of unity in E. Let U_n be the set of all n-th roots of unity in E. Then $1 \in U_n$ and, for each $\mu, \nu \in U_n$,

$(\mu\nu)^n = \mu^n\nu^n = 1$; hence, by the Corollary of Theorem 2.3.6, U_n is a subgroup of E^*, of order n. By Theorem 4.4.6, U_n is cyclic. ■

Definition 4.11.2: Let n be a positive integer, and let F be a field of characteristic 0 or of characteristic $p > 0$, with $p \nmid n$. Then any generator, ζ, of the (cyclic) group, U_n, of all n-th roots of unity in a splitting field E of $x^n - 1$ over F is a *primitive n-th root of unity*.

As a consequence of Theorem 4.11.1, we now have the following result.

Corollary: If n is a positive integer, F a field of characteristic 0 or of characteristic $p > 0$, with $p \nmid n$, then there is a primitive n-th root of unity contained in some extension field of F.

Under the hypotheses of Definition 4.11.2, by Exercise 2.3.13, the number of primitive n-th roots of unity, for any given n, is $\phi(n)$, where ϕ is the Euler Phi-Function.

Example:

In the complex number field, for each positive integer n, $\zeta = e^{2\pi i/n} = \cos\frac{2\pi}{n} + i\sin\frac{2\pi}{n}$ is a primitive n-th root of unity, since

$$[\zeta] = \{e^{2\pi ih/n} \mid h = 0, \dots, n-1\} = U_n$$

consists of (the) n distinct n-th roots of 1 in $\mathbb{C}$. In the complex plane, by de Moivre's Theorem, the points representing the ζ^h $(h = 0, \dots, n-1)$ are equally spaced around the unit circle, hence form the vertices of a regular n-gon. The primitive n-th roots of unity are the ζ^h with gcd $(h, n) = 1$. For example, the sixth roots of unity are as shown in Figure 15. The primitive sixth roots of unity are $\zeta = e^{2\pi i/6}$ and $\zeta^5 = e^{10\pi i/6} = e^{5\pi i/3} = \zeta^{-1}$.

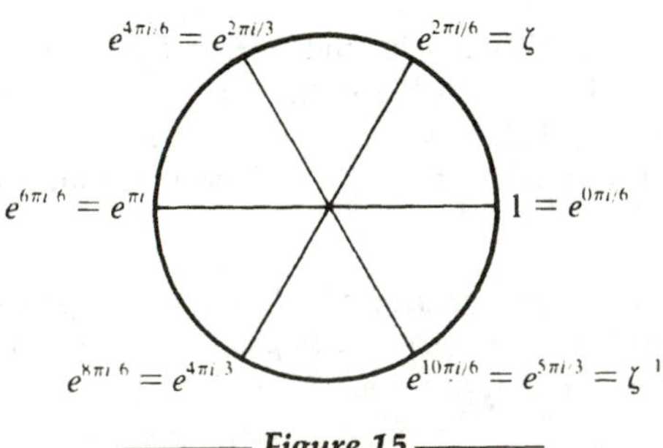

Figure 15

In the following, we explore further the properties of roots of unity in the complex number field.

Definition 4.11.3: For n a positive integer, let $\zeta_1, \ldots, \zeta_{\varphi(n)}$ be the primitive n-th roots of unity in $\mathbb{C}$. Then the polynomial

$$\Phi_n = \prod_{i=1}^{\phi(n)} (x - \zeta_i)$$

in $\mathbb{C}[x]$ is the *n-th cyclotomic polynomial.*

The word *cyclotomic* means "circle cutting," or "circle dividing," referring to the fact that the n n-th roots of unity divide the circumference of the unit circle into n equal arcs.

We shall prove that, for each positive integer n, Φ_n has integer coefficients, is irreducible over $\mathbb{Q}$, and is thus the minimal polynomial, over $\mathbb{Q}$, of each of the primitive n-th roots of unity in $\mathbb{C}$.

The following result regarding primitive polynomials (see Definition 3.9.1) will be helpful.

Lemma: Let r, s, t be polynomials in $\mathbb{Q}[x]$ such that $r = st$, with r monic in $\mathbb{Z}[x]$ and s monic in $\mathbb{Q}[x]$. Then s and t are both primitive polynomials in $\mathbb{Z}[x]$.

Proof: Note that, since $r \in \mathbb{Z}[x]$ is monic, r is primitive. As in Section 3.9, proof of Lemma 3, we write $s = \frac{a}{b}\bar{s}$, $t = \frac{c}{d}\bar{t}$, where $a, b, c, d \in \mathbb{Z}$ and $\bar{s}, \bar{t}$ are primitive in $\mathbb{Z}[x]$. Then

$$bdr = ac\bar{s}\bar{t}.$$

By Section 3.9, Lemmas 1 and 2, since r is primitive, we conclude that r is an associate of $\bar{s}\bar{t}$ in $\mathbb{Z}[x]$, i.e., $r = \pm\bar{s}\bar{t}$. If we choose $a, b, c, d \in \mathbb{Z}$ such that $\bar{s}$ and $\bar{t}$ have positive leading coefficients then, clearly, $r = \bar{s}\bar{t}$. Since r is monic, so are $\bar{s}$ and $\bar{t}$. But then, from $s = \frac{a}{b}\bar{s}$, we conclude that $\frac{a}{b}$ is the leading coefficient of s; hence $\frac{a}{b} = 1$ and $s = \bar{s}$. From $r = st$, with r, s monic, we have t monic, hence, similarly, $t = \bar{t}$. Thus, $s, t \in \mathbb{Z}[x]$, both primitive. ■

Theorem 4.11.2: For n a positive integer, the n-th cyclotomic polynomial Φ_n has integer coefficients and is irreducible over $\mathbb{Q}$.

Proof: If $\zeta \in \mathbb{C}$ is a primitive n-th root of unity, then the primitive n-th roots of unity, in $\mathbb{C}$, are the $\phi(n)$ numbers ζ^h, where $\gcd(h, n) = 1$, $1 \leq h < n$. By Definition 4.11.3,

$$\Phi_n = \prod_{\substack{1 \leq h < n \\ \gcd(h, n) = 1}} (x - \zeta^h).$$

Let $f = m_{\zeta/\mathbb{Q}}$. Then $f \in \mathbb{Z}[x]$, for: since ζ is a zero of $x^n - 1$, $m_{\zeta/\mathbb{Q}} | x^n - 1$ in $\mathbb{Q}[x]$. Thus, there is a polynomial g in $\mathbb{Q}[x]$ such that

$$x^n - 1 = fg. \tag{1}$$

By the preceding Lemma, since $x^n - 1$ is primitive in $\mathbb{Z}[x]$ and f is monic in $\mathbb{Q}[x]$, we conclude that f, g are both (monic) primitive polynomials in $\mathbb{Z}[x]$.

We next show that the zeros of $f = m_{\zeta/\mathbb{Q}}$ are precisely the $\phi(n)$ primitive n-th roots of unity. From this, it will follow that $\Phi_n = f \in \mathbb{Z}[x]$, irreducible over $\mathbb{Q}$.

First note that, if η is a zero of f, then there is a $\mathbb{Q}$-isomorphism of $\mathbb{Q}(\zeta)$ onto $\mathbb{Q}(\eta)$, sending ζ to η. This implies that ζ and η have the same order in the multiplicative group $\mathbb{C}^*$. But then η has order n, and is thus a primitive n-th root of unity.

It remains to show that *every* primitive n-th root of unity is a zero of f, i.e., that ζ^h is a zero of f for each integer h $(1 \leq h < n)$ with $\gcd(h, n) = 1$. Since h is a product of primes not dividing n, it suffices to prove that, for each zero, η, of f and each prime p not dividing n, η^p is also a zero of f. Successive application of this result, starting with ζ, will then yield the conclusion that ζ^h is a zero of f.

We proceed by contradiction. Suppose that, for p a prime, $p \nmid n$, η^p is *not* a zero of f for some zero, η, of f. Since η^p is clearly a zero of $x^n - 1$, we have, from (1),

$$0 = f(\eta^p)g(\eta^p).$$

Since $f(\eta^p) \neq 0$, we conclude that $g(\eta^p) = 0$. Put another way, η is a zero of the polynomial $l(x) = g(x^p) \in \mathbb{Q}[x]$. But then $f = m_{\eta/\mathbb{Q}}$ divides l in $\mathbb{Q}[x]$:

$$g(x^p) = l = fq \tag{2}$$

for $q \in \mathbb{Q}[x]$. Since l is monic in $\mathbb{Z}[x]$ and f is monic, we have $q \in \mathbb{Z}[x]$, again by the preceding Lemma.

To cope with $g(x^p)$, we now "reduce modulo p." By Exercise 4.1.8, the canonical ring epimorphism of $\mathbb{Z}$ onto $\mathbb{Z}_p = \mathbb{Z}/p\mathbb{Z}$ extends to an epimorphism of $\mathbb{Z}[x]$ onto $\mathbb{Z}_p[x]$. For $t \in \mathbb{Z}[x]$, we denote by $\tilde{t}$ the polynomial in $\mathbb{Z}_p[x]$ corresponding to t. From (2), we obtain

$$\tilde{g}(x^p) = \tilde{f}\tilde{q}. \tag{3}$$

But $\tilde{g}(x^p) = [\tilde{g}(x)]^p$ (see Exercise 11.11), and so

$$[\tilde{g}(x)]^p = \tilde{f}\tilde{q}. \tag{4}$$

If $\tilde{d} \in \mathbb{Z}_p[x]$ is an irreducible divisor of $\tilde{f}$, then $\tilde{d} | \tilde{g}$. Hence, from

$$\bar{1}x^n - \bar{1} = \widetilde{x^n - 1} = \tilde{f}\tilde{g}, \tag{5}$$

(bars denoting residue classes modulo p in $\mathbb{Z}_p$), we infer that $\tilde{d}$ is a divisor of $\bar{1}x^n - \bar{1}$ of multiplicity at least 2. This implies that the polynomial $\bar{1}x^n - \bar{1}$ has a zero in common with its derivative, $(n\bar{1})x^{n-1}$. But this is impossible since $p \nmid n$, hence $(n\bar{1})x^{n-1}$ is not the 0-polynomial.

We conclude that η^p *is* a zero of f. It follows that the zeros of f are precisely the primitive n-th roots of unity, i.e., the numbers ζ^h, $\gcd(h, n) = 1$. Since f and Φ_n are both monic, separable polynomials, we have $f = m_{\zeta/\mathbb{Q}} = \Phi_n = \prod_{\substack{1 \leq h < n \\ \gcd(h, n) = 1}} (x - \zeta^h)$.

Thus, $\Phi_n \in \mathbb{Z}[x]$, and Φ_n is irreducible over $\mathbb{Q}$. ∎

Corollary 1: For n a positive integer, the n-th cyclotomic polynomial Φ_n is the minimal polynomial over $\mathbb{Q}$ of each primitive n-th root of unity in $\mathbb{C}$.

Corollary 2: If $\zeta \in \mathbb{C}$ is a primitive n-th root of unity, then $[\mathbb{Q}(\zeta):\mathbb{Q}] = \phi(n)$, and $\mathbb{Q}(\zeta)/\mathbb{Q}$ is a Galois extension.

Proof: Since $m_{\zeta/\mathbb{Q}} = \Phi_n$, $[\mathbb{Q}(\zeta):\mathbb{Q}] = \deg \Phi_n = \phi(n)$, and the $\mathbb{Q}$-conjugates of ζ are all powers of ζ, hence contained in $\mathbb{Q}(\zeta)$. It follows that $\mathbb{Q}(\zeta)$ is normal over $\mathbb{Q}$, hence a Galois extension of $\mathbb{Q}$. ■

Now, note that, if U_n is the set of *all* n-th roots of unity in $\mathbb{C}$ ($n \in \mathbb{Z}^+$), then each $\eta \in U_n$ has multiplicative order d for some divisor d of n, and is thus a primitive d-th root of unity. Conversely, if $d|n$, $d > 0$, then every (primitive) d-th root of unity is also an n-th root of unity. The set, U_n, can therefore be partitioned as follows: $U_n = \bigcup_{\substack{d|n \\ d>0}} P_d$, where P_d is the set of all *primitive* d-th roots of unity. From this, it follows that

$$x^n - 1 = \prod_{\substack{d|n \\ d>0}} \Phi_d.$$

The cyclotomic polynomials can thus be computed recursively:

$$\Phi_n = \frac{x^n - 1}{\prod_{\substack{d|n \\ 0 \leq d < n}} \Phi_d}.$$

Obviously, $\Phi_1 = x - 1$. Hence

$$\Phi_2 = \frac{x^2 - 1}{x - 1} = x + 1,$$

$$\Phi_3 = \frac{x^3 - 1}{x - 1} = x^2 + x + 1,$$

$$\Phi_4 = \frac{x^4 - 1}{(x - 1)(x + 1)} = x^2 + 1,$$

$$\Phi_5 = \frac{x^5 - 1}{x - 1} = x^4 + x^3 + x^2 + x + 1,$$

$$\Phi_6 = \frac{x^6 - 1}{(x - 1)(x + 1)(x^2 + x + 1)} = x^2 - x + 1,$$

$$\Phi_7 = \frac{x^7 - 1}{x - 1} = x^6 + x^5 + x^4 + x^3 + x^2 + x + 1,$$

$$\Phi_8 = \frac{x^8 - 1}{(x - 1)(x + 1)(x^2 + 1)} = x^4 + 1,$$

$$\Phi_9 = \frac{x^9 - 1}{(x - 1)(x^2 + x + 1} = x^6 + x^3 + 1,$$

$$\Phi_{10} = \frac{x^{10} - 1}{(x - 1)(x + 1)(x^4 + x^3 + x^2 + x + 1)} = x^4 - x^3 + x^2 - x + 1, \quad \text{etc.}$$

Note: The least positive integer n such that Φ_n has a coefficient of absolute value greater than 1 is $n = 105$. In fact, for $n < 385$, the coefficients of Φ_n do not exceed 2 in absolute value. It is therefore all the more striking that, by a theorem of I. Schur, there exist cyclotomic polynomials with coefficients of arbitrarily large absolute value. (See [34]: Emma Lehmer, *On the Magnitude of the Coefficients of the Cyclotomic Polynomial*, Bull. AMS 42 (1936).)

Exercises 4.11

True or False

1. For each positive integer n, the n-th cyclotomic polynomial is

$$x^{n-1} + x^{n-2} + \cdots + x + 1.$$

2. If $\zeta \neq 1$ in $\mathbb{C}$ is an n-th root of 1, then ζ has $\phi(n)$ distinct $\mathbb{Q}$-conjugates.

3. For each positive integer n and each prime p, the polynomial $x^n - 1$ in $\mathbb{Z}_p[x]$ is separable over $\mathbb{Z}_p$.

4. Every root of unity is a primitive k-th root of unity for some positive integer k.

5. $\cos 72° + i \sin 72°$ is a primitive tenth root of unity.

4.11.1. Let F be a field and let $n > 1$ be an integer. Prove: if some extension field of F contains n distinct roots of unity, then either Char $F = 0$, or Char $F = p > 0$, with $p \nmid n$.

4.11.2. For p a prime, prove that $\Phi_p(x + 1) \in \mathbb{Q}[x]$ is an Eisenstein polynomial and hence deduce the irreducibility of Φ_p over $\mathbb{Q}$. (Recall Exercise 3.9.6.)

4.11.3. Prove that, for p prime, k a positive integer, $\Phi_{p^k}(x) = \Phi_p(x^{p^{k-1}})$. Use this result to determine Φ_{16}, Φ_{25}, and Φ_{27}.

4.11.4. Let $n \geq 1$ be an integer, and let $\zeta \in \mathbb{C}$ be a primitive n-th root of unity.

(1) Prove that the Galois group of the n-th cyclotomic polynomial Φ_n over $\mathbb{Q}$ is equal to $G(\mathbb{Q}(\zeta)/\mathbb{Q})$.

(2) Find the elements of $G(\mathbb{Q}(\zeta)/\mathbb{Q})$ and show that $G(\mathbb{Q}(\zeta)/\mathbb{Q})$ is isomorphic to $\mathbb{Z}_n^*$.

(3) Find the elements of $G(\mathbb{R}(\zeta)/\mathbb{R})$ and show that $G(\mathbb{R}(\zeta)/\mathbb{R})$ is isomorphic to a subgroup of $\mathbb{Z}_n^*$.
(Hint: for $n > 2$, $\phi(n)$ is even.)

4.11.5. Let F be a field, n a positive integer not divisible by Char F, and let ζ be a primitive n-th root of 1 contained in an extension field of F. Prove that $F(\zeta)/F$ is a Galois extension, with $G(F(\zeta)/F)$ isomorphic to a subgroup of $\mathbb{Z}_n^*$ (hence abelian).

4.11.6. Express the ninth roots of unity as powers of e and represent them in the complex plane. Identify the primitive ninth roots of unity.

4.11.7. Find the cyclotomic polynomial Φ_9 and use it to express the ninth roots of unity in terms of square roots and cube roots.

4.11.8. Find the cube roots of -1 and represent them in the complex plane.

4.11.9. Let $\theta \neq \pm 1$ be a sixth root of 1. Prove that θ is a primitive sixth root of 1 if and only if θ is a cube root of -1.

4.11.10. Prove that, for $n > 2$, n not prime, the polynomial

$$f(x) = x^{n-1} + x^{n-2} + \cdots + x + 1$$

is reducible over $\mathbb{Q}$. (Conclude that there is no integer k such that $f(x + k)$ is an Eisenstein polynomial.)

4.11.11. Let p be a prime and let h be a polynomial in $\mathbb{Z}_p[x]$. Prove that $[h(x)]^p = h(x^p)$.

4.11.12. Assuming as known that every non-negative real number has a real n-th root, for each positive integer n, prove: every complex number has n complex n-th roots. (Hint: use de Moivre's Theorem.)

4.11.13. Let $n > 1$ be an odd integer. Prove that $\Phi_{2n}(x) = \Phi_n(-x)$.

4.11.14. Let p, q be distinct primes.

(1) Prove that $\Phi_{pq}(x) = \dfrac{\Phi_q(x^p)}{\Phi_q(x)} = \dfrac{\Phi_p(x^q)}{\Phi_p(x)}$.

(2) Conclude that $\Phi_p(x)\ \Phi_q(x^p) = \Phi_q(x)\ \Phi_p(x^q)$.

Radical Extensions I

(Note: Results from Section 2.13 are used in this section.)

In order for a polynomial $f \in F[x]$ to be solvable by radicals (in the sense of Definition 4.12.1), its splitting field must be contained in a special kind of extension: an extension obtainable by successive adjunction of radicals.

In order to include the case where Char $F = p > 0$, we must generalize the notion of "radical" somewhat beyond the familiar idea of root extraction.

Definition 4.12.1: An element α in an extension field of a field F is a *radical* over F if it satisfies one of the following conditions:

(1) α is a zero of some polynomial $x^n - a \in F[x]$, where Char $F \nmid n$;

(2) α is a zero of some polynomial $x^p - x - a \in F[x]$, where Char $F = p$.

An extension K/F is a *radical extension* if it has a *radical tower*, i.e., if there are subfields K_i $(i = 0, \ldots, t)$ and elements α_i $(i = 1, \ldots, t)$ of K such that

$$F = K_0 \subset K_1 \subset \cdots \subset K_t = K,$$

where $K_i = K_{i-1}(\alpha_i)$, α_i a radical over K_{i-1} $(i = 1, \ldots, t)$.

A separable polynomial $f \in F[x]$ is *solvable by radicals* over F if a splitting field E of f over F is contained in a radical extension of F.

Corollary: If $F \subset E \subset K$ and $\alpha \in K$ is a radical over F, then α is also a radical over E.

Proof: If α is a zero of $x^n - a \in F[x]$, with Char $F \nmid n$, then since $x^n - a \in E[x]$ and Char E = Char F does not divide n, α is a radical over E.

If α is a zero of $x^p - x - a \in F[x]$, with Char $F = p$, then since $x^p - x - a \in E[x]$ and Char E = Char $F = p$, α is a radical over E. ■

We have already observed that, if Char $F \nmid n$, then the adjunction to F of a primitive n-th root of unity produces a Galois extension with abelian (but not necessarily cyclic) Galois group over F (see Exercise 4.11.5). We now show that, *if F contains a primitive n-th root of unity* (where Char $F \nmid n$), then the adjuction to F of a zero of any polynomial $x^n - a \in F[x]$ produces an extension with cyclic Galois group over F.

Theorem 4.12.1: Let F be a field and let n be a positive integer such that Char $F \nmid n$. Suppose that F contains a primitive n-th root of unity. Let $a \in F$, and let α be a zero (in an extension field of F) of the polynomial $x^n - a$. Then:

(1) $F(\alpha)/F$ is a Galois extension;
(2) $m_{\alpha/F} = x^k - b \in F[x]$, where $k = \min\{t \in \mathbb{Z}^+ \mid \alpha^t \in F\}$, and $k \mid n$;
(3) $G(F(\alpha)/F)$ is cyclic, of order k.

Proof: *(1)* Let $\zeta \in F$ be a primitive n-th root of unity. Then

$$\alpha, \zeta\alpha, \ldots, \zeta^{n-1}\alpha$$

are distinct zeros of $x^n - a$ in $F(\alpha)$. Thus, $x^n - a$ is a separable polynomial, and $F(\alpha) = F(\alpha, \zeta\alpha, \ldots, \zeta^{n-1}\alpha)$ is a splitting field of $x^n - a$ over F, hence a Galois extension of F.

(2) Let $k = \deg m_{\alpha/F}$. Then there are k integers i_j $(0 \le i_j \le n-1; j = 1, \ldots, k)$ such that the F-conjugates of α are the k elements $\alpha\zeta^{i_j}$, hence $m_{\alpha/F} = \prod_{j=1}^{k} (x - \alpha\zeta^{i_j})$.

Since the constant term of $m_{\alpha/F}$ is $(-\alpha)^k \prod_{j=1}^{k} \zeta^{i_j} \in F$, we have $\alpha^k \in F$. Thus, α is a zero of $x^k - b$ for some $b \in F$, and so $m_{\alpha/F} \mid x^k - b$. Since $\deg m_{\alpha/F} = k$, we conclude that $m_{\alpha/F} = x^k - b$.

It remains to show that $k \mid n$. For each $j = 1, \ldots, k$, $b = (\alpha\zeta^{i_j})^k = b(\zeta^{i_j})^k$, whence $(\zeta^{i_j})^k = 1$. Thus, the ζ^{i_j}, $j = 1, \ldots, k$, are k distinct k-th roots of unity, Since they form a cyclic group of order k, at least one of the ζ^{i_j} is a primitive k-th root of unity—call it η. But $o(\eta) = k$ and $\eta^n = 1$; hence $k \mid n$.

(3) The zeros of $x^k - b$ are $\alpha, \alpha\eta, \ldots, \alpha\eta^{k-1}$. Let ϕ be the F-automorphism of $F(\alpha)$ defined by: $\phi(\alpha) = \alpha\eta$. Then

$$\begin{aligned}
\phi^2(\alpha) &= \phi(\alpha\eta) = \phi(\alpha)\phi(\eta) = \alpha\eta^2 \\
\phi^3(\alpha) &= \phi(\alpha\eta^2) = \phi(\alpha)\phi(\eta^2) = \alpha\eta^3 \\
&\vdots \\
\phi^{k-1}(\alpha) &= \alpha\eta^{k-1}.
\end{aligned}$$

Since $|G(F(\alpha)/F)| = [F(\alpha) : F] = k$, we conclude that $G(F(\alpha)/F) = \{\iota, \phi, \ldots, \phi^{k-1}\} = [\phi]$, a cyclic group of order k, where $k|n$. ∎

Corollary: If $E = F(\alpha)$, where $m_{\alpha/F} = x^k - b$, Char $F \nmid k$, and F contains a primitive k-th root of unity, then E/F is a Galois extension and $G(E/F)$ is cyclic of order k.

We now investigate the second type of radical allowed in Definition 4.12.1.

Theorem 4.12.2: Let F be a field of characteristic $p > 0$ and let $f = x^p - x - a \in F[x]$. If $E = F(\alpha)$, where α is a zero of f, then E/F is a Galois extension, with $G(E/F)$ cyclic, of order 1 or p.

Proof: For each $i = 0, \ldots, p - 1$ in $\mathbb{Z}_p$, we have $i^p = i$, by Fermat's Little Theorem (Corollary 5 of Theorem 2.5.2); hence

$$(\alpha + i)^p - (\alpha + i) - a = \alpha^p + i^p - \alpha - i - a = \alpha^p - \alpha - a = 0.$$

Thus, $\alpha, \alpha + 1, \ldots, \alpha + (p - 1)$ are p distinct zeros of $f = x^p - x - a$, all located in $F(\alpha)$. But then f is separable over F and $E = F(\alpha) = F(\alpha, \alpha + 1, \ldots, \alpha + (p - 1))$ is a splitting field for f over F, hence a Galois extension of F. If $\alpha \in F$, then $E = F$ and $|G(E/F)| = 1$. Suppose $\alpha \notin F$. Let $k = \deg m_{\alpha/F}$. Then $1 < k \leq p$ and

$$m_{\alpha/F} = \prod_{j=1}^{k} (x - (\alpha + i_j)),$$

where $0 \leq i_j \leq p - 1$. The coefficient of x^{k-1} in $m_{\alpha/F}$ is equal to $-k\alpha - \sum_{j=1}^{k} i_j \in F$. If $k < p$, this implies that $\alpha \in F$, contrary to hypothesis. Thus, $k = p$, and $m_{\alpha/F} = x^p - x - a$. But then $[E : F] = |G(E/F)| = p$, and so $G(E/F)$ is cyclic. ∎

Corollary: If K/F is a radical extension, then K/F is separable.

Proof: By Definition 4.12.1, there is a field tower

$$F = K_0 \subset K_1 \subset \cdots \subset K_t = K$$

such that, for each $i = 1, \ldots, t$, $K_i = K_{i-1}(\alpha_i)$, where α_i is a radical over K_{i-1}. For given i, suppose first that α_i is zero of $x^{n_i} - a_i \in K_{i-1}[x]$, Char $F \nmid n_i$. If $a_i \neq 0$, then $x^{n_i} - a_i$ and its derivative $n_i x^{n_i - 1}$ have no common zeros (since Char $K_{i-1} \nmid n_i$). By Theorem 4.8.1, $x^{n_i} - a_i$ is separable over K_{i-1}. But then $m_{\alpha_i/K_{i-1}}$, and hence α_i, is separable over K_{i-1}, and so $K_i = K_{i-1}(\alpha_i)$ is a separable extension of K_{i-1}, by Corollary 3 of Theorem 4.8.4. If $a_i = 0$, then $\alpha_i^{n_i} = 0$; hence $\alpha_i = 0$, $K_i = K_{i-1}$, and K_i/K_{i-1} is separable.

Next suppose that α_i is a zero of $x^p - x - a_i \in K_{i-1}[x]$, where Char $F = p$. By Theorem 4.12.2, $x^p - x - a_i$ is separable over K_{i-1}. But then $m_{\alpha_i/K_{i-1}}$, α_i, and $K_i = K_{i-1}(\alpha_i)$ are separable over K_{i-1}.

Since separability is transitive (Corollary 2, Theorem 4.8.4), it follows that K/F is a separable extension. ∎

Our next objective is to prove that, if E/F is a finite Galois extension such that E is contained in a radical extension, then $G(E/F)$ is solvable. We need two lemmas.

Lemma 1: Let E/F be a finite Galois extension and let ζ be an element (contained in some extension of E) that is algebraic and separable over F. Then $E(\zeta)/F(\zeta)$ is a Galois extension with $G(E(\zeta)/F(\zeta))$ isomorphic to a subgroup of $G(E/F)$.

Proof: Since ζ is separable over F, $E(\zeta)/F$ is a separable extension, hence so is $E(\zeta)/F(\zeta)$. (See Theorem 4.8.4, Corollary 3, and Lemma, p. 275). If ϕ is an $F(\zeta)$-monomorphism of $E(\zeta)$ into a normal closure of $E(\zeta)/F(\zeta)$, then ϕ is an F-monomorphism that fixes ζ. By Theorem 4.7.4, since E/F is normal, $\operatorname{Im} \phi_E = E$, whence $\operatorname{Im} \phi = E(\zeta)$. But then, by Theorem 4.7.4, it follows that $E(\zeta)/F(\zeta)$ is normal and is, thus, a Galois extension. For each $\sigma \in G(E(\zeta)/F(\zeta))$, let $\mu(\sigma) = \sigma'$, the F-automorphism of E that agrees with σ on E. Then $\mu : G(E(\zeta)/F(\zeta)) \to G(E/F)$ is, clearly, a group homomorphism. If $\sigma \in \operatorname{Ker} \mu$, then $\sigma|_E = \iota_E$; since σ also fixes ζ, it is therefore the identity on $E(\zeta)$. Thus. μ is a monomorphism, and so $G(E(\zeta)/F(\xi))$ is isomorphic to a subgroup of $G(E/F)$. ■

Lemma 2: If L/F is a finite radical extension and N is a normal closure of L/F, then N/F is a finite radical Galois extension.

Proof: Let

$$F = L_0 \subset L_1 \subset \cdots \subset L_t = L$$

be a radical tower for L/F. For each i $(1 \le i \le t)$, we have $L_i = L_{i-1}(\alpha_i)$ where α_i is a radical over L_{i-1}, in the sense of Definition 4.12.1. Then $L = F(\alpha_1, \dots, \alpha_t)$, and a normal closure, M, of L/F can be obtained by forming a splitting field, over L, hence over F, of the polynomial $f = \prod_{i=1}^{t} m_{\alpha_i/F}$ (see the proof of Theorem 4.7.5). Since L/F is separable (by the Corollary of Theorem 4.12.2), each $m_{\alpha_i/F}$ is separable over F, hence f is separable over F, and so M/F is a finite Galois extension. More generally, for each $j = 1, \dots, t$, the splitting field, M_j, in M, of $f_j = \prod_{i=1}^{j} m_{\alpha_i/F}$ is the normal closure, in M, of L_j/F, with M_j/F a Galois extension, and we have the field tower

$$F = M_0 \subset M_1 \subset \cdots \subset M_t = M. \tag{1}$$

This tower can be refined to a radical tower for M/F. For each j $(1 \le j \le t)$, let $\alpha_{j1} = \alpha_j, \alpha_{j2}, \dots, \alpha_{jk_j}$ be the F-conjugates of α_j in M_j. Then

$$M_{j-1} \subset M_{j-1}(\alpha_{j1}) \subset M_{j-1}(\alpha_{j1})(\alpha_{j2}) \subset \cdots \subset M_{j-1}(\alpha_{j1})(\alpha_{j2}) \cdots (\alpha_{jk_j}) = M_j.$$

Since α_{j1} is a radical over $L_{j-1} \subset M_{j-1}$, α_{j1} is a radical over M_{j-1} (see the Corollary of Definition 4.12.1). For each $1 = 2, \dots, k_j$, there is an F-isomorphism θ_1 of $F(\alpha_{j1})$ to $F(\alpha_{jl})$ sending α_{j1} to α_{jl}. Since M_j/F is normal, θ_l extends to an

F-automorphism, θ_1, of M_j, sending α_{j1} to α_{jl}. Since α_{j1} is a radical over L_{j-1}, we conclude that $\alpha_{jl} = \theta_1\alpha_{j1}$ is a radical over $\theta_1 L_{j-1}$. From $\theta_1 L_{j-1} \subset M_{j-1} \subset M_{j-1}(\alpha_{j1}, \ldots, \alpha_{jl-1})$, it follows that α_{jl} is a radical over $M_{j-1}(\alpha_{j1}, \ldots, \alpha_{jl-1})$. Thus, (2) forms part of a radical tower for M/F. Using (2), for $j = 1, \ldots, t$, we can refine (1) to a radical tower for M/F. Thus, M/F is a finite radical Galois extension. Since any normal closure, N, of L/F is F-isomorphic to M, it, too, has the required properties. ∎

We are now ready to prove one half of the Fundamental Theorem on Radical Extensions.

Theorem 4.12.3: Let E/F be a finite Galois extension such that E is contained in a radical extension K of F. Then $G(E/F)$ is a solvable group.

Proof: Let E/F be a finite Galois extension such that E is contained in a radical extension K of F, and let N be a normal closure of K. By Lemma 2, N/F is a finite radical Galois extension. Let

$$F = F_0 \subset F_1 \subset \cdots \subset F_t = N \tag{1}$$

be a radical tower for N/F, with $F_i = F_{i-1}(\alpha_i)$ $(i = 1, \ldots, t)$, where either α_i is a zero of some polynomial $x^{n_i} - a_i \in F_{i-1}[x]$, Char $F \nmid n_i$, or α_i is a zero of some polynomial $x^p - x - a_i \in F_{i-1}[x]$, $p =$ Char F. In the latter case, let $n_i = 1$. Define $n = \prod_{i=1}^{t} n_i$. By the Corollary of Theorem 4.11.1, since Char $F \nmid n$, there is a primitive n-th root, ζ, of unity contained in some extension of N. By Lemma 1, $N(\zeta)/F(\zeta)$ is a Galois extension, with Galois group $G(N(\zeta)/F(\zeta))$ isomorphic to a subgroup of $G(N/F)$. Furthermore, $N(\zeta)/F$ is a Galois extension. For, since $N(\zeta)/F(\zeta)$ and $F(\zeta)/F$ are separable, so is $N(\zeta)/F$. Since N/F is normal and the F-conjugates of ζ are powers of ζ in $N(\zeta)$, the F-conjugates of all elements of $N(\zeta)$ are in $N(\zeta)$. Hence $N(\zeta)/F$ is normal and is, thus, a Galois extension.

Consider the field tower

$$F(\zeta) = F_0(\zeta) \subset F_1(\zeta) \subset \cdots \subset F_t(\zeta) = N(\zeta). \tag{2}$$

For each $i = 1, \ldots, t$, we have $F_i(\zeta) = F_{i-1}(\zeta)(\alpha_i)$. If α_i is a zero of $x^{n_i} - a_i \in F_{i-1}[x] \subset F_{i-1}(\zeta)[x]$, with Char $F \nmid n_i$, then $n_i | n$ and $\zeta^{n/n_i} \in F_{i-1}(\zeta)$ is a primitive n_i-th root of unity. By Theorem 4.12.1, it follows that $F_i(\zeta)/F_{i-1}(\zeta)$ is a Galois extension with cyclic Galois group $G(F_i(\zeta)/F_{i-1}(\zeta))$. If α_i is a zero of $x^p - x - a_i \in F_{i-1}[x] \subset F_{i-1}(\zeta)[x]$, with Char $F = p$, then, by Theorem 4.12.2 and Lemma 1, $F_i(\zeta)/F_{i-1}(\zeta)$ is, again, a Galois extension with cyclic Galois group.

Corresponding to the field tower (2), we have the following chain of subgroups of $G(N(\zeta)/F(\zeta))$:

$$\begin{aligned} G(N(\zeta)/F(\zeta)) &= G(N(\zeta)/F_0(\zeta)) \supset \cdots \supset G(N(\zeta)/F_t(\zeta)) \\ &= G(N(\zeta)/N(\zeta)) = \{\iota\}. \end{aligned} \tag{3}$$

This is a normal series since the normality of the extension $F_i(\zeta)/F_{i-1}(\zeta)$ (by Theorem 4.10.1(3)) implies

$$G(N(\zeta)/F_i(\zeta)) \lhd G(N(\zeta)/F_{i-1}(\zeta))$$

$(i = 1, \ldots, t)$. For each $i = 1, \ldots, t$,

$$G(N(\zeta)/F_{i-1}(\zeta))/G(N(\zeta)/F_i(\zeta)) \cong G(F_i(\zeta)/F_{i-1}(\zeta));$$

hence (3) is a solvable normal series, and $G(N(\zeta)/F(\zeta))$ is a solvable group.

Now, $F(\zeta)/F$ is a Galois extension with abelian Galois group (see Exercise 4.11.5), and

$$G(F(\zeta)/F) \cong G(N(\zeta)/F)/G(N(\zeta)/F(\zeta)).$$

By Theorem 2.14.3(3), it follows that $G(N(\zeta)/F)$ is solvable. Finally, since $G(E/F)$ is a homomorphic image of $G(N(\zeta)/F$, it follows (by Theorem 2.14.3(2) that $G(E/F)$ is solvable. ∎

Corollary: If $f \in F[x]$ is a separable polynomial which is solvable by radicals over F, then the Galois group of f over F is solvable.

Proof: For $f \in F[x]$ a separable polynomial, any splitting field E of f over F is a finite Galois extension (by Theorem 4.10.2). If f is solvable by radicals, then E is contained in a radical extension of F. By Theorem 4.12.3, $G(E/F)$ is a solvable group. ∎

Before we examine Galois groups of specific polynomials, it will be useful to relate our work to Galois' original point of view.

Lemma: Let $f \in F[x]$ be a separable polynomial of degree n. Then the Galois group of f over F is isomorphic to a subgroup of the symmetric group S_n.

Proof: Let E be a splitting field of f over F, and let $\alpha_1, \ldots, \alpha_m$ be the distinct zeros of f in E. Then $E = F(\alpha_1, \ldots, \alpha_m)$. If $\phi \in G(E/F)$, then, for each $i = 1, \ldots, m$, $\phi(\alpha_i) = \alpha_j$ for some j, $1 \leq j \leq m$. It follows that ϕ induces a permutation, ϕ', on the set $A = \{\alpha_1, \ldots, \alpha_m\}$. Let S_A be the group of all permutations on A. Define $\mu : G(E/F) \to S_A$ by: $\mu(\phi) = \phi'$ (where ϕ' is the permutation induced on A by ϕ). Then μ is well-defined and, clearly, preserves products. If $\phi \in \operatorname{Ker} \mu$, then $\phi' = \iota_A$; hence, ϕ fixes all of the α_i as well as all elements of F, and is thus the identity on $E = F(\alpha_1, \ldots, \alpha_m)$. It follows that μ is a monomorphism of $G(E/F)$ into S_A. Since $G(E/F)$ is the Galois group of f over F and S_A is isomorphic to the symmetric group, S_m, we conclude that $G(E/F)$ is isomorphic to a subgroup of S_m. But $m \leq n$, hence S_m is isomorphic to a subgroup of S_n, and so $G(E/F)$ is isomorphic to a subgroup of S_n. ∎

Example:

We have seen (Theorem 2.14.2) that S_n is not a solvable group when $n \geq 5$. To find a polynomial which is not solvable by radicals, it thus suffices to find a polynomial with Galois group isomorphic to S_n for some $n \geq 5$.

Consider the polynomial $f = x^5 - 80x + 5 \in \mathbb{Q}[x]$. By the Eisenstein criterion, f is irreducible over $\mathbb{Q}$ (Theorem 3.9.2). Since (by Rolle's Theorem of calculus) the interval between two real zeros of f must contain at least one zero of f', we examine f' to find an upper bound for the number of real zeros of f. Since

$$f' = 5x^4 - 80 = 5(x-2)(x+2)(x^2+4)$$

has just two real zeros 2 and -2, f has *no more than* three real zeros.

Since $f(0) = 5$, $f(1) = -74$, $f(3) = 8$, and $f(-4) = -699$, the Intermediate Value Theorem of calculus implies that f has a zero in each of the open intervals $(-4, 0)$, $(0, 1)$, and $(1, 3)$. It follows that f has *exactly* three real zeros, hence two imaginary zeros. (We assume as known that $\mathbb{C}$ is algebraically closed, i.e., every polynomial with coefficients in $\mathbb{C}$ splits over $\mathbb{C}$.)

Let E be the splitting field of f in $\mathbb{C}$. If α is one of the zeros of f in E, then $[\mathbb{Q}(\alpha):\mathbb{Q}] = 5 = \deg f$ (since f is irreducible over $\mathbb{Q}$). Hence $5 \mid [E:\mathbb{Q}]$, and so $5 \mid |G(E/\mathbb{Q})|$. By Theorem 2.15.3, $G(E/\mathbb{Q})$ has a subgroup, hence an element, of order 5. By the Lemma, $G(E/\mathbb{Q})$ is isomorphic to a subgroup, H, of S_5. Thus, H contains an element of order 5, i.e., a 5-cycle.

The mapping $\psi: \mathbb{C} \to \mathbb{C}$ such that, for each $z \in \mathbb{C}$, $\psi(z) = \bar{z}$ (the complex conjugate of z) is an $\mathbb{R}$-automorphism of $\mathbb{C}$, which induces an $\mathbb{R}$-monomorphism of E into $\mathbb{C}$, hence an $\mathbb{R}$-automorphism of E. (Note that every $\mathbb{R}$-automorphism is also a $\mathbb{Q}$-automorphism.) If β is one of the imaginary zeros of f, then, since f has real coefficients, $f(\bar{\beta}) = f(\psi\beta) = f^{\psi}(\psi\beta) = \psi(f(\beta)) = \psi(0) = 0$. Thus, the imaginary zeros of f are β and its complex conjugate, $\bar{\beta} = \psi\beta$. Clearly, $\psi^2 = \iota$. On the set, A, of all zeros of f in E, ψ thus induces the transposition $\psi' = (\beta, \bar{\beta})$ which, under any one of the obvious isomorphisms between S_A and S_n (see Exercise 2.6.7), corresponds to a transposition in H. Thus, H contains a 5-cycle and a transposition. By Exercise 2.6.14, it follows that $H = S_5$, which is not a solvable group (see Theorem 2.14.2). But then $G(E/\mathbb{Q})$ is not a solvable group; hence, by Theorem 4.12.3, f is not solvable by radicals over $\mathbb{Q}$.

The result illustrated in the example generalizes readily to the following.

Theorem 4.12.4: If $f \in \mathbb{Q}[x]$ is an irreducible polynomial of prime degree $p \geq 5$, and if f has exactly $p - 2$ real zeros, then f is not solvable by radicals over $\mathbb{Q}$.

We leave the proof as an exercise. (see Exercise 4.12.1.).

Exercises 4.12

True or False

1. If $f \in F[x]$ is a separable polynomial with Galois group isomorphic to S_n for some $n \geq 5$, then f is not solvable by radicals over F.
2. If $f \in F[x]$ is a separable polynomial of degree $n \geq 5$, and $E \supset F$ is a splitting field for f over F, with $[E:F] = n!$, then f is not solvable by radicals over F.

3. If $f \in F[x]$ is a separable polynomial of degree $n \geq 5$ and $E \supset F$ is a splitting field of f over F, with $[E : F] = \frac{n!}{2}$, then f is not solvable by radicals over F.
4. If $f \in F[x]$ is a separable polynomial of degree $n \geq 5$, then f is not solvable by radicals over F.
5. The polynomial $x^5 - x^2 + x - 1$ is solvable by radicals over $\mathbb{Z}_3$.

4.12.1. Prove Theorem 4.12.4, i.e., prove: if $f \in \mathbb{Q}[x]$ is an irreducible polynomial of prime degree $p \geq 5$ and if f has exactly $p - 2$ real zeros, then the Galois group of f over $\mathbb{Q}$ is isomorphic to S_p, whence f is not solvable by radicals over $\mathbb{Q}$.

4.12.2. Prove that the polynomial

$$g = 6x^7 - 14x^6 - 14x^3 + 42x^2 - 7$$

is not solvable by radicals over $\mathbb{Q}$.

4.12.3. Prove that the polynomial $f = x^{10} + 4x^2 + 2$ is solvable by radicals over $\mathbb{Z}_5$, in the sense of Definition 4.12.1.

4.12.4. Let $\zeta \in \mathbb{C}$ be a primitive twelfth root of unity, and let $F = \mathbb{Q}(\zeta)$. Let $\alpha = \sqrt[12]{25}$.

(1) Find $m_{\alpha/F}$.
(2) Find a generator for the Galois group $G(F(\alpha)/F)$ and list its elements.
(3) Draw the lattice of subgroups of $G(F(\alpha)/F)$ and the lattice of intermediate fields of $F(\alpha)/F$.

4.12.5. In the preceding exercise, is $\mathbb{Q}(\alpha)/\mathbb{Q}$ a Galois extension? Is $F(\alpha)/\mathbb{Q}$ a Galois extension?

4.12.6. Let F be a field of characteristic 0, or of *odd* characteristic $p > 0$. Prove that every extension E/F, with $[E : F] = 2$, is a radical extension.

4.12.7. Let F be a field of characteristic 2. Prove or disprove: every extension of degree 2 of F is a radical extension.

4.12.8. Assuming the Fundamental Theorem of Algebra (which states that every polynomial of positive degree in $\mathbb{C}[x]$ splits over $\mathbb{C}$), prove that every polynomial in $\mathbb{R}[x]$ is solvable by radicals over $\mathbb{R}$.

4.12.9. Let $f \in \mathbb{Q}[x]$ be a cubic polynomial, with splitting field E over $\mathbb{Q}$. If $[E : \mathbb{Q}] = 3$, prove that $E/\mathbb{Q}$ does *not* have a radical tower.

Nonetheless, as we shall see, f *is* solvable by radicals over $\mathbb{Q}$. Why is this not a contradiction?

4.12.10. Generalize Exercise 4.12.9 by replacing 3 with an arbitrary odd prime p.

4.12.11. For each prime $p \geq 5$, the following construction, due to Richard Brauer, guarantees the existence of an irreducible polynomial of degree p, in $\mathbb{Q}[x]$, not solvable by radicals over $\mathbb{Q}$:

Let $n_1 < \cdots < n_{p-2}$ be even integers, and let m be a positive integer. Let

$$g(x) = (x^2 + m)(x - n_1) \ldots (x - n_{p-2})$$

and let $f(x) = g(x) - 2$. Prove that, for m sufficiently large, f is an irreducible polynomial with exactly $p - 2$ real zeros.

Conclude that f is not solvable by radicals over $\mathbb{Q}$. (See Theorem 4.12.4 and Exercise 4.12.1.) (cf. [7]: N. Jacobson, *Lectures in Abstract Algebra*, Vol. III, p. 107.)

4.13

Radical Extensions II

(Note: Results from Section 2.14, are used in this section.)

Our aim here is to prove the converse of Theorem 4.12.3. By the Lemma on p. 121, every solvable group has a normal series whose factors are cyclic, of prime order. We thus begin by investigating Galois extensions with cyclic Galois groups.

Definition 4.13.1: A finite Galois extension E/F is a *cyclic extension* if $G(E/F)$ is a cyclic group.

We have already seen that (under the conditions of Theorem 4.12.1 or Theorem 4.12.2), the adjunction of a single radical produces a cyclic extension. To obtain the converses of Theorems 4.12.1 and 4.12.2, we require the concepts of "norm" and "trace."

Definition 4.13.2: Let E/F be a finite Galois extension, with Galois group $G(E/F) = \{\sigma_1 = \iota, \sigma_2, \ldots, \sigma_n\}$. Then, for each $\alpha \in E$, the *norm*, $N_{E/F}(\alpha)$, *of* α over F is given by: $N_{E/F}(\alpha) = (\sigma_1\alpha)(\sigma_2\alpha)\ldots(\sigma_n\alpha)$.

Theorem 4.13.1: Let E/F be a finite Galois extension of degree n, with $G = G(E/F)$.

(1) For each $\alpha \in E$, $N_{E/F}(\alpha) \in F$.

(2) For $\alpha \in E$ and $\sigma \in G$, $N_{E/F}(\alpha) = N_{E/F}(\sigma\alpha)$.

(3) For $\alpha, \beta \in E$, $N_{E/F}(\alpha\beta) = N_{E/F}(\alpha)N_{E/F}(\beta)$.

(4) For $\alpha \in F$, $N_{E/F}(\alpha) = \alpha^n$.

(5) $N_{E/F}(0) = 0$, $N_{E/F}(1) = 1$ and $N_{E/F}(\alpha^{-1}) = [N_{E/F}(\alpha)]^{-1}$ for each $\alpha \neq 0$ in E.

(6) For $\alpha \neq 0$ in E and $\sigma \in G(E/F)$, $N_{E/F}\left(\dfrac{\alpha}{\sigma\alpha}\right) = 1$.

(7) If $\alpha \in E$ and $[F(\alpha):F] = k$, then $N_{E/F}(\alpha) = (-1)^n a_0^{n/k}$, where a_0 is the constant term of $m_{\alpha/F}$.

Proof: *(1)* If $G = \{\iota = \sigma_1, \sigma_2, \ldots, \sigma_n\}$, then (using Theorem 2.2.4 and Example 1.3.19) it is easy to see that

$$G = \{\sigma\sigma_1, \sigma\sigma_2, \ldots, \sigma\sigma_n\} = \{\sigma_1\sigma, \sigma_2\sigma, \ldots, \sigma_n\sigma\} \quad \text{for each} \quad \sigma \in G.$$

Thus, if $\alpha \in E$ and $\sigma \in G$, then

$$\sigma(N_{E/F}(\alpha)) = \sigma\left(\prod_{i=1}^{n} \sigma_i\alpha\right) = \prod_{i=1}^{n} (\sigma\sigma_i)(\alpha) = \prod_{i=1}^{n} \sigma_i\alpha = N_{E/F}(\alpha).$$

But then $N_{E/F}(\alpha) \in F$ (the fixed field of G).

(2) For $\alpha \in E$, $\sigma \in G$,

$$N_{E/F}(\sigma\alpha) = \prod_{i=1}^{n} \sigma_i(\sigma\alpha) = \prod_{i=1}^{n} (\sigma_i\sigma)(\alpha) = \prod_{i=1}^{n} \sigma_i\alpha = N_{E/F}(\alpha).$$

(3) For $\alpha, \beta \in E$,

$$N_{E/F}(\alpha\beta) = \prod_{i=1}^{n} \sigma_i(\alpha\beta) = \prod_{i=1}^{n} \sigma_i\alpha\ \sigma_i\beta = \prod_{i=1}^{n} \sigma_i\alpha \prod_{i=1}^{n} \sigma_i\beta = N_{E/F}(\alpha)N_{E/F}(\beta).$$

(4) For $\alpha \in F$, $\sigma_i\alpha = \alpha$ for each $i = 1, \ldots, n$, hence

$$N_{E/F}(\alpha) = \prod_{i=1}^{n} \sigma_i\alpha = \prod_{i=1}^{n} \alpha = \alpha^n.$$

(5) By (4), $N_{E/F}(0) = 0^n = 0$ and $N_{E/F}(1) = 1^n = 1$. If $\alpha \neq 0$, then $\sigma_i\alpha \neq 0$ for each $i = 1, \ldots, n$; hence $N_{E/F}(\alpha) \neq 0$. From $1 = N_{E/F}(\alpha\alpha^{-1}) = N_{E/F}(\alpha)N_{E/F}(\alpha^{-1})$ follows $N_{E/F}(\alpha^{-1}) = [N_{E/F}(\alpha)]^{-1}$.

(6) For $\alpha \neq 0$ in F and $\sigma \in G$, since $\sigma\alpha \neq 0$, we have

$$\begin{aligned} N_{E/F}\left(\frac{\alpha}{\sigma\alpha}\right) &= N_{E/F}(\alpha(\sigma\alpha)^{-1}) = N_{E/F}(\alpha)N_{E/F}((\sigma\alpha)^{-1}) = N_{E/F}(\alpha)[N_{E/F}(\sigma\alpha)]^{-1} \\ &= N_{E/F}(\alpha)[N_{E/F}(\alpha)]^{-1} = 1. \end{aligned}$$

(7) If $[F(\alpha) : F] = k$, then $k|n$ and α has k distinct F-conjugates. Labeling the elements of G so that $\sigma_1\alpha, \ldots, \sigma_k\alpha$ are distinct, we have

$$m_{\alpha/F} = \prod_{i=1}^{k} (x - \sigma_i\alpha).$$

Hence $\prod_{i=1}^{k} \sigma_i\alpha = (-1)^k a_0$, where a_0 is the constant term of $m_{\alpha/F}$. Let $H = G(E/F(\alpha))$. Then $|H| = [E : F(\alpha)] = \frac{n}{k}$. For $\sigma, \tau \in G$, we have $\sigma\alpha = \tau\alpha \Leftrightarrow \sigma^{-1}\tau\alpha = \alpha \Leftrightarrow \sigma^{-1}\tau \in H \Leftrightarrow \sigma H = \tau H$. Thus, for each $i = 1, \ldots, k$, $\sigma\alpha = \sigma_i\alpha$ if and only if σ belongs to the left coset $\sigma_i H$ of H in G. It follows that, for each $i = 1, \ldots, k$, exactly $\frac{n}{k}$ of the automorphisms in G map α to $\sigma_i\alpha$. But then

$$N_{E/F}(\alpha) = \prod_{i=1}^{n} \sigma_i\alpha = \left(\prod_{i=1}^{k} \sigma_i\alpha\right)^{n/k} = [(-1)^k a_0]^{n/k} = (-1)^n a_0^{n/k}. \qquad \blacksquare$$

For E/F a finite Galois extension, Theorem 4.13.1 implies that the mapping $\alpha \mapsto N_{E/F}(\alpha)$ defines a group homomorphism, $N_{E/F}$, of E^* into F^*, and that $\alpha/\sigma\alpha \in \text{Ker } N_{E/F}$ for each $\alpha \in E^*$ and each $\sigma \in G(E/F)$. The next theorem states that, in the special case where E/F is a cyclic extension, with σ a generator of $G(E/F)$, *every* element of Ker $N_{E/F}$ is of the form $\alpha/\sigma\alpha$ for some $\alpha \in E^*$.

Theorem 4.13.2 (Hilbert's Theorem 90—Multiplicative Version): Let E/F be a finite cyclic extension, with $G(E/F) = [\sigma]$. Then, for $\beta \in E$, $N_{E/F}(\beta) = 1$ if and only if $\beta = \alpha/\sigma\ \alpha$ for some $\alpha \in E$.

Proof: Suppose $[E : F] = n$, hence $|G(E/F)| = n$. Let $\beta \in E$, with $N_{E/F}(\beta) = 1$. By Lemma 2, p. 282, the F-automorphisms $\iota, \sigma, \sigma^2, \ldots, \sigma^{n-1}$ are linearly independent over E. Hence, in particular, the linear combination, τ, of $\iota, \sigma, \ldots, \sigma^{n-1}$ defined by $\tau = \iota + \beta\sigma + (\beta \cdot \sigma\beta)\sigma^2 + \cdots + (\beta \cdot \sigma\beta \cdot \sigma^2\beta \cdot \cdots \cdot \sigma^{n-2}\beta)\sigma^{n-1}$ is not the 0-endomorphism of E (since $\beta \neq 0$). Thus, there is some element, y, in E such that $\tau(y) \neq 0$. Let $\alpha = \tau(y)$. Then

$$\alpha = y + \beta\,\sigma(y) + (\beta \cdot \sigma\beta)\sigma^2(y) + \cdots + (\beta \cdot \sigma\beta \cdot \sigma^2\beta \cdot \cdots \cdot \sigma^{n-2}\beta)\sigma^{n-1}(y),$$

hence

$$\begin{aligned}\sigma(\alpha) &= \sigma y + \sigma\beta \cdot \sigma^2(y) + (\sigma\beta \cdot \sigma^2\beta)\sigma^3(y) + \cdots + \overbrace{(\sigma\beta \cdot \sigma^2\beta \cdot \cdots \cdot \sigma^{n-1}\beta)}^{N(\beta)/\beta}\overbrace{\sigma^n(y)}^{y} \\ &= \sigma(y) + \sigma\beta \cdot \sigma^2(y) + (\sigma\beta \cdot \sigma^2\beta)\sigma^3(y) + \cdots + (1/\beta)y.\end{aligned}$$

But then $\beta \cdot \sigma(\alpha) = \alpha$. Since $\alpha = \tau(y) \neq 0$, we have $\sigma\alpha \neq 0$. Hence $\beta = \dfrac{\alpha}{\sigma\alpha}$, as required.

The converse is part (6) of Theorem 4.13.1. ∎

This enables us to prove the converse of Theorem 4.12.1:

Theorem 4.13.3: Let E/F be a finite cyclic Galois extension of degree n, with Char $F \nmid n$, such that F contains a primitive n-th root of unity. Then $E = F(\alpha)$, where $m_{\alpha/F} = x^n - a \in F[x]$.

Proof: Suppose $G(E/F) = [\sigma]$. Let $\zeta \in F$ be a primitive n-th root of unity. Then $N_{E/F}(\zeta) = \zeta^n = 1$. By Theorem 4.13.2, there is some $\alpha \in E$ such that $\zeta = \dfrac{\alpha}{\sigma\alpha}$. But then $1 = \zeta^n = \dfrac{\alpha^n}{(\sigma\alpha)^n}$; hence $\sigma(\alpha^n) = (\sigma\alpha)^n = \alpha^n$. This implies that $\sigma^i(\alpha^n) = \alpha^n$ for each $i = 0, \ldots, n-1$, and so α^n is fixed by each F-automorphism of E. It follows that $\alpha^n \in F$. Let $a = \alpha^n$. Then α is a zero of $x^n - a \in F[x]$, as required. Since the elements α, $\sigma\alpha = \alpha\zeta^{-1}$, $\sigma^2\alpha = (\sigma\alpha)\zeta^{-1} = \alpha\zeta^{-2}, \ldots, \sigma^{n-1}\alpha = \alpha\zeta^{-(n-1)}$ are all distinct, α has n distinct F-conjugates in $F(\alpha)$. Thus, $[F(\alpha):F] \geq n$. But $F \subset F(\alpha) \subset E$, and $[E:F] = n$; hence $[F(\alpha):F] \leq n$. It follows that $[F(\alpha):F] = n$, and so $E = F(\alpha)$. Since $\deg m_{\alpha/F} = n$ and $m_{\alpha/F} \mid x^n - a$, we conclude that $m_{\alpha/F} = x^n - a$. ∎

To deal with the case where Char $F > 0$, we require the additive analog of the norm of an extension.

Definition 4.13.3: Let E/F be a finite Galois extension, with Galois group $G(E/F) = \{\sigma_1 = \iota, \sigma_2, \ldots, \sigma_n\}$. Then, for each $\alpha \in E$, the *trace*, $\mathrm{Tr}_{E/F}(\alpha)$, of α over F is given by:

$$\mathrm{Tr}_{E/F}(\alpha) = \sum_{i=1}^{n} \sigma_i\alpha.$$

The following theorem is analogous to Theorem 4.13.1.

Theorem 4.13.4: Let E/F be finite Galois of degree n, with $G = G(E/F)$.

(1) For each $\alpha \in E$, $\mathrm{Tr}_{E/F}(\alpha) \in F$.

(2) For $\alpha \in E$ and $\sigma \in G$, $\mathrm{Tr}_{E/F}(\alpha) = \mathrm{Tr}_{E/F}(\sigma\alpha)$.

(3) For $\alpha, \beta \in E$, $\mathrm{Tr}_{E/F}(\alpha + \beta) = \mathrm{Tr}_{E/F}\alpha + \mathrm{Tr}_{E/F}\beta$.

(4) For $\alpha \in F$, $\text{Tr}_{E/F}\alpha = n\alpha$.
(5) $\text{Tr}_{E/F}(0) = 0$, $\text{Tr}_{E/F}(1) = n$ and $\text{Tr}_{E/F}(-\alpha) = -\text{Tr}_{E/F}(\alpha)$ for each $\alpha \in E$.
(6) For each $\alpha \in E$ and $\sigma \in G$, $\text{Tr}_{E/F}(\alpha - \sigma\alpha) = 0$.
(7) If $\alpha \in E$ and $[F(\alpha) : F] = k$, then $\text{Tr}_{E/F}(\alpha) = -\frac{n}{k} a_{k-1}$, where a_{k-1} is the coefficient of x^{k-1} in $m_{\alpha/F}$.

Since the proof is completely analogous to the proof of Theorem 4.13.1, we leave it as an exercise (Exercise 4.13.4).

For E/F a finite Galois extension, Theorem 4.13.4 implies that the mapping $\alpha \mapsto \text{Tr}_{E/F}(\alpha)$ defines a group homomorphism, $\text{Tr}_{E/F}$, of $\langle E, + \rangle$ into $\langle F, + \rangle$, and that $\alpha - \sigma\alpha \in \text{Ker Tr}_{E/F}$ for each $\alpha \in E$ and each $\sigma \in G(E/F)$. The next theorem proves that, in the special case where E/F is a cyclic extension, with σ a generator of $G(E/F)$, every element of $\text{Ker Tr}_{E/F}$ is of the form $\alpha - \sigma\alpha$ for some $\alpha \in E$.

Theorem 4.13.5 (Hilbert's Theorem 90—Additive Version): Let E/F be a finite cyclic extension, with $G(E/F) = [\sigma]$. Then, for $\beta \in E$, $\text{Tr}_{E/F}(\beta) = 0$ if and only if $\beta = \alpha - \sigma\alpha$ for some $\alpha \in E$.

Proof: Suppose $[E : F] = n$, hence $G(E/F) = \{\iota, \sigma, \ldots, \sigma^{n-1}\}$. Let $\beta \in E$, with $\text{Tr}_{E/F}(\beta) = 0$. By the Artin-Dedekind Lemma, p. 282, the F-automorphisms $\iota, \sigma, \ldots, \sigma^{n-1}$ are linearly independent over E. Hence, in particular, there is some element $\gamma \in E$ such that $\text{Tr}_{E/F}\gamma = \gamma + \sigma\gamma + \sigma^2\gamma + \cdots + \sigma^{n-1}\gamma \neq 0$. Let

$$\alpha = \frac{1}{\text{Tr}_{E/F}(\gamma)}[\beta\,\sigma(\gamma) + (\beta + \sigma(\beta))\sigma^2(\gamma) + \cdots + (\beta + \sigma(\beta) + \cdots + \sigma^{n-2}(\beta))\sigma^{n-1}(\gamma)].$$

Then

$$\begin{aligned}\sigma(\alpha) &= \frac{1}{\text{Tr}_{E/F}(\gamma)}[\sigma(\beta)\sigma^2(\gamma) + (\sigma(\beta) + \sigma^2(\beta))\sigma^3(\gamma) \\ &\quad + \cdots + \overbrace{(\sigma(\beta) + \sigma^2(\beta) + \cdots + \sigma^{n-1}(\beta))}^{\text{Tr}_{E/F}(\beta) - \beta = -\beta}\overbrace{\sigma^n(\gamma)}^{\gamma}] \\ &= \frac{1}{\text{Tr}_{E/F}(\gamma)}[\sigma(\beta)\sigma^2(\gamma) + (\sigma(\beta) + \sigma^2(\beta))\sigma^3(\gamma) + \cdots + (\sigma(\beta) \\ &\quad + \sigma^2(\beta) + \cdots + \sigma^{n-2}(\beta))\sigma^{n-1}(\gamma) - \beta\gamma].\end{aligned}$$

But then

$$\begin{aligned}\alpha - \sigma\alpha &= \frac{1}{\text{Tr}_{E/F}(\gamma)}[\beta(\gamma + \sigma(\gamma) + \cdots + \sigma^{n-1}(\gamma))] \\ &= \frac{1}{\text{Tr}_{E/F}(\gamma)}\beta\text{Tr}_{E/F}(\gamma) = \beta.\end{aligned}$$

The converse is part (6) of Theorem 4.13.4. ■

This enables us to prove the converse of Theorem 4.12.2:

Theorem 4.13.6: If Char $F = p > 0$ and E/F is a cyclic extension of degree p over F, then there is some $\alpha \in E$ such that $E = F(\alpha)$ and $m_{\alpha/F} = x^p - x - a \in F[x]$.

Proof: Suppose $G(E/F) = [\sigma]$. By Theorem 4.13.4 (4), $\mathrm{Tr}_{E/F}(1) = p \cdot 1 = 0$. By Theorem 4.13.5, there is some $\alpha \in E$ such that $1 = \alpha - \sigma(\alpha)$. Since $\alpha - \sigma(\alpha) \neq 0$, we have $\alpha \notin F$. Since $[E : F] = p$, a prime, we conclude that $[F(\alpha) : F] = p$, and so $E = F(\alpha)$. From $1 = \alpha - \sigma(\alpha)$, we have $\sigma(\alpha) = \alpha - 1$; hence $\sigma(\alpha^p - \alpha) = (\sigma(\alpha))^p - \sigma(\alpha) = (\alpha - 1)^p - (\alpha - 1) = \alpha^p - 1 - \alpha + 1 = \alpha^p - \alpha$. But then $\alpha^p - \alpha$ is in the fixed field, F, of $G(E/F) = [\sigma]$. Hence there is some $a \in F$ such that α is a zero of $x^p - x - a$. Since $[F(\alpha) : F] = p$, we conclude that $m_{\alpha/F} = x^p - x - a$. ∎

We are now ready to obtain the converse of Theorem 4.12.3 and thus complete the Fundamental Theorem on Radical Extensions.

Theorem 4.13.7 (Fundamental Theorem on Radical Extensions): Let E/F be a finite Galois extension. Then E is contained in a radical extension of F if and only if $G(E/F)$ is a solvable group.

Proof: We have already proved (Theorem 4.12.3): if E is contained in a radical extension of F, then $G(E/F)$ is solvable.

Conversely, suppose that $G = G(E/F)$ is solvable. Let

$$G = G_0 \rhd G_1 \rhd \cdots \rhd G_t = \{\iota\} \tag{1}$$

be a composition series for G, with G_{i-1}/G_i cyclic of prime order p_i $(i = 1, \ldots, t)$, and let

$$F = F_0 \subset F_1 \subset \cdots \subset F_t = E \tag{2}$$

be the corresponding field tower. Then F_i/F_{i-1} is a normal, separable, hence Galois, extension of degree $p_i = |G_{i-1}/G_i|$, for each $i = 1, \ldots, t$. If Char $F = 0$, let $n = \prod_{i=1}^{t} p_i$. If Char $F = p > 0$, let $n = \prod_{\substack{p_i \neq p \\ 1 \leq i \leq t}} p_i$. Then there is a primitive n-th root, ζ, of unity contained in some extension of E. By Lemma 1, p. 300, $E(\zeta)/F(\zeta)$ is a Galois extension, with $G(E(\zeta)/F(\zeta))$ isomorphic to a subgroup of $G(E/F)$. In fact, in the field tower

$$F(\zeta) = F_0(\zeta) \subset F_1(\zeta) \subset \cdots \subset F_t(\zeta) = E(\zeta), \tag{3}$$

again by Lemma 1, p. 300, $F_i(\zeta)/F_{i-1}(\zeta)$ is a Galois extension, with $G(F_i(\zeta)/F_{i-1}(\zeta))$ isomorphic to a subgroup of $G(F_i/F_{i-1})$, hence to a subgroup of G_{i-1}/G_i, for each $i = 1, \ldots, t$. Thus, $F_i(\zeta)/F_{i-1}(\zeta)$ is a cyclic extension, of degree p_i, or of degree 1, for each $i = 1, \ldots, t$. If $[F_i(\zeta) : F_{i-1}(\zeta)] = 1$, then, trivially, $F_i(\zeta) = F_{i-1}(\zeta)(\alpha_i)$, where (for example) $\alpha_i = 1$, a zero of $x - 1 \in F_{i-1}(\zeta)$. If $[F_i(\zeta) : F_{i-1}(\zeta)] = p_i \neq p$, then, by Theorem 4.13.3, $F_i(\zeta) = F_{i-1}(\zeta)(\alpha_i)$, where $m_{\alpha_i/F_{i-1}(\zeta)} = x^{p_i} - a_i$ for some $a_i \in F_{i-1}(\zeta)$. Note that ζ^{n/p_i} is a primitive p_i-th root of unity, contained in $F(\zeta) \subset F_{i-1}(\zeta)$. If $[F_i(\zeta) : F_{i-1}(\zeta)] = p_i = p$, then, by Theorem 4.13.6,

$F_i(\zeta) = F_{i-1}(\zeta)(\alpha_i)$, where $m_{\alpha_i/F_{i-1}(\zeta)} = x^p - x - a_i$ for some $a_i \in F_{i-1}(\zeta)$. It follows that (3) is a radical tower for $E(\zeta)/F(\zeta)$. But then

$$F \subset F(\zeta) \subset F_1(\zeta) \subset \cdots \subset F_t(\zeta) = E(\zeta) \tag{4}$$

is a radical tower for $E(\zeta)/F$, and so E is contained in a radical extension of F. ∎

Corollary 1: A separable polynomial $f \in F[x]$ is solvable by radicals over F if and only if the Galois group of f over F is a solvable group.

Corollary 2: If $f \in F[x]$ is a separable polynomial of degree $n \leq 4$, then f is solvable by radicals over F.

Proof: By the Lemma on p. 302, the Galois group of f over F is isomorphic to a subgroup of S_n, hence to a subgroup of S_4. Since S_4 is solvable, every subgroup of S_4 is solvable; hence the Galois group of f over F is solvable, and so f is solvable by radicals over F. ∎

Many individual polynomials of degree > 4 are, indeed, solvable by radicals. For example, any polynomial $x^n - a$ over a field F (Char $F \nmid n$) is obviously solvable by radicals since each of its zeros *is* a radical. However, we shall see that the general polynomial (see Definition 4.14.2) of degree $n > 4$ is *not* solvable by radicals over its coefficient field. This implies the impossibility of finding *general formulas* for the zeros of polynomials of degree $n > 4$. Precisely how to *define* a general polynomial of degree n over a field F is not at all obvious; we address ourselves to this question in the next section.

Exercises 4.13

True or False

1. If $f \in F[x]$ is a separable polynomial, with splitting field E of odd degree over F, then f is solvable by radicals over F.
2. If $f \in F[x]$ is a separable polynomial with Galois group D_n (where D_n is the dihedral group consisting of the symmetries of a regular n-gon), then f is solvable by radicals over F.
3. The polynomial $x^5 - 2$ is solvable by radicals over $\mathbb{Q}$.
4. The polynomial $x^5 - 2$ is solvable by radicals over $\mathbb{Z}_5$.
5. The Galois group of $x^{10} - 3$ over $\mathbb{Q}$ is cyclic.

4.13.1. Let $F = \mathbb{Q}$, $E = \mathbb{Q}(\sqrt{2})$. Find $N_{E/F}(\alpha)$ and $\mathrm{Tr}_{E/F}(\alpha)$, where
(1) $\alpha = \sqrt{2}$;
(2) $\alpha = 3$;
(3) $\alpha = 3 + \sqrt{2}$.

4.13.2. Let $F = \mathbb{Q}$, $E = \mathbb{Q}(\sqrt[4]{2})(i)$. Find $N_{E/F}(\alpha)$ and $\mathrm{Tr}_{E/F}(\alpha)$, where
(1) $\alpha = \sqrt[4]{2}$;
(2) $\alpha = i$;

(3) $\alpha = \sqrt[4]{2}\, i$;
(4) $\alpha = \sqrt{2}$;
(5) $\alpha = 1 + i + \sqrt{2}$.

4.13.3. Let $F = \mathbb{R}$, $E = \mathbb{C}$. For $a, b \in \mathbb{R}$, find $N_{E/F}(a + bi)$ and $\mathrm{Tr}_{E/F}(a + bi)$.

4.13.4. Prove Theorem 4.13.4.

4.13.5. Express the zeros of the polynomial $f = x^6 + 4x^3 + 8$ in $\mathbb{Q}[x]$ using only radicals of rational numbers. Factor f over $\mathbb{Q}$.

4.13.6. Find the Galois group, G, of the polynomial $f = x^6 + 4x^3 + 8$ over $\mathbb{Q}$. Let E be the splitting field, in $\mathbb{C}$, of f over $\mathbb{Q}$. Draw the subgroup lattice of G, and the lattice of subfields intermediate between $\mathbb{Q}$ and E. Find a primitive element for $E/\mathbb{Q}$.

4.13.7. Prove that, for each positive integer n, the polynomial $f = x^{2n} + bx^n + c$ in $\mathbb{Q}[x]$ is solvable by radicals over $\mathbb{Q}$. Prove that the Galois group of f over $\mathbb{Q}$ has order at most $2n^2\phi(n)$.

4.13.8. Let n be a positive integer. Prove that the polynomial $x^{2n} + bx^n + d^n$, with $b, d \in \mathbb{Q}$, has Galois group of order $\leq 2n\phi(n)$.

4.13.9. By the theorem of Thompson and Feit (see p. 119), every finite group of odd order is solvable. Use this result to prove: if F is a field, and if $f \in F[x]$ is a polynomial whose Galois group over F contains no F-automorphism $\sigma = \sigma^{-1}$, then f is solvable by radicals over F. (Is the converse true?)

4.13.10. Let $f \in F[x]$ be a polynomial whose Galois group over F has order 6. Prove that f is solvable by radicals over F. (See Exercise 2.6.16.)

4.13.11. Let p, q be primes, $q \not\equiv 1 \bmod p$, and let E/F be a Galois extension of degree pq. Suppose that the characteristic of F is equal to neither p nor q, and that F contains a primitive pq-th root of unity. Prove that there is some $\alpha \in E$ such that $E = F(\alpha)$ and $\alpha^{pq} \in F$. (See Exercise 2.15.9.)

4.14

Transcendence Sets and the Definition of a General Polynomial of Degree n

We now approach the problem: given a field F, for what positive integers n do there exist general formulas (involving only finitely many field operations and root extractions) that will yield the zeros of any polynomial f of degree n in $F[x]$?

Clearly, there can be no such formulas for given F and n if there exists an individual polynomial of degree n in $F[x]$ which is not solvable by radicals over F (e.g., for $F = \mathbb{Q}$ and $n = 5$). On the other hand, even if *every individual* polynomial of degree n in $F[x]$ were known to be solvable by radicals over F, there would be no reason to assume that a *single* formula would solve *all* such polynomials.

This suggests that, in order to formalize the notion of "general formulas," for the zeros of polynomials in $F[x]$, we would do well to transcend the point of view that deals only with polynomials *in* $F[x]$. We need, instead, to define (for given F and n)

a polynomial that can serve as a pattern for all polynomials of degree n in $F[x]$, and whose solution might serve as a "formula" for the solution of all such polynomials. The coefficients of a "general polynomial of degree n over F" will be "algebraically independent" indeterminates over F.

To begin with, we generalize the notions of algebraicity and transcendence.

Definition 4.14.1: Let F be a field, $K \supset F$ an extension field of F, and let $\{\alpha_1, \ldots, \alpha_n\}$ $(n \geq 1)$ be a subset of K. Then $\{\alpha_1, \ldots, \alpha_n\}$ is *algebraically dependent* over F if $f\{\alpha_1, \ldots, \alpha_n\} = 0$ for some $f \neq 0$ in $F[x_1, \ldots, x_n]$. If $f\{\alpha_1, \ldots, \alpha_n\} \neq 0$ holds for all $f \neq 0$ in $F[x_1, \ldots, x_n]$, then $\{\alpha_1, \ldots, \alpha_n\}$ is *algebraically independent* over F. In this case, $\{\alpha_1, \ldots, \alpha_n\}$ forms a *transcendence set* over F. We shall find it convenient to regard the empty set as a transcendence set over F.

Remark 1: If $\{\alpha_1, \ldots, \alpha_n\}$ is a transcendence set over F, then each α_i is transcendental over F. For, if α_i is algebraic over F, then there is a non-zero polynomial f_i in $F[x_i]$ such that $f_i(\alpha_i) = 0$. Clearly, f_i can be regarded as a (non-zero) polynomial in $F[x_1, \ldots, x_i, \ldots, x_n]$, where all terms involving x_j $(j \neq i)$ occur with 0 coefficient only. Thus, $f_i(\alpha_1, \cdots, \alpha_n) = 0$, and so the set $\{\alpha_1, \ldots, \alpha_n\}$ is algebraically dependent over F. Contradiction!

However, *the transcendence over F of each* α_i $(i = 1, \ldots, n)$ *does not suffice* to make the set $\{\alpha_1, \ldots, \alpha_n\}$ a transcendence set over F! For example, the set $\{\pi, 2\pi\}$ consists of two real numbers, each transcendental over $\mathbb{Q}$, but it is *not* a transcendence set over $\mathbb{Q}$ since $f(\pi, 2\pi) = 0$, where $f = 2x_1 - x_2 \in \mathbb{Q}[x_1, x_2]$.

Remark 2: As in the case of polynomials in a single indeterminate, we can define substitution homomorphisms with domain $F[x_1, \ldots, x_n]$ $(n \geq 1)$. If $\alpha_1, \ldots, \alpha_n \in K$, $K \supset F$, then

$$\phi_{\alpha_1, \ldots, \alpha_n/F} : F[x_1, \ldots, x_n] \to K$$

sends each $f(x_1, \ldots, x_n)$ to the corresponding field element $f(\alpha_1, \ldots, \alpha_n)$. The mapping $\phi_{\alpha_1, \ldots, \alpha_n/F}$ clearly preserves operations. It is then apparent that the Corollary of Definition 4.2.1 generalizes to the following Corollary of Definition 4.14.1:

$\{\alpha_1, \ldots, \alpha_n\} \subset K$ is algebraically dependent over F if and only if $\text{Ker}\ \phi_{\alpha_1, \ldots, \alpha_n/F} \neq \{0\}$;

$\{\alpha_1, \ldots, \alpha_n\} \subset K$ is a transcendence set over F if and only if $\text{Ker}\ \phi_{\alpha_1, \ldots, \alpha_n/F} = \{0\}$.

Remark 3: In view of Remark 2, if $\{u_1, \ldots, u_n\}$ is a transcendence set over a field F, then $\phi_{u_1, \ldots, u_n/F}$ defines an isomorphism of $F[x_1, \ldots, x_n]$ onto $F[u_1, \ldots, u_n]$. We may thus regard $F[u_1, \ldots, u_n]$ as simply a polynomial domain, over F, in the (algebraically independent) indeterminates $u_1, \ldots, u_n$. Indeed, if $\alpha_1, \ldots, \alpha_n \in E$ for some extension E of F, then there is a *substitution homomorphism*

$$\phi_{\alpha_1, \ldots, \alpha_n/F} : F[u_1, \ldots, u_n] \to E,$$

sending u_i to α_i $(i = 1, \ldots, n)$.

If $\{u_1, \ldots, u_n\}$ forms a transcendence set over F, we denote by $F(u_1, \ldots, u_n)$ the quotient field of the integral domain $F[u_1, \ldots, u_n]$.

Definition 4.14.2: Let F be a field, $\{u_1, \ldots, u_n\}$ a transcendence set over F. Then the polynomial

$$g_n = x^n + u_1 x^{n-1} + \cdots + u_{n-1}x + u_n$$

in $F(u_1, \ldots, u_n)[x]$ is a *general* (or *generic*) *polynomial of degree n over F*.

Example:

For $F = \mathbb{Q}$ and $n = 2$, $\{u_1, u_2\}$ a transcendence set over $\mathbb{Q}$, we have $g_2 = x^2 + u_1 x + u_2 \in \mathbb{Q}(u_1, u_2)[x]$. Let E be a splitting field of g_2 over $\mathbb{Q}(u_1, u_2)$. Since the numbers

$$\beta_{1,2} = \frac{-u_1 \pm \gamma}{2} \tag{1}$$

serve as zeros of g_2 if $\gamma^2 = u_1^2 - 4u_2$ $(\gamma = \beta_1 - \beta_2 \in E)$, we have $E = \mathbb{Q}(u_1, u_2)(\beta_1, \beta_2) = \mathbb{Q}(u_1, u_2)(\gamma)$. If $\gamma \in \mathbb{Q}(u_1, u_2)$, then $\gamma = \dfrac{s(u_1, u_2)}{t(u_1, u_2)}$ for some $s, t \in \mathbb{Q}[x_1, x_2]$, whence

$$\gamma^2 = \left[\frac{s(u_1, u_2)}{t(u_1, u_2)}\right]^2 = u_1^2 - 4u_2$$

and so

$$[s(u_1, u_2)]^2 = [t(u_1, u_2)]^2(u_1^2 - 4u_2).$$

Since the highest power to which u_2 occurs is even on the left-hand side and odd on the right-hand side, this contradicts the algebraic independence of $\{u_1, u_2\}$ over $\mathbb{Q}$. It follows that $\gamma \notin \mathbb{Q}(u_1, u_2)$; hence $[E : \mathbb{Q}(u_1, u_2)] = 2$ and g_2 is irreducible over $\mathbb{Q}(u_1, u_2)$.

Formula (1), more familiarly written

$$\beta_{1,2} = \frac{-u_1 \pm \sqrt{u_1^2 - 4u_2}}{2},$$

enables us to solve every polynomial $f = x^2 + \alpha_1 x + \alpha_2$ in $\mathbb{Q}[x]$ by substituting α_1 for u_1 and α_2 for u_2.

More generally, if a formula can be found for the zeros of a general polynomial $g_n \in F(u_1, \ldots, u_n)[x]$, then every polynomial $f = x^n + \alpha_1 x^{n-1} + \cdots + \alpha_{n-1}x + \alpha_n$ in $F[x]$ can be solved by substitution of α_i for u_i $(i = 1, \ldots, n)$, provided these substitutions do not entail division by zero. (See Exercise 4.16.2)

On the other hand, if, for given F and n, a general polynomial g_n proves to be not solvable by radicals, then no formulas exist that will solve *every* polynomial in $F[x]$ (such formulas would "ignore" any algebraic relations that might hold between the coefficients $\alpha_1, \ldots, \alpha_n$ of specific polynomials in $F[x]$, and would therefore solve any general polynomial g_n with algebraically independent coefficients $u_1, \ldots, u_n$).

The historical problem regarding the existence of "formulas" thus reduces to the question: *for what fields F and positive integers n is a general polynomial* $g_n \in F(u_1, \ldots, u_n)[x]$ *(defined in Definition 4.14.2) solvable by radicals over* $F(u_1, \ldots, u_n)$?

Exercises 4.14

True or False

1. If $u_1, u_2, \ldots, u_n$ are transcendental over F, then $\{u_1, \ldots, u_n\}$ is a transcendence set over F.
2. If u is transcendental over F, then $x^2 + ux + 3u^2$ is a general polynomial of degree 2 over F.
3. Since π and e are algebraically independent over $\mathbb{Q}$, $x^2 + ex + \pi$ is a general polynomial of degree 2 over $\mathbb{Q}$.
4. For $\{u_1, u_2\}$ a transcendence set over $\mathbb{Q}$, the quadratic formula provides the solution of $x^2 + u_1x + u_2$, by radicals, over $\mathbb{Q}$.
5. $x^3 + \sqrt[3]{2}\,x^2 + ex + \pi$ is a general polynomial of degree three over $\mathbb{Q}$.

4.14.1. Prove: if F is a field and S is an algebraically independent set over F, then every subset of S is algebraically independent over F.

4.14.2. Prove that $\mathbb{Q}[\pi]/(\pi^2 - 2)\mathbb{Q}[\pi]$ is a field isomorphic to $\mathbb{Q}(\sqrt{2})$.

4.14.3. Prove that $\mathbb{Q}(e)$ is isomorphic to $\mathbb{Q}(\pi)$ (assuming as known that π and e are transcendental over $\mathbb{Q}$).

4.14.4. Prove that $\mathbb{Q}(\pi, e)$ is not isomorphic to $\mathbb{Q}(\pi)$ (assuming as known that $\{\pi, e\}$ is a transcendence set over $\mathbb{Q}$).

4.14.5. Prove that $\mathbb{Q}(\pi)$ is isomorphic to $\mathbb{Q}(\sqrt{\pi})$.

4.14.6. Let $\{t_1, t_2\}$ be a transcendence set over a field F, and let

$$h_2 = (x - t_1)(x - t_2) \in F(t_1, t_2)[x].$$

Note that $h_2 = x^2 - (t_1 + t_2)x + t_1t_2$. Let $u_1 = -(t_1 + t_2)$, $u_2 = t_1t_2$. Prove that $\{u_1, u_2\}$ is a transcendence set over F, and conclude that h_2 is a general polynomial of degree 2 over F. (This result will be generalized in the next section.)

4.14.7. Let F be a field, n a positive integer, and let $\{u_1, \ldots, u_n\}$ and $\{\tilde{u}_1, \ldots, \tilde{u}_n\}$ be transcendence sets over F. Let $g_n = x^n + u_1x^{n-1} + \cdots + u_n$ and let $\tilde{g}_n = x^n + \tilde{u}_1x^{n-1} + \cdots + \tilde{u}_n$. (Thus, g_n and $\tilde{g}_n$ are both general polynomials of degree n over F.)

Prove that the Galois group, G, of g_n over $F(u_1, \ldots, u_n)$ is isomorphic to the Galois group, $\tilde{G}$, of $\tilde{g}_n$ over $F(\tilde{u}_1, \ldots, \tilde{u}_n)$. It is for this reason that one usually speaks simply of "the" general polynomial of degree n.)

4.14.8. Let $\{u, v\}$ be a transcendence set over $\mathbb{Z}_2$. If $F = \mathbb{Z}_2(u, v)$ and $E = \mathbb{Z}_2(u, v)(\alpha, \beta)$, where $\alpha^2 = u$ and $\beta^2 = v$, prove that E/F is not a simple extension. What is the number of intermediate fields K such that $F \subset K \subset E$? (See Exercises 4.8.4–4.8.6.)

4.15

Symmetric Functions and the Unsolvability of a General Polynomial of Degree $n > 4$

Definition 4.15.1: Let $\{t_1, \ldots, t_n\}$ be a transcendence set over a field F. Then an element, q, of $F(t_1, \ldots, t_n)$ is a *symmetric function* of $t_1, \ldots, t_n$ if, for every permutation $\sigma \in S_n$, $q(t_1, \ldots, t_n) = q(t_{\sigma(1)}, \ldots, t_{\sigma(n)})$.

Example:

Of special interest to us will be the *elementary symmetric polynomials* $s_1, \ldots, s_n \in F(t_1, \ldots, t_n)$ defined by:

$$s_1 = \sum_{i=1}^{n} t_i$$

$$s_2 = \sum_{1 \le i < j \le n} t_i t_j$$

$$s_3 = \sum_{1 \le i < j < k \le n} t_i t_j t_k$$

$$\vdots$$

$$s_n = \prod_{i=1}^{n} t_i.$$

For example, if $n = 3$, we have $s_2 = t_1 t_2 + t_1 t_3 + t_2 t_3$. If $\sigma = (132) \in S_3$, then $\sigma(s_2) = t_3 t_1 + t_3 t_2 + t_1 t_2 = s_2$. In general, $\sigma(s_i) = s_i$ for each $i = 1, \ldots, n$ and each $\sigma \in S_n$ since the application of σ to the subscripts $1, \ldots, n$ causes each term of s_i to be mapped to some term of s_i. Thus, the s_i are, indeed, symmetric functions in $t_1, \ldots, t_n$. We shall see that *every* symmetric function $q(t_1, \ldots, t_n) \subset F(t_1, \ldots, t_n)$ is expressible as a rational function in $s_1, \ldots, s_n$, i.e., $q(t_1, \ldots, t_n) = \dfrac{f(s_1, \ldots, s_n)}{g(s_1, \ldots, s_n)}$ $(f, g \in F[x_1, \ldots, x_n])$.

Lemma: Let F be a field and let $\{t_1, \ldots, t_n\}$ be a transcendence set over F. If $\sigma \in S_n$, then the mapping $\phi : F(t_1, \ldots, t_n) \to F(t_1, \ldots, t_n)$ defined by $\phi\left(\dfrac{f(t_1, \ldots, t_n)}{g(t_1, \ldots, t_n)}\right) = \dfrac{f(t_{\sigma 1}, \ldots, t_{\sigma n})}{g(t_{\sigma 1}, \ldots, t_{\sigma n})}$ for each $f, g \in F[x_1, \ldots, x_n]$, $g \neq 0$, is an F-automorphism of $F(t_1, \ldots, t_n)$. (Thus, every permutation on the set $\{t_1, \ldots, t_n\}$ extends to an F-automorphism of $F(t_1, \ldots, t_n)$.)

Proof: For each $\sigma \in S_n$, since $\{t_{\sigma 1}, \ldots, t_{\sigma n}\} = \{t_1, \ldots, t_n\}$ is a transcendence set over F, the substitution homomorphism

$$\phi_{t_{\sigma 1}, \ldots, t_{\sigma n}/F} : F[t_1, \ldots, t_n] \to F(t_1, \ldots, t_n)$$

is an F-monomorphism with image $F[t_{\sigma 1},\ldots,t_{\sigma n}] = F[t_1,\ldots,t_n]$; its extension to the quotient field $F(t_1,\ldots,t_n)$ is the required F-automorphism, ϕ, of $F(t_1,\ldots,t_n)$. ■

Theorem 4.15.1: Let F be a field, and let $\{t_1,\ldots,t_n\}$ be a transcendence set over F. Let $h_n = \prod_{i=1}^{n} (x - t_i) \in F(t_1,\ldots,t_n)[x]$. Then

(1) $h_n = x^n - s_1 x^{n-1} + s_2 x^{n-2} - \cdots + (-1)^n s_n$, where s_i is the i-th elementary symmetric polynomial in $t_1,\ldots,t_n$.

(2) The Galois group of h_n over $F(s_1,\ldots,s_n)$ is isomorphic to the symmetric group S_n.

(3) h_n is irreducible over $F(s_1,\ldots,s_n)$.

Proof: *(1)* Multiplying out the factors of h_n, we obtain $x^n + \alpha_1 x^{n-1} + \cdots + \alpha_n$, where $\alpha_i = (-1)^i s_i$ for each $i = 1,\ldots,n$.

(2) By (1), $h_n \in F(s_1,\ldots,s_n)[x]$, Since $\{t_1,\ldots,t_n\}$ is a transcendence set over F, the t_i are obviously distinct; hence, h_n is separable over F, and thus separable over $F(s_1,\ldots,s_n)$. Let E be a splitting field of h_n over $F(s_1,\ldots,s_n)$, and let G be the Galois group of E over $F(s_1,\ldots,s_n)$. By the Lemma, every permutation of the set $\{t_1,\ldots,t_n\}$ extends to an F-automorphism, ϕ, of $F(t_1,\ldots,t_n)$, which leaves each of the elementary symmetric polynomials $s_1,\ldots,s_n$ fixed. Now, $E = F(s_1,\ldots,s_n)(t_1,\ldots,t_n) = F(t_1,\ldots,t_n)$. Hence the $n!$ permutations on $\{t_1,\ldots,t_n\}$ induce $n!$ distinct $F(s_1,\ldots,s_n)$-automorphisms of E, and so $|G| \geq n!$. On the other hand, by the Lemma, p. 302, since E is a splitting field of an n-th degree polynomial over $F(s_1,\ldots,s_n)$, $|G| \leq n!$. It follows that $|G| = n!$, and that G is isomorphic to S_n.

(3) Suppose h_n is reducible over $F(s_1,\ldots,s_n)$. Then, for each $i = 1,\ldots,n$, the polynomial $m_{t_i/F(s_1,\ldots,s_n)}$ has degree $k_i < n$; hence t_i has only k_i possible images under any $F(s_1,\ldots,s_n)$-automorphism of E. This implies that some permutations of $\{t_1,\ldots,t_n\}$ do not extend to $F(s_1,\ldots,s_n)$-automorphisms of E, whence $|G| < n!$, contradicting (2). It follows that h_n is irreducible over $F(s_1,\ldots,s_n)$. ■

Theorem 4.15.1 enables us to give a very elegant proof of the following classical result:

Corollary (Fundamental Theorem on Symmetric Functions): Let F be a field, $\{t_1,\ldots,t_n\}$ a transcendence set over F. Then every symmetric function in $F(t_1,\ldots,t_n)$ is expressible as a rational function $\dfrac{f(s_1,\ldots,s_n)}{g(s_1,\ldots,s_n)}$ $(f, g \in F[x_1,\ldots,x_n])$, where the s_i $(i = 1,\ldots,n)$ are the elementary symmetric polynomials in $t_1,\ldots,t_n$.

Proof: Since every permutation of the t_i extends to an automorphism contained in $G = G[F(t_1,\ldots,t_n)/F(s_1,\ldots,s_n)]$, every symmetric function in $F(t_1,\ldots,t_n)$ lies in the fixed field of G. By Theorem 4.10.1, the fixed field of G is $F(s_1,\ldots,s_n)$, hence every symmetric function is an element of $F(s_1,\ldots,s_n)$. ■

Recall that the polynomial h_n of Theorem 4.15.1 was defined by

$$h_n = \prod_{i=1}^{n} (x - t_i),$$

where $\{t_1, \ldots, t_n\}$ is a transcendence set over a field F. Two obvious questions regarding h_n may already have occurred to you.

(1) Do the coefficients of h_n form a transcendence set over F, i.e., is h_n a general polynomial of degree n over F?

(2) If $g_n = x^n + u_1 x^{n-1} + \cdots + u_n$ is a general polynomial of degree n over F, with zeros $t_1, \ldots, t_n$ in a splitting field over $F(u_1, \ldots, u_n)$, is $\{t_1, \ldots, t_n\}$ a transcendence set over F, i.e., does g_n satisfy the conditions imposed on h_n in Theorem 4.15.1?

The answer to each of these questions turns out to be "yes" (see the proof of Theorem 4.15.1, and Exercise 4.15.4), and Theorem 4.15.1 thus implies that the Galois group of g_n over $F(u_1, \ldots, u_n)$ is isomorphic to S_n.

Some preliminary work is required.

Definition 4.15.2: If F is a subfield of K and $\{w_1, \ldots, w_n\} \subset K$ $(n \geq 0)$ is a transcendence set over F such that $K/F(w_1, \ldots, w_n)$ is algebraic, then $\{w_1, \ldots, w_n)$ is a *transcendence base* for K/F.

It can be shown that every extension has a transcendence base, and all transcendence bases for a given extension have the same cardinality. For our purposes, it suffices to consider *finitely generated* extensions, i.e., extensions K/F such that $K = F(\beta_1, \ldots, \beta_n)$ for some $\beta_1, \ldots, \beta_n \in K$.

Theorem 4.15.2: Let $K = F(R)$, where R is a finite subset of K. Then:

(1) K/F has a finite transcendence base $S \subset R$ such that $[K : F(S)]$ is finite;

(2) if T is another transcendence base for K/F, then S and T have the same cardinality.

Proof: *(1)* The set R, being finite, has a maximal (finite) subset S, possibly empty, such that S is a transcendence set over F. If $S = \varnothing$, then K/F is algebraic; hence $S = \varnothing$ is a transcendence base for K/F, and $[K : F]$ is finite since each element of R is algebraic over F. If $S \neq \varnothing$, label the elements $\beta_1, \ldots, \beta_n$ of R so that $S = \{\beta_1, \ldots, \beta_k\}(1 \leq k \leq n)$. If $k = n$, then $[K : F(S)] = 1$. Suppose $k < n$. Then each $\beta \in R \backslash S$ is algebraic over $F(S) = F(\beta_1, \ldots, \beta_k)$. For, otherwise, $f(\beta) \neq 0$ for each non-zero polynomial $f \in F(\beta_1, \ldots, \beta_k)[x]$. This implies that $h(\beta_1, \ldots, \beta_k, \beta) \neq 0$ for each non-zero polynomial $h \in F[x_1, \ldots, x_k, x_{k+1}]$, and so $\{\beta_1, \ldots, \beta_k, \beta\}$ is a transcendence set over F, contrary to the maximality of S. By Exercise 4.4.9, it follows that $K/F(S) = F(S)(\beta_{k+1}, \ldots, \beta_n)/F(S)$ is a finite extension. But then $K/F(S)$ is algebraic, hence S is a transcendence base with the required properties.

(2) If $S = \varnothing$, then every transcendence base for K/F is empty since K/F is algebraic. If $S \neq \varnothing$, suppose $T = \{\gamma_1, \ldots, \gamma_h\}$ is another transcendence base for

K/F. Then γ_1 is algebraic over $F(S)$, hence the set $S \cup \{\gamma_1\}$ is algebraically dependent over F. Since $\{\gamma_1\}$ is a transcendence set over F, $S \cup \{\gamma_1\}$ has a maximal algebraically independent subset containing γ_1, i.e., a subset $S_1 \cup \{\gamma_1\}$, where $S_1 \subsetneqq S$. If $\beta \in S \backslash S_1$, then $S_1 \cup \{\gamma_1\} \cup (\beta)$ is algebraically dependent over F, whence β is algebraic over $F(S_1 \cup \{\gamma_1\})$. But then $F(S)$ is algebraic over $F(S_1 \cup \{\gamma_1\})$, and from $K/F(S)$ algebraic we infer that $K/F(S_1 \cup \{\gamma_1\})$ is algebraic, hence $S_1 \cup \{\gamma_1\}$ is a transcendence base for K/F.

We next form $S_1 \cup \{\gamma_1, \gamma_2\}$, an algebraically dependent set. Since $\{\gamma_1, \gamma_2\}$ is a subset of a transcendence set, it is itself a transcendence set. Hence $S_1 \cup \{\gamma_1, \gamma_2\}$ has a maximal algebraically independent subset of the form $S_2 \cup \{\gamma_1, \gamma_2\}$, $S_2 \subsetneqq S_1$. Again, if $\beta \in S_1 \backslash S_2$, then since $S_2 \cup \{\gamma_1, \gamma_2\} \cup \{\beta\}$ is algebraically dependent, β is algebraic over $F(S_2 \cup \{\gamma_1, \gamma_2\})$, whence $K/F(S_2 \cup \{\gamma_1, \gamma_2\})$ is algebraic, and $S_2 \cup \{\gamma_1, \gamma_2\}$ is a transcendence base for K/F. Continuing in this manner, we can replace elements of S by the elements γ_i, $i = 1, \ldots, h$, until we exhaust the γ_i. This implies that $h \leq k$.

In the same way, exchanging γ's for β's, we can prove the opposite inequality, and so $h = k$. ■

Remark: Statement (2) of Theorem 4.15.2 can be worded more strongly: if T is *any* transcendence base for K/F, then T and S have the same (finite) cardinality. (See Exercise 4.15.1.).

We are now ready to determine the Galois group of a general polynomial of degree n over its coefficient field.

Theorem 4.15.3: Let F be a field, n a positive integer. Let $\{u_1, \ldots, u_n\}$ be a transcendence set over F, and let E be a splitting field over $F(u_1, \ldots, u_n)$ of the general polynomial

$$g_n = x^n + u_1 x^{n-1} + \cdots + u_n.$$

Then g_n is a separable polynomial, irreducible over $F(u_1, \ldots, u_n)$, and $G(E/F(u_1, \ldots, u_n))$ is isomorphic to S_n.

Proof: Over E, the polynomial g_n factors into linear factors:

$$g_n = \prod_{i=1}^{n} (x - t_i),$$

and so $E = F(u_1, \ldots, u_n)(t_1, \ldots, t_n)$. (At this stage, we do not even know that the t_i are distinct.) Since the u_i are expressible in terms of the t_i (in fact, $u_i = (-1)^i s_i(t_1, \ldots, t_n)$, where $s_i \in F[x_1, \ldots, x_n]$ is the i-th elementary symmetric polynomial), we have $E = F(t_1, \ldots, t_n)$. Consider the field tower

$$F \subset F(u_1, \ldots, u_n) \subset F(t_1, \ldots, t_n) = E.$$

Since $\{u_1, \ldots, u_n\}$ forms a transcendence set over F, and $E/F(u_1, \ldots, u_n)$ is algebraic, $\{u_1, \ldots, u_n)$ is a transcendence base for E/F. Suppose $\{t_1, \ldots, t_n\}$ is algebraically dependent over F. Then, by Theorem 4.15.2(1), $\{t_1, \ldots, t_n\}$ contains a transcendence base for E/F having $m < n$ elements. This contradicts Theorem 4.15.2(2). Thus, $\{t_1, \ldots, t_n\}$ is a transcendence set over F, and our polynomial

$g_n = x^n + u_1 x^{n-1} + \cdots + u_n = \prod_{i=1}^{n} (x - t_i)$ satisfies the conditions we imposed on the polynomial h_n of Theorem 4.15.1, with $u_i = (-1)^i s_i(t_1, \ldots, t_n)$ for each $i = 1, \ldots, n$. Theorem 4.15.1 immediately yields the result that g_n is irreducible over $F(u_1, \ldots, u_n)$ (which is clearly equal to $F(s_1, \ldots, s_n)$), and that $G(E/F(u_1, \ldots, u_n))$ is isomorphic to S_n. Since $\{t_1, \ldots, t_n\}$ is a transcendence set over F, the t_i are certainly distinct, and so g_n is separable. ■

Theorem 4.15.4 (Solvability Criterion for a General Polynomial of Degree n): Let F be a field, $n \in \mathbb{Z}^+$, and let $\{u_1, \ldots, u_n\}$ be a transcendence set over F. Then the general polynomial $g_n = x^n + u_1 x^{n-1} + \cdots + u_n$ is solvable by radicals over $F(u_1, \ldots, u_n)$ if and only if $n \leq 4$. (Equivalently, but more interestingly: g_n is *not* solvable by radicals over $F(u_1, \ldots, u_n)$ if and only if $n \geq 5$.)

Proof: This result follows immediately from Theorems 4.15.3, 4.13.7 (Corollary), and 2.14.2. ■

Remark: An obvious historical question arising at this point is: how could Abel obtain this theorem without the benefit of Galois Theory? For those who wish to see for themselves, an English translation of Abel's 1824 paper is printed in [23], Chapter 8. Galois' original works are reprinted in [13]

Exercises 4.15

True or False

1. Some day, some really smart person will find a way to solve a general polynomial of degree $n \geq 5$ by radicals.
2. If g_n is a general polynomial of degree $n \geq 1$ over a field F, then the Galois group of g_n over F is isomorphic to S_n.
3. Since π and e are algebraically independent over $\mathbb{Q}$, the zeros of the polynomial $x^2 + \pi x + e$ form a transcendence set over $\mathbb{Q}$.
4. A polynomial of degree 6 cannot be solved by radicals.
5. Any polynomial of degree ≤ 4, with coefficients in a field, can be solved by radicals.

4.15.1. If T is a transcendence base for K/F, consisting of k elements ($k \geq 0$ in $\mathbb{Z}$), prove that any $k + 1$ elements of K are algebraically dependent over F. (Conclude that the existence of a finite transcendence base for K/F precludes the existence of an *infinite transcendence base* for K/F, (i.e., of an infinite subset W of K such that every finite subset of W is a transcendence set over F, and $K/F(W)$ is algebraic,) Hence conclude: if K/F has a finite transcendence base, then all transcendence bases of K/F have the same cardinality. (Using set theory, this result can be shown to hold also for extensions having an infinite transcendence base.)

4.15.2. Let G be a finite group. Prove that there are fields F and K such that K/F is finite Galois, with Galois group isomorphic to G.

4.15.3. Assume as known that π and e are algebraically independent over $\mathbb{Q}$. Let $s_1 = \pi^2 + e^3$, $s_2 = \pi^2 e^3$. Find the Galois group of the polynomial $(x - \pi^2)(x - e^3)$ over $\mathbb{Q}(s_1, s_2)$.

4.15.4. Let F be a field and let $\{u_1, \ldots, u_n\}$ be a subset of some extension field of F. Let $g = x^n + u_1 x^{n-1} + \cdots + u_n \in F(u_1, \ldots, u_n)[x]$, and suppose that

$$g = \prod_{i=1}^{n} (x - t_i)$$

in a splitting field E for g over $F(u_1, \ldots, u_n)$.

Prove that $\{u_1, \ldots, u_n\}$ is a transcendence set over F if and only if $\{t_1, \ldots, t_n\}$ is a transcendence set over F. (Note: one of these implications was established in the proof of Theorem 4.15.3.)

4.15.5. Let F be a field, n a positive integer and g_n a general polynomial of degree n over F. If $u_1, \ldots, u_n$ are the coefficients of g_n and $t_1, \ldots, t_n$ are the zeros of g_n, prove that every permutation of the set $\{t_1, \ldots, t_n\}$ extends to an $F(u_1, \ldots, u_n)$-automorphism of $F(t_1, \ldots, t_n)$.

4.15.6. For n a positive integer, let g_n be a general polynomial of degree n over a field F. Find a composition series for the Galois group of $F(t_1, \ldots, t_n)/F(u_1, \ldots, u_n)$ (notation as in the preceding exercise):

(1) for $n = 2$;
(2) for $n = 3$;
(3) for $n = 4$;
(4) for $n \geq 5$.

4.15.7. With t_i, u_i defined as in the two preceding exercises ($i = 1, \ldots, u$) and $n \geq 3$, prove that there is a subfield M of $F(t_1, \ldots, t_n)$ such that:

(a) $[M : F(u_1, \ldots, u_n)] = 2$,
and
(b) $M = F(u_1, \ldots, u_n)(\delta)$, where $\delta = \prod_{1 \leq i < j \leq n} (t_i - t_j)$, and $\delta^2 \in F(u_1, \ldots, u_n)$. ($\delta^2$ is the discriminant of g_n—see Definition 4.16.1.)

(Hint: Recall that a permutation σ on $\{1, 2, \ldots, n\}$ is even if and only if $\prod_{1 \leq i < j \leq n} (t_{\sigma i} - t_{\sigma j}) = \prod_{1 \leq i < j \leq n} (t_i - t_j)$—see the proof of Theorem 2.6.4.)

4.16

Solution of a General Polynomial of Degree $n \leq 4$ by Radicals

For F any field, u_1 transcendental over F, the general polynomial $g_1 = x + u_1$ has one zero: $-u_1$.

For F a field, with Char $F = 0$ or Char $F = p > 2$, $\{u_1, u_2\}$ a transcendence set over F, the general polynomial $g_2 = x^2 + u_1 x + u_2$ has zeros $x_1 = \dfrac{-u_1 + \delta}{2}$ and $x_2 = \dfrac{-u_1 - \delta}{2}$, where $\delta^2 = u_1^2 - 4u_2 \in F(u_1, u_2)$. Note that $x_1 - x_2 = \delta$. Write $\Delta = \delta^2 = (x_1 - x_2)^2 = u_1^2 - 4u_2$, the discriminant of g_2.

Definition 4.16.1: Let f be a polynomial of degree $n \geq 2$ in $K[x]$, K a field. If $f = \prod_{i=1}^{n} (x - x_i) \in E[x]$, where E is a splitting field of f over K ($x_1, \ldots, x_n$ not necessarily distinct), then $\Delta = \prod_{1 \leq i < j \leq n} (x_i - x_j)^2$ is the *discriminant* of f over K.

If $g_n \in F(u_1, \ldots, u_n)[x]$ is a general polynomial of degree n over F ($\{u_1, \ldots, u_n\}$ a transcendence set over F), with splitting field E over $F(u_1, \ldots, u_n)$, then the Galois group G of g_n over $F(u_1, \ldots, u_n)$ is isomorphic to S_n, by Theorem 4.15.3. The zeros $x_1, \ldots, x_n$ of g_n in E are all distinct and form a transcendence set over F. If, with each $\sigma \in G$, we associate the permutation, σ, in S_n such that $\sigma x_i = x_{\sigma i}$ for each $i = 1, \ldots, n$, the resulting correspondence is an isomorphism of G onto S_n. Let $\delta = \prod_{1 \leq i < j \leq n} (x_i - x_j)$. Then $\delta^2 = \Delta$, the discriminant of g_n. By the proof of Theorem 2.6.4, if $\sigma(\delta) = \prod_{1 \leq i < j \leq n} (x_{\sigma i} - x_{\sigma j})$ for $\sigma \in S_n$, then $\sigma(\delta) = \delta$ if and only if σ is an even permutation (i.e., $\sigma \in A_n$), and $\sigma(\delta) = -\delta$ if and only if σ is an odd permutation. It follows that Δ is in the fixed field, $F(u_1, \ldots, u_n)$, of G and δ is in the fixed field of H, the subgroup of G corresponding to the alternating group A_n. If Char $F \neq 2$, then $\sigma(\delta) = -\delta \neq \delta$, hence δ is *not* in the fixed field of G, and so $[F(u_1, \ldots, u_n)(\delta) : F(u_1, \ldots, u_n)] = 2$.

(However, if $f = x^n + a_1 x^{n-1} + \cdots + a_n \in F[x]$, with zeros $x_1, \ldots, x_n$ in a splitting field of f over F, then $\delta = \prod_{1 \leq i < j \leq n} (x_i - x_j)$ may well be in F. For example, the zeros of a quadratic $x^2 + a_1 x + a_2 \in \mathbb{R}[x]$ are real if the discriminant $a_1^2 - 4a_2$ has square roots in $\mathbb{R}$ and imaginary otherwise.)

We now examine the general cubic,

$$g_3 = x^3 + u_1 x^2 + u_2 x + u_3 \tag{1}$$

over a field F, with Char $F = 0$ or Char $F > 3$.

Let E be a splitting field of g_3 over $F(u_1, u_2, u_3)$, with x_1, x_2, x_3 the zeros of g_3 in E. Then

$$E = F(u_1, u_2, u_3)(x_1, x_2, x_3) = F(x_1, x_2, x_3)$$

(since the u_i are in $F(x_1, x_2, x_3)$).

To simplify the solution, set $y = x + \frac{1}{3}u_1 \in E$. Then $x = y - \frac{1}{3}u_1$, and $g_3(x) = y^3 + py + q$, where

$$p = -\frac{1}{3}u_1^2 + u_2$$

$$q = \frac{2}{27}u_1^3 - \frac{1}{3}u_1 u_2 + u_3.$$

Write $f_3(y) = y^3 + py + q$. Then the zeros of f_3 are

$$y_1 = x_1 + \frac{1}{3}u_1,\ y_2 = x_2 + \frac{1}{3}u_1,\ y_3 = x_3 + \frac{1}{3}u_1 \tag{2}$$

Since $u_1 = -(x_1 + x_2 + x_3)$ and the x_i are all distinct, it follows that the y_i are distinct elements of $E = F(x_1, x_2, x_3)$. Thus, f_3 is separable over $F(p, q)$ and splits over E. Let $\bar{E}$ be the splitting field, in E, of f_3 over $F(p, q)$. Then $\bar{E} = F(p, q)(y_1, y_2, y_3) = F(y_1, y_2, y_3)$. By Theorem 4.15.3, the Galois group G of g_3 over $F(u_1, u_2, u_3)$ is isomorphic to S_3. The six distinct $F(u_1, u_2, u_3)$-automorphisms of E induce six distinct $F(p, q)$-automorphisms of $\bar{E}$. Thus, the Galois group $\bar{G} = G(\bar{E}/F(p, q))$ of f_3 over $F(p, q)$ has order at least six. Since deg $f_3 = 3$, $\bar{G}$ is isomorphic to a subgroup of S_3, hence isomorphic to S_3. (See Figure 16.)

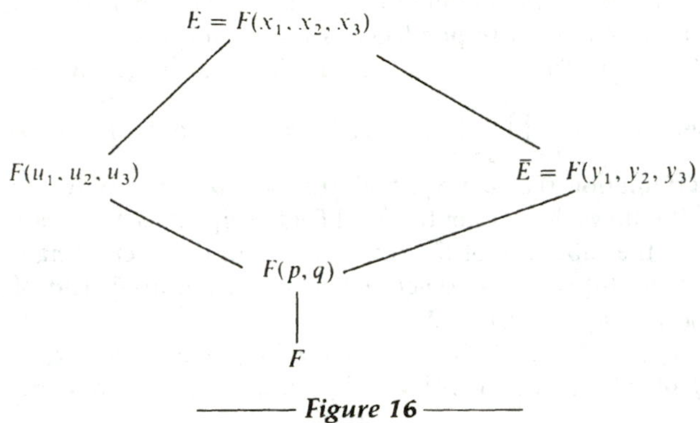

Figure 16

The polynomials g_3 and f_3 have the same discriminant

$$\Delta = \prod_{1 \le i < j \le n} (x_i - x_j)^2 = \prod_{1 \le i < j \le n} (y_i - y_j)^2.$$

It can be verified (cf. [9]) that

$$\Delta = -4p^3 - 27q^2.$$

Corresponding to the normal series

$$G \rhd H \rhd \{\iota\},$$

where G is the Galois group of f_3 over $F(p, q)$ and H is the subgroup corresponding to A_3 in S_3, we have the field tower

$$F(p, q) \subset F(p, q)(\delta) \subset E,$$

where $\delta = \prod_{1 \le i < j \le n} (y_i - y_j)$, and $[E : F(p, q)(\delta)] = 3$.

Since the characteristic of E is not a divisor of 3, there is an extension of E that contains a primitive cube root, ω, of 1. Indeed, we may write $\omega = \dfrac{-1 + \sqrt{-3}}{2}$ where $\sqrt{-3}$ represents either one of the square roots of -3, necessarily located in $E(\omega)$. Then $\omega^2 = \dfrac{-1 - \sqrt{-3}}{2}$.

To solve the cubic f_3, we now employ the *Lagrange resolvents*:

$$(1, y_1) = y_1 + y_2 + y_3 = 0$$
$$(\omega, y_1) = y_1 + \omega y_2 + \omega^2 y_3 \qquad (2)$$
$$(\omega^2, y_1) = y_1 + \omega^2 y_2 + \omega y_3$$

located in $E(\omega)$.

Solving equations (2) for y_1, y_2, y_3 (the determinant of (2) being $3(\omega^2 - \omega) = -3\sqrt{-3} \neq 0$), one obtains

$$y_1 = \frac{1}{3}[(\omega, y_1) + (\omega^2, y_1)]$$
$$y_2 = \frac{1}{3}[\omega^2(\omega, y_1) + \omega(\omega^2, y_1)] \qquad (3)$$
$$y_3 = \frac{1}{3}[\omega(\omega, y_1) + \omega^2(\omega^2, y_1)]$$

It can be shown (cf. [9]) that

$$(\omega, y_1)^3 = -\frac{27}{2}q + \frac{3}{2}\sqrt{-3}\,\delta$$
$$(\omega^2, y_1)^3 = -\frac{27}{2}q - \frac{3}{2}\sqrt{-3}\,\delta,$$

and that $(\omega, y_1) \cdot (\omega^2, y_1) = -3p$. Cube roots must therefore be chosen so as to satisfy this condition. With cube roots appropriately chosen, writing $\delta = \sqrt{\Delta}$, one obtains

$$(\omega, y_1) = \sqrt[3]{-\frac{27}{2}q + \frac{3}{2}\sqrt{-3\Delta}}$$
$$(\omega^2, y_1) = \sqrt[3]{-\frac{27}{2}q - \frac{3}{2}\sqrt{-3\Delta}}, \qquad (4)$$

where

$$\Delta = -4p^3 - 27q^2.$$

Substitution in (3) yields the zeros, y_i, of f_3, from which the zeros, x_i, of g_3 can be obtained using $x_i = y_i - \frac{1}{3}u_1 \quad (i = 1, 2, 3)$.

For any *specific* polynomial $x^3 + a_1x^2 + a_2x + a_3 \in F[x]$, the roots can be found by substituting a_i for $u_i \quad (i = 1, 2, 3)$.

Example 1:

Let $g = x^3 + 3x^2 + x + 1 \in \mathbb{Q}[x]$. Set $x = y - \frac{1}{3} \cdot 3 = y - 1$. Then

$$\begin{aligned} g(x) = f(y) &= (y-1)^3 + 3(y-1)^2 + y - 1 + 1 \\ &= y^3 + 3y - 6y + y - 1 + 3 \\ &= y^3 - 2y + 2. \end{aligned}$$

With $p = -2$, $q = 2$, we obtain

$$\triangle = -4(-2)^3 - 27 \cdot 2^2 = 32 - 108 = -76,$$

$$(\omega, y_1)^3 = -27 + \frac{3}{2}\sqrt{228} = -27 + 3\sqrt{57},$$

$$(\omega^2, y_1)^3 = -27 - \frac{3}{2}\sqrt{228} = -27 - 3\sqrt{57}.$$

In order for $(\omega, y_1)(\omega^2, y_1) = -3 \cdot p = 6$ to hold, it suffices, in this case, to choose both cube roots to be the real cube roots:

$$\sqrt[3]{-27 + 3\sqrt{57}} \cdot \sqrt[3]{-27 - 3\sqrt{57}} = \sqrt[3]{27^2 - 9 \cdot 57} = 6,$$

where $\sqrt[3]{\ }$ denotes the real cube root in each case.

Using (3), we thus obtain

$$y_1 = \frac{1}{3}\left[\sqrt[3]{-27 + 3\sqrt{57}} + \sqrt[3]{-27 - 3\sqrt{57}}\right]$$

$$y_2 = \frac{1}{3}\left[\omega^2 \sqrt[3]{-27 + 3\sqrt{57}} + \omega \sqrt[3]{-27 - 3\sqrt{57}}\right]$$

$$y_3 = \frac{1}{3}\left[\omega \sqrt[3]{-27 + 3\sqrt{57}} + \omega^2 \sqrt[3]{-27 - 3\sqrt{57}}\right]$$

as the zeros of f, and $x_1 = y_1 - 1$, $x_2 = y_2 - 1$, $x_3 = y_3 - 1$ as the zeros of the original polynomial g. Let E be the splitting field, in $\mathbb{C}$, of g, hence of f, over $\mathbb{Q}$. Since f is an Eisenstein polynomial, f and g are both irreducible over $\mathbb{Q}$, whence $3 \mid [E : \mathbb{Q}]$. Also, since

$$\mathbb{Q} \subset \mathbb{Q}(\sqrt{-76}) \subset E,$$

$2 \mid [E : \mathbb{Q}]$. It follows that $6 \mid [E : \mathbb{Q}]$. Since $[E : \mathbb{Q}] \leq 6$, this implies $[E : \mathbb{Q}] = 6$ and $G(E/\mathbb{Q})$ is isomorphic to S_3.

Example 2:

Let $f = y^3 - 3y + 1$. Since $f(1) \neq 0$ and $f(-1) \neq 0$, f is irreducible over $\mathbb{Q}$. We have $p = -3$, $q = 1$, $\Delta = -4 \cdot (-27) - 27 = 108 - 27 = 81$,

$$(\omega, y_1)^3 = -\frac{27}{2} + \frac{3}{2}\sqrt{-3 \cdot 81} = 27\omega = 27e^{2\pi i/3},$$

$$(\omega^2, y_1)^3 = -\frac{27}{2} - \frac{3}{2}\sqrt{-3 \cdot 81} = 27\omega^2 = 27e^{4\pi i/3}.$$

We need to choose cube roots so that $(\omega, y_1)(\omega^2, y_1) = -3p = 9$. Choosing $(\omega, y_1) = 3e^{8\pi i/9}$ and $(\omega^2, y_1) = 3e^{10\pi i/9}$, and writing $\sqrt[3]{\omega} = e^{8\pi i/9}$ and $\sqrt[3]{\omega^2} = e^{10\pi i/9}$, we obtain

$$y_1 = \sqrt[3]{\omega} + \sqrt[3]{\omega^2} = e^{8\pi i/9} + e^{10\pi i/9}$$

$$y_2 = \omega^2 \sqrt[3]{\omega} + \omega \sqrt[3]{\omega^2} = e^{20\pi i/9} + e^{16\pi i/9}$$

$$y_3 = \omega \sqrt[3]{\omega} + \omega^2 \sqrt[3]{\omega^2} = e^{14\pi i/9} + e^{22\pi i/9}.$$

Despite the appearance of imaginary radicals in these expressions, all three of the zeros of f are real! For, $f(-2) = -1$, $f(0) = 1$, $f(1) = -1$, and $f(2) = 3$; hence, by the Intermediate Value Theorem, f has a zero on each of the open intervals $(-2, 0)$, $(0, 1)$, and $(1, 2)$. (See Exercise 4.16.5.)

If E is the splitting field of f over $\mathbb{Q}$, then $G(E/\mathbb{Q})$ is isomorphic to A_3. For:

$$\delta = \sqrt{\Delta} = \prod_{1 \le i < j \le n} (y_i - y_j) = 9 \in \mathbb{Q},$$

hence every $\sigma \in G(E/\mathbb{Q})$ fixes δ, and thus corresponds to an even permutation in S_3.

These two examples illustrate the following result.

Theorem 4.16.1: Let F be a field, with Char $F = 0$, or Char $F > 3$. Let f be an irreducible cubic in $F[x]$, with Galois group G over F, zeros y_1, y_2, y_3 and discriminant $\Delta = \delta^2$, $\delta = (y_1 - y_2)(y_1 - y_3)(y_2 - y_3)$. Then G is isomorphic to S_3 if and only if $\delta \notin F$; and G is isomorphic to A_3 if and only if $\delta \in F$.

Proof: In any case, G is isomorphic to a subgroup of S_3. Let E be a splitting field for f over F. Since f is irreducible over F, $[E : F(\alpha)] = 3$ for any zero $\alpha \in E$ of f, hence $3 \mid |G|$. Thus, either $G \cong S_3$ or $G \cong A_3$. Let Δ be the discriminant of f. If Δ has no square root in F, then $[F(\delta) : F] = 2$, hence $2 \mid |G|$, and so $G \cong S_3$. If Δ has square roots in F, then $\delta = \prod_{1 \le i < j \le n} (y_i - y_j) \in F$ is fixed under each $\sigma \in G$. Thus, the restriction of σ to $\{y_1, y_2, y_3\}$ induces an even permutation on the y_i. By the Lemma, p. 302, it follows that G is isomorphic to A_3. ■

Remark: For a cubic polynomial f with real coefficients, the zeros y_1, y_2, y_3 of f are either all real, or one real and two conjugate imaginary. In the latter case,

$$\delta = (y_1 - y_2)(y_1 - y_3)(y_2 - y_3)$$

is pure imaginary, and so $\Delta = \delta^2$ is negative. Hence, (ω, y_1) and (ω^2, y_1) are both real (from (4)), and the only imaginary radical occurring in (3) is $\sqrt{-3}$.

On the other hand, if the y_i are all real, then δ is real, hence $\Delta = \delta^2 \ge 0$. If the zeros are distinct, then $\Delta > 0$; hence, from (4), (ω, y_1) and (ω^2, y_1) are both imaginary, and formulas (3) express the three real zeros y_1, y_2, y_3 in terms of imaginary radicals.

Next, for F a field with Char $F = 0$ or Char $F > 3$, and $\{u_1, \ldots, u_4\}$ a transcendence set over F, let $g_4 = x^4 + u_1x^3 + u_2x^2 + u_3x + u_4 \in F(u_1, \ldots, u_4)[x]$ be the general quartic over F. We first eliminate the cubic term by setting $x = y - \frac{1}{4}u_1$, thus obtaining

$$f_4(y) = y^4 + py^2 + qy + r$$

$(p, q, r \in F(u_1, \ldots, u_4))$.

If E is a splitting field of g_4 over $F(u_1, u_2, u_3, u_4)$ and $x_1, \ldots, x_4$ are the zeros of g_4 in E, then $y_i = x_i + \frac{1}{4}u_1$ $(i = 1, \ldots, 4)$ are the (distinct) zeros of f_4, located in E since $u_1 = \sum_{i=1}^{4} x_i$. Let $\bar{E}$ be the splitting field in E of f_4 over $F(p, q, r)$. In analogy to the general cubic, we have

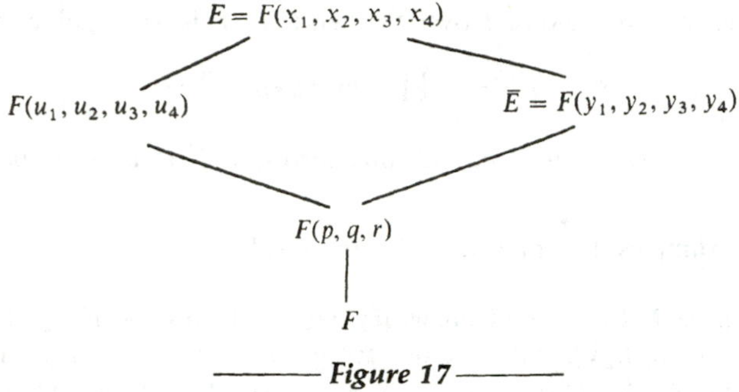

Figure 17

and it is easy to show that the Galois group $\bar{G}$ of f_4 over $F(p, q, r)$ is isomorphic to S_4.

Corresponding to the normal series $S_4 \rhd A_4 \rhd V_4 \rhd \{\iota\}$, where $V_4 = \{\iota, (12)(34), (13)(24), (14)(23)\}$, there is a field tower

$$F(p, q, r) \subset F_1 \subset F_2 \subset \bar{E}$$

where F_1 and F_2 are the fixed fields, respectively, of the subgroups of $\bar{G}$ corresponding to A_4 and V_4.

Consider the elements $z_1, z_2, z_3 \in \bar{E}$ given by

$$z_1 = (y_1 + y_2)(y_3 + y_4)$$

$$z_2 = (y_1 + y_3)(y_2 + y_4)$$

$$z_3 = (y_1 + y_4)(y_2 + y_3).$$

These elements are, clearly, in F_2. Let h be the polynomial in $\bar{E}[y] \subset E[y]$ defined by

$$h = (y - z_1)(y - z_2)(y - z_3).$$

The coefficients of this cubic, h, are equal to plus or minus the elementary symmetric polynomials in z_1, z_2, z_3. But every permutation of y_1, y_2, y_3, y_4 merely permutes z_1, z_2, z_3; hence every $\sigma \in \bar{G}$ fixes the coefficients of h. It follows that

$h \in F(p, q, r)[y]$. The polynomial h is called the *resolvent cubic* of g_4. It can be shown (see [9]) that

$$h = y^3 - 2py^2 + (p^2 - 4r)y + q^2.$$

By solving the cubic equation $h(y) = 0$, one obtains z_1, z_2 and z_3. Next, using

$$0 = (y_1 + y_2) + (y_3 + y_4) \quad \text{and} \quad z_1 = (y_1 + y_2)(y_3 + y_4),$$
$$0 = (y_1 + y_3) + (y_2 + y_4) \quad \text{and} \quad z_2 = (y_1 + y_3)(y_2 + y_4),$$
$$0 = (y_1 + y_4) + (y_2 + y_3) \quad \text{and} \quad z_3 = (y_1 + y_4)(y_2 + y_3),$$

one obtains the formulas

$$y_1 = \tfrac{1}{2}[\sqrt{-z_1} + \sqrt{-z_2} + \sqrt{-z_3}],$$
$$y_2 = \tfrac{1}{2}[\sqrt{-z_1} - \sqrt{-z_2} - \sqrt{-z_3}],$$
$$y_3 = \tfrac{1}{2}[-\sqrt{-z_1} - \sqrt{-z_2} + \sqrt{-z_3}],$$
$$y_4 = \tfrac{1}{2}[-\sqrt{-z_1} + \sqrt{-z_2} - \sqrt{-z_3}],$$

where the square roots must be chosen such that $\sqrt{-z_1}\sqrt{-z_2}\sqrt{-z_3} = -q$. The zeros of the general quartic g_4 are $x_i = y_i - \frac{1}{4}u_1$, $i = 1, \ldots, 4$.

Example:

Let $f = y^4 + 2y^2 + 2y + \frac{5}{4} \in \mathbb{Q}[x]$. We have $p = 2$, $q = 2$, $r = \frac{5}{4}$. Let $h = y^3 - 2py^2 + (p^2 - 4r)y + q^2 = y^3 - 4y^2 - y + 4 = (y - 4)(y - 1)(y + 1)$. The zeros of h are $z_1 = 4$, $z_2 = 1$, $z_3 = -1$. We need to choose square roots of -4, -1, and 1 such that the product of the square roots is $-q = -2$. Since $2i \cdot i \cdot 1 = -2$, we may use the square roots $2i$, i, and 1, and thus obtain

$$y_1 = \tfrac{1}{2}[2i + i + 1] = \tfrac{1}{2}(1 + 3i),$$
$$y_2 = \tfrac{1}{2}[2i - i - 1] = \tfrac{1}{2}(-1 + i),$$
$$y_3 = \tfrac{1}{2}[-2i + i - 1] = \tfrac{1}{2}(-1 - i),$$
$$y_4 = \tfrac{1}{2}[-2i - i + 1] = \tfrac{1}{2}(1 - 3i).$$

(A different choice of square roots, such as $2i$, $-i$, -1, would have yielded merely a different labeling of the zeros.)

Exercises 4.16

True or False

1. The best way to solve any cubic is to use the cubic formulas.
2. If $f \in \mathbb{Q}[x]$ has 3 real zeros, then the zeros of f can be expressed in terms of real radicals.
3. The discriminant, Δ, of the general polynomial $g_n = x^n + u_1x^{n-1} + \cdots + u_n$ over a field F is contained in $F(u_1, \ldots, u_n)$.

4. The discriminant, Δ, of a polynomial $f \in F[x]$ is contained in F.
5. The square root, δ, of the discriminant of a polynomial $f \in F[x]$ is fixed by every even permutation of the zeros of f.

4.16.1. With notation as defined on page 321, prove:
(a) $F(x_1, x_2, x_3) \neq F(y_1, y_2, y_3)$, and
(b) $F(p, q) \neq F(u_1, u_2, u_3)$.

4.16.2. Let F be a field of characteristic 2. Find a solution (by radicals) of the general polynomial g_2 over F. Is every polynomial of degree 2 in $F[x]$ solvable by radicals? (Recall Exercise 4.12.7.)

4.16.3. Solve the cubic

$$x^3 + 3x^2 + 4x + 3 = 0.$$

4.16.4. Solve the quartic

$$y^4 + \frac{1}{2}y^2 + 3y + \frac{37}{16} = 0.$$

4.16.5. In Example 2, p. 324, simplify the expressions for y_1, y_2, y_3 to show that

$$y_1 = 2\cos \pi/9 \sim -1.88$$

$$y_2 = 2\cos 2\pi/9 \sim 0.35$$

$$y_3 = 2\cos 4\pi/9 \sim 1.53.$$

Check these values by substituting in the equation.

4.17 Classical Constructions II; The Fundamental Theorem of Algebra

(Note: Results from Section 2.15 are used in this section.)

In Section 4.5, we proved the Fundamental Theorem on Constructibility (Theorem 4.5.1), which implies that a constructible real number must be algebraic, of degree a power of 2, over $\mathbb{Q}$. (Corollary, Theorem 4.5.1.)

Using the Fundamental Theorem of Galois Theory (Theorem 4.10.1) and Theorem 2.15.4 on p-groups, we now obtain a simple sufficient condition for constructibility.

Theorem 4.17.1: Let ρ be a real number, algebraic over $\mathbb{Q}$, and let E be the splitting field, in $\mathbb{C}$, of $m_{\rho/\mathbb{Q}}$ over $\mathbb{Q}$. If $E \subset \mathbb{R}$ and $[E:\mathbb{Q}]$ is a power of 2, then ρ is constructible over $\mathbb{Q}$.

Proof: Since Char $\mathbb{Q} = 0$, $E/\mathbb{Q}$ is separable. By Theorem 4.7.3, $E/\mathbb{Q}$ is normal. Thus, $E/\mathbb{Q}$ is a Galois extension, with $[E:\mathbb{Q}] = 2^t$ for some integer $t \geq 0$, and the Galois group $G = G(E/\mathbb{Q})$ has order 2^t. By Theorem 2.15.6, there is a normal series

$$G = G_t \rhd G_{t-1} \rhd \cdots \rhd G_0 = \{\iota\},$$

where $|G_i| = 2^i$, hence $|G_i/G_{i-1}| = 2$, for each $i = 1, \ldots, t$. If we denote by E_i the subfield of E corresponding to G_{t-i} ($i = 0, \ldots, t$), we have

$$\mathbb{Q} = E_0 \subset E_1 \subset \cdots \subset E_t = E \subset \mathbb{R},$$

with $[E_i : E_{i-1}] = 2$ for each $i = 0, \ldots, t$ (see Theorem 4.10.1). Since $\rho \in E$, Theorem 4.5.1 implies that ρ is constructible over $\mathbb{Q}$. ∎

We are now in a position to prove the Fundamental Theorem on Constructibility of regular polygons. Clearly, a regular n-gon ($n > 2$) is constructible (i.e., its vertices are constructible points) if and only if $\cos \frac{2\pi}{n}$ is a constructible real number. The following Lemma and the formulas derived in Exercises 4.17.5 and 4.17.6 for the Euler Phi Function will play a major role in the proof.

Lemma: For n a positive integer, the real number $\rho = \cos 2\pi/n$ is constructible if and only if $\phi(n)$ is a power of 2.

Proof: Let $\zeta = \cos 2\pi/n + i \sin 2\pi/n$. Then ζ is a primitive n-th root of unity and $\mathbb{Q}(\zeta)/\mathbb{Q}$ is a Galois extension of degree $\phi(n)$. (See Corollary 2, Theorem 4.11.2.) The Galois group $G = G(\mathbb{Q}(\zeta)/\mathbb{Q})$ consists of $\phi(n)$ $\mathbb{Q}$-automorphisms δ_j, each sending ζ to ζ^j for some positive integer j, $1 \leq j \leq n$, $\gcd(j, n) = 1$. (See Exercise 4.11.4.) Since $\rho = \cos 2\pi/n = 1/2(\zeta + \bar{\zeta}) = 1/2(\zeta + \zeta^{-1})$, a non-trivial $\mathbb{Q}$-automorphism δ_j in G fixes $\rho \Leftrightarrow 1/2(\zeta + \zeta^{-1}) = 1/2(\zeta^j + \zeta^{-j}) \Leftrightarrow \cos 2\pi/n = \cos 2\pi j/n \Leftrightarrow j = n - 1 \Leftrightarrow \sigma_j(\zeta) = \zeta^{n-1} = \bar{\zeta}$. Thus, $G(\mathbb{Q}(\zeta)/\mathbb{Q}(\rho))$ consists of the two automorphisms ι and σ_{n-1}. It follows that $[\mathbb{Q}(\zeta) : \mathbb{Q}(\rho)] = 2$; hence $[\mathbb{Q}(\rho) : \mathbb{Q}] = \phi(n)/2$.

Now, if $\rho = \cos 2\pi/n$ is constructible over $\mathbb{Q}$, then (by the Corollary of Theorem 4.5.1), $[\mathbb{Q}(\rho) : \mathbb{Q}] = \phi(n)/2$ is a power of 2, hence $\phi(n)$ is a power of 2.

Conversely, suppose $\phi(n) = [\mathbb{Q}(\zeta) : \mathbb{Q}]$ is a power of 2. Since $\mathbb{Q}(\zeta)/\mathbb{Q}$ is a normal extension of $\mathbb{Q}$, containing ρ, $\mathbb{Q}(\zeta)$ contains all $\mathbb{Q}$-conjugates of ρ, hence contains the splitting field, E, of $m_{\rho/\mathbb{Q}}$. Moreover, $E \subset \mathbb{R}$. For, every $\mathbb{Q}$-automorphism, τ, of $\mathbb{Q}(\rho)$ extends to a $\mathbb{Q}$-automorphism σ_j of the normal extension $\mathbb{Q}(\zeta)$ of $\mathbb{Q}$; hence, each $\mathbb{Q}$-conjugate, $\tau(\rho)$, of ρ is equal to $1/2(\zeta^j + \zeta^{-j}) = 1/2(\zeta^j + \overline{\zeta^j}) = \cos 2\pi j/n \in \mathbb{R}$. From $\mathbb{Q} \subset E \subset \mathbb{Q}(\zeta)$, we conclude that $[E : \mathbb{Q}]$ is a divisor of $\phi(n) = [\mathbb{Q}(\zeta) : \mathbb{Q}]$, hence a power of 2. But then, by Theorem 4.17.1, $\rho = \cos 2\pi/n$ is constructible over $\mathbb{Q}$. ∎

Definition 4.17.1: A prime p is called a *Fermat prime* if it is of the form

$$p = 2^m + 1$$

for some positive integer m.

Remark 1: If p is a *prime* of the form

$$p = 2^m + 1$$

(m a positive integer), then $m = 2^t$ for some positive integer t (see Exercise 4.17.1).

Remark 2: To date, the only known Fermat primes are the numbers $2^{2^t} + 1$ for $t = 0, 1, 2, 3, 4$.

Theorem 4.17.2: Let $n > 2$ be an integer. Then a regular n-gon is constructible if and only if

$$n = 2^m p_1 \dots p_s$$

where $m \geq 0$ and the p_i are distinct Fermat primes.

Proof: By the preceding Lemma, a regular n-gon is constructible if and only if $\phi(n)$ is a power of 2. If $n = 2^m p_1^{\alpha_1} \dots p_s^{\alpha_s}$, where $m \geq 0$, $\alpha_i \geq 1$, p_i distinct odd primes ($i = 1, \dots, s$), then (by Exercise 4.17.5), $\phi(n) = 2^{m-1} p_1^{\alpha_1 - 1} \dots p_s^{\alpha_s - 1} (p_1 - 1) \dots (p_s - 1)$ in case $m \geq 1$, and $\phi(n) = p_1^{\alpha_1 - 1} \dots p_s^{\alpha_s - 1} (p_1 - 1) \dots (p_s - 1)$ in case $m = 0$. It follows that $\phi(n)$ is a power of 2 if and only if $\alpha_i = 1$ and $p_i - 1$ is a power of 2, for each $i = 1, \dots, s$, i.e., if and only if

$$n = 2^m p_1 \dots p_s$$

($m \geq 0$, p_i distinct Fermat primes). ■

Example:

Since $5 = 2^2 + 1$, it is possible to construct a regular pentagon.

Let $\zeta = \cos \frac{2\pi}{5} + i \sin \frac{2\pi}{5}$. Then $m_{\zeta/\mathbb{Q}} = \Phi_5 = x^4 + x^3 + x^2 + x + 1$, the fifth cyclotomic polynomial. From $\zeta^4 + \zeta^3 + \zeta^2 + \zeta + 1 = 0$, multiplying by ζ^{-2}, we have

$$\begin{aligned} 0 &= \zeta^2 + \zeta + 1 + \zeta^{-1} + \zeta^{-2} = \zeta^2 + 2 + \zeta^{-2} + \zeta + \zeta^{-1} - 1 \\ &= (\zeta + \zeta^{-1})^2 + (\zeta + \zeta^{-1}) - 1. \end{aligned}$$

Thus, $\zeta + \zeta^{-1}$ is a zero of the quadratic polynomial $x^2 + x - 1 \in \mathbb{Q}[x]$, hence equal to one of $\frac{-1 + \sqrt{5}}{2}, \frac{-1 - \sqrt{5}}{2}$. Since $\zeta^{-1} = \bar{\zeta} = \cos \frac{2\pi}{5} - i \sin \frac{2\pi}{5}$, we have $\zeta + \zeta^{-1} = 2 \cos \frac{2\pi}{5} > 0$. Hence $\zeta + \zeta^{-1} = \frac{-1 + \sqrt{5}}{2}$, and thus $\cos \frac{2\pi}{5} = \frac{-1 + \sqrt{5}}{4}$.

The construction may now proceed as follows:

Obtain $\sqrt{5}$ as the hypotenuse of a right triangle with legs 1 and 2. Subtract 1 and divide by 4 (cf. p. 258). This gives $\cos \frac{2\pi}{5}$.

Lay off $\cos \frac{2\pi}{5} = OP$ on a radius OP_1 of a unit circle, as shown. Erect a perpendicular at P to obtain a second vertex P_2 of the inscribed regular pentagon. Lay off $P_1 P_2$ along the circumference to obtain the remaining vertices of the pentagon. (See Figure 18.)

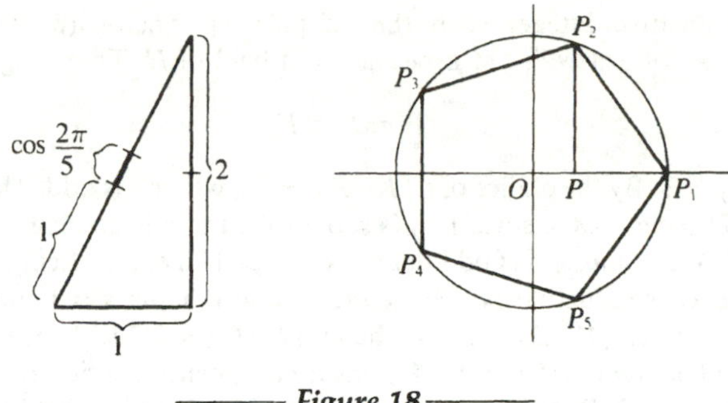

Figure 18

Having completed the pentagon, we can proceed to construct any regular n-gon, where $n = 2^m \cdot 5$, $m \geq 1$, simply by successive bisections.

Theorem 4.17.2 was proved by Gauss in 1796, without benefit of Galois Theory. We conclude with another of Gauss' great theorems, first proved by Gauss in his doctoral dissertation (1799), twelve years before Galois' birth: the Fundamental Theorem of Algebra. Despite its name, it is a theorem on complex numbers and cannot be proved by purely algebraic means, without making use of at least some simple results of *real* analysis. Many proofs of this theorem are known—Gauss himself contributed several. Our proof makes essential use of Galois Theory. The only analytic tool we use is the Intermediate Value Theorem on continuous functions of which the following Lemma is an easy consequence.

Lemma 1: Every polynomial of odd degree in $\mathbb{R}[x]$ has at least one real root.

We leave the proof as an exercise (Exercise 4.17.9).

We also assume as known:

Lemma 2: Every complex number has complex square roots.

This is a special case of Exercise 4.11.12 in which results based on de Moivre's Theorem on complex numbers were to be used. While de Moivre's Theorem yields very simply that $z = r(\cos\theta + i\sin\theta)$ ($r > 0$ in $\mathbb{R}$, $\theta \in \mathbb{R}$) has square roots $\pm r^{1/2}\left(\cos\frac{\theta}{2} + i\sin\frac{\theta}{2}\right)$, it is not *necessary* to use de Moivre's Theorem to prove Lemma 2. (See Exercise 4.17.8.)

Theorem 4.17.3 (Fundamental Theorem of Algebra): Every polynomial f of positive degree in $\mathbb{C}[x]$ splits over $\mathbb{C}$.

Proof: Let f be a polynomial of positive degree in $\mathbb{C}[x]$ and let E be a splitting field for f over $\mathbb{C}$. Since $\mathbb{R} \subset \mathbb{C} \subset E$, there is a normal closure, K, of $E/\mathbb{R}$. Since Char $\mathbb{R} = 0$, $K/\mathbb{R}$ is a Galois extension of even degree, since $[\mathbb{C} : \mathbb{R}] = 2$. Let t be the largest positive integer such that $2^t | [K:\mathbb{R}]$. Then $G(K/\mathbb{R})$ has a Sylow 2-subgroup, H, of order 2^t. Let L be the fixed field of H. Then

$$\mathbb{R} \subset L \subset K,$$

with $[K:L] = 2^t$. By the choice of t, $[K:\mathbb{R}] = 2^t s$, where s is odd. Thus, $[L:\mathbb{R}] = s$ is odd. By Theorem 4.8.5, since $L/\mathbb{R}$ is separable, there is an element, α, in L such that $L = \mathbb{R}(\alpha)$. Then $m_{\alpha/\mathbb{R}}$ has odd degree s. If $s > 1$, then, by Lemma 1, $m_{\alpha/\mathbb{R}}$ has a zero in $\mathbb{R}$, hence is reducible over $\mathbb{R}$—contradiction! Thus, $s = 1$, and $[K:\mathbb{R}] = 2^t$. From $\mathbb{R} \subset \mathbb{C} \subset K$, $[\mathbb{C}:\mathbb{R}] = 2$, we have $[K:\mathbb{C}] = 2^{t-1}$; hence $G(K/\mathbb{C})$ is a 2-group. By Theorem 2.15.6, $G(K/\mathbb{C})$ has a composition series all of whose $t-1$ factors have order 2. But then, by Theorem 4.10.1, there is a field tower

$$\mathbb{C} = K_0 \subset K_1 \subset \cdots \subset K_{t-1} = K,$$

where $[K_i : K_{i-1}] = 2$ for each $i = 1, \ldots, t-1$. If $t - 1 \geq 1$, then $K_1 = \mathbb{C}(\alpha_1)$ where $\alpha_1 \in K_1$, with $\alpha_1^2 \in \mathbb{C}$ (see Exercise 4.8.10). By Lemma 2, $\alpha_1 \in \mathbb{C}$. But then $[K_1 : K_0] = [K_1 : \mathbb{C}] = 1 \neq 2$. Contradiction! It follows that $t - 1 = 0$, hence $[K:\mathbb{C}] = 2^0 = 1$ and $K = \mathbb{C}$. From $\mathbb{C} \subset E \subset K$, we have $E = \mathbb{C}$, and so f splits over $\mathbb{C}$. ■

Exercises 4.17

True or False

1. Some day, someone will figure out how to construct a regular 11-gon.
2. There are infinitely many known integers n such that a regular n-gon is constructible.
3. A regular 680-gon is constructible.
4. Every finite extension of $\mathbb{R}$ is isomorphic to $\mathbb{C}$.
5. $\mathbb{C}$ has no infinite extensions.

4.17.1. Prove: if $p = 2^m + 1$ is prime $(m \geq 1)$, then m is a power of 2.

4.17.2. Construct a regular decagon.

4.17.3. Express sin 72° in terms of radicals over $\mathbb{Q}$.

4.17.4. Based on the five known Fermat primes, how many odd integers n are there such that a regular n-gon is constructible?

4.17.5. Let $\phi : \mathbb{Z}^+ \to \mathbb{Z}^+$ be the Euler Phi Function, defined by: $\phi(n)$ is the number of elements in the set $\{m \in \mathbb{Z}^+ | m \leq n \text{ and } \gcd(m, n) = 1\}$ $(n \in \mathbb{Z}^+)$. Prove:

(1) If p is prime, then $\phi(p) = p - 1$.

(2) If p is prime, $k \geq 1$, then $\phi(p^k) = p^k - p^{k-1}$.

(3) If $m, n \in \mathbb{Z}^+$ and $\gcd(m, n) = 1$, then $\phi(mn) = \phi(m)\phi(n)$.

4.17.6. Using the preceding exercise, prove: if $n = p_1^{k_1} \ldots p_t^{k_t}$, p_i distinct primes, $k_i \geq 1$ $(i = 1, \ldots, t)$, then

$$\phi(n) = \prod_{i=1}^{t} \left(p_i^{k_i} - p_i^{k_i - 1} \right) = \prod_{i=1}^{t} p_i^{k_i - 1}(p_i - 1) = n \prod_{i=1}^{t} \left(1 - \frac{1}{p_i} \right).$$

4.17.7. Prove that $\phi(n)$ is even for all $n > 2$.

4.17.8. Without using de Moivre's Theorem, prove that every complex number has complex square roots.

4.17.9. Use the Intermediate Value Theorem on continuous real-valued functions to prove that every polynomial of odd degree in $\mathbb{R}[x]$ has a zero in $\mathbb{R}$.

4.17.10. Prove that the only algebraic extension of $\mathbb{C}$ is $\mathbb{C}$ itself.

Bibliography

Abstract Algebra (General)

1. Birkhoff, G., and MacLane, S. *A Survey of Modern Algebra.* 4th ed. New York: Macmillan, 1977.
2. Fraleigh, J. B. *A First Course in Abstract Algebra.* 3rd ed. Reading, MA: Addison-Wesley, 1983.
3. Goldhaber, J. K., and Ehrlich, G. *Algebra.* New York: Macmillan, 1970. Rev. ed. Huntington, NY: Krieger, 1980.
4. Herstein, I. N. *Topics in Algebra.* 2nd ed. New York: Wiley, 1975.
5. Hungerford, T. W. *Algebra.* New York: Holt, Rinehart and Winston, 1974.
6. Jacobson, N. *Basic Algebra I.* San Francisco: W. F. Freeman & Co., 1974.
7. Jacobson, N. *Lectures in Abstract Algebra, Vols. I, II, III.* Princeton, NJ: Van Nostrand, 1951, 1960, 1964.
8. Lang, S. *Algebra.* Reading, MA: Addison Wesley, 1965.
9. Van der Waerden, B. L. *Moderne Algebra.* Berlin: Springer, 1931. Engl. transl.: *Algebra.* New York: Ungar, 1949.

Galois Theory

10. Artin, E. *Galois Theory.* Notre Dame, IN: University of Notre Dame, 1944.
11. Gaal, L. *Classical Galois Theory with Examples.* 4th ed. New York: Chelsea, 1988.
12. Stewart, I. *Galois Theory.* London: Chapman and Hall, 1973.
13. Verriest, G. *Oeuvres d'Évariste Galois—É. Galois et la Theorie des Équations Algebriques.* Paris: Gauthier-Villars, 1951.

Group Theory

14. Feit, W., and Thompson, J. *Solvability of Groups of Odd Order.* Pacific Journal of Mathematics, Vol. 13, 1962.
15. Fuchs, L. *Abelian Groups.* Oxford: Pergamon Press, 1960.
16. Gorenstein, D. *Classifying Finite Simple Groups.* Bulletin of the American Mathematical Society, Vol. 14, No. 1, pp. 1–98.
17. Hall, M. *Theory of Groups.* New York: Macmillan, 1961.
18. Kurosh, A. E. *The Theory of Groups.* 2nd Engl. ed. New York: Chelsea, 1960.
19. Scott, W. R. *Group Theory.* Englewood Cliffs, NJ: Prentice-Hall, 1964.

History and Philosophy of Mathematics

20. Aaboe, A. *Episodes from the Early History of Mathematics.* New Haven, CT: Yale University, 1964.

21. Bell, E. T. *Men of Mathematics*. New York: Simon and Schuster, 1937.
22. Cajori, F. *History of Mathematics*. 4th ed. New York: Chelsea, 1980.
23. Calinger, R. *Classics of Mathematics*. Oak Park, IL: Moore Publishing Co., 1982.
24. Dieudonné, J. *Pour l'Honneur de l'Esprit Humain*. Paris: Hachette, 1987.
25. Newman, J. R. *The World of Mathematics*. New York: Simon and Schuster, 1956.
26. Russell, B. *Introduction to Mathematical Philosophy*. New York: Macmillan 1920; Allen and Unwin, 1960.

Linear Algebra

27. Anton, H. *Elementary Linear Algebra*. 4th ed. New York: Wiley, 1984.
28. Curtis, C. W. *Linear Algebra*. New York: Springer, 1984.
29. Hoffman, K., and Kunze, R. *Linear Algebra*. 2nd ed. Englewood Cliffs, NJ: Prentice Hall, 1971.
30. Leon, S. J. *Linear Algebra with Applications*. New York: Macmillan, 1980.

Number Theory

31. Adams, W. W., and Goldstein, L. J. *Introduction to Number Theory*. Englewood Cliffs, NJ: Prentice Hall, 1976.
32. Bressoud, D. M. *Factorization and Primality Testing*. New York: Springer, 1989.
33. Hardy, G. H., and Wright, E. M. *An Introduction to the Theory of Numbers*. 4th ed. Oxford University Press, 1960.
34. Lehmer, E. *On the Magnitude of Coefficients of the Cyclotomic Polynomial*. Bulletin of the American Mathematical Society 42 (1936).
35. Pollard, H., and Diamond, H. G. *The Theory of Algebraic Numbers*. 2nd ed. Carus Mathematical Monograph # 9, Mathematical Association of America, 1975.
36. Rose, H. E. *A Course in Number Theory*. Oxford Science Publications, 1988.
37. Shanks, D. *Solved and Unsolved Problems in Number Theory*. 3rd ed. New York: Chelsea, 1985.
38. Weinberger, P. *On Euclidean Rings of Algebraic Integers*. Analytic Number Theory, Proceedings of Symposia in Pure Mathematics, Vol. XXIV, pp. 321–332, American Mathematical Society, 1973.

Ring Theory

39. Herstein, I. N. *Noncommutative Rings*. Carus Mathematical Monograph # 15, Mathematical Association of America, 1973.
40. McCoy, N. H. *Rings and Ideals*. Carus Mathematical Monographs # 8, Mathematical Association of America, 1968.
41. Samuel, P. P. *About Euclidean Rings*. Journal of Algebra 19 (1971), pp. 282–301.

Set Theory and Foundations

42. Cohen, L. W., and Ehrlich, G. *The Structure of the Real Number System*. Princeton, NJ: Van Nostrand, 1963; rev. ed. Huntington, NY: Krieger, 1977.
43. Enderton, H. B. *Elements of Set Theory*. New York: Academic Press, 1977.
44. Halmos, P. *Naive Set Theory*. Princeton, NJ: Van Nostrand, 1960.
45. Suppes, P. *Axiomatic Set Theory*. Princeton, NJ: Van Nostrand, 1960.

Index

Abel, Niels Hendrik (1802–1829), 3
abelian group, 34
abstract group, 58
al-Khowarizmi, Mohammed ibn Musa, 2
algebraic
  element, 243
  extension, 250
algebraically closed extension, 232
algebraically independent set, 312
alternating group, 75
Artin, Emil (1898–1962)
Artin-Dedekind Lemma, 282
associate, 163
automorphism, 87, 140
  inner, 105
Axiom of Choice, 13

basis, 199
binary operation, 27
  associative, 28
  commutative, 28
binary relation, 7
  order, 7
  partial order, 7
  reflexive, 7
  symmetric, 7
  transitive, 7
Brauer, Richard (1901–1977), 304

Cardano, Girolamo (1501–1576), 2
Cartesian product, 6
Cayley, Alfred Lord (1821–1895)
Cayley table, 36
Cayley's Theorem, 77
center, 85
centralizer, 125
characteristic, 240
characteristic polynomial, 230
class equation, 125
co-domain, 12
column space, 223
commutative diagram, 97
commutative ring, 133
commutator, 106
  additive, 150
commutator subgroup, 106
companion matrix, 237
composition factors, 121
composition of functions, 12
composition series, 121
congruence, 23
conjugacy class of subsets, 126, 127
conjugate class, 85, 124
conjugate elements, 85
conjugate subsets, 126
constructibility
  of a point, 258
  of a real number, 258
  of a regular polygon, 329, 330
coordinate vector, 212
Correspondence Theorem, 119
coset, 60
cycle, 69
cycle structure, 85
cyclic extension, 305
cyclic group, 47
cyclotomic polynomial, 293

Dedekind, Richard (1831–1916), 4, 282
degree
  of an extension, 250
  of a field element, 251
  of a polynomial, 158
determinant, 215
diagonalizable matrix, 231
dihedral group, 78
dimension of a vector space, 202
direct product of groups, 109
  internal, 110
discriminant, 75
divisibility, 20, 163
division algorithm
  for integers, 19
  for polynomials, 159
division ring, 133
domain of a function, 12
duplication of a cube, 257, 260

e
  transcendence of, 243
eigenvalue
  of a linear operator, 229
  of a matrix, 230
eigenvector, 229, 230
eigenspace, 232
Eisenstein's Irreducibility Criterion, 187
elementary symmetric function, 315
element of a set, 4
endomorphism
  of groups, 87
  of rings, 140
equivalence class, 8
equivalence relation, 7
Euclid (c. 330–275 B.C.)
Euclidean domain, 174
Euclidean norm, 174
Euclid's Lemma, 22
Euler, Leonhard (1707–1783)
Euler Phi Function, 65, 332
extension field, 238
  algebraic, 238
  algebraically closed, 232
  cyclic, 305
  finite, 250
  Galois, 281
  normal, 266
  radical, 297
  separable, 273
  simple, 242
  transcendental, 250

F-conjugate, 266
F-homomorphism, 264
factor group, 94
factor set, 8
factors of a normal series, 119
Fermat, Pierre Auguste de (1601–1665)
Fermat prime, 329
Fermat's Little Theorem, 64
Ferrari, Lodovico (1522–c. 1560), 2
Ferro, Scipione del (1465–1526), 2
field, 133
  algebraically closed, 232
  prime, 239
  of quotients, 180
field tower, 251
finite extension, 250
fixed field, 281
function, 12
  codomain of a function, 12
  domain of a function, 12
  identity function, 13
  image of a function, 12
  inverse function, 13
  range of a function, 12
Fundamental Partition Theorem, 9
Fundamental Theorem
  of Algebra, 331
  of Arithmetic, 22
  of Galois Theory, 285
  of Homomorphism
    for Groups, 97
    for Rings, 148
  on Constructibility, 259
  on Radical Extensions, 309
  on Symmetric Functions, 316

Galois, Évariste (1811–1832), 3
  correspondence, 285
  extension, 281
  group, 281, 286
Gauss, Karl Friedrich (1777–1855), 331
Gauss' Lemma, 183
Gaussian integers, 162, 176
general cubic equation, 2, 321
general linear group, 34
general polynomial, 313
general quartic equation, 2, 326
geometric groups, 78

golden section, 262
greatest common divisor, 20, 171
greatest lower bound, 11
group, 3, 33
  abelian, 34
  abstract, 58
  alternating, 75
  of bijections, 35
  cyclic, 47
  dihedral, 78
  with operators, 190
  of permutations, 35, 66
  primary, 114
  solvable, 119
  symmetric, 35
groupoid, 33

Hermite, Charles (1822–1901), 243
Hilbert, David (1862–1943), 4
Hilbert's Theorem 90
  additive, 308
  multiplicative, 306
homomorphism, 87, 140, 191, 193, 207

ideal, 140
  maximal, 151
  prime, 153
  principal, 161
idempotent, 138
identity element, 29, 132
identity function, 13
identity matrix, 30, 214
image, 12
index, 63
induction, 18, 19
inner automorphism, 105
integers, 16
integral domain, 133
intersection, 5
inverse element, 31
inverse function, 13
inverse image, 88
inverse matrix, 214
involutory, 138, 237
irreducible element, 164
irreducible polynomial, 164
isometry, 78
isomorphism, 55, 87, 140

Jordan-Hölder Theorem, 121

kernel of a homomorphism, 88, 141
Klein, Felix (1849–1925), 3

Lagrange, Joseph Louis (1736–1813)
Lagrange resolvent, 323
Lagrange's Theorem, 63
lattice, 11
least common multiple, 25, 171
least upper bound, 11
left-regular representation, 78
Lindemann, Ferdinand (1852-1939), 243
linear dependence, 197
linear independence, 197
linear operator, 207
  invertible, 208
  non-singular, 207
  singular, 207
linear transformation, 207
Liouville, Joseph (1809–1882)
local ring, 155

M-group, 190
M-homomorphism, 191
M-subgroup, 190
mapping, 12
matrix, 213, 214
  diagonalizable, 231
  inverse, 214
  of a linear transformation, 218
  non-singular, 224
  partitioned, 218
  singular, 224
  transition, 225
matrix addition, 214
matrix multiplication, 214
matrix units, 144
maximal ideal, 151
minimal polynomial, 244
module, 192
monomorphism, 87, 140
motion, 78
multiplicative set, 181

nilpotent, 138, 237
Noether, Emmy (1882–1935), 4
non-singular matrix, 224
norm
  Euclidean, 174
  of a field element, 305
normal closure, 268
normal extension, 266

normalizer, 127
normal series, 119
  solvable, 119
normal subgroup, 82
nullity, 207, 223
nullspace, 207, 223

orbit, 68
order
  of an element, 51
  of a group, 34
order relation, 7
ordered basis, 212
ordered integral domain, 147
ordered pair, 6

partial order relation, 7
partition, 8
permutation, 35
  even, 74
  odd, 74
pi (π), 243
polynomial, 158
polynomial ring, 158
positive cone, 147
power set, 5
primary group, 114
prime, 20, 166
prime field, 239
prime ideal, 153
primitive element, 278, 279
primitive polynomial, 183
primitive root of unity, 292
principal ideal, 161
principal ideal domain, 161

quaternion group, 86
quaternions, 135
quotient group, 94

radical, 297
  extension, 297
  of an ideal, 147
  tower, 297
rank
  of a linear operator, 207
  of a matrix, 223
reflection, 79
reflexive binary relation, 7
regular pentagon, 330, 331
regular polygon, 78, 329, 330
relatively prime, 21, 172
remainder, 19, 174
residue class, 23, 141
residue class ring, 141
resolvent cubic, 327
right regular representation, 78
ring, 132
  Boolean, 150
  commutative, 133
  division, 133
  matrix, 214
  polynomial, 158
  of quotients, 182
  simple, 143
  von Neumann regular, 228
  with identity, 132
  without zero divisors, 133
root of unity, 52, 291
rotation, 79

scalar, 194
  multiplication, 194
Schreier Refinement Theorem, 121
semigroup, 33
separability degree, 276
separable
  element, 273
  extension, 273
  polynomial, 273
set, 4
  denumerable, 26
  empty, 4
  finite, 26
  infinite, 26
  power, 5
similarity of matrices, 226
simple
  group, 102
  ring, 143
solvability
  of a group, 119
  by radicals, 319
span, 196
special linear group, 45
splitting field, 248
squaring of a circle, 257, 260
standard ordered basis, 217
Steinitz, Ernst (1871-1928), 4
Structure Theorem for Finite Abelian Groups, 113, 117
subfield, 238

subgroup, 43
  characteristic, 106
  cyclic, 46
  commutator, 106
  invariant, 106
  normal, 82
  proper, 44
  Sylow, 126
submodule, 193
subring, 136
subspace, 196
Sylow, Ludvig Mejdell (1832–1918), 125
  Sylow subgroup, 126
  Sylow Theorems, 125, 128
symmetric binary relation, 7
symmetric function, 315
symmetric group, 35
symmetries
  of an equilateral triangle, 79
  of a regular polygon, 78
  of a square, 80

trace, 307
transcendence base, 317
transcendence set, 312
transcendental
  element, 243
  extension, 250
transitive binary relation, 7
transposition, 73
trisection of an angle, 257, 260

union, 5
unique factorization domain, 168
unit, 134
unit group, 134
unit-regular, 228
unitary, 192

vector, 194
vector space, 194

well-ordering of $\mathbb{Z}^+$, 17

zero divisors, 133
zero matrix, 214
zero ring, 135
Zorn's Lemma, 155